Advances in Computational Intelligence and Autonomy for Aerospace Systems

Advances in Computational Intelligence and Autonomy for Aerospace Systems

EDITED BY

John Valasek

Timothy C. Lieuwen, Editor-in-Chief
Georgia Institute of Technology
Atlanta, Georgia

Published by
American Institute of Aeronautics and Astronautics, Inc.
12700 Sunrise Valley Drive, Suite 200, Reston, VA 20191-5807

Cover images, clockwise from left:
Deep Space 1.
Courtesy NASA/JPL-Caltech.
NASA's Ikhana aircraft.
Credits: NASA Photo/Carla Thomas
NASA's Mars 2020 Rover Artist's Concept #3.
Image Credit: NASA/JPL-Caltech

American Institute of Aeronautics and Astronautics, Inc., Reston, VA.

ISBN 978-1-62410-478-7

TABLE OF CONTENTS

Preface . . . . **xi**

SECTION I INTELLIGENT SPACE SYSTEMS

Chapter 1 Advances in Intelligent Systems for Space Applications . . 1

Christopher Bowman, *Data Fusion & Neural Networks (DF&NN);* Christopher Tschan, *Aerospace Corporation*; Paul Zetocha, *AFRL/RV*

Nomenclature 1
1.1 Introduction 2
1.2 Building Data-Driven and Goal-Driven Computational Intelligent Systems 8
1.3 Specific Intelligent System Space Applications 14
1.4 Summary 43
References 43

Chapter 2 Advances in Functional Fault Modeling for the Design and Operation of Space Systems 47

Kevin J. Melcher, *NASA Glenn Research Center, Cleveland, Ohio 44135, USA;* William A. Maul and Amy K. Chicatelli, *Vantage Partners, LLC, Brookpark, Ohio 44142, USA*; Gordon Aaseng, Eric Barszcz and Ann Patterson-Hine, *NASA Ames Research Center, Moffet Field, CA 94035, USA*; Barton D. Baker, *NASA Marshall Space Flight Center, Huntsville, AL 35808, USA*

Nomenclature 47
2.1 Introduction 49
2.2 TEAMS Overview and Example Models 54
2.3 Advances in FFM Development, Implementation, and Application . . 63
2.4 The Future of FFM 103
2.5 Concluding Remarks 104
Acknowledgements 104
References 105

SECTION II INTELLIGENT VALIDATION & VERIFICATION METHODS

Chapter 3 Toward Safe Intelligent Unmanned Aircraft Using Formal Methods and Runtime Monitoring 107

Christoph Torens and Florian-Michael Adolf, *German Aerospace Center (DLR), Institute of Flight Systems, Braunschweig, 38108, Germany*

SECTION IV INTELLIGENT FLIGHT CONTROL

PREFACE

Why a new book on intelligent aerospace systems? Although it has been 29 years since the publication of *Machine Intelligence and Autonomy for Aerospace Systems* (AIAA, 1989), no formal and widely accepted definition exists of what exactly makes an aerospace system intelligent and autonomous. A linear combination of eight characteristics that are likely to be inherent were listed in *Advances in Intelligent and Autonomous Aerospace Systems* (AIAA, 2012). But it has been only six years since the publication of that book. The book you are reading started as a 2nd Edition of it, in which the work contained therein would be updated with perhaps a few new chapters added. However, the field is changing so rapidly that the researchers of the work published six years ago were not merely updating their work, but rather significantly changing their work and pursuing new directions as new algorithms and approaches were developed, such as Deep Learning, and new applications from other domains provided insight and results, such as connected vehicles. In addition to new approaches, directions, and applications, the importance of formal methods and Validation & Verification (V&V), which are needed for useful and practical application of intelligent systems to space, air, and ground vehicles, is gaining momentum. Clearly a new book was in order rather than a 2nd Edition.

It is enlightening to review the scope and content of the two previous books and to compare them to the present book to see how the field has progressed in the last 30 years. The first book focused exclusively on spacecraft that were currently anticipated from contemporary requirements and applications on future space missions. It was envisioned that these would present building blocks for further development and advances. Machine Intelligence and Human Machine Interaction aspects were emphasized as a means to increase the technical feasibility and economic feasibility of spacecraft systems. As defined in that work, these aspects described systems that accept human commands, conduct reasoning, and then perform manipulative tasks. It is interesting that Expert Systems and Knowledge Based Systems were featured so prominently in this earlier work, since they did not see nearly the widespread use that was envisioned for them. On the other hand, Information Acquisition for single and multi-sensor systems was identified as an important emerging area, and this is certainly true today and for the foreseeable future. A rough indicator of the level of maturation of the intelligent systems of a quarter century ago can be inferred from the fact that only one of the 12 chapters of the book featured a system or application that used closed-loop intelligent or autonomous control. That application was the control of a deformable spacecraft system.

It is clear that the field had matured to a level in which closed-loop control is increasingly pervasive as seven of the 12 chapters of the second book, published 23 years later, feature closed-loop intelligent or autonomous control. That book is organized into sections titled Intelligent Flight Control, Intelligent Propulsion and Health Management, and Intelligent Planning and Multi-Agent Systems. But many of the chapters have distinct technical elements which could have easily justified their inclusion in a different section, highlighting the interdisciplinary nature of recent intelligent systems. The applications addressed in this book span flapping Micro Air Vehicles to generic transport aircraft to propulsion control to architectures for Integrated Health Management and Health Monitoring. In addition, a variety of methods are presented to support and conduct missions ranging from autonomous soaring to cooperative teams of UAVs to space exploration missions to air vehicle search and target tracking.

When I edited *Advances in Intelligent and Autonomous Aerospace Systems* my goal was to provide both the aerospace researcher and the practicing aerospace engineer with an exposition on the latest innovative methods and approaches that focuses on intelligent and autonomous aerospace systems. That goal has not changed for this book. The chapters are written by leading researchers in this field, and include ideas, directions, and recent results on current intelligent aerospace research issues with a focus reflected by the four sections the book is organized into: Space Systems, Validation & Verification Methods, Health Monitoring, and Flight Control. A broad spectrum of methods and approaches are presented in the nine chapters, including Data-Driven and Goal-Driven Computational Intelligence; Functional Fault Modeling; Runtime Verification and Monitoring; Artificial Immune System; Model Reference Adaptive Control and Nonlinear Dynamic Inversion Adaptive Control; Semisupervised Learning; and Adaptive Architectures.

Once again it is my sincere hope that the reader will find the book as useful and rewarding as we have in writing it.

John Valasek

College Station, TX
July 2018

CHAPTER 1

Advances in Intelligent Systems for Space Applications

Christopher Bowman*
Data Fusion & Neural Networks (DF&NN)

Christopher Tschan*
Aerospace Corporation

Paul Zetocha*
AFRL/RV

NOMENCLATURE

AbNet	abnormality network
ACS	satellite attitude control subsystem
ACU	abnormal (space) catalog update
ADCV	abnormality detection classification viewer
AFRL	Air Force Research Laboratory
AFSCN	Air Force Satellite Control Network
ALPS	AFSCN Link Protection System
ANOM	suite of abnormality detection tools
AOCS	satellite attitude and orbit control subsystem
AoR	area of responsibility
ARCADE	Advanced Research Collaboration and Development Environment
BFN	Bayesian fusion node
CA	context assessment
CACM	context assessment and conformity management
ClassCat	classification category application
CAOSD+CA	continuous anomalous situation discriminator + conjunction analysis
C/NOFS	Communications/Navigation Outage Forecasting System
DF&NN	Data Fusion & Neural Networks, a small technology company
DF&RM	data fusion and resource management
DNN	dual node network
DRM	detection rate management
DST	disturbance storm time index
EPS	satellite electrical power subsystem

*Ph.D.

GAIM	Global Assimilative Ionospheric Model
GEO	geostationary Earth orbit
GOES	Geostationary Operational Environmental Satellite
HAV	historical abnormality visualization, which uses a specific file format
HEO	high Earth orbit
IPP	ionospheric penetration point
JDL	Joint Directors of Laboratories, a historical organization that supported data fusion technology development
KB	knowledge base
LEO	low Earth orbit
MEO	medium Earth orbit
MOPS	measures of performance
NN	neural network
PA	performance assessment
PAPM	performance assessment and process management
RMS	root mean square
RSO	resident space object
SAS	satellite as a sensor
SDA	space domain awareness
SEAES	Spacecraft Environmental Anomalies Expert System
SGP4	simplified general perturbations orbit propagation model for near-Earth objects
SDP4	simplified general perturbations orbit propagation model for deep-space objects
SOH	system state of health variables
SNR	signal to noise ratio
SpWx	space weather
SSA	space situational awareness
SSN	space surveillance network
SSUSI	Special Sensor Ultraviolet Spectrographic Imager
TEC	total electron content
TemPats	temporal pattern recognition application
TLE	two-line Keplerian (space object) element set
TrnSat	tool for creating an abnormality detection application using satellite state-of-health telemetry data
UDOP	user defined operational picture

1.1 INTRODUCTION

1.1.1 SCOPE AND OBJECTIVES

This chapter defines an intelligent system and its functional roles. Section 1.2 decomposes problem types and describes when intelligent system data-driven or goal-driven solutions should be used versus model-driven systems. Then key capabilities needed for intelligent systems are described with examples given from an intelligent abnormality detection and characterization system. Section 1.3 describes three specific instantiations of intelligent systems for space applications as follows:

- Prediction of the Effects of Space Weather on Satellites
- Abnormal Space Object Orbital Event Detection, Characterization, and Prediction
- Prediction of Global Positioning System (GPS) Signal Outage Due to Scintillation

These applications serve as a tutorial for the data-driven application of intelligent systems as well as a summary of the performance evaluation results achieved by intelligent systems. These intelligent system developments have been implemented on historical "big data" sets [1–4], as follows:

1. Many years of GOES State of Health (SOH) data overlapping with greater than 10 years of space weather from NOAA
2. Eight years of space catalog historical data from spacetrack.org
3. Many months of dual-band GPS signal to noise ratio (SNR) overlapping with space weather scintillation data

The reader will come away with an appreciation for the following:

- When to apply intelligent systems
- Benefits of intelligent systems
- How to infuse these systems into existing Data Fusion & Resource Management (DF&RM) systems
- What capabilities intelligent systems can offer
- Specific examples of intelligent system capabilities for space applications
- How to evaluate their performance
- How to enable these systems to maintain and improve themselves automatically

1.1.2 WHAT IS AN INTELLIGENT SYSTEM?

> *Intelligent system*: A system that discovers solutions toward achieving goals/desires in (usually difficult) problems given access to relevant inputs. This is done without being given the problem solution, such as rules and equation solutions based on models.

Intelligent systems can be centralized or distributed in a hierarchy, or autonomous with or without communications. Intelligent systems can work alone, in groups, and be teamed with humans. Intelligent systems are used for harder problems where the states of the system are not known. Intelligent systems detect unanticipated conditions to meet the user goal for difficult problems. The resulting intelligent systems can be centralized or distributed in a hierarchy or operate teamed with humans on-the-loop or autonomously, given access to relevant heterogeneous upstream data. The intelligent system can learn correlations in background phenomenologies to detect abnormal entity signature patterns and to predict future behavior. Prediction is a nasty problem: Most solutions assume that past behavior can be used to predict future behavior given similar conditions, but how can "black swans" (i.e., novel and unforeseen entities/events) be predicted? One approach is to look for abnormal behavior (e.g., abnormal correlations or abnormal signatures), then flag that something is amiss. Such behaviors can be correlated to historical signatures. The problem, however, is that there will be a tendency to put the new behavior into one of the historical classes. The problem of detecting a novel abnormality and its possible precursors, all with unknown signatures can be addressed with Independent Component Analysis (ICA) and deep Multi-Start Residual Training (MSRT) learning approaches since these approaches learn normal behavior then score the extent of the abnormality. Intelligent system examples include unknown abnormality detection, discovering correlations in phenomenologies, learned adaptive control, and coming up with reasoned judgment to predict events. The intelligent systems described here can be standalone or used to complement rule-based and first principles approaches.

1.1.3 FUNCTIONAL ROLES FOR INTELLIGENT SYSTEMS

The most common roles for intelligent systems are as a subset or combination of the five functional levels of the Data Fusion & Resource Management (DF&RM) Dual Node Network (DNN) technical architecture. These five levels have been defined in [5] based upon the DF&RM DNN technical architecture, which is an extension of the Joint Directors of Laboratories (JDL) fusion model. These dual resource-response management functional levels are depicted in Fig. 1.1 and described in Table 1.1.

To accomplish DF&RM intelligent systems are typically integrated into data and management mining, which are interlaced with each DF&RM level. The difference between these functions is summarized as follows:

- *Data mining* discovers and models aspects of the data input to each fusion node
- *Data fusion* combines data/information to estimate/predict the entity states at each fusion node
- *Management mining* "dual" discovers and models implications of commands to each RM node
- *Resource management* uses fusion products to plan/control "dual" resource responses at each RM node

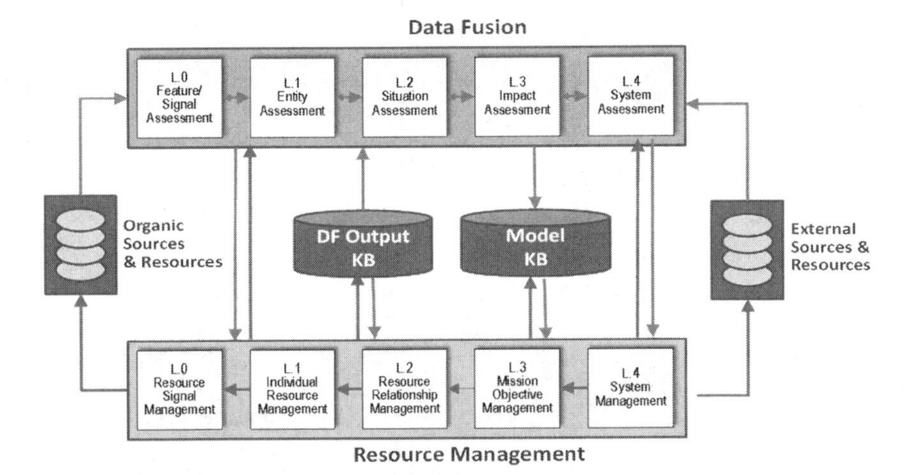

Fig. 1.1 DF&RM Dual Node Network (DNN) technical architecture.

DF&RM solutions as intelligent systems—or otherwise—are implemented as a network of interlaced DF&RM nodes. Fig. 1.2 shows an example of data mining nodes (see the middle row of functional boxes) interlaced with a data fusion node network (the bottom row of fusion nodes). This sample fusion node network shows only one fusion node per DNN fusion level. Typically, there are many fusion nodes per DNN level, which can be partitioned over time, input sources/ sensors, entity types, and other criteria. This partitioning is done to accomplish as much as possible as early as possible in the fusion node network.

These DF&RM nodes are implemented within and across the DF&RM functional levels described earlier to achieve the knee-of-the-curve in performance versus cost, with both reducing as additional decomposition is imposed. Each fusion node at each functional level performs all the classical data fusions, shown as follows:

- *Data preparation*: Assuring consistency in format, spatio-temporal and measurement framework, and confidence;
- *Data association*: Generating, evaluating, and selecting hypotheses of model scope (i.e., of the range of phenomena to be explained by the model)
- *State estimation*: Estimating and predicting the distribution and dependencies of characteristics and behavior of given entities or entity classes

The data fusion network is typically implemented as a fan-in tree with feedback minimized to achieve necessary performance. The management nodes perform dual fan-out functions and are interlaced with the fusion nodes to meet requirements. Figure 1.3 shows a sample interlacing of "dual" DF&RM nodes.

TABLE 1.1 DF&RM DUAL NODE NETWORK (DNN) FUNCTIONAL LEVELS

DF&RM	RM Level	RM Level Description	DF Level	DF Level Description
Level 0	Signal Management	Control individual response signals and actions for specific resources	Signal/ Feature Assessment	Detect/characterize/predict individual entity signals and features from available sources
Level 1	Independent Resource Scheduling	Task continuous and discrete individual resource responses (e.g., resource modes, attacks, countermeasures)	Entity Assessment	Detect/estimate/predict continuous parametric (e.g., kinematics, signature) and discrete attributes (e.g., ID, cause) of entity track states
Level 2	Resource Relationship Management	Task/control resource relationships (e.g., aggregation, coordination, de-confliction) among resource responses	Situation Assessment	Detect/estimate/predict relationships (e.g., aggregation, causal, command/control, coordination, adversarial relationships) among entity states
Level 3	Mission Objective Management	Establish/modify the objective of level 0, 1, 2 action, response, or relationship states based upon mission planning	Mission Impact Assessment	Predict/estimate the impact of signal, entity, or relationship states based upon course-of-action analysis
Level 4	System Management	Task/control/adjudicate DF&RM processes, and system engineering for distributed consistency, process performance, contextual conformity, problem-to-solution space mappings, and data mining model discovery	System Assessment	Detect/estimate/predict the performance, consistency, and context assessment of the level 0/1/2/3 outputs against truth, distributed processes, and relevant mission context

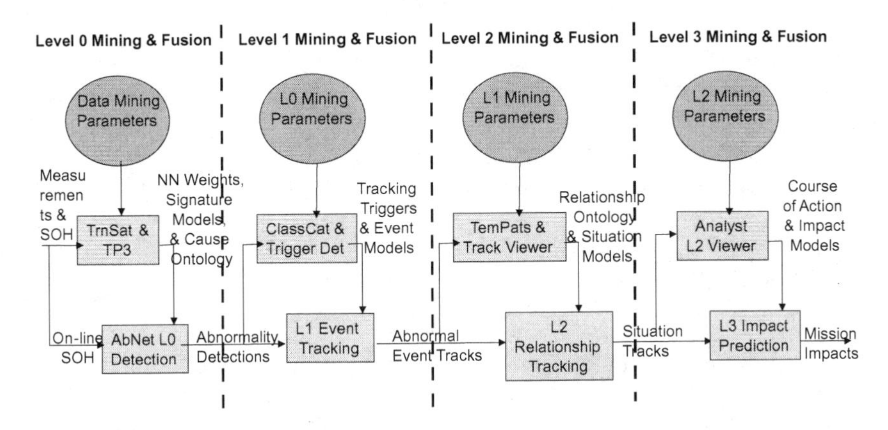

Fig. 1.2 Data mining interlaced with fusion node network.

An overlooked role of intelligent systems is to support a DF&RM system Context Assessment & Conformity Management (CACM). CACM methods opportunistically exploit data sources that are not used in the baseline DF&RM system to improve its robustness [6]. CA methods find relevant data in external data sources, and create and refine the baseline DF&RM products and the situational models. CA methods also measure the conformity of this relevant context data with the baseline fusion system products. CM methods decide what to do based upon the context assessment results, including flagging discrepancies,

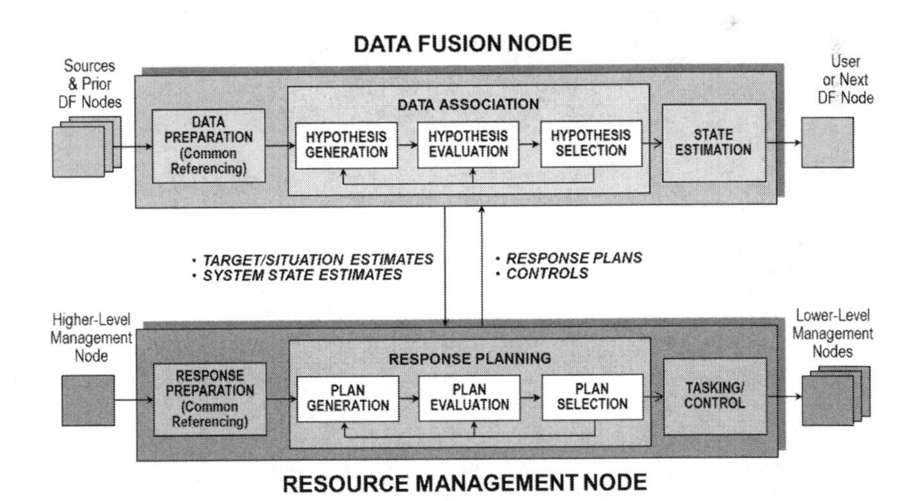

Fig. 1.3 Intelligent systems perform DF&RM in interlaced "dual" nodes.

modifying baseline DF&RM products, improving models, and managing response resources.

Intelligent systems are beginning to play a significant role in automatically learning to characterize burgeoning large volumes of external nontraditional "big data" and flag operationally relevant behaviors to enable their comparison with baseline data fusion systems products and models (e.g., taxonomies and ontologies). Such relevant external context data can improve the quantity, quality, availability, timeliness, and diversity of the baseline fusion system sources and therefore can improve prediction, estimation accuracy, and robustness at all levels of fusion. Performance Assessment and Process Management (PAPM) methods [7] are being used to manage intelligent system learning and operational parameters, to further improve their performance.

1.2 BUILDING DATA-DRIVEN AND GOAL-DRIVEN COMPUTATIONAL INTELLIGENT SYSTEMS

Intelligent systems are not the solution to every problem. This section describes the role for intelligent systems (i.e., when intelligent system solutions are recommended).

1.2.1 ROLE FOR INTELLIGENT SYSTEMS

We describe alternative solution techniques that find and characterize patterns to solve problems in diverse big data sets. The steps from data to information, to knowledge, to wisdom are as follows:

1. Leveraging the available diverse source data to understand patterns in currently evolving situations in space (i.e., entity and event relationships) yields information.
2. Understanding patterns in information with comparisons to truth yields knowledge.
3. Understanding general principles from knowledge, to predict consequences upon which actions are decided, yields wisdom.

Each step along this path is more difficult and provides corresponding benefits. We segment these problems into four parts progressing from easier to the more difficult detection and characterization sub-problems. Current SSA tools being developed are working on the first step to provide information from diverse data sources for the characterization of space objects and the identification of specific events of interest. A "divide & conquer" approach to this entity and event relationship detection and characterization problem can include four problem types, as shown in Fig. 1.4.

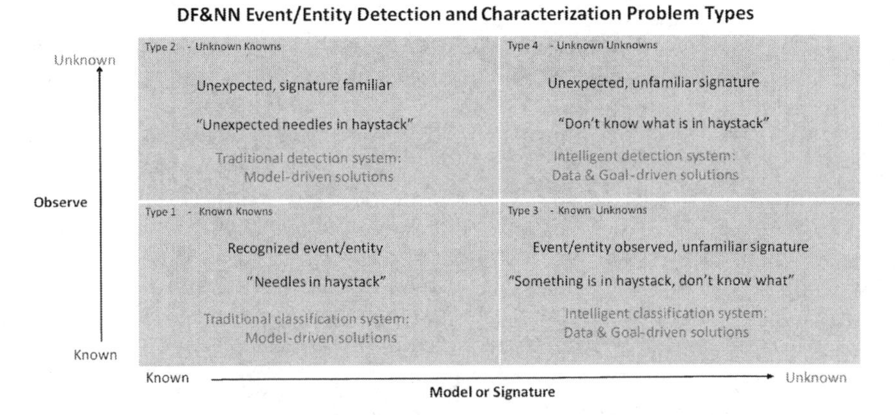

Fig. 1.4 DF&NN problem-to-solution space mappings guidelines.

The first two problem types are effectively solved using model-based solutions where we know the models for the objects and can derive solution criteria. For type 3, where models are unknown and classification of observation data is known, data-driven classification tools are applicable to learn the classification of these given signatures. Goal-driven techniques are used to maintain the system as new known signature types appear. For type 4, where models and entity classifications are not available, these problems are segmented into static and dynamic problems. The static problems are primarily driven by the historical data, such as for abnormality detection problems that have primarily fixed goals, and the dynamic problems are much more goal-driven, such as for adaptive systems where the goals can be fixed or dynamic. The solutions to these goal-driven problems typically have reduced performance (e.g., accuracy) as the problems progress from model-driven, to data-driven, and to goal-driven system solutions, as shown in Fig. 1.5.

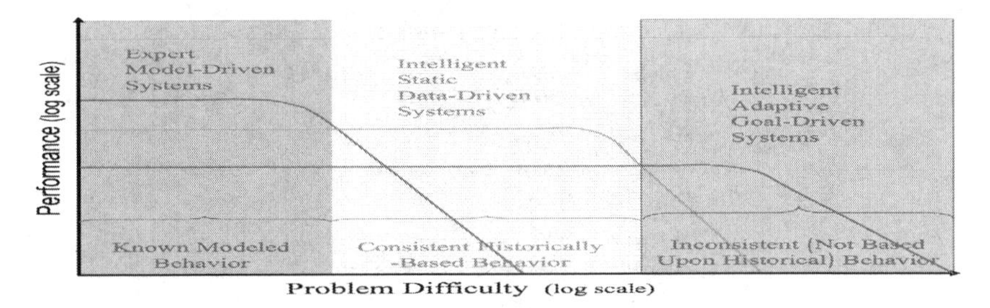

Fig. 1.5 Problem-to-solution space capability mapping depends upon problem difficulty.

This figure compares how intelligent system solutions provide higher affordable performance as opposed to developing solutions with hard-to-derive models (e.g., from model-driven to data-driven). This chart also shows the improved performance of intelligent goal-driven systems for hard dynamic problems. For type 4 problems, intelligent systems detect a range from unanticipated entities to meet user goals, to difficult multi-spectral entity detection problems. This is done without knowing the problem solution, such as expert rules and physics-based solutions based on models, because no such valid models can be affordably developed for type 4 problems. The resulting intelligent systems can be centralized or distributed in a hierarchy, operate in collaboration with humans on-the-loop, or autonomously, given access to relevant heterogeneous upstream data. The intelligent system learns correlations in background phenomenologies, so as to detect abnormal entity signature patterns.

1.2.2 DEVELOPMENT OF INTELLIGENT DATA-DRIVEN ABNORMALITY DETECTION SYSTEMS

When problem-to-solution space mapping indicates that an intelligent system is appropriate, the developer must survey available applications. Data-driven intelligent system solutions may need to include many of the following capabilities:

- Learn normal historical pattern behavior and correlations in the data over time
- Detect unexpected unknown abnormal temporal behavior and abnormal data correlations in real time even if the inputs values are within normal limits
- Cluster abnormal signatures for characterization
- Provide adaptive online abnormality suppression and enhancement
- Provide tools to manage detection rates
- Track abnormal clusters to create event tracks and characterizations
- Provide detection sensitivity analysis, and event track and characterization accuracy performance assessment

Not all of these will be needed in a typical problem. In developing the selected subset of these it is preferred that the intelligent system limit its dependence upon vigilance parameters, not be based only on statistical limits, avoid unsupervised approaches where possible, not use a fixed architecture, apply Occam's razor principle (i.e., use the simplest solution as possible), and use balanced training sets (i.e., avoid training on large sets of redundant similar normal behaviors and balance these with the smaller normal behavior subsets of the data), remove most of the bad data, limit uninteresting detections on rare data, ease expert naming of abnormality detections, and enable user detection sensitivity adaptation.

Some intelligent system tools already exist to address data-driven intelligent system problems. The ANOM (Anomaly) intelligent system developed by Data Fusion & Neural Networks (DF&NN), LLC is used in application examples in

Section 1.3. ANOM enables humans on-the-loop to improve performance. Examples of the tools illustrated in Section 1.3 include:

- **ClassCat:** An abnormal event classification category naming tool that names the possible causes of abnormal clusters and their confidences using historical context, detection signature similarity, and cause naming displays, so that future similar detection signature causes are named and user responses defined. The result is that these named clusters can be ignored (or not) depending on human system operator preferences.
- **Adaptive Online Abnormality Suppression:** The online adaptive abnormality suppression and enhancement service GUI enables the human system operator to easily and immediately get rid of persistent known and understood abnormality detections that they no longer want to see, while maintaining accurate classifications for other abnormality signatures occurring under these persistent undesired abnormality detections conditions. The reason for persistent undesired abnormality detections includes new phenomenology in the data being evaluated that were not present with the intelligent system machine learning was conducted.
- **Smoking Gun:** The ability to statistically rank cause and effect confidences from two sets of events (e.g., system abnormalities and environment or equipment events) given user cause and effect temporal, spatial, or event type conditions and allowed causality constraints and boundaries. This includes presentation to the user for tailoring of the constraints and selection of the cause and effects to be used online.
- **Retraining and Testing File Saving and Processing:** The ability to flag times containing undesired abnormality detection intervals so that they can be explicitly included for future "normal" behavior retraining that meets user retraining requirements. This also includes saving the test data and users requirements for adequate testing to allow promotion of the neural networks for online operations.
- **Offline Component Retraining:** The abnormality detection components are retrained on new system input data after a user-specified period of time. The retraining can also be performed if the average RMS error sensitivity increases beyond the desired noise floor. If the system is commanded into a new operational mode, then components will be trained again instead of retrained. This retraining significantly increases the sensitivity of the online abnormality detection tools to new fault patterns because the abnormality detection thresholds can be reduced.
- **Known Abnormality Detection Enhancement:** This retraining uses the known abnormal times as well as the normal historical data. This tool trains the components with an increased error on the outputs for these times. This modification can be adjusted to achieve the desired level of abnormality

detection enhancement for the subtle known abnormality signatures. The result is a single set of components that have been trained to detect both known and unknown abnormal signatures. This achieves a hybrid approach containing both supervised and unsupervised machine learning.

- **Detection Rate Management:** The Detection Rate Management (DRM) retraining capability enables the user to control the number of abnormalities reported by ANOM under "mostly normal" conditions via improved component training. This is accomplished by the tool, which automatically adjusts the abnormality detection parameters to achieve the analyst-selected average frequency of abnormality detections over a selected mostly normal historical data set. The individual measurand sensitivity values are automatically adjusted for those measurand variables that have the largest number of undesired abnormality detections compared to the size of the increase in the measurand sensitivity value. This capability enables an online detection rate management that will adapt the sensitivity of ANOM automatically online (e.g., a type of automated gain control that requires user approval). The output is the modified measurand sensitivity file for real-time abnormality detection, used to achieve the desired overall detection rate.
- **Historical Similarity Assessment:** The similarity assessment capability enables the user of the real-time or offline Abnormality Detection Classification Viewer (ADCV) to find similar historical abnormal cluster detection signatures. This is done by processing the Historical Abnormality Visualization (HAV) file over the selected historical time interval. These capabilities are used to provide the supervised feedback needed for further component retraining (e.g., for increasing the sensitivity for measurands that have too low sensitivity).

1.2.3 DEVELOPMENT OF INTELLIGENT GOAL-DRIVEN ABNORMALITY DETECTION SYSTEMS

In this section, we examine automated retraining determination, offline dynamic normal SOH behavior retraining, and retraining testing and online promotion.

Based upon user goals, the ANOM abnormality detection tools just described have been used to create neural networks (NNs) that can be automatically retrained on dynamically changing patterns. ANOM automatically determines when it is time to retrain, what to retrain on, what to test on, and when to promote the new NN architecture and weights per entity. This adaptive normal behavior retraining capability significantly reduces false abnormality scores of the online abnormality detection tools to new normal patterns. The NNs not only detect abnormal static measurand patterns, but also detect abnormalities in their correlations across input measurands and their temporal variations over user-selected time windows of interest. The NNs preprocess the incoming data to generate the standard deviation, linear regression slope, and frequency variations of each time varying input variable over one or many user-specified time windows for each. These are entered into the NNs along with the current variable values during normal behavior learning and during online real-time operations. The static and temporal abnormalities are detected in real time for each time batch of input measurands.

NNs are computational units implemented either in software or hardware that are modeled after the neuronal structure of the mammalian cerebral cortex. The neural network architecture (i.e., placement of nodes and connections) and the connection weights are optimized and trained using the selected training algorithm, based upon the number of training examples versus the number of patterns to be learned, such as the efficient back-propagation learning (e.g., Riedmiller & Braun algorithm) used for big data. Once these NNs are trained on normal behavior, then similar normal behavior will be predictable and score low. An example of the functional flow for such goal-driven abnormality detection is shown in Fig. 1.6. In one space data application, ANOM has automatically generated and maintained over 350,000 neural networks. This adaptive normal behavior recognition capability has to eliminate ringing abnormal scores of the online abnormality detection tools to new normal patterns. In this application, the user defines how long space object abnormal space catalog events are allowed to get before they need to be terminated by automated retraining and promotion of new NNs.

A management example of goal-driven driven processing is spacecraft smart structure L-band antennae adaptive control learning antennae vibration modes online [11]. In this application the L-band antennae system traditional controller can cause the antennae to go unstable when the antennae is unfolded in space to change its vibration modes, by as much as 20 percent. The neural controller is given the goal of maintaining vibration levels below a threshold. The neural controller modifies the traditional controller commands in real time and feeds back the effects of these control perturbations so as to learn the commands necessary to meet the vibration goals.

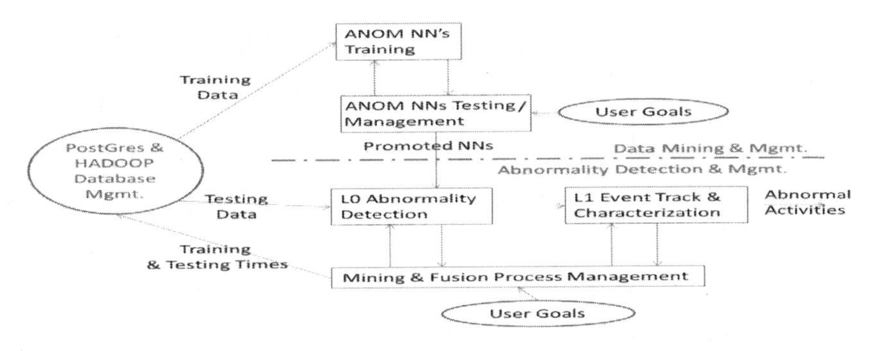

Fig. 1.6 ANOM intelligent system goal-driven solution functional partitioning.

In summary, the ANOM intelligent system level zero abnormality detection fusion tools perform the following for the applications in Section 1.3:

- Learning measurand temporal patterns and parameter correlations over time offline
- Detecting unanticipated abnormal dynamic behaviors in real time
- Finding and showing similar historical behaviors with drilldown to the raw inputs
- Adapting the system based upon user feedback
- Automatically adapting to dynamic systems normalcy changes based upon user goals to determine:
    - When to retrain
    - What to retrain on
    - What to test on
    - When to promote the updated application to online operations

1.3 SPECIFIC INTELLIGENT SYSTEM SPACE APPLICATIONS

In this section, we present three intelligent system space applications: prediction of the effects of space weather on satellites; abnormal space object orbital event detection, characterization, and prediction; and prediction of GPS outages due to ionospheric scintillation.

1.3.1 PREDICTION OF THE EFFECTS OF SPACE WEATHER ON SATELLITES

This intelligent system application exploits existing data to provide comprehensive situational awareness of the effects on satellite behavior due to space weather. It is based on intelligent software including: ANOM Detection, Smoking Gun, and Abnormality Detection Classification Viewer (ADCV). These tools are applied to historical Geostationary Operational Environmental Satellite (GOES) 11 State of Health and Space Environment data to include the Spacecraft Environmental Anomalies Expert System (SEAES) data for GOES 11. AFRL computed SEAES hazard quotients [26]. The results presented are the outcome of multiple years of effort and a decade-long vision. The team had to learn how to "exploit existing data to provide comprehensive situational awareness." With this intelligent system software, that vision is becoming a reality.

1.3.1.1 COLLABORATION

This work would not have been possible without cooperation and collaboration of multiple parties, including:

- Aerospace Corporation
    - Transformed GOES data into a format SAS could use

    - Created solar min and solar max SAS applications for GOES 10, 11, 12
    - Created SAS abnormality results to be used by Smoking Gun for GOES 10, 11, 12
- AFRL/RV
    - Created the space weather data files for GOES 10, 11, 12 to be used by Smoking Gun
- Data Fusion & Neural Networks
    - Created/evolved the Smoking Gun and ADCV tools
    - Performed correlation of the GOES abnormalities and space weather events
    - Determined initial results and performed drill down to individual mnemonic signatures

Previous work had been anomaly-centric. When an anomaly occurred, the anomaly resolution team tried to determine if the anomaly was caused by space weather. This new work takes a different approach. The first step is to detect the unknown-unknowns abnormal states in historical spacecraft telemetry. The second is to correlate the abnormalities with historical space weather phenomena. This approach has the benefit of providing a previously unprecedented level of situational awareness. This includes historical insight into low-level effects of space weather and actual telemetry signatures for each mnemonic affected by a specific space weather phenomenon. We initially used, and are reporting here, the process and tool use for GOES 10–12 telemetry because these satellites are retired. The reader can expect the tool use and space weather effects results to be similar for the current constellation of civil GOES satellites as well as non-weather satellites.

1.3.1.2 SPACE WEATHER ABNORMALITY ATTRIBUTION BACKGROUND AND NEED

Domain Awareness forms the framework for operations, planning, and decision making. Space Domain Awareness (SDA) brings knowledge of the operational space environment, its supporting ground elements and links, and the projection of its future status. To achieve effective space situational awareness, the SDA system architecture must provide the decision maker and user the right data, information, tools, and decision aids, at the right time. Using a net-centric service-oriented data fusion approach allows a rapid assessment of the situation, capitalizes on many available data sources, and adapts to situations in a timely manner. Relevant information can be gathered across a broad range of sources. Once this data is identified, it must be developed into actionable information for the decision maker.

Space assets are susceptible to numerous anomalous conditions, including space weather events, Radio Frequency Interference (RFI), proximity operations, conjunctions, bus failures, and other satellite anomalies. As a result, space system operators need an easy-to-use methodology for segregating the distinct causes of anomalous conditions. In addition, space system operators also need a simple system for collaborating and coordinating their investigation of anomalous conditions.

A semi-automated satellite mission re-planning system capable of detecting, reporting, and characterizing abnormalities and then recommending space-based asset responses will be integral to U.S. space mission assurance. This system will run continuously, monitoring the relationships between space asset events, the potential for satellite system outages, and the ongoing missions relying on space-based assets. The system will provide immediate input to the human-in-the-loop in the form of a series of satellite mission re-planning and response options.

There are five functional levels of SDA (e.g., incident detection/causality characterization, event tracking/characterization, event relationship assessment, mission impact prediction, and process and context assessments) and five dual response management levels within which the automated response decision aids of interest reside. These five levels have been described based upon the DF&RM DNN technical architecture, which is an extension of the JDL fusion model [6, 7, 9, 27, 28].

The Air Force Research Laboratory (AFRL) requires tools to develop, validate, transition, and operationalize Space Weather Attribution research technologies and incorporate them into civilian or military prototypes as risk reduction activities. The objective is to affordably integrate these capabilities into a net-centric information technology infrastructure. Activities include development of space weather attribution technologies, performance assessments, services that utilize real sources of space systems data such as SOH and SNR signatures, and authoritative space weather sources.

When doing attribution of satellite abnormalities, it is important to rule out environment causes (i.e., the interaction of the satellite with the surrounding space weather). The problem is that the current SSA community does not have a methodology to detect, characterize, and verify space weather effects on satellites. Notably, no automated tools are available to provide affordable real-time space weather attribution across all satellites. Also, no traceback of abnormality signatures to similar historical abnormality signatures is provided.

This effort addresses these problems by using over 10 years of real satellite SOH and space weather data at Geosynchronous Earth Orbit (GEO) to search for correlations of space weather events with the automatically detected abnormal events in each data set. High energy particles and electromagnetic (EM) radiation associated with space weather events can cause significant downtime and permanent degradation of satellite capabilities. An important in situ means of measuring space weather events in the geosynchronous domain is the set of GOES satellites, which measure X-rays and fluxes of energetic charged particles. At present it is difficult to leverage this data for real-time resource management and situational assessment. One roadblock is the difficulty in detecting (let alone predicting) the level of space weather conditions in real time and attributing the corresponding effects of the space weather conditions as measured on one satellite onto other satellites that are spatially and/or temporally removed from the measurement location. While ongoing efforts have had some success in correlating behavioral anomalies in satellite systems with space weather events on the same satellite, directly linking potential cause and effect on separate satellites require new data analysis approaches such as the DF&NN Temporal Patterns (TEMPATS) detection tool set.

Data assimilation models used in space weather research can provide a means to extrapolate sparse direct measurements to a full coverage estimation of the space environment, but these models require detailed physical models, as well as significant expertise, to configure and operate. In addition, they provide only a proxy estimate that then has to be linked to the potentially tagged abnormalities on the satellites of interest. In this effort, we research and develop an alternative non-parametric, statistical modeling approach that is more data driven. The goal is to build a simpler and more direct approach to identifying likely causes and effects as measured on spatially and/or temporally distant satellites. The techniques have been drawn from predictive modeling and data mining, including discretization of orbit regions and construction of independent variables that capture various temporal lags. Regression and related techniques have been used to identify the combination of derived variables that can most accurately predict space weather conditions and correlations with satellite system abnormalities in GEO, LEO, and MEO regimes. This combination of tools has been assessed in their ability to predict and characterize satellite system abnormalities in real time from fused space weather measurements—providing valuable environment versus man-made cause insight for satellite operator decision support.

The result should enable satellite operators, engineers, and leadership to have better SDA and thus the confidence to quickly and better choose response courses of actions. The automated anomaly characterization should drastically shorten the response time and increase the survivability of the satellites. This capability can be applied to new and existing satellite systems and fused with Satellite as a Sensor (SAS) abnormality detections and characterizations, navigation signal abnormalities on GPS, and radio frequency intrusions in the AFSCN Link Protection System (ALPS) reception bands.

1.3.1.3 APPROACH

This section describes the data, the tools, and how the tools are used. The real data processed includes:

- Abnormalities detected in GOES state-of-health telemetry
  - GOES 10 (2007–2008)
  - GOES 11 (2001–2008)
  - GOES 12 (2001–2008)
- Chronological space weather phenomena tailored to satellite's location in GEO from SEAES

  - Outputs of SEAES include: 1) surface charging, 2) internal charging, 3) energetic ion "single events," and 4) radiation dose
- The tools applied are as follows:
  - Smoking Gun to correlate satellite abnormalities and space weather phenomena; it has undergone numerous revisions over the past 10 years
  - ADCV to drill down and see telemetry signatures

Our satellite abnormality to space weather correlation quantification process begins with Smoking Gun, used to find correlations worth investigating (see Fig. 1.7). Second, we use ADCV to drill down and examine the signatures as shown in Fig. 1.8. More specifically starting with Fig. 1.7, the GUI loads spacecraft abnormality output (i.e., effects) detections from the DF&NN ANOM tools and the external space weather events that need to becorrelated (i.e., possible causes). The space weather events come from either the ANOM tools applied to NOAA space measurements (e.g., energetic electrons, protons, etc.) or other space weather detection tools such as SEAES. Many years of data can be loaded. The data can be filtered by any of the values shown on the top row. The filtered data can then be sorted by any of the columns in the next row. A common filter is by the number of abnormalities during the listed space weather event (as in Fig. 1.7) another is by the correlation Rate Ratio showing the ratio of the abnormalities per hour during the event divided by those not during the space weather event. For each space weather event of interest in the

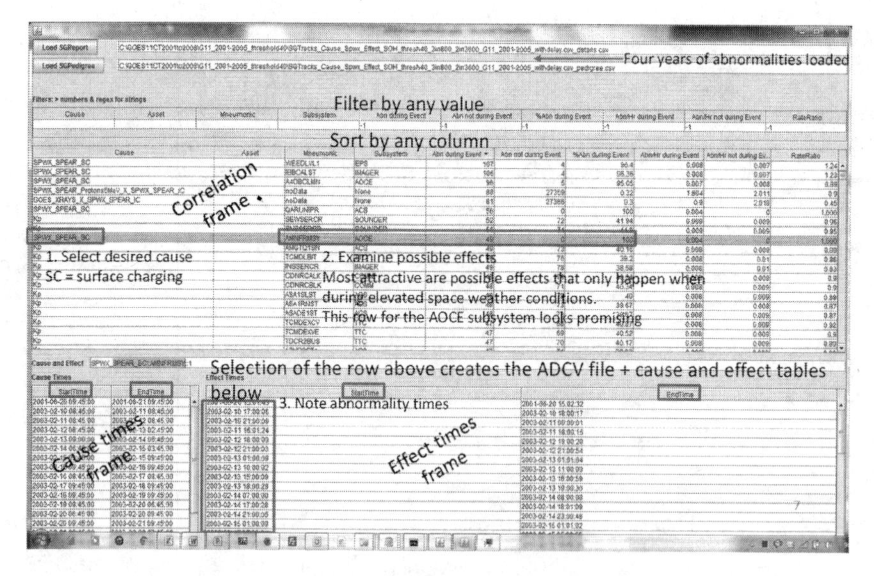

Fig. 1.7 A sample of how Smoking Gun is used.

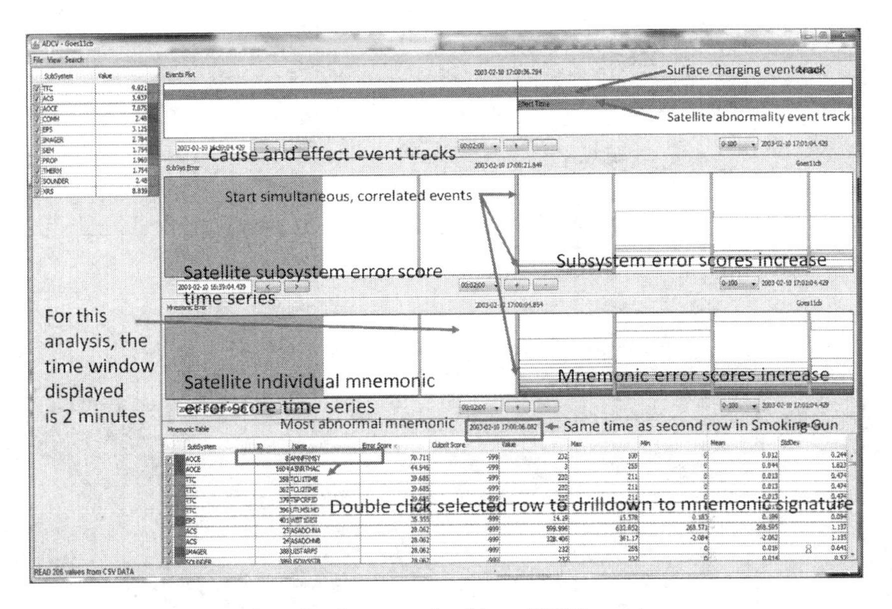

Fig. 1.8 An example of how ADCV is used.

top table, the Smoking Gun visualization bottom table shows the drilldown of all the cause event start and end times with all the correlated effect event start and end times.

The selection of a row in the Smoking Gun top table creates the ADCV visualization in Fig. 1.8. ADCV shows the satellite detected abnormality details with drilldown to the input satellite abnormal SOH strip charts over any selected time window. The subsystem abnormality scores are shown on the top left. The rest of the screen shows three strip charts (i.e., abnormal SOH and space weather events, satellite subsystem abnormality scores, and individual satellite mnemonic error scores) over the selected time window and the table of the ranked list of measurands at the selected time point of interest in the window. This mnemonic (i.e., SOH measurand and temporal variables) table shows the mnemonic subsystem, ID number, mnemonics name, its abnormality error score as compared to normal from ANOM (i.e., range 0 to 100), culprit name and confidence as to the measurand that has caused the abnormality, the current value of the mnemonic as used by ANOM (e.g., value at the last time it came in or −999 if no valid value was available), then the four statistics of the mnemonic over the ANOM training set (max, min, mean, and standard deviation).

Figure 1.9 shows the selection of one of the mnemonics in the table (in this case, the highest abnormality scoring) and the strip chart that pops up when double clicked, showing the values of that mnemonic in the raw input data.

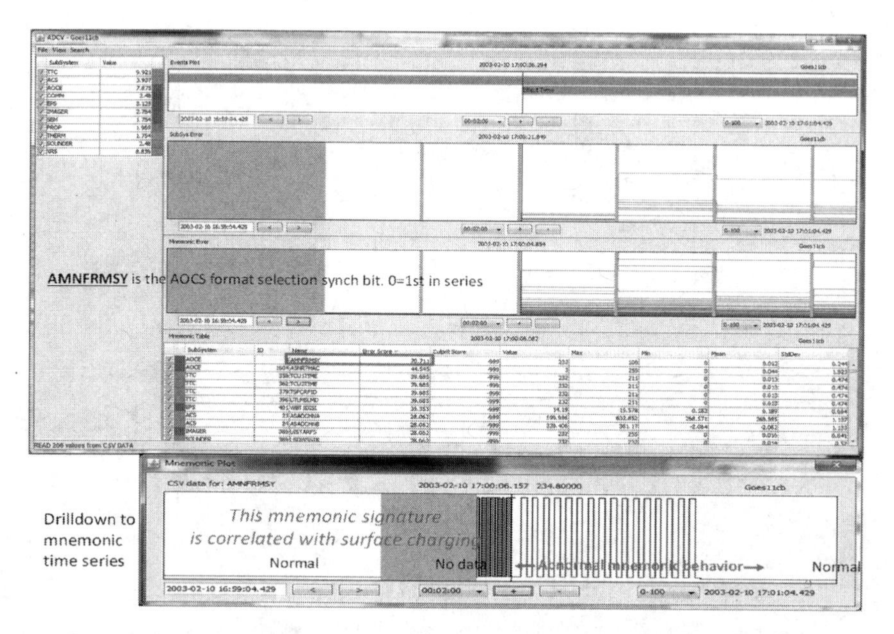

Fig. 1.9 A sample of the results to be described summary and recommendations.

The GOES 11 results found correlations in three of the subsystems investigated. These are summarized as follows:

- AOCE: Same abnormal Format Selection Synch Bit signatures repeat, but *only during surface charging*. The affected mnemonic quickly returns to normal behavior.
- ACS: Interesting that thruster duty cycles might change only during surface charging events. Mnemonics monitoring other thrusters (not shown) exhibited similar behavior.
- EPS: Rapid fluctuations in battery discharge current during surface charging events (not shown) probably is not desirable.

This is the first time we have seen repeatable signatures in telemetry that have high correlation scores with space weather phenomena. The time scale for these effects is relatively small (on the order of seconds). While interesting, these results should not be considered definitive. This capability should be applied to additional spacecraft that are currently providing operational support so that spacecraft operators and engineers can assess the value of this capability.

1.3.1.4 CONCLUSIONS

DF&NN provided real-data-based Smoking Gun and ADCV prototype tool demonstrations of GOES satellite SOH abnormality correlations with nearby abnormal space weather at GEO. These results can be fused with other sources using the DF&NN Bayesian Fusion Node (BFN) web services.

Benefits are automated space weather satellite abnormality attribution, satellite anomaly forecast, visualization with drilldown to engender trust, multiple source integration, and validation tools to characterize abnormal SOH and space weather events detected online at operational sites using the DF&NN patented BFS ANOM data and goal-driven abnormality detection and characterization system [2, 29–31]. The competitive advantages of the abnormality characterization technology are in its affordability (derived from the data-driven normal pattern learning software); its ability to automatically detect and characterize unexpected abnormal signatures; and its extendibility/reusability, derived from the Dual Node Network (DNN) DF&RM technical architecture. Operational prototypes of this capability are being evaluated at two satellite operations sites and these ANOM tools have been applied offline to over 100 different large real data sets on over 200 combined years of data.

1.3.1.5 RECOMMENDATIONS

Our next step is to secure AFRL direction on near term way forward priorities. These may include the following:

- Obtain substantial documentation of results prior to get GOES' subject matter expert (SME) feedback
- Continue documenting possible space weather signatures on GOES 11, other GOES, and other satellites. Question is whether to go deep on GOES 11 or analyze results for other GOES satellites (to see if space weather vulnerabilities of GOES 11 are also seen on other satellites)
- Focus on other satellite SOH and space weather correlations
- Coordinate with the space weather, satellite manufacturer, and satellite operations community to get feedback

Future research directions recommended include the following:

1. Go deeper on GOES 11 historical data
   - Document all signatures for each mnemonic/space weather phenomena correlation on GOES 11
   - Examine all correlations on GOES 11 to include detection of temporal correlations using TEMPATS
2. Repeat the process for other GOES satellites (10 and 12)
   - Determine if there are any similarities between signatures seen on more than one satellite

- Determine if there are differences between space weather correlations during solar max and solar min

3. Ask GOES SME(s) for perspective on findings
4. Perform similar analysis for other satellite constellations once results are affirmed
5. Include existing DMSP data and upcoming NOAA data on test to Smoking Gun for LEO satellites
6. Introduce automation to perform the analysis and save hours of human SME analysis
7. Give feedback on accuracy of the current space environment event thresholds
8. Quantify probability of space weather effects given space weather event correlations
9. Identify operational applications to support AFRL entity assessment initiatives

1.3.2 ABNORMAL SPACE OBJECT ORBITAL EVENT DETECTION, CHARACTERIZATION, AND PREDICTION

The second intelligent system application is the Abnormal Catalog Update (ACU). ACU is a goal-driven enterprise-level client–server system that uses neural networks to analyze and detect abnormal events and incorrect satellite track taggings in space catalog updates of the two-line Keplerian element (TLE) records. Data derived from historical TLE records for each object are used to train individual neural networks to discover and learn expected deviations between an object's predicted and observed orbital state. These trained neural networks are then used to find unexpected deviations in new TLE records resulting from an abnormal orbital maneuver, from an erroneous tagging, or from other causes. To accurately detect abnormalities across a wide range of objects, the neural network architectures in ACU are optimized for each Resident Space Object (RSO) to avoid dependencies on individual orbital states or specific propagation models. The ACU then characterizes and tracks the sequences of abnormal catalog updates for each RSO.

1.3.2.1 FINDING THE UNKNOWN-UNKNOWNS ABNORMAL SPACE CATALOG BEHAVIORS

Built upon a distributed, enterprise-level client–server architecture, the Abnormal Catalog Update (ACU) is a data fusion analysis service that provides the ability to automatically assess orbital catalog data across tens of thousands of objects, flagging and characterizing abnormal TLE records. ACU is an intelligent turnkey system that is user-goal driven. ACU determines on its own for each RSO when to retrain, what to retrain on, what to test on, and when to promote the new neural networks adapted to each RSOs new dynamic behavior. The resulting ACU products are used as context to provide improved characterizations for the baseline conjunction prediction system tracks. Timely detection and characterization of unexpected space object behavior is of vital importance. Unfortunately, online and historical data of this type are not available for the vast majority of RSOs. The ACU system is based on the premise that unanticipated changes in an object's orbital state can be detected by learning normal behavior in order to flag unknown abnormal behavior. ACU is designed to serve as an initial warning system capable of detecting abnormalities across the entire space catalog for every catalog update. Specifically, ACU is a turnkey enterprise-level system for automatically detecting and characterizing abnormal space catalog updates across thousands of space objects using the TLEs.

An important objective of the SSA mission is to provide end users with the most accurate information about the location, status, activities, and intents of all objects in space. The space catalog contains the most current location information about all known objects. The ACU service that has been developed by DF&NN for AFRL/RV provides an enhanced awareness of the state of all known objects by monitoring the entire space catalog in real time. Detections and characterizations can be presented to end users through a variety of interfaces, including the fusion system UDOP (user-defined operating picture), an interactive, cross-platform Rich Internet Application Interface residing in a web browser, and as XML output over HTTP.

This new system is implemented as a web service that is compatible with the JMS architecture and has been deployed in the AFRL/RV Advanced Research Collaboration and Development Environment (ARCADE). ACU is a multi-user system that is scalable, extensible, supports a variety of user levels, and permits an analyst to customize the detection thresholds, reporting filters, and other analysis parameters for individual objects. ACU is designed to fuse the outputs from other SSA tools, such as maneuver detection, conjunction prediction, space weather sensors, etc. This capability will enable operators to quickly assess the state of all objects, identify those with suspect behavior, and drill down for detailed information to determine possible courses of action. Information provided by ACU can assist operators in making decisions about intent and danger-level of objects, changes in object configuration, location, and mission, Space Surveillance Network (SSN) tracking performance evaluation, space weather effects, conjunction prediction response, and SSN sensor tasking.

Because TLEs are readily available, they have been used as a data source in a wide variety of space-related problems, such as detecting satellite thrust maneuvers, estimating collision probabilities, and even calibrating thermospheric density models. However, care must be exercised when employing computational methods that depend on assumptions regarding the accuracy of the TLE orbital parameters, as the precision of these parameters is uncertain and varies both between objects and over time for a given object. The next version of ACU will use more accurate ephemeris data.

1.3.2.2 ACU ABNORMAL TLE DETECTION AND CHARACTERIZATION

One analytical approach that has proven successful in situations where data is noisy or missing is to use neural networks. Like the brain, NNs can automatically learn normal behavior and then generalize to detect previously not encountered situations, performing well even when the incoming data is noisy, incomplete, or inaccurate. As a consequence, NNs have been employed in such ill-structured domains such as the classification of satellites, detection of changes in and segmentation of satellite imagery, optimization of satellite broadcast schedules, and detection and prediction of space weather events.

In ACU, NNs are used to first learn the behavior of each object in the space catalog using a training data set derived from historical TLEs for that object. ACU then uses this knowledge to subsequently detect unexpected unknown deviations from this learned normal behavior in incoming TLEs (e.g., abnormal maneuvers). NNs perform in two different modes: *training* and *testing*. Training is an adaptive process by which an NN modifies its internal structure based on a set of examples, so that it provides a desired response when presented with a set of input stimuli. Once trained, NNs are able to quickly detect a new pattern from online observations without any prior knowledge of the pattern itself.

The vast diversity of objects within the space catalog, combined with a varying sampling rate and suspect precision of the TLEs themselves, necessitates that ACU construct a sequential collection of NNs dedicated to each object in the catalog. Each NN is responsible for analyzing a sliding window of chronologically ordered TLE data for that specific object, and trained using the TLE data immediately preceding the beginning of its testing period. As part of the training data, normal station keeping maneuvers and other normal variations (e.g., measurement noise) are learned by the NNs.

The training success of NNs depends on the number of examples available. Because different objects in the catalog are sampled at different rates, the lengths of training periods and testing periods are determined by the number of data records rather than by a fixed unit of time. As a result, NNs for objects that are sampled more frequently will have training and testing periods that are shorter in duration than objects with low sampling rates. As new TLE data for a given object are incorporated into ACU each day, ACU attempts to analyze the new data using an existing, already trained NN. However, if the testing period for the current NN has expired, then a new NN is created and trained automatically.

The SGP4/SDP4 orbital model is used to generate ephemeris data for each TLE. This data is used as the input parameters for each NN. The error between the NN output and observed values of these measurands is used to calculate a single root mean square (RMS) error value for each TLE record. The time series of these RMS error values is then evaluated to flag significant changes in RMS error values, indicating an abnormal event. The results are summarized on the ACU query dashboard shown in Fig. 1.10.

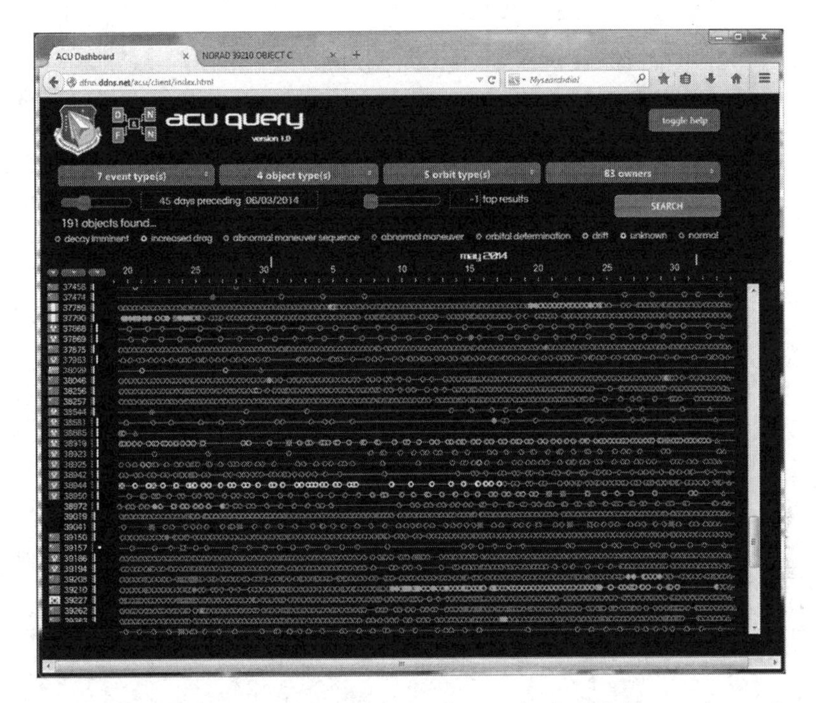

Fig. 1.10 Sample ACU query dashboard shows characterization of the top abnormal catalog updates.

The user specifies which of the following they want displayed:

- 7 event types (e.g., abnormality maneuvers, abnormal maneuvers sequences, drift, orbital determination, increased drag, unknown)
- 4 object types (e.g., satellite, rocket booster, debris, undetermined)
- 5 orbit types (e.g., GEO, MEO, LEO, HEO, undetermined)
- 83 owners (e.g., ESA, USA, India, Iran, China, Russia, etc.)

The user can also specify the time period of interest and the number of top results that they want to see. When the user finds an RSO of interest they can double click on it to drill-down to see the characterization confidences; the top contributing TLEs abnormality scores; the strip plots of the TLEs; and the strip plots of the latitude, longitude, and altitude over the selected historical time periods of interest. These strip charts show every catalog update so that changes in update rate are readily apparent. The user can also mouse over some of the characterizations to drill down for more details such as a list of possible confusers for a satellite observation miss-tagging meaning that the NORAD TLE was actually represents a different object. An example is shown in Fig. 1.11. To focus the display, the user can select which of the TLEs and locations they want in each strip chart.

An overview of the Continuous Anomalous Orbital Situation Discriminator + Conjunction Analysis (CAOSD+CA), ACU, and other sources fusion node network is shown in Fig. 1.12. The figure depicts how ACU fits into a more complete space object management system. The ACU Bayesian Fusion Node (BFN) is used to track and characterize ACU detection results. Its ACU tracks are fused with the CAOSD+CA and other sources to enable the user to focus on the specific conjunctions of interest.

The BFN logic can be expanded online by the human analyst. Examples of this logic include the following:

- CAOS-D proximity report or intelligence can initiate event (e.g., a conjunction prediction)
- Subsequent proximity reports and intelligence reports are associated by event time and satellite and used to update the conjunction track

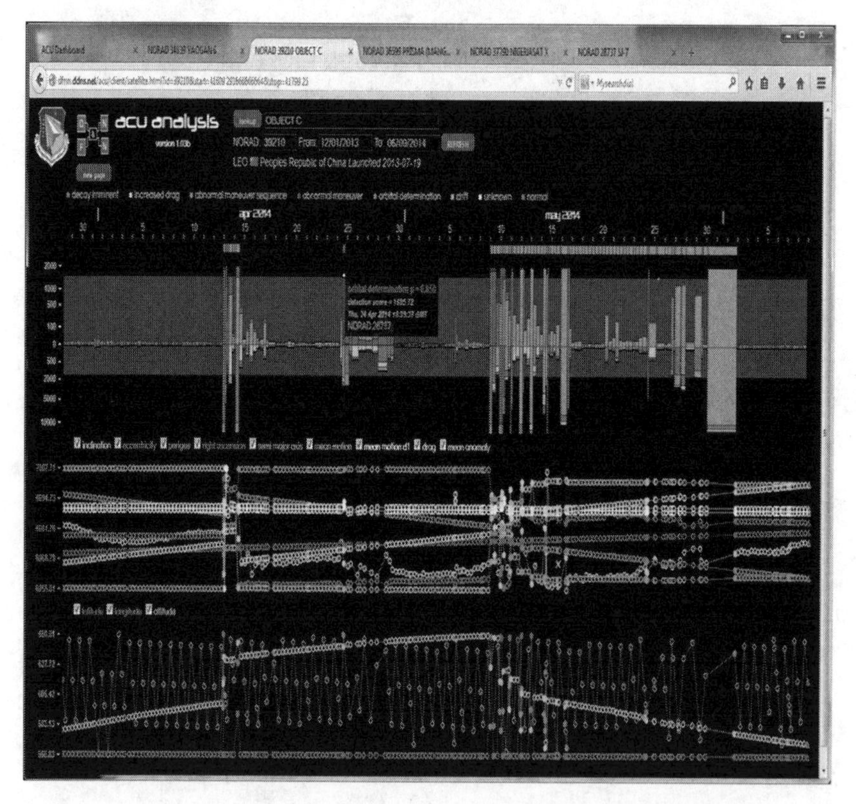

Fig. 1.11 Sample ACU drilldown for an abnormal RSO catalog update.

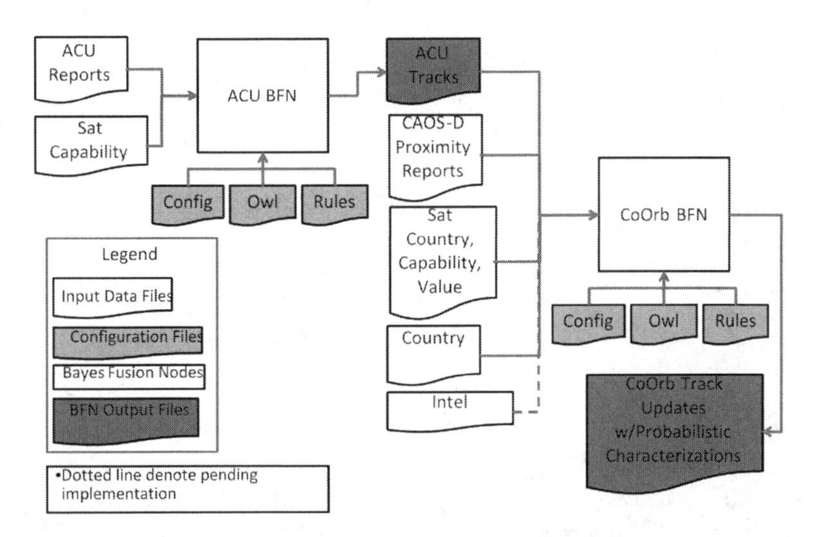

Fig. 1.12 ACU, conjunction prediction, and other sources fusion node network.

- Tracks expire when the last estimate of the event end time is past
- Distance thresholds are defined to characterize the conjunction (e.g., <10 km = “near”, 10–50 km = “far”, >50 km = “none”)
- Relative speed is used for additional conjunction characterization (e.g., fast = “Destructive”)
- RSO type and capabilities plus recent ACU tracks with maneuver characterizations improve conjunction cause characterization (e.g., “Accidental” vs. “Proximity Operations” with culprit maneuver detections)
- Blue/Red/Grey (i.e., friendly, not friendly, neutral) status, satellite value improve intent characterization (e.g., “non-cooperative” vs. “Test”)
- Intelligence can improve characterization of events (e.g., if Jamming resulting in low SNR, then P(non-cooperative) increases

The result is timelier and cost effective with more significant conjunction detection and characterization.

1.3.2.3 SYSTEM ARCHITECTURE

Given that the space catalog contains thousands of objects that are currently in orbit, and that each of these objects is associated with a growing list of NNs within ACU, efficiently handling the logistics of procuring the daily updates of TLE data, generating and training the NNs, and then testing for abnormal events is a critical operational requirement. There are currently over 280 K

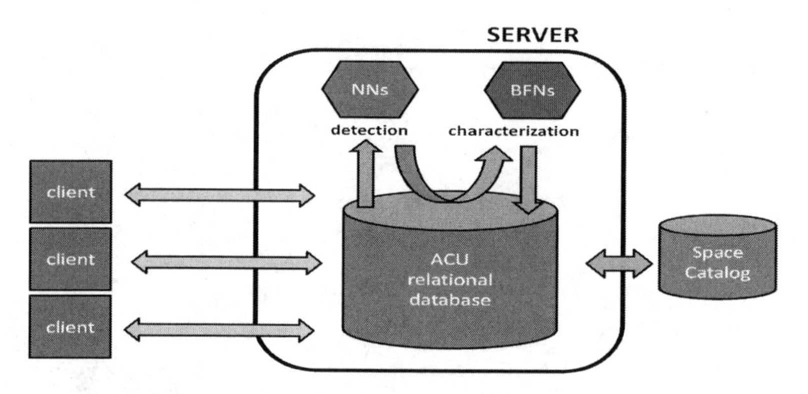

Fig. 1.13 Overview of ACU system architecture.

NNs available to create the ACU user products from 2008 to the present. To accomplish this, ACU utilizes an enterprise system based on a client–server model, shown in Fig. 1.13. Specifically, ACU consists of four key components: a back-end relational database, a suite of analysis modules managed by a server-side application server, a web server for delivering the client use interface, and a client-side Rich Internet Application (RIA).

1.3.2.4 ACU PERFORMANCE ASSESSMENT RESULTS

The historical data scenario used for the initial performance assessment of ACU follows:

- 6/1/2010–1/15/2011 space catalog with 18,000 resident space objects (RSOs)
- 28.7 million TLE records processed since 2008 to learn normal and flag abnormal TLE updates
- 280,000 neural networks trained to operate on line for recent and historical abnormal TLE flagging
- 18,000 Bayesian Fusion Nodes for abnormal event characterization
- Baseline CAOS-D conjunctions:
    - Events of interest defined by user:
        - 2010 active satellites with 870 satellites in <1000 km orbit
        - All proximity events with <10 km closest range flagged
        - Slow relative velocity events with <100 km closest range flagged (5–10% of total)
    - Conjunctions predicted with 10-day horizon, 1 run/day, 570 satellites involved
    - 565,000 conjunction reports generated with 100,000 BFN possible conjunction tracks

  - 1700 “interesting” BFN conjunction tracks with p(“no proximity event”) <0.5
- ACU detections
  - 65 abnormal satellite updates out of ~20,000 reports total
  - 28 BFN characterized event tracks

Fig. 1.14 shows how much ACU helps focus the user from over 500,000 of conjunction predictions to proximity events on interest with confidences. The figure shows the proximity operations probabilities with and without ACU. The lightest and darkest highlighted (non-zero) events are where P(proximity operations) has improved due to BFN tracking.

Fig. 1.15 also shows the proximity detection quantitative performance results for late 2010 that would have been achieved had this system been operational then. The track characterization error is lowered with the addition of ACU, which allows the user to focus on the conjunctions of highest mission interest. This performance assessment was accomplished using the DF&NN DNN Level 4 Performance Assessment and Process Management (PAPM) software that automatically computes the user specified Measures of Performance (MOPS) during PA.

The PM was then used to optimize the co-orbital proximity BFN detection settings for both the baseline fusion system and the ACU context augmented system. Both runs highlight more optimal settings than the analyst's starting fusion algorithm settings. In this case the PAPM cut track-wise characterization error from 0.45 to 0.31 to 0.12 error. This was accomplished using the Pareto optimal fronts that the PAPM interactions find. These Pareto optimal fronts were

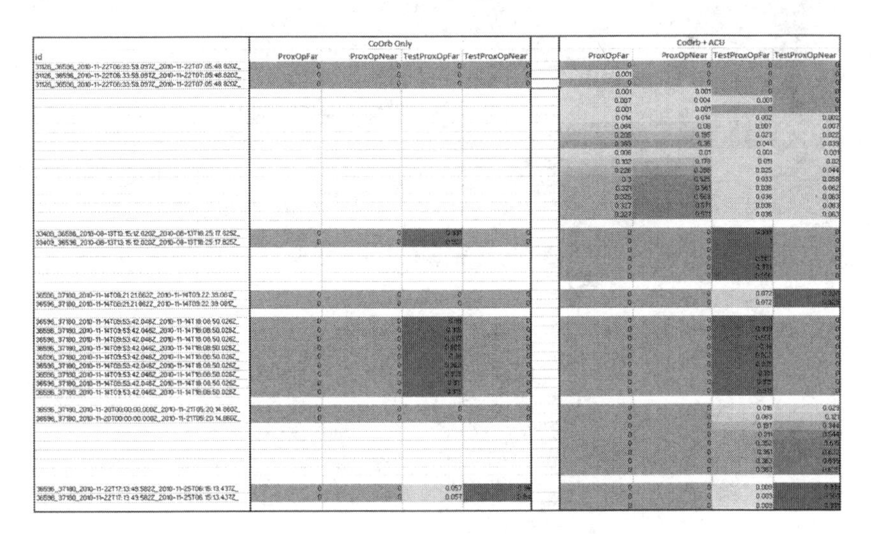

Fig. 1.14 Quantitative performance for co-orbit conjunction only versus with ACU fused.

MOPS	CP BFN – with/ACU	CP BFN – CAOSD only
Pd (% of Truth Events Detected | % of Truth Minutes Detected)	100.0% | 98.5%	100.0% | 98.5%
Pfa (% of Fusion Events Unmatched to Truth | % of Fusion Minutes in Unmatched Fusion Events| % of Fusion Minutes in Unmatched Fusion Events + those in Matched Fusion Events outside of Truth Event span)	26.7% | 0.4% | 19.3%	26.7% | 0.4% | 19.3%
Truth Event Totals (# Truth events | tot mins in all truth events | avg mins per truth event)	5 | 116501 | 23300	5 | 116501.0 | 23300.2
Fusion Event Totals (# Fusion events | tot mins in all fusion events | avg mins per fusion event)	15 | 142255 | 9483	15 | 142255.7 | 9483.7
Matched/Detection Events Totals (#events | tot mins | avg mins)	5 | 114733 | 22946	5 | 114733.2 | 22946.6
False Alarm Fusion Message Totals (#events | tot mins | avg mins)	4 | 508 | 127	4 | 508.7 | 127.2
Missed Truth Event Totals -- False Negatives (#events | tot mins | avg mins)	0	0
Missed Truth Event Minutes (w/late starting or early ending fusion events)	1,767	1767
Unmatched Fusion Event Minutes (including early starting or late ending fusion events)	27,522	27522.5
Avg Delay in start of fusion tracks compared to truth tracks for matches (mins)	-1204	-1204
Avg Delay in end of fusion tracks compared to truth tracks for matches (mins)	6488	6488.9
Characterization err (# | min | max | avg | time weighted average)	11 | 0 | 1 | 0.124 | 0.006	11 | 0 | 1 | 0.182 | 0.016
Truth Time (secs | % of Total)	6990060 | 1.58%	6990060 | 1.58%
Fusion Time (secs | % of Total)	8535342 | 1.93%	8535342 | 1.93%

Fig. 1.15 Co-orbital proximity performance assessment Comparison of results with and without ACU for late 2010.

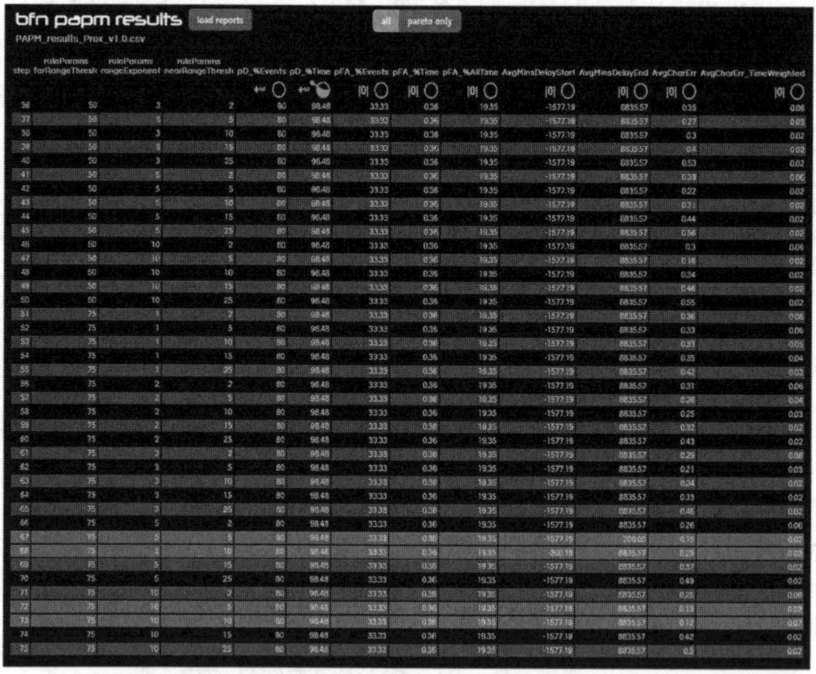

bfn papm results load reports all pareto only

PAPM_results_Prox_v1.0.csv

step	ruleParams farRangeThresh	ruleParams rangeExponent	ruleParams nearRangeThresh	pD_%Events	pD_%Time	pFA_%Events	pFA_%Time	pFA_%AllTime	AvgMinsDelayStart	AvgMinsDelayEnd	AvgCharErr	AvgCharErr_TimeWeighted
36	50	3	2	80	98.48	33.33	0.36	19.35	-1577.19	8835.57	0.35	0.06
37	50	3	5	80	98.48	33.33	0.36	19.35	-1577.19	8835.57	0.27	0.03
38	50	3	10	80	98.48	33.33	0.36	19.35	-1577.19	8835.57	0.3	0.02
39	50	3	15	80	98.48	33.33	0.36	19.35	-1577.19	8835.57	0.4	0.03
40	50	3	25	80	98.48	33.33	0.36	19.35	-1577.19	8835.57	0.53	0.02
41	50	5	2	80	98.48	33.33	0.36	19.35	-1577.19	8835.57	0.33	0.06
42	50	5	5	80	98.48	33.33	0.36	19.35	-1577.19	8835.57	0.22	0.02
43	50	5	10	80	98.48	33.33	0.36	19.35	-1577.19	8835.57	0.31	0.02
44	50	5	15	80	98.48	33.33	0.36	19.35	-1577.19	8835.57	0.44	0.02
45	50	5	25	80	98.48	33.33	0.36	19.35	-1577.19	8835.57	0.56	0.02
46	50	10	2	80	98.48	33.33	0.36	19.35	-1577.19	8835.57	0.3	0.06
47	50	10	5	80	98.48	33.33	0.36	19.35	-1577.19	8835.57	0.18	0.02
48	50	10	10	80	98.48	33.33	0.36	19.35	-1577.19	8835.57	0.34	0.02
49	50	10	15	80	98.48	33.33	0.36	19.35	-1577.19	8835.57	0.46	0.02
50	50	10	25	80	98.48	33.33	0.36	19.35	-1577.19	8835.57	0.55	0.02
51	75	1	2	80	98.48	33.33	0.36	19.35	-1577.19	8835.57	0.36	0.06
52	75	1	5	80	98.48	33.33	0.36	19.35	-1577.19	8835.57	0.33	0.06
53	75	1	10	80	98.48	33.33	0.36	19.35	-1577.19	8835.57	0.33	0.05
54	75	1	15	80	98.48	33.33	0.36	19.35	-1577.19	8835.57	0.35	0.04
55	75	1	25	80	98.48	33.33	0.36	19.35	-1577.19	8835.57	0.42	0.03
56	75	2	2	80	98.48	33.33	0.36	19.35	-1577.19	8835.57	0.31	0.06
57	75	2	5	80	98.48	33.33	0.36	19.35	-1577.19	8835.57	0.26	0.04
58	75	2	10	80	98.48	33.33	0.36	19.35	-1577.19	8835.57	0.25	0.03
59	75	2	15	80	98.48	33.33	0.36	19.35	-1577.19	8835.57	0.32	0.02
60	75	2	25	80	98.48	33.33	0.36	19.35	-1577.19	8835.57	0.43	0.02
61	75	3	2	80	98.48	33.33	0.36	19.35	-1577.19	8835.57	0.29	0.06
62	75	3	5	80	98.48	33.33	0.36	19.35	-1577.19	8835.57	0.21	0.03
63	75	3	10	80	98.48	33.33	0.36	19.35	-1577.19	8835.57	0.24	0.02
64	75	3	15	80	98.48	33.33	0.36	19.35	-1577.19	8835.57	0.33	0.02
65	75	3	25	80	98.48	33.33	0.36	19.35	-1577.19	8835.57	0.46	0.02
66	75	5	2	80	98.48	33.33	0.36	19.35	-1577.19	8835.57	0.26	0.06
67	75	5	5	80	98.48	33.33	0.36	19.35	-1577.19	200.00	0.15	0.02
68	75	5	10	80	98.48	33.33	0.36	19.35	-200.19	8835.57	0.25	0.03
69	75	5	15	80	98.48	33.33	0.36	19.35	-1577.19	8835.57	0.37	0.02
70	75	5	25	80	98.48	33.33	0.36	19.35	-1577.19	8835.57	0.49	0.02
71	75	10	2	80	98.48	33.33	0.36	19.35	-1577.19	8835.57	0.25	0.06
72	75	10	5	80	98.48	33.33	0.36	19.35	-1577.19	8835.57	0.13	0.03
73	75	10	10	80	98.48	33.33	0.36	19.35	-1577.19	8835.57	0.12	0.07
74	75	10	15	80	98.48	33.33	0.36	19.35	-1577.19	8835.57	0.42	0.02
75	75	10	25	80	98.48	33.33	0.36	19.35	-1577.19	8835.57	0.5	0.02

Fig. 1.16 Integrated PA and PAPM reduces parameter search space for "fair" assessments.

compared to the other sensitivity analysis results and shown to the user using the BFN PAPM results display, such as shown in Fig. 1.16 where the light grey rows are the Pareto optimal fronts. The visualizations allow drilldown to the ACU events, which with the ACU visualizations allow drilldown to the characterization confidences, the TLE values, and the latitude/longitude/altitude strip plots, as described in Kraus et al. [25].

1.3.2.5 SUMMARY

The ACU system is the only known system able to provide unknown abnormality detections with characterizations based upon the entire space catalog in real time. "Normal" TLE behavior changes dynamically over time and across objects. The quality and frequency of sampling data varies both across objects and over time. The objects themselves are a heterogeneous group in all types of orbits (e.g., MEO, LEO, GEO, and HEO). To solve this problem, ACU exploits the learning capabilities of neural networks. NNs are trained to learn the expected normal TLE update behavior. These trained NNs for each object are then used to detect and characterize unexpected unknown abnormal deviations in the TLE incoming data. Based upon user goals ACU automatically determines when to retrain, what to retrain on, what to test on, and when to promote new NNs for each RSO in the space catalog.

1.3.3 PREDICTION OF GPS OUTAGES DUE TO IONOSPHERIC SCINTILLATION

The final intelligent system application we illustrate performs prediction of GPS site-satellite pair outages due to scintillation and is important for high confidence GPS operations. This application shows how we have improved the prediction confidences for GPS scintillation worldwide. We do this by combining abnormal GPS SNR detections for many hundreds of GPS satellite and receiver site pairs, plus by using space weather data as context. Adding this global data context assessment to the traditional S4 smoothed high SNR variance detections reduces the number of significant GPS SNR event tracks and provide more confident cause characterizations.

1.3.3.1 SPACE WEATHER ATTRIBUTION BACKGROUND AND NEEDS

Situational Awareness forms the framework for operations, planning, and decision making. Space Situational Awareness (SSA) brings knowledge of the operational space environment, its supporting ground elements and links, and the projection of its future status. To achieve effective space situational awareness, the SSA system architecture must provide the decision maker and user the right data, information, tools, and decision aids at the right time. Using a net-centric service-oriented data fusion approach (i.e., fusion services) allows a rapid assessment of the situation, capitalizes on many available data sources, and adapts to situations in a timely manner. Information will need to be gathered across a broad range of DOD, civil, and commercial sources. Once mission relevant data is identified, it must be fused into actionable information for the decision maker.

Because space assets are susceptible to numerous anomalous conditions, including space weather events, communications faults, Radio Frequency Interference (RFI), conjunctions, bus failures, and other satellite anomalies, the decision makers will need a distributed satellite resource management system to effectively accomplish their space missions. Near real-time integrated SSA methods are needed to detect and distinguish between environmental, man-made, and equipment conditions and provide real-time response recommendations to evolving conditions. Distributed satellite resource management promises continued access to space capabilities so as to maintain mission-critical information after space asset degradation. Currently, in the event that space-based assets suffer an outage or abnormality, the responsibility falls to the humans-in-the-loop to follow checklist procedures to restore operations. Unfortunately, these procedural checklists are time-consuming and are not always optimized with consideration of the need to maintain Space Situational Awareness. Also, systems used to compensate for satellites suffering outages may not achieve the restoration of service with sufficient time to adequately support ongoing missions.

This section focuses on the need for anomaly attribution and prediction for the GPS. Current mission needs include intelligence collection and maintaining satellite communications to all military forces operating in the area of responsibility. There are five functional levels of SSA (e.g., incident detection/causality, event tracking/characterization, event relationship assessment, mission impact prediction, and system process assessments) and five dual response management levels within which the automated response decision aids of interest reside. These five levels are based upon the Data Fusion & Resource Management Dual Node Network (DNN) technical architecture, which is an extension of the JDL fusion model, see references [6, 26, and 28].

1.3.3.2 BASELINE GPS ABNORMAL SNR DETECTION

The core dataset used to detect and predict GPS constellation outages was dual channel SNR data for each satellite to site GPS broadcast around the world collected in late 2010 and early 2011. The DF&NN ANOM abnormality detection system was used to learn normal SNR behavior from a large number of GPS sites and corresponding GPS satellites. The ANOM system was trained on data from September/October 2010 and then used to detect abnormal signatures for March 2011. This Proof of Concept (POC) prototype focused on characterizing and predicting the abnormalities during March 2011. During March 2011, there were 62 million SNR samples that were filtered down to 73,000 abnormal communications (as identified by ANOM) then down to 1086 abnormal event tracks for given satellite/site pairs (identified by the DF&NN ANOM Event Tracker), and finally to 141 situation tracks (identified by the DF&NN BFN web service).

The DF&RM DNN technical architecture was used to develop the multiple source fusion system shown in Fig. 1.17. This fusion system processes the data through ANOM, event tracking, situation tracking, and scintillation prediction. To determine the fusion logic, it was necessary to build a better understanding of the SNR data through visualizations and analyses. One analysis involved mapping out the spatial distribution of the Ionospheric Pierce Points (IPP), defined as the point where the line of site between a satellite and site crosses the ionosphere at 350 km altitude. Fig. 1.18 shows a frame from an animation of all the IPPs colored and sized by the ANOM SNR abnormality score. Dark grey squares denote communications with abnormal SNR characteristics detected. As evidenced, the IPPs are in clusters overhead the ground sites—in fact, other analyses demonstrated that almost all IPPs occur within ± 10 degrees of the site locations. This finding is important because it constrains the range of ionospheric sampling that the GPS constellation performs, and thus simplifies the fusion logic responsible for highlighting scintillation events. The ionospheric modeling and situation characterization in this case is essentially discretized into regions corresponding to the ionospheric "umbrella" over each site.

Another analysis clarified the association between likely scintillation effects and local time. While subject matter experts (SMEs) informed us that scintillation effects are primarily seen at night, data analysis helped highlight a two-tiered effect between late evening and early morning hours. The two-tiered association of ANOM identified abnormalities with local time is shown in Fig. 1.19. Percentages are displayed, with the bars of each shade (corresponding to events exceeding a given abnormality threshold) adding to 1.0. The various heights of the bars

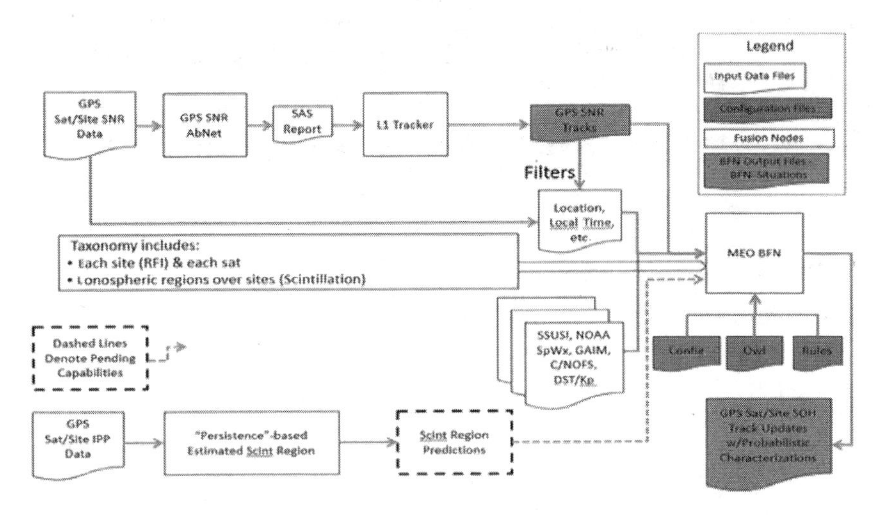

Fig. 1.17 Fusion node network for abnormal GPS SNR detection, characterization, and prediction.

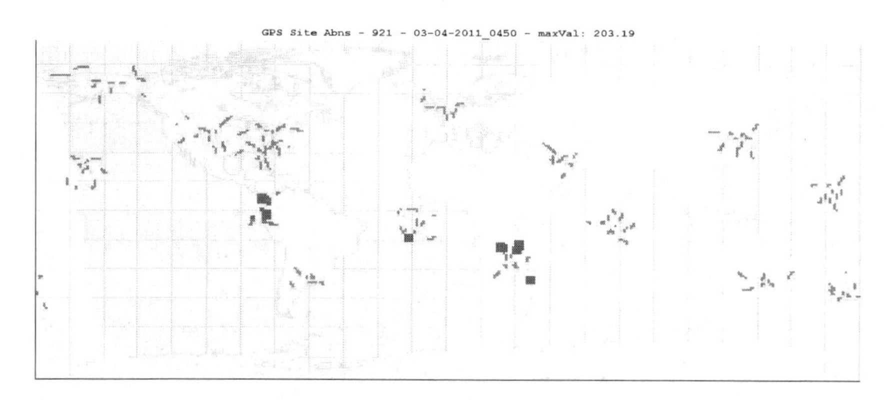

Fig. 1.18 Ionospheric Pierce Points on the line of sight for all communications between GPS sites and satellites during a 1-hour period in March 2011.

highlight the skewed distribution of abnormalities across local time. Abnormalities over 25 are clearly concentrated in the hours from 8 pm to 12 am, while the early morning (from 12 am to 3 am) hours still have elevated rates, but ones that are much reduced from the late evening.

These effects and various other SME insights and intuitions helped specify the fusion logic for the BFN. For the GPS POC, the taxonomy used 67 possible situations: 16 sites (e.g., ground site radio frequency interference), 35 satellites (e.g., single event upsets or other satellite problems), and 16 Scintillation Regions (350 km alt, ± 10 degrees from sites). Each event track fused into a situation track heightens probability of situation characterizations corresponding to the relevant site and satellite. The probability of the scintillation regions overhead being the "cause" of the situation is raised if the local time during the event start is during the night (and especially if it's between 2000 and 2400 local time), nearby sites ($<$4000 km distant) are simultaneously experiencing SNR events,

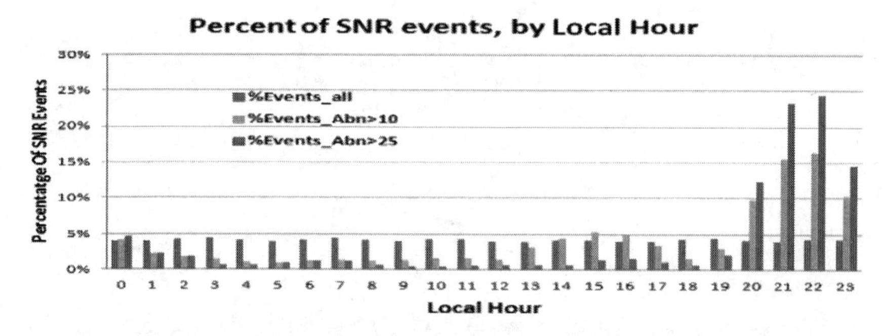

Fig. 1.19 Two-tiered association of ANOM identified abnormalities with local time.

and the site is <30 degrees of latitude from the equator. An event track is associated with an ongoing situation track if the Situation Probability Vector (SPV) for the event track is sufficiently similar (cosine similarity) to that of the situation track. If the SPVs are too dissimilar, then the event track will initiate a new situation track. These simple rules serve to highlight overlapping data in multiple event tracks as the likely cause of the abnormalities. If, for example, many satellites have abnormal SNR when communicating with QUI (Quito, Ecuador) during local day, the resulting situation track will attribute the abnormality to the problems at the QUI site (e.g., caused by RFI), while suppressing the probabilities of issues with the involved satellites and the scintillation regions.

1.3.3.3 VIEWING BFN SITUATIONS

We applied the Bayesian Fusion Node (BFN) service to track the ANOM abnormal SNR detections and track the space weather sources (see references 4, 5, and 6). The same BFN application (see Fig. 1.20) was used to implement the multi-source fusion. The only change was to the runtime configuration files specifying the message parsing and pre-processing, the situation taxonomy, and the fusion logic. The BFN is a flexible high-level fusion application that can be deployed in varied contexts and accommodate changes to both data and decision logic without requiring re-compilation, re-deployment, and developer intervention. Benefits are automated SpWx attribution, forecast, visualization, integration,

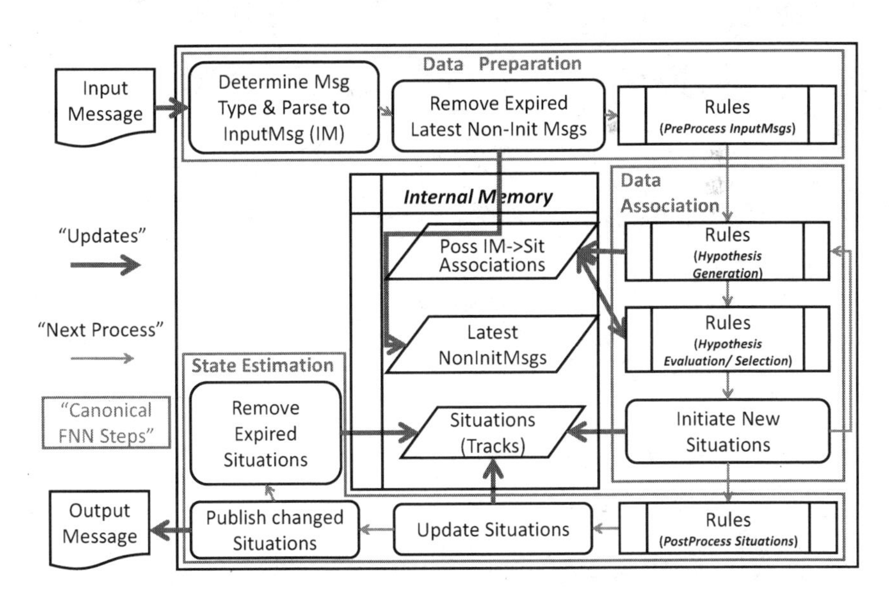

Fig. 1.20 Bayes Fusion Node (BFN) functional partitioning is based upon the DNN technical architecture.

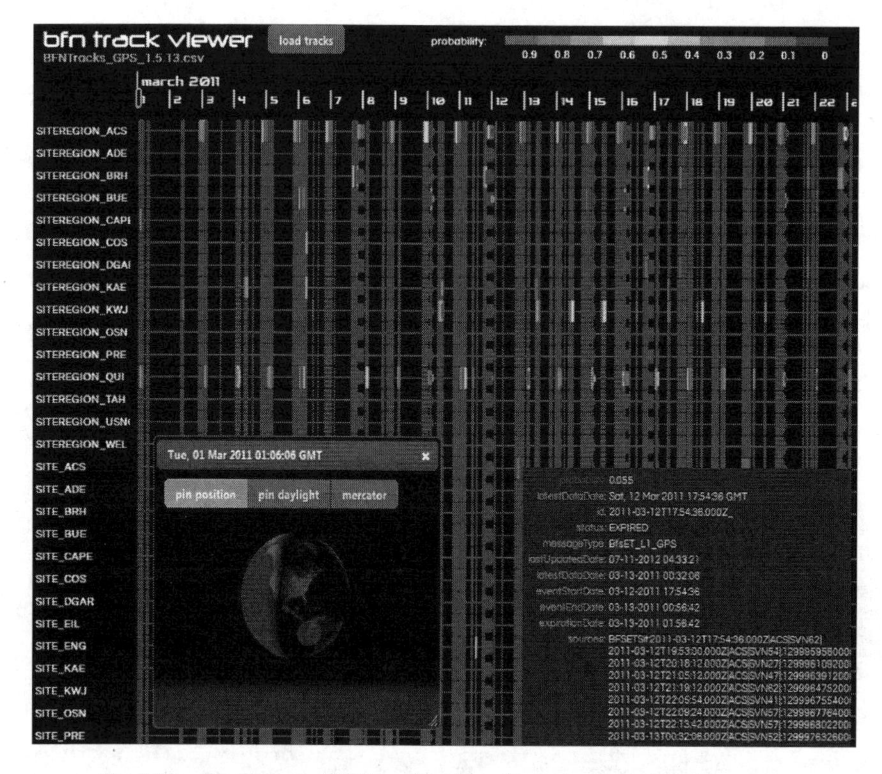

Fig. 1.21 Screenshot of BFN track viewer displaying GPS situation tracks.

and validation tools to characterize abnormal SOH and SpWx events detected online at operational sites using the DF&NN patented ANOM system.

The 141 situation tracks generated during this period were visualized using the BFN Track Viewer (BFNTV). The BFNTV is currently implemented for use in a web browser (via JavaScript). It consumes CSV files produced by the BFN application that contain all the BFN track updates that would be published in a real-time deployment. The screenshot of the BFNTV shown in Fig. 1.21 displays some GPS track information. The possible situation characterizations/causes in the taxonomy are listed on the left side (67 possible). The tracks are displayed as widenings in the horizontal lines corresponding to the characterizations, and high probability characterizations for a given track are shown by coloring the relevant track line. Time can be animated by dragging the time indicator along the timeline, which rotates the day/night shading on the globe or on the alternate mercator visualization. Likely scintillation regions are highlighted on the globe when the time indicator is over a situation with high probabilities of scintillation activity. A popup with the situation details (including a list of the fused event tracks) is shown when the mouse dwells over a track—this popup will be an entry point for drill through to event (L1), abnormality (L0), and raw SNR data implementation. Using the BFNTV enables visual identification of interesting patterns in the data that seem likely to be due to scintillation activity. The scintillation regions over the sites ASC (Ascension Island), QUI, KWJ (Kwajalein), and to a lesser extent, KAE (Keena Point, HI), which is higher latitude, were most active during the analysis period (especially March 3–March 23, 2011). There was a robust sequence of scintillation region activity that appears to follow sunset going east to west every night for a week (March 10, March 13–18). The sequence of scintillation regions that repeats is ASC then QUI (then sometimes KAE) then KWJ—a progression likely to reflect actual scintillation activity. The BFN logic filters highlight these scintillation effects and the BFNTV helps to visualize and animate the repeating pattern each night.

1.3.3.4 GPS "CONSTELLATION AS SENSOR"

The initial GPS analysis used only the temporal and site/satellite identifier information from the GPS SNR to perform SpWx attribution. We then added a new type of scintillation region estimation and new external data sources to increase the characterization accuracy and introduce predictive capabilities into the GPS fusion framework.

Figures 1.22 and 1.23 are frames from a movie IPPs from satellite/site communications with abnormal SNR. Each black square is an instantaneous abnormal SNR event; each dark grey dot represents an abnormal SNR event where the geomagnetic latitude and the local time are in the relevant range for likely scintillation caused abnormalities. The regions are calculated in a data-driven fashion, by constructing a minimal bounding box (dark grey box) that contains all of the

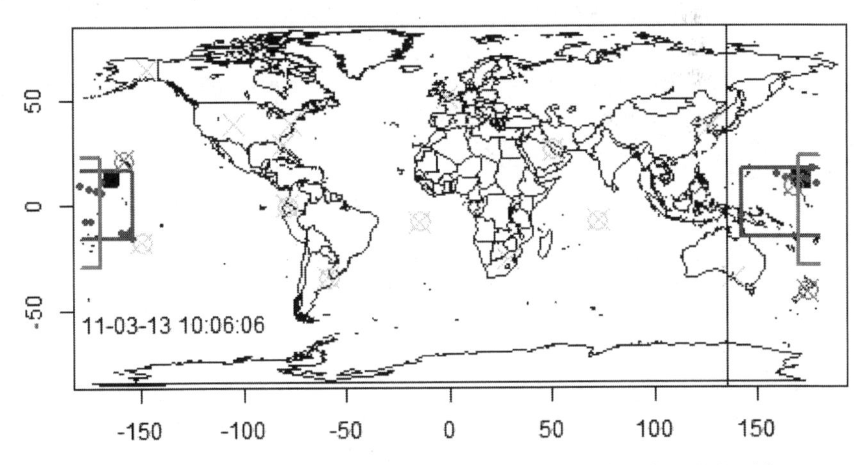

Fig. 1.22 Worldwide GPS SNR scintillation behavior prediction baseline SNR with context (BFS SNR, dark grey and SSUSI, light grey).

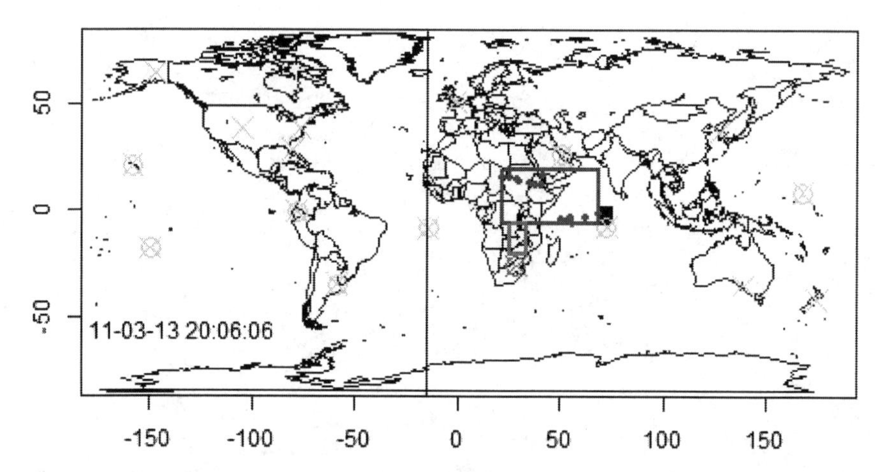

Fig. 1.23 Eight hours later the "predicted" abnormalities are confirmed/reinforced at DGAR and South Africa.

qualifying IPPs translated to follow sunset as if they were samples of a persisting ionospheric region. The IPPs for each frame are generated by persisting the likely scintillation caused IPPs and moving them to constant heliocentric coordinates, correcting for geomagnetic inclination at their new position. The persistence of each "active" IPP starts with some default value (e.g., 6 hours) and is increased every time a new SNR abnormality has an IPP that is nearby the propagated position of the previous IPP sample. Likewise, the persistence of the active IPPs is decremented when the inferred region fails to cause abnormal SNR communications.

Normal (i.e., not flagged abnormal) communications could be factored into the persistence in an exhaustive approach by considering the IPPs at all-time points, but for ease of calculation an abnormality-based approach was used instead. The failure to cause abnormalities is judged by the region being proximal to sites without observing abnormal IPPs. These simple approaches implement the heuristic that a scintillation region is still relevant when it is triggering new abnormal SNR IPP samples. The persistent region can be used to adjust confidences for characterization of ongoing events and, more important, can be used to predict future events in a relatively intuitive and simple manner.

The widespread, direct sampling–driven "modeling" of the ionosphere is unique to this effort. The dynamics of the estimated scintillation regions, including their apparently accurate persistence over time and tracking of the sunset, have not been previously demonstrated in a data-driven manner. This approach thus provides novel insight into ionospheric dynamics and their impacts on GPS functioning. These insights could be of direct interest to the scientific community and could be used to validate and tune ionospheric modeling efforts like GAIM that are used by the Air Force to estimate and predict possible decreases in GPS accuracy for mission planning.

1.3.3.5 DRILL DOWN INTO EXTERNAL DATA SOURCES TO SUPPORT GPS

Several other sources of data were integrated via the BFN and tested for association to the abnormal GPS SNR reports. These data included:

- Communications/Navigation Outage Forecasting System (C/NOFs)
- Global Assimilative Ionospheric Model, Total Electron Content (GAIM-TEC)
- Special Sensor Ultraviolet Spectrographic Imager (SSUSI) from DMSP
- NOAA Space Weather event reports, especially X-ray events
- Disturbance Storm Time index (DST) from publicly available sources

Drilldown plots for GPS SNR are intended to support rapid analysis and validation of the GPS SNR abnormalities and the L1 event tracks. In addition, they provide an intuitive tool for browsing the information content of the available SpWx data sources. The BFN track viewer interface has been altered to provide rudimentary drilldown capabilities from L2 tracks to these plots files, to allow a near-online drilldown experience for SMEs. These plots were programmatically generated for every time period spanned by L1 SNR abnormality tracks or any Communication/Navigation Outage Forecasting System (C/NOFS) events. Separate plots are generated for every satellite communicating with the site during the period, to allow for comparison of the various satellite communications at various azimuths and elevations. The bottom four plots display the raw signal data and the two plots above display the L0 and L1 ANOM detected abnormality events and tracks. The drilldown shown in Fig. 1.24 appears when double clicking an event of interest in Fig. 1.21. The top three panels display the Disturbance Storm Time (DST), Special Sensor Ultraviolet Spectrographic Imager (SSUSI) bubble detected nearby, and the C/NOFS events each temporally registered.

The C/NOFS data was only available for a given site when the satellite was in the appropriate location. The measured time periods totaled ~80 for each of the 4 sites investigated (ACS, KWJ, QUI, and DGAR, Diego Garcia). The ionospheric SMEs manually reviewed the C/NOFS data for each of these periods in order to identify C/NOFS signatures that were likely to support low, medium, or high probability of scintillation for GPS transmissions to the given site. These probabilities were investigated in aggregate and for each site and time period measured. Plots were automatically generated for all C/NOFS abnormality periods to enable rapid validation and review by SMEs.

The GAIM-TEC data was generated from the Global Assimilation of Ionospheric Measurements (GAIM) model using relatively coarse granularity and without the Full Physics (FP) modeling. The Total Electron Content (TEC) density information was used to derive gradients, maximum and minimums of the TEC density in the region over the sites of interest. In addition, rates of change of these gradients and other TEC derived variables were calculated to attempt to capture dynamics in the ionosphere. This data was then investigated

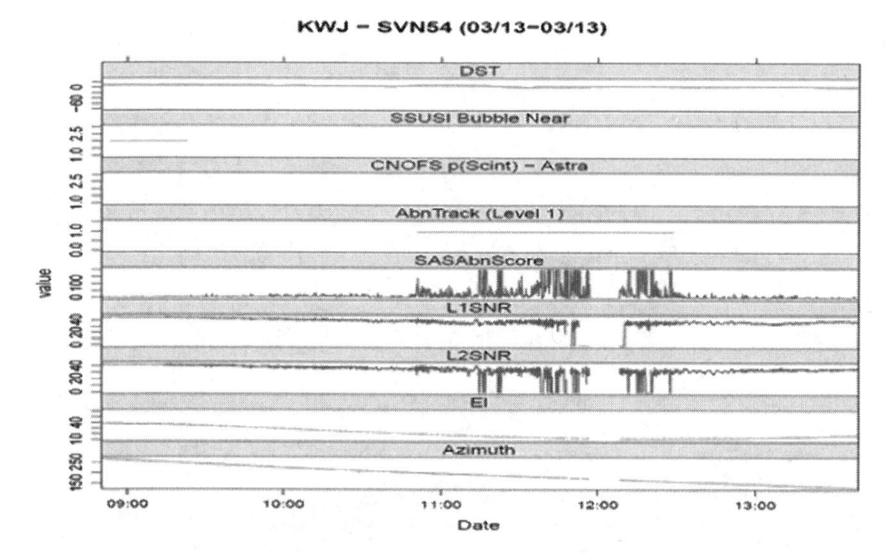

Fig. 1.24 Drilldown visualization of "raw" GPS relevant SNR data.

in aggregate and for each period of interest over each site during periods with either elevated CNOFS and/or abnormal SNR at that site.

The SSUSI "depletion regions/ionospheric bubbles" data provides a 3D point cloud of ion "depleted" regions associated with "ionospheric bubbles" that are thought to cause scintillation effects. These regions were essentially flattened to a fixed height of 350 km. Figures 1.22 and 1.23 provide examples of the association between the estimated scintillation regions (in dark grey) and the SSUSI depletion regions (in light grey). Like CNOFS, the SSUSI instrument samples a small subset of the ionosphere in a given time, so the regions reported and analyzed reflected times and places when there was a measurement that reported scintillation bubbles. Non-sampled and periods without bubbles were both reflected as a lack of bubbles.

In viewing a video made up of frames like Figures 1.22 and 1.23, the association between the SSUSI ionospheric bubbles and the regions identified by the IPPs of abnormal SNR samples is clear. The SSUSI bubbles are often overlapping or adjacent "forward" (i.e., immediately westward) of IPP defined regions. This association offers exciting possibilities to enhance the predictive capabilities of the data-driven approach as well as refining characterization estimates and source bias estimates. If development of the data-driven ionospheric scintillation estimation and prediction approach receives further funding, SSUSI will be the primary external data source that we seek to incorporate. Likewise, future evaluations of external data will focus on sources that measure similar physical phenomena.

Global reports and indices such as NOAA SpWx reports and DST were downloaded from publicly available sources and were aggregated and analyzed for meaningful correlation to the abnormal SNR events. Figures 1.25 and 1.26 show similar investigations into the association between SNR abnormalities and X-ray and DST levels, respectively. In each case the frequency and intensity of SNR abnormalities is plotted on the vertical axis while the potentially associated environmental measurement is plotted along the horizontal axis. X-rays and DST are both discretized into bins, and then counts of abnormalities seen during the periods of the relevant environmental conditions are plotted in the vertical axis. In both plots, no clear relationship is apparent, at least for the period analyzed (March 2011).

The conclusions from reviewing the full set of plots, and aggregate analyses of several of the data sources, was that ANOM abnormalities *all* were declared to be valid detections of ionospheric scintillation events by the SMEs. Further review is needed to identify whether there are any "true" missed detections of events. Of the data sources included in the plots (C/NOFS, GAIM TEC, SSUSI, NOAA SpWx, and DST), only SSUSI seemed to provide significant additional information for characterization and scintillation prediction. C/NOFS events did seem to have a nontrivial correlation with the ANOM identified abnormalities and some visually identified raw SNR signal fluctuation, but the association was relatively weak. Of these data sources, the SSUSI data is the only external data source that showed

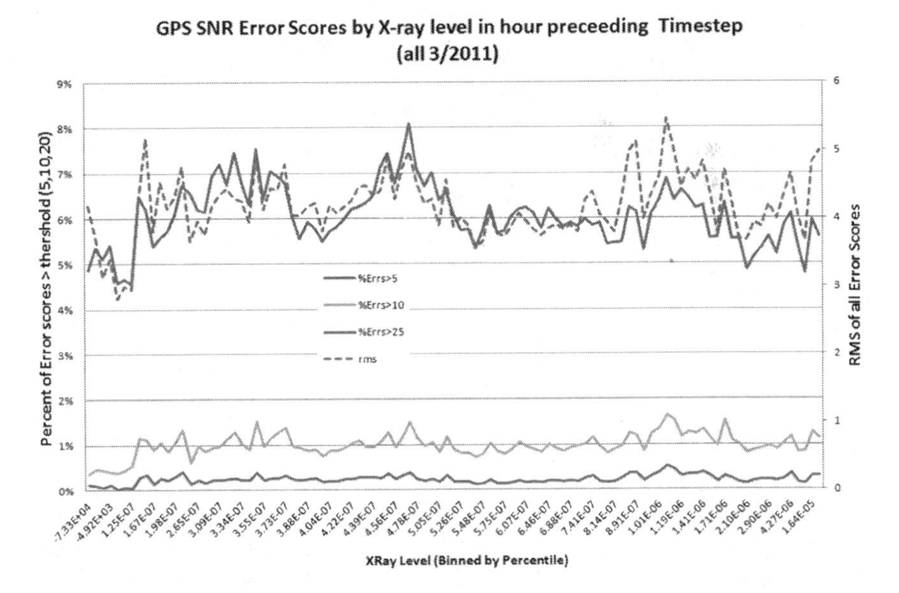

Fig. 1.25 1-8 angstrom X-ray levels as measured by GOES15 plotted against number of SNR abnormalities seen at various severity levels and the RMS of all the abnormality scores.

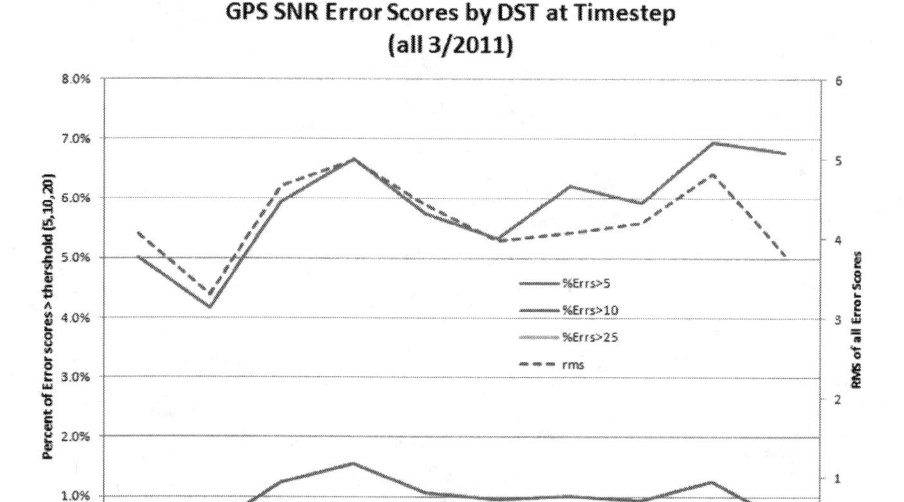

Fig. 1.26 DST levels plotted against number of SNR abnormalities seen at various severity levels and the RMS of all the abnormality scores.

enough correlation to the abnormal SNR measurements to suggest a significant gain from keeping it incorporated in the BFN.

1.3.3.6 GPS SCINTILLATION PREDICTION SUMMARY

There are five functional levels of SSA and five dual response management levels within which the automated response decision aids of interest reside. These five levels are based upon the Data Fusion & Resource Management (DF&RM) Dual Node Network (DNN) technical architecture which is an extension of the JDL fusion model [6, 27, 28]. Because space assets are susceptible to numerous anomalous conditions decision makers need a distributed satellite resource management system to effectively accomplish their space mission [31]. Near real-time integrated SSA methods are needed to: detect mission impacting events and distinguish between environmental, equipment, and man-made acts (both intentional and unintentional). SSA also needs to predict these events and provide real-time response recommendations to evolving unexpected scenarios. Distributed satellite resource management promises continued access to space capabilities so as to maintain mission-critical information after space-based asset degradation. A semi-automated satellite mission re-planning system capable of confirming and characterizing abnormalities and then recommending space-based asset responses will be integral to the future improved use of space. The response recommendation decisions can be provided by the dual resource management processes described in the DNN technical architecture.

This chapter has described the methodology and software developed to affordably improve the robustness of distributed fusion systems by dynamic intelligent use of new context sources that are external to baseline operational systems that are costly to modify. Our context assessment tools provide a graduated range of context-based feedback to significantly improve operations. Affordable model-driven and data-driven data mining methods to discover unknown models from non-traditional and "big data" sources [7, 9], are used to automatically learn entity behaviors and correlations with baseline operational fusion products. This chapter has described our GPS SNR abnormality detection and scintillation software development and test results.

1.4 SUMMARY

The role for intelligent systems in solving DF&RM and mining problems has been described. We have described the DF&RM DNN technical are architecture that defines the components, interfaces, and engineering methodology to solve these problems. Hybrid model, data, and goal-driven solutions are expected as the problems range from simple to complex. The role for intelligent systems has been described as for the harder problems where affordable solution models do not exist. When models do exist, they provide the preferred solution.

Near real-time integrated SSA methods are needed to detect mission impacting events and distinguish between environmental, equipment, and man-made acts (both intentional and unintentional). SSA also needs to predict these events and provide real-time response recommendations to evolving unexpected scenarios. Distributed satellite resource management, promises continued access to space capabilities so as to maintain mission-critical information after space-based asset degradation. We have described three specific instantiations of intelligent systems for space applications that will support space operations.

The intelligent systems developed for space data applications shown in this chapter were successful and illustrate the increasing potential for such systems for space applications.

REFERENCES

[1] Bowman, C., and Steinberg, A., "Adaptive Context Assessment and Context Management for Multi-INT Fusion," National Symposium on Sensor and Data Fusion (NSSDF), NGA, Oct. 27–30, 2014.

[2] Bowman, C., Haith, G., Tschan, C., and Zetocha, P., "The Search for Signatures of Space Weather Effects," *AIAA SciTech 2016*, San Diego, CA, Jan. 4–8, 2016.

[3] Bowman, C., and Tschan, C., "Goal-Driven Automated Dynamic Retraining for Space Weather Abnormality Detection," *AIAA Space 2013 Conference Proceedings*, May, 2013.

[4] Bowman, C., "Abnormal Orbital Event Detection, Characterization, and Prediction," *AIAA InfoTech 2015*, Orlando, FL, Jan. 5–9, 2015.

[5] Bowman, C., "GPS Scintillation Outage Prediction," *AIAA SciTech 2015* at Orlando, FL, Jan. 5–9, 2015.

[6] Steinberg, A., Bowman, C., Haith, G., Morefield, C., Morefield, M., and Blasch, E., "Adaptive Context Assessment and Context Management," *17th International Conference on Information Fusion (Fusion2014)*, Oct. 2014.

[7] Bowman, C., "Process Assessment and Process Management for Intelligent Data Fusion & Resource Management Systems," *AIAA Space 2012*, Pasadena, CA. Sept. 2012.

[8] Kraus, B., and Bowman, C., "Detecting Abnormal Space Catalog Updates," *2012 AIAA InfoTech@Aerospace Conference*, Garden Grove, CA, June 2012.

[9] Bowman, C., and Tschan, C., "Data-Driven & Goal-Driven Computational Intelligence for Autonomy and Affordability," *2012 AIAA InfoTech@Aerospace Conference*, Garden Grove, CA, June 2012.

[10] Haith, G., and Bowman, C., "Data-Driven Performance Assessment and Process Management for Space Situational Awareness," *AIAA Infotech@Aerospace Conference*, Atlanta, GA, April 2010.

[11] Bowman, C., "Spacecraft Smart Structure Neural Network Adaptive Control," *2nd Government Neural Network Applications Workshop Proceedings*, U.S. Army, Huntsville, Alabama, Sept. 10–12, 1991.

[12] Kelecy, T. et al., "Satellite Maneuver Detection Using Two-line Element (TLE) Data," *AMOS Technical Conference Proceedings*, Wailea, HI, Sept. 2007.

[13] Bérend, N., "Estimation of the Probability of Collision Between Two Catalogued Orbiting Objects," *Advances in Space Research*, Vol. 23, Issue 1, 1999, pp. 243–247.

[14] Levit, C., and Marshall, W., "Improved Orbit Predictions Using Two-line Elements," *Eighth US/Russian Space Surveillance Workshop Space Surveillance Detecting and Tracking Innovation*, Maui, HI, April 2010.

[15] LaPorte, F., and Sasot, E., "Operational Management of Collision Risks for LEO Satellites at CNE," *Space Operations Communicator*, Vol. 5, No. 4, Oct.–Dec. 2008.

[16] Lei, C. et al., "An Analytic Method of Collision Detection for Active Spacecrafts," *55th International Astronautical Congress*, Vancouver, Canada, 2004.

[17] Doornbos, E. et al., "Use of Two-Line Element Data for Thermosphere Neutral Density Model Calibration," *Advances in Space Research*, Vol. 41, Issue 7, 2008, pp. 1115–1122.

[18] Cauquy, M. A., Roggemann, M. C., and Schulz, T. J., "Distance-Based and Neural-Net-Based Approaches for Classifying Satellites using Spectral Measurements," *Optical Engineering* Vol. 45, No. 3, 2006, 036201.

[19] Poelman, C. J., and Meltzer, S. R., "Spacecraft Identification by Multispectral Signature Analysis Using Neural Networks," Phillips Laboratory Technical Report PL-TR-97-1053, 1997.

[20] Dentamaro, A. V., and Phan, D. D., "Test of Neural Network Techniques Using Simulated Dual-band Data of LEO Satellites," *AMOS Technical Conference Proceedings*, Maui, HI, Sept. 2010.

[21] Awed, M., "An Unsupervised Artificial Neural Network Method for Satellite Image Segmentation," *International Arab Journal of Information Technology*, Vol. 7, No. 2, April 2010.

[22] Helmy, A. K., and El-Tawee, Gh.S., "Neural Network Change Detection Model for Satellite Images Using Textural and Spectral Characteristics," *American Journal of Engineering and Applied Sciences*, Vol. 3 No. 4, 2010, pp. 604–610.

[23] Funabiki, N., and Nishikawa, S., "A binary Hopfield neural-network approach for satellite broadcast scheduling problems," *IEEE Transactions on Neural Networks*, Vol. 8, No. 2, 1997.

[24] Boberg, F. et al., "Real Time Kp Predictions from Solar Wind Data Using Neural Networks," *Physics and Chemistry of the Earth, Part C: Solar, Terrestrial & Planetary Science*," Vol. 25, No. 4, 2000, pp. 275–280.

[25] Kraus, B., and Bowman, C. et al., "Detecting Abnormal Space Catalog Updates" *AIAA InfoTech Aerospace Conference*, Garden Grove, CA, June 2012.

[26] O'Brien, T. P., "A Spacecraft Environmental Anomalies Expert System for Geosynchronous Orbit," *Space Weather*, 7, S09003, doi:10.1029/2009SW000473, SEAES-GEO, 2009.

[27] Bowman, C. L., and Steinberg, A. S., *Handbook of Multi-Sensor Data Fusion*, CRC Press, Boca Raton, 2009, Chaps. 3, 22.

[28] Bowman, C. L., "The Dual Node Network (DNN) Data Fusion & Resource Management (DF&RM) Architecture," *AIAA Intelligent Systems Conference*, Chicago, Sept. 20–22, 2004.

[29] Haith, G., Bowman, C., and Tschan, C., "Statistical Methods for Correlating Space Environment Measurements and Effects Amongst Multiple GEO Satellites," *2010 AIAA Infotech Aerospace Conference*, Atlanta, GA, April 2010.

[30] Haith, G., Bowman, C., Tschan, C., and Soderlund, P., "Toward an Automated Situation Assessment for Abnormal Behavior of Satellites in Low Earth Orbit (LEO)," *2010 AIAA Infotech Aerospace Conference*, Atlanta, GA, April 2010.

[31] Bowman, C. L., and Haith, G., "Engineering Resource Management Solutions by Leveraging Dual Data Fusion Solutions," *2010 AIAA Infotech Aerospace Conference*, Atlanta, GA, April 2010.

CHAPTER 2

Advances in Functional Fault Modeling for the Design and Operation of Space Systems

Kevin J. Melcher*

NASA Glenn Research Center, Cleveland, Ohio 44135, USA

William A. Maul† and Amy K. Chicatelli‡

Vantage Partners, LLC, Brookpark, Ohio 44142, USA

Gordon Aaseng§, Eric Barszcz¶ and Ann Patterson-Hine**

NASA Ames Research Center, Moffet Field, CA 94035, USA

Barton D. Baker††

NASA Marshall Space Flight Center, Huntsville, AL 35808, USA

NOMENCLATURE

AC	Abort Condition
ACAWS	Advanced Caution and Warning System
ACU	Actuator Control Unit
ADIO	Analog to Digital Input/Output
AGSM	Advanced Ground Systems Maintenance
AIAA	American Institute of Aeronautics and Astronautics
AMO	Autonomous Mission Operations
AMPS	Advanced Modular Power Supply
AT	Abort Trigger
BHS	Basic Hydraulic System
C3R	Command, Control, Communications, and Range
COTS	Commercial-off-the-shelf
cRIO	Compact Real-time Input Output

*Team Lead, Systems Health Management Methods for Space Exploration, Intelligent Control and Autonomy Branch, MS 77-1, AIAA Associate Fellow.

†SLS IVFM Technical Team Lead, 3000 Aerospace Parkway/VPL-3, AIAA Member.

‡Aerospace Engineer, Vantage Partners LLC, VPL-3, AIAA Senior Member.

§Computer Scientist, Intelligent Systems Division, MS 269-1.

¶Computer Engineer, MS 269-1, non-member.

**Fault Management Engineer, MS 269-1, AIAA Senior Member.

††Aerospace Flight Systems Engineer, MSFC/EV43, non-member.

CSV	Comma Separated Variable
C&W	Caution and Warning
DMC	D-Matrix Comparator
EFFBD	Enhanced Functional Flow Block Diagram
EFT-1	Exploration Flight Test 1
EPS	Electrical Power System
ETA	Extended Testability Analysis
FDIR	Fault Detection, Isolation, and Recovery
FFBD	Functional Flow Block Diagram
FFM	Functional Fault Model or Functional Fault Modeling
FIR	Failure Impacts Reasoner
FM	Fault Management
FMEA	Failure Modes and Effects Analysis
GCM	Generic Component Model
GEMINI	Generic Model Instantiator
GH2	Gaseous Hydrogen
GHe	Gaseous Helium
GSDO	Ground Systems Development and Operations
Hz	Hertz
ICM	Instantiated Component Model
ID	Identifier
ISHEM	Integrated System Health Engineering and Management
ISHM	Integrated System Health Management
IVFM	Integrated Vehicle Failure Model
IVHM	Integrated Vehicle Health Management
LCC	Launch Commit Criteria
LRU	Line Replaceable Unit
MBSE	Model Based Systems Engineering
MBSU	Main Bus Switching Unit
M&FM	Mission and Fault Management
NA	Not Available
NASA	National Aeronautics and Space Administration
PDU	Power Distribution Unit
PHM	Prognostics and Health Management
PLB	Programmable Load Banks
RM	Redundancy Management
SBS	System Breakdown Structure
SHM	System Health Management
SLS	Space Launch System
TEAMS	Testability Engineering and Maintenance System
TRL	Technology Readiness Level
VBA	Visual Basic for Applications
VERA	Verification Analysis

VHM	Vehicle Health Management
V&V	Verification and Validation

SYMBOLS

°C	Degrees Celsius
°F	Degrees Fahrenheit
°K	Degrees Kelvin
D_{cov}	Detection Coverage (%)
FM_{miss}	Number of Failure Modes not detected during testability analysis
FM_{total}	Total Number of Failure Modes used during testability analysis

2.1 INTRODUCTION

An important element of intelligent and autonomous systems is Integrated Systems Health Management (ISHM) and its variations, for example, Integrated System Health Engineering and Management (ISHEM), Integrated Vehicle Health Management (IVHM), System Health Management (SHM), Vehicle Health Management (VHM), and Prognostics and Health Management (PHM). ISHM has been defined as "the capabilities of a system that preserve the system's ability to function as intended" [1]. This means that ISHM protects system function to ensure that system goals are achieved. The ISHM discipline is applicable over the entire system life cycle, from early system design through operation [2], and as such, requires a systems perspective. A recent, more detailed discussion of ISHM [3]—including its scope, its history, and its terminology—is recommended as a foundation for the work described in this chapter. The terminology discussion is likely to be particularly useful to readers as it clarifies the use of ISHM terms and places them in a broad theoretical context.

In intelligent and autonomous systems, ISHM protects a system's functionality by providing a means to self-assess its health state [4], by managing redundancy (and loss thereof), and by informing the system's planning and control functions regarding loss of capability due to internal system failures. An important subset of ISHM is Fault Management (FM)—the set of operational capabilities that perform their primary functions when the nominal system design is unable to keep state variables within acceptable bounds [3]. FM operates as a set of meta-control loops that predict, detect, and respond to existing or prospective failures, maintaining or returning the system to a controllable state [5].

One FM technology that has been gaining acceptance in recent U.S. National Aeronautics and Space Administration (NASA) applications is the Functional Fault Model (FFM). As will be shown, this acceptance is evidenced by the maturation of FFMs from a research topic to a mid-Technology Readiness Level (TRL) technology that embodies a number of design analysis and real-time diagnostic assessment capabilities.

An FFM is an abstract failure space representation that defines the effect propagation paths of critical failure modes associated with a given system. These models embody information typically found in a Failure Modes and Effects (FMEA) analysis, in system design diagrams, and in other system design documentation such as Concepts of Operation. Further, because they capture effect propagation paths as well as the failure modes and detection mechanisms, FFMs can provide insight into the safety and reliability characteristics of the systems they model. Examples of FFM applications and demonstrations can be found in the aerospace research community and specifically in references [6–12]. The types of applications vary from ground-based diagnostics and routine maintenance procedures to real-time fault detection, isolation, and recovery. A more in-depth description of FFMs is provided in Section 2.1.1.

This chapter presents advances in FFM technology that have occurred during NASA programs over the period from 2006 through 2016. Throughout this chapter, advances in FFMs will be discussed generically. However, to facilitate the discussion and present tutorial FFM examples, a commercial software product, the Testability Engineering and Maintenance System (TEAMS) [13] from Qualtech Systems, Inc., will be consistently used. This software has been used by a number of recent NASA projects [6–12] to develop FFMs. Further, although the FFM advances described in this chapter are specific to TEAMS-based FFMs, the general objectives, purpose, and "lessons learned" can be translated to other FFM modeling techniques.

The most recent FFM advances included the development of cross-program FFM modeling conventions and practices [14], the expansion of analytical tools [6, 15–18], and library-based modeling approaches [19]. The new analytical tools provide more efficient verification of FFMs and their associated FM requirements, whereas library-based modeling approaches can result in reduced model development time. To achieve these improvements, FFMs must maintain a level of consistency during development that enables efficient revision and integration of the models with external processing and reporting software tools, real-time diagnostic architectures, and other FFMs. Under the Ground System Development and Operations program, the Space Launch System Program, and the Orion Program, a set of cross-program FFM modeling conventions [14] was established to address these issues. The utility of these new capabilities is highlighted by the fact that reporting capabilities of several of them have been integrated by Qualtech Systems, Inc., into the commercial version of their TEAMS software.

The chapter is organized as follows. Throughout the remainder of Section 2.1, fundamental FFM concepts and topics are presented. These topics provide an understanding of FFMs, how NASA is using and advancing FFMs, and some key benefits and limitation of FFMs. The intent in presenting this information is to provide a technical foundation for the FFM novice. In Section 2.2, an overview of the TEAMS software is presented along with an introduction to two example FFMs. These discussions provide a basis for many of the examples used to illustrate the FFM advances found in Section 2.3. Section 2.3 describes advances in the development of FFMs in the context of the current NASA FFM development process. Section 2.4 speaks to the future of FFMs, including current challenges, suggestions for future work, and the anticipated impact of FFMs on the development and implementation of future space systems. The chapter closes by discussing conclusions, acknowledging key support for the chapter, and providing a list of relevant references.

2.1.1 WHAT IS A FUNCTIONAL FAULT MODEL?

Functional fault models are a subset of functional models. The NASA Systems Engineering Handbook [20] describes the use of Functional Flow Block Diagrams (FFBD) and Enhanced Functional Flow Block Diagrams (EFFBD) for defining system functions and depicting the nominal time sequence of functional events including control flows and data flows.

FFMs complement FFBDs and EFFBDs by representing a system's failure space. An FFM represents the failure effect propagation paths between the origination points of failure modes and the observation points within a system. It can be used to provide a diagnostic of the modeled system. FFMs contain a significant amount of information about the system, including its design, operation, and off-nominal behavior.

In general, models are developed for many purposes and take different forms, ranging from simple spreadsheets to sophisticated physics-based simulations that model complex multi-domain systems. FFMs tend to be in the middle of this spectrum. FFMs are hierarchical models that can represent very complex structures if the system being modeled has many subsystems and components but the failure effects being propagated are represented at a qualitative level. For example, consider a component representing a helium tank that has a leak as one of its failure modes. The failure effect would be represented by propagating a low helium pressure signal, which may be detected downstream by a pressure sensor and a "test" for "low helium pressure." The qualitative nature of the test—a representation of the logic used to detect the failure effect—allows various analyses to be performed in the absence of detailed physical characteristics, such as a pipe's inner diameter and length.

The ability to model subsystems and components at various levels of fidelity with a qualitative description of failure effects allows FFMs to mature along with the system design while providing useful analyses of the current design or proposed changes. This primarily involves performing fault isolation to determine whether a proposed sensor suite will be able to isolate faults to a Line Replaceable Unit (LRU). After the design phase, the matured FFM can be used during testing and operations to diagnose systems and help guide troubleshooting, thereby utilizing the FFM throughout the life of the system from development (i.e., Phase A) through operations and sustainment (i.e., Phase E) [20].

2.1.2 NASA'S ADVANCEMENT OF FUNCTIONAL FAULT MODELS

Two goals in NASA's advancement of FFMs are to benefit system engineering design and development processes and provide diagnostic and decision support for system operations. These goals are motivated by a desire to improve safety and to reduce the life-cycle cost of crewed space systems.

A driving issue is the scale of the FFM [21]. The model scale affects the time and number of modelers required for model development, the time and resources needed to test and verify the model(s), the approach used to present the model and its output to stakeholders, and the location of the resulting diagnostic reasoner (i.e., onboard spacecraft or ground-based). Other issues revolve around FFM support for mission decision processes and the development of frameworks to support onboard and ground-based diagnostic reasoners.

A number of aerospace projects at NASA have used or are using a Commercial Off-The-Shelf (COTS) Functional Fault Modeling tool called TEAMS from Qualtech Systems, Inc. Over the years, NASA modelers have developed a number of in-house software tools to augment the TEAMS software suite. Although some aspects of these tools are specific to TEAMS, the general concepts apply to building, testing, and using any FFM.

As an example of the scale of TEAMS models being developed at NASA, the model of the Electrical Power System (EPS) for the Orion EFT-1 test flight contained approximately 3500 failure modes and 2200 tests [10]. Development of these large models using the TEAMS GUI can be extremely tedious and time consuming. To address that issue, NASA has been developing software tools to reduce the development time and the number of human-touch errors typically associated with large FFMs. One tool [15] provides a means to easily make sweeping changes in the model. It provides techniques for semiautomatically placing and connecting components in an FFM from source material such as channelization spreadsheets. Another tool is a template library tool [19] for often-used components such as valves and cards. These tools are briefly discussed later in the chapter.

To assist in integration, test, and verification processes, NASA developed a set of FFM modeling conventions [14, 22]. NASA also developed a software tool that can be used to check a given model's compliance with these conventions and identify various modeling errors [18]. Further, an analysis and reporting tool [16, 17] is also available to post processes FFM output and generate reports for review by subsystem experts. Additionally, a software tool is currently under construction that uses the FFM model to automate the generation of interface control documentation for subsystem models.

Operationally there is a nontrivial amount of software needed to integrate a diagnostic reasoner into a mission control environment. Typically, the steps involve 1) collecting telemetered data required by tests within the FFM; 2) defining and implementing processes for data with different sampling rates, and for missing or bad data; 3) executing code to generate a pass/fail for applicable tests; 4) communicating with the diagnostic reasoner to get a diagnosis; and 5) presenting diagnostic output to users. Additionally, NASA uses the diagnosis as input to a Failure Impacts Reasoner (FIR), to provide information to mission operators. The operational phase typically has real-time requirements on the amount of time allowed to generate a diagnosis or an FIR result. Such a framework, the Advanced Caution and Warning System (ACAWS), has been developed at NASA, and future plans are to implement the ACAWS to support the Exploration Mission 1 and Exploration Mission 2 flights.

2.1.3 BENEFITS OF FUNCTIONAL FAULT MODELS

FFMs can provide benefits to intelligent and autonomous systems—both to system operations and to system engineering and design processes used to develop these systems.

The most obvious benefit of FFMs is the ability of FFMs to support the real-time detection and diagnosis of failures, which enables state self-awareness, the autonomous self-determination of a system's state. State self-awareness is required for determination of the achievability of mission goals that require specific onboard capabilities. FFMs provide a computationally efficient and cost-effective means of detecting and isolating system failures during system operation. The associated failure data can then be used by higher-level reasoning algorithms to assess loss of capability and, subsequently, the likelihood that a mission or goal can be achieved.

Another benefit is that FFMs may be used to identify unmet requirements for off-nominal operation (e.g., abort conditions, caution and warning conditions, launch commit criteria, and fault isolation) early in the systems engineering process when the design can be corrected at lower cost [23]. To achieve this benefit, FFMs are developed using qualitative, rather than quantitative, relationships to describe the behavior of a system in its failure space. Properly designed qualitative FFMs are able to perform the same analysis as those with quantitative relationships and, as the system design matures, qualitative models may be converted to quantitative models to support system operations and maintenance.

Current applications of FFMs are targeting human-rated spaceflight systems that are part of the infrastructure required for human crews to travel to the planet Mars. The large communication time delays associated with this mission can reduce the effectiveness of ground-based mission operations teams' ability to support the crew. Automated onboard diagnosis and isolation of critical failures, and response to those failures, should significantly improve crew safety and mission success.

2.2 TEAMS OVERVIEW AND EXAMPLE MODELS

2.2.1 TEAMS OVERVIEW

This section briefly introduces TEAMS, a suite of software tools from Qualtech Systems, Inc. [13], as a representative FFM tool. The suite contains four main components that are used to create and analyze FFMs (TEAMS Designer), perform real-time diagnostics (TEAMS-RT), provide guided troubleshooting for maintenance (TEAMATE), and provide a database for collecting and analyzing fleet data (TEAMS-RDS). Here, an overview of the modeling and analysis capabilities provided by TEAMS Designer is presented.

At the heart of TEAMS FFMs is a dependency matrix often called the D-Matrix. Figure 2.1 provides a notional example of this matrix. Rows of the D-Matrix represent failure modes and columns represent tests. If a failure mode has a failure effect that propagates and is detectable by a test, the cell at the intersection of the failure mode row and test column will contain a one (1). Otherwise, the cell will contain a zero (0). The series of ones and zeros in a given row are the detection signature for the failure mode associated with that row.

Analyses of information contained in the D-Matrix provide assessments of the system's diagnostic capabilities. A unique detection signature allows the detected

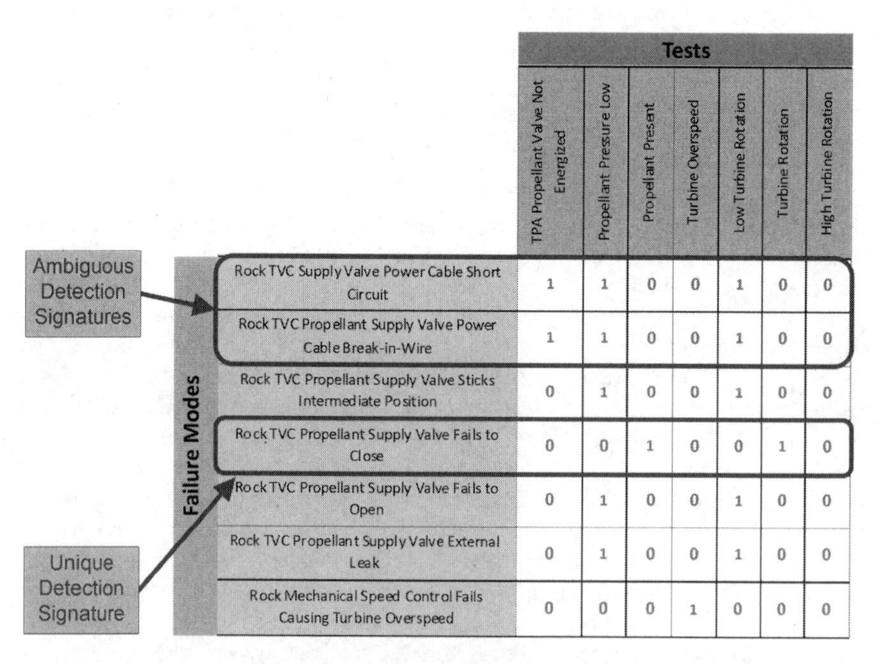

Failure Modes	TPA Propellant Valve Not Energized	Propellant Pressure Low	Propellant Present	Turbine Overspeed	Low Turbine Rotation	Turbine Rotation	High Turbine Rotation
Rock TVC Supply Valve Power Cable Short Circuit	1	1	0	0	1	0	0
Rock TVC Propellant Supply Valve Power Cable Break-in-Wire	1	1	0	0	1	0	0
Rock TVC Propellant Supply Valve Sticks Intermediate Position	0	1	0	0	1	0	0
Rock TVC Propellant Supply Valve Fails to Close	0	0	1	0	0	1	0
Rock TVC Propellant Supply Valve Fails to Open	0	1	0	0	1	0	0
Rock TVC Propellant Supply Valve External Leak	0	1	0	0	1	0	0
Rock Mechanical Speed Control Fails Causing Turbine Overspeed	0	0	0	1	0	0	0

Fig. 2.1 Notional dependency matrix.

failure effects to be isolated (i.e., tracked back) to a single failure mode. However, nonunique detection signatures result in ambiguity that limits the isolation of detected failure effects to an ambiguity group, where an ambiguity group is a group of failure modes that all have the same detection signature.

As an example of these diagnostic concepts, row 4 in Fig. 2.1 shows that the failure mode *Rock TVC Propellant Supply Valve Fails to Close* has a unique detection signature. Its failure effects can be detected by the combination of tests *Propellant Present* and *Turbine Rotation.* Because no other failure mode can be detected using those two tests in combination, the failure effects can be isolated unambiguously to that failure mode. On the other hand, rows 1 and 2 show that the failure modes *Rock TVC Supply Valve Power Cable Short Circuit* and *Rock TVC Propellant Supply Valve Power Cable Break-in-Wire* have matching detection signatures. The failure effects of each of these failure modes is detected by the combination of tests *TPA Propellant Valve Not Energized, Propellant Pressure Low,* and *Low Turbine Rotation.* In this case, there is not enough information to isolate the failure effects to a single failure mode.

The lowest level of failure sources are comprised of individual failure modes. However, it is common to perform analyses where the failure sources are at a higher level in the model hierarchy, for example, an LRU—a unit which, should it fail, is designed to be replaced in the field. Determining whether or not a given component meets the criteria for designation as an LRU has to do with the component's accessibility, its interfaces, and the capability to unambiguously isolate critical failure modes originating in the component. FFMs allow system engineers to address the latter concern. Analysis of an FFM's D-Matrix can determine if the test signature associated with each of the component's failure modes can be isolated to the LRU or, instead, has ambiguity with failure modes originating outside the LRU. If any of the test signatures are ambiguous, the component's LRU designation should be questioned.

A simple TEAMS model, Fig. 2.2 is composed of failure sources whose failure effects propagate as "signals" along paths to tests that detect those signals. Failure sources are represented by modules that emit signals representing failure effects. The darker hatched module in Fig. 2.2 represents a broken heating element that emits a signal (failure effect) called "no heat." Tests are collected into groups at "test points," denoted in Fig. 2.2 by the circle labeled *Temp-Tests.* A test point is typically associated with a sensor. For example, a temperature sensor may have an associated test point with several tests: off-scale low test, low temperature test, high temperature test, and off-scale high test.

Paths are implemented using links, switches, and AND nodes. Links provide the actual connections between nodes in the model. Switches allow for propagation paths to be changed based on the configuration of the system. For example, a spacecraft orbiting the moon may be operating on power from the solar panels most of the time and from batteries when in the moon's shadow. It would be incorrect to identify the solar panels as failed simply because the spacecraft is in the moon's shadow. Changing a switch setting when entering and

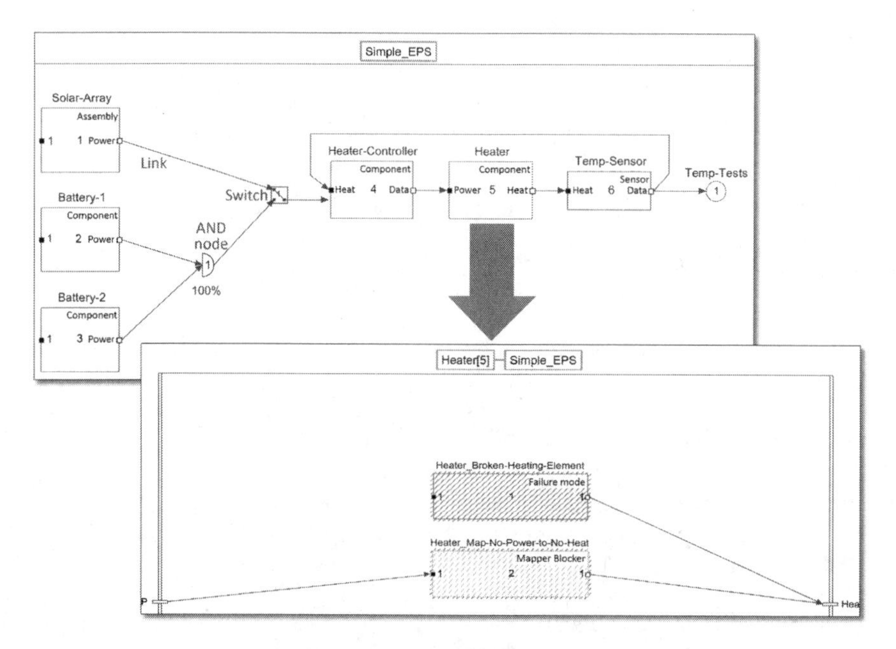

Fig. 2.2 Simple TEAMS model.

exiting the shadow would allow the proper (battery or solar panel) failure effects to propagate. Multiple switch settings, "switch modes," may be collected together as a "system mode." Then the user only has to specify the system mode and all switch states will be set automatically. Used to model system redundancy, AND nodes are characterized by the number of input links and a threshold percentage. If failure effects (signals) exist on a number of input links greater than or equal to the threshold percentage, the signals will be propagated. Otherwise, the signals are blocked and do not propagate further. For example, if the two batteries on the left in Fig. 2.2 operate in parallel and either is sufficient to power the system, then both batteries must be lost before power to the system is lost.

Because a TEAMS model is hierarchical, a module may contain links, switches, AND nodes, test points, and lower-level modules, as shown in the expanded view of the heater in Fig. 2.2. All links are directional. Input ports to a module are on the left-hand side of the module and output ports are on the right-hand side.

TEAMS does not perform physics-based simulation of the failure effects along the paths in the model. However, it does allow the user to transform (i.e., map) one failure effect into another to emulate physical effects at an abstract level. An example is shown in the heater module of Fig. 2.2 where an input failure effect of *No Power* is transformed into the output failure effect of *No Heat* by the lighter hatched module labeled *Heater_Map-No-Power-to-No-Heat*. In addition to being able to map one signal into another, the user may also block specified signals at a module.

There are many more features of TEAMS models than can be described in this overview. For more information about the suite of TEAMS tools and TEAMS modeling, see reference [13].

Two relatively simple FFMs are used to facilitate the discussion throughout the remainder of this chapter. The first FFM represents a Basic Hydraulic System (BHS) model, and the second an Advanced Modular Power Supply (AMPS) for a space system. Each of these FFMs has features that make it more suited than the other model for illustrating certain key points.

Assessments of the system's diagnostic capabilities are performed in a process called testability analysis with the FFMs. For this analysis, the user selects the model configuration defining functional flow paths within the model. The user can also restrict the failure modes and sensors tests to be included in the analysis. During the analysis, each failure mode is evaluated individually. The result of the testability analysis is a dependency matrix indicating the tests that detect the functions propagating from each failure mode.

2.2.2 BASIC HYDRAULIC SYSTEM FFM

Under NASA's Ares I Project, FFM model developers created a series of supporting tools to assist model development and reporting. To advance and demonstrate FFM tool capabilities, smaller generic systems with minimally required complexity were defined and associated FFMs established. One such model, the BHS, is representative of the actuator portion of a Thrust Vector Control (TVC) system. The BHS model contains two redundant independent hydraulic power sources and two actuators. The actuators work in tandem, gimbaling a rocket engine to direct the thrust. One actuator gimbals the engine along the yaw axis of the vehicle's reference frame and the other along the pitch axis.

This BHS model reflects the FFM approach used for similar models under the Ares I Project. It contains components and features common for dynamic fluid systems and addresses some advanced FFM modeling techniques such as the representation of system-level outcomes as special test points. The model adheres to the most currently employed modeling conventions and practices (Section 2.3.2.3) and will be utilized in Section 2.3 to describe advances in the development, implementation, and application of FFMs that support analytical verification of system design requirements. A brief description of the BHS model follows.

A notional block diagram displaying the fluid flow through the simple hydraulic system represented in the BHS model is shown in Fig. 2.3. The entire system contains two independent hydraulic supply circuits that are cross-strapped to provide primary and backup power to two actuators. Each hydraulic supply circuit is pressurized by a turbine-pump assembly driven by a generic gaseous propellant [e.g., gaseous helium (GHe) or gaseous hydrogen (GH2)], indicated by the dotted flow lines. The pump pressurizes the hydraulic fluid (indicated by double-walled flow lines), which flows through check valve and filter components to a pressure-selector valve and then on to support the actuator load. Each actuator possesses a pressure-selector valve to which the actuator's primary and backup hydraulic supply and return lines are connected. If the primary power circuit to an actuator fails (i.e., loses significant hydraulic pressure) then the selector valve will switch over to the backup hydraulic power source. The hydraulic return portion of the circuit (indicated by solid flow lines) contains a reservoir that provides a low pressurization through mechanical means to support boot-strap power-up of the system. Electric power flow to the propellant supply valve and actuator is represented by the dashed lines.

Figure 2.4 shows the location of sensors in the BHS system. Sensors are indicated by circles and ovals. For each hydraulic circuit, electrical current indicates that the propellant supply valve is energized and the valve's position measurement is available. The BHS also provides a measurement of the inlet propellant pressure to the turbine and the shaft speed of the turbine-pump assembly. Within the hydraulic fluid circuit, the system contains pressure measurements along the supply-side as well as the return-side and a differential pressure across the hydraulic filter. The position of each selector valve, indicating the

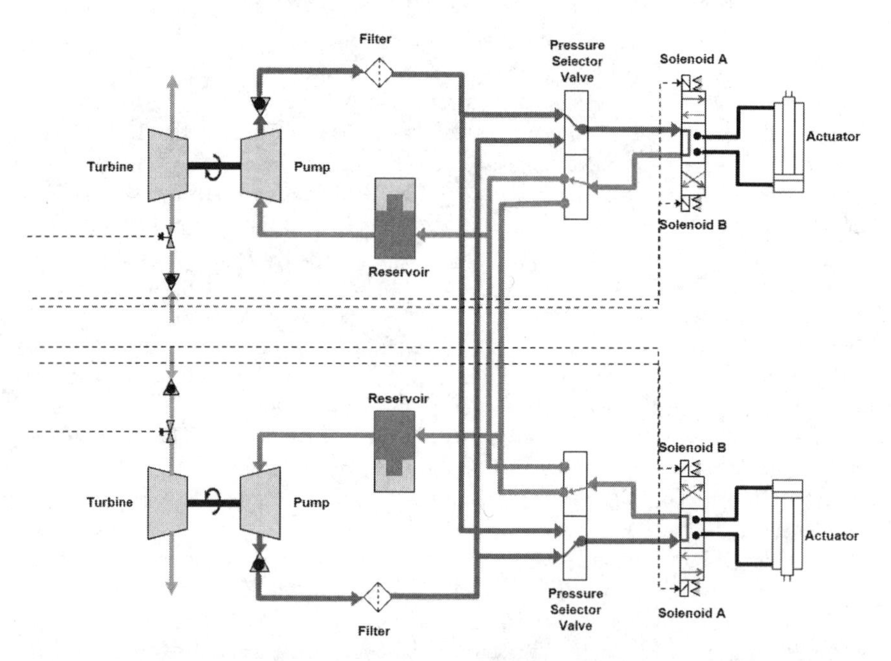

Fig. 2.3 A notional view of the electrical and fluid flows for the basic hydraulic system.

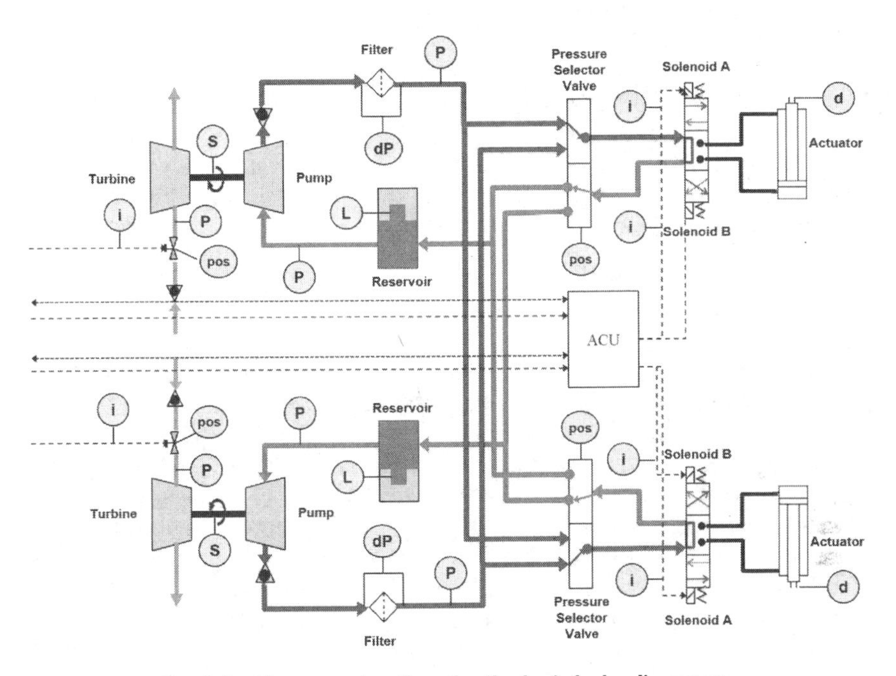

Fig. 2.4 The sensor locations for the basic hydraulic system.

position of primary or backup, is also provided. The redundant electrical power supply to the actuators is distributed through a simple actuator control unit (ACU) to the actuator, driving the actuator in either position. Current measurements for the actuators are provided, as well as the actuator displacement measurement.

There are a number of features within this basic model that make it useful for advancing the development of support tools and analysis processes. The BHS model contains 132 failure modes, including those defining faults for the 34 sensors within the model. Failures indicating loss of propellant supply to the turbine-pump assemblies or loss of electrical power to the valves and actuators are also represented in the BHS FFM. The BHS model utilizes a series of mapper-blocker modules to transition the failure effects of one property (e.g., *loss of electrical power*) into another property effect (e.g., *loss of hydraulic pressure*). The BHS model also contains system-level effects defined as test points, intended to enable a combined analysis with low-level sensor detectable effects propagated from failure modes (e.g., *low pressure*) and higher-level system effects (e.g., *loss of actuator function* or *complete loss of vehicle control*). Each of these features is intended to expose the applied tool or process to realistic system representations.

2.2.3 ADVANCED MODULAR POWER SUPPLY FFM

NASA has developed an AMPS as an engineering prototype and analysis system for spacecraft electrical power systems [24]. The AMPS project developed, demonstrated, and evaluated key modular power technologies to minimize non-recurring development costs and to reduce recurring costs associated with integration, mission operations, and maintenance. To achieve these goals and advance the development of AMPS systems, the project included an FFM-based diagnostic system. The hardware and associated diagnostic capabilities are herein described as a basis for the discussion of advances in FFM for operational support (Section 2.3.4.2).

The AMPS consists of Main Bus Switching Units (MBSUs) and Power Distribution Units (PDUs), with interfaces to power sources and electrical equipment. The MBSUs receive power from a simulated Solar Array or from an actual lithium ion battery. The AMPS is designed so that any electrical equipment can receive power via multiple redundant sources of power. Each of the units contain power sensors, remotely controlled switches, and a Compact Real-time Input Output (cRIO) processor. This processor performs analog to digital conversions and engineering unit conversions of the sensor data and accepts and routes external commands to equipment. An external AMPS Power Controller receives data from each of the component's processors, while operator interface applications display data and accept commands to turn switches on and off, configure setpoints, simulate failures, etc.

The AMPS MBSUs and PDUs contain switches, power supplies, busses, a cRIO processor, sensors, and structural components. The AMPS FFM models these components, their failure modes, and relevant electrical loads. The electrical loads can be simulated with Programmable Load Banks (PLB) or actual electrical equipment can be used if available. An overview of the AMPS system is shown in Fig. 2.5. In the system, an electrical load is connected to each of the PDU output channels. For illustrative purposes only, however, three loads are shown in Fig. 2.5. These loads represent three typical classes of loads that include critical loads, requiring redundant power inputs to a single load; high-power loads, requiring more power than a single power channel can provide; and normal loads, requiring a single power input.

To use the FFM for operational diagnosis, the TEAMS (hereafter referred to generically as the "FFM development software") diagnostic reasoner is used along with interface and data processing software. The reasoner loads the run-time model at initialization and the interface/data processing software obtains data from the system, performs tests on that data, and then packages the data for transmission back to the reasoner. Results are returned from the reasoner and distributed to displays, to other reasoners, to diagnostic logs/databases, to other applications, or to end-users.

Prior to use by the reasoner, sampled real-time data must be analyzed and converted to actionable knowledge that can be used by the diagnostic reasoner.

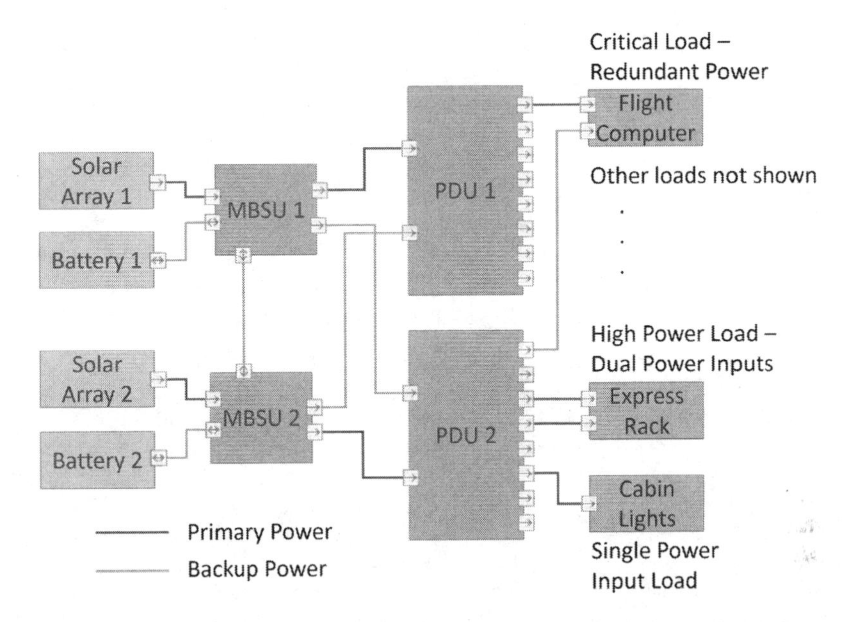

Fig. 2.5 The AMPS architecture indicating the primary components and the electrical power connections between them.

To do this, specialized interface and data processing software is used to apply diagnostic tests embodied in the FFM to the data and classify the result of each test as pass, fail, or unknown. The diagnostic reasoner operates on results of the classification process.

The AMPS model includes failures requiring different approaches to testing and diagnosis. The failure types that have been modeled are:

Component total failures. The MBSU and PDU components could suffer a failure that results in a complete loss of function. Either a failure of the cRIO, or a loss of power to the cRIO, could result in the absence of power or data outputs, that is, a "dead box" failure. These failures are detected by the absence of data from the component, and because no data is available, isolation of the specific fault is not possible without manual testing inside the component.

Switch failures. Both the MBSU and PDU contain power switches that control the power inputs and outputs, and the switches contain sensors that measure the actual state of each power switch. Two common failure modes of a switch are a "fail open" condition, in which the switch fails to provide power when commanded, and a "fail closed" condition, in which the switch fails to turn off when commanded. Switch failures are detected by a discrepancy between the commanded state and the actual measured state of a switch and can be supported by electrical measurements such as current or voltage.

Short circuits. A short circuit on a line is typically characterized by a current spike followed by the tripping of a circuit breaker. The current spike will be brief, no more than a few milliseconds. Because the current sensors in the AMPS system are only read once per second, it is probable that the current data will look normal—loads can be switched on and off—and the only indication of the short circuit will be an indication that the circuit breaker has tripped. Distinguishing between an actual short circuit and a "false trip," in which a breaker trips due to a switch malfunction, requires sensor sampling rates much higher than available in the AMPS system.

Load overcurrent. An overcurrent of an electrical load is characterized by current rising over a period of time above a peak threshold, and possibly exceeding the circuit's overcurrent trip threshold. For example, a pump with a degrading bearing could require increasing power to overcome friction forces in the pump, and if the degradation is significant enough, could draw more power than the circuit is capable of providing. Operating with more load on a circuit than allowed, such as connecting too much equipment to a power strip, could also induce an overcurrent and a switch trip. For AMPS diagnosis, if current above a load threshold is observed for at least a few seconds prior to a switch trip, a load overcurrent fault is diagnosed.

The sensors and data used to detect and isolate the AMPS failures include:

Loss of data. The failure of a PDU or MBSU results in no data from the component. In the AMPS system, when data is lost, nothing is received from the component. In many systems, however, data will continue to be received but each element may be labeled as static, or processing may need to be performed to detect that none of the data is updating as expected.

Switch states. Sensors provide the actual open or closed states of switches, and component software provides the commanded state of switches. AMPS software includes a mismatch indication when the commanded state and the actual state differ.

Switch trip indications. If a switch has tripped due to an overcurrent, a trip sensor in the switch sets a trip indication.

Voltage sensors. Voltage measurements at the PDU inputs from the MBSUs are used by the diagnostic model to determine if power is being delivered by the individual MBSUs. A voltage of 60 volts and above is understood to indicate normal power, and a voltage below 60 volts is understood to indicate that the MBSU is not delivering power. Normally power is either close to 120 volts, indicating good power, or near 0 volts (allowing for sensor "noise"), indicating no power.

Current sensors. Amperage is measured on each PDU and MBSU output channel and is used to detect over-current faults.

The AMPS model represents failures typical of operational spacecraft power systems. The model contains about 70 failure modes and uses about 70 data elements. The model was constructed to diagnose failures for which sufficient data is available to isolate the failures, so with a single fault there is no diagnostic ambiguity.

2.3 ADVANCES IN FFM DEVELOPMENT, IMPLEMENTATION, AND APPLICATION

Successful FFM applications result from following a proven set of procedures that have systematically evolved and been defined by experienced modelers. In this section, advances in FFM are presented in the context of the FFM development, implementation, and application process [23] shown in Fig. 2.6. The discussion begins with an overview of the FFM development, implementation, and application process. This is followed by a detailed discussion of each step of the process. As part of this discussion, advances in FFM are described where they are first used in the process.

The FFM development process shown in Fig. 2.6 is an iterative process with four primary phases: 1) knowledge acquisition, 2) conceptual design, 3) implementation and verification, and 4) application. Although these phases are shown in series, the phases are inherently coupled. Shown as the initial phase, knowledge acquisition is typically an ongoing activity throughout the entire FFM process. This is especially true if qualitative FFMs are developed early in the system design cycle and evolve with the system design. As this evolution takes place, new or revised design knowledge is periodically generated and must be incorporated into the FFM. Examples of this knowledge are 1) review board approved design changes and 2) enhanced understanding of the system application as a result of verification and validation. The second phase, conceptual design, involves defining the architecture and constructing the FFM. Knowledge gained from Phase 1 is initially flowed into this phase; however, subsequent modifications to the FFM may be required due to evolving requirements, system design data, and results obtained during verification and validation. Under the third phase, implementation and verification, the initial FFM is developed, verified, and validated. As with prior phases, evolving system requirements and knowledge are likely to require that the FFM be periodically revised, reverified, and revalidated. The fourth phase, application, involves using the FFM to conduct analyses supporting the systems engineering development process and online detection and isolation of failures. In an effective systems engineering process, results from this phase will inevitably feed back into the system design. As the system design matures, the FFM will reflect that as well and require fewer revisions and less maintenance.

Although specific aerospace applications may require modification of this generally accepted process, the primary functions outlined here are commonly practiced by the majority of current FFM modelers. The following sections provide more details of these four key steps in the FFM process.

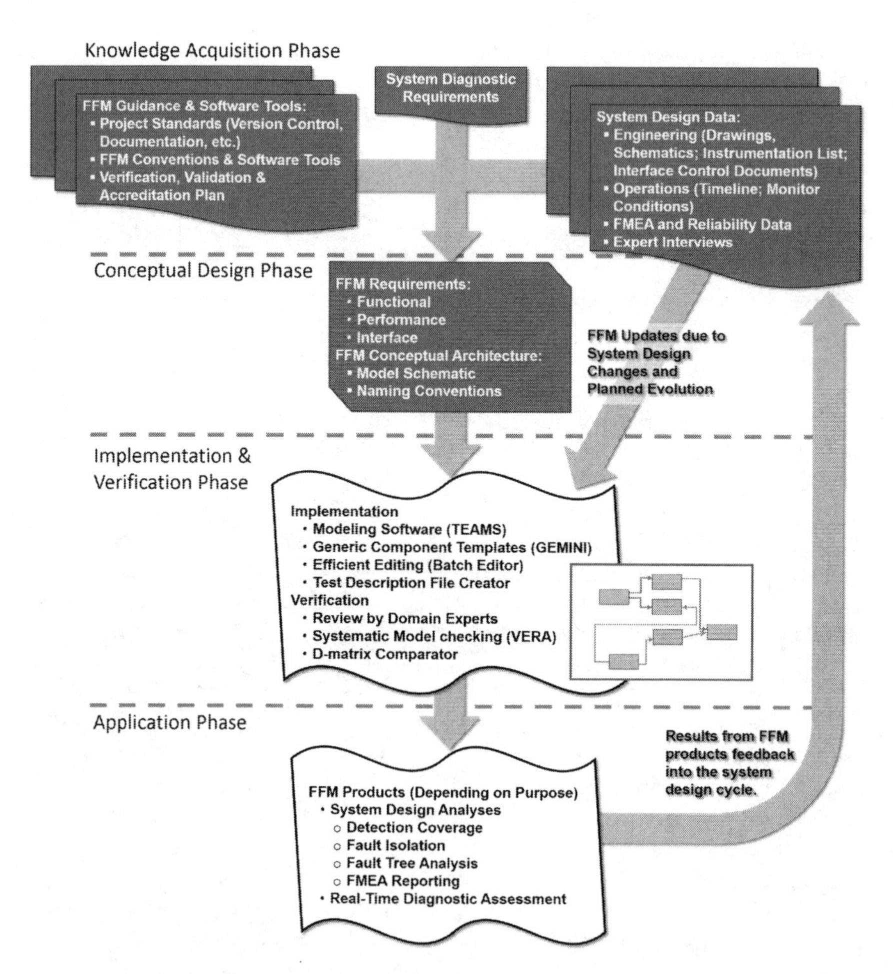

Fig. 2.6 Functional Fault Model development process.

2.3.1 KNOWLEDGE ACQUISITION

Note in Fig. 2.6 that the knowledge acquisition phase includes two main components: 1) system design data, and 2) FFM guidance and software tools. Connecting these are the overall system diagnostic requirements that flow down through the whole FFM model development process. As described next, knowledge acquisition for system design pertains to the application that is being modeled, and knowledge acquisition for FFM guidance and software tools pertains to the FFM development process.

System Design Data: For the first phase in the FFM development process, collecting information about the system under consideration for the FFM application is obviously an important initial step in the model development process. Documentation that is typically generated for a project can be found in several forms, such as:

- Concept of operations
- FMEA
- Schematics and instrumentation lists
- Interface definition and control
- Ground systems and maintenance procedures

The above items are not intended to be an exhaustive list, but they do illustrate the comprehensive set of information and data that might be needed when building the FFM concerning engineering, operations, and safety and mission assurance knowledge. In addition to written information, domain experts offer a source of information that can be acquired through informal as well as formal project meetings. Each setting offers a different type of opportunity in which to learn and gather information for the FFM application. For example, informal meetings with system designers may provide practical modeling information, whereas project design reviews are responsible for approving changes to the design that will affect the final FFM product. If the FFM is to be applied to hardware, information about the testbed and facilities will be important as these may be incorporated into the model.

If the FFM application is based on new technology or concepts, then there may be a lack of system design information. In this case, research can be performed to obtain the domain information that can support the development of the FFM. The means to gather information can largely be accomplished via the Internet where web-based search engines and data bases may provide usable information. In that same manner, the American Institute of Aeronautics and Astronautics (AIAA), the Institute of Electrical and Electronics Engineers, the National Aeronautics and Space Administration (NASA), and other organizations maintain their own websites that contain information specific to their sponsored activities and research. Even with an exhaustive literature search, there can still be applicable information that is not readily obtainable because it is proprietary. In such cases where the FFM application is similar to a previous one, reusing, modifying, or just examining the previously developed model will provide insight that can be adapted to the current application.

FFM Guidance and Software Tools: In addition to system design information that is collected to model the application under consideration, the FFM development and modeling process will have domain information that needs to be identified as well. Knowledge acquisition for the FFM development and modeling process will involve recognizing domain information, whether it is collected in documentation or informally practiced by established FFM development teams. Conventions and standards that define modeling best practices, as well as plans for accreditation and Verification and Validation (V&V) procedures, are needed for successful FFM development and must be established if they are not already defined. In addition, configuration management for the FFM application will track its development, and organization of the associated system design information will facilitate modeling accuracy. Depending on the scope of the project, configuration management may be handled in a formal or informal manner. Software utilities that facilitate modeling, analyzing, and verifying the FFM application will also need to be identified. These tools aid the FFM development team and provide consistency within the modeling environment. From the knowledge acquisition process, the following documentation and software utilities are typically identified for the FFM development process:

- Conventions and standards for modeling
- Standards and plans for accreditation and V&V
- Configuration management plan
- Software utilities for consistent modeling, editing, and verification checking

As before, with the system design domain information, this list is not intended to be all-encompassing, but to provide examples of the typical types of domain information that will be utilized during the FFM development process. When domain information is lacking from the systems engineering design and modeling process perspectives, the FFM development process can lack direction and consequently may not deliver a product that the customer has confidence in. When the FFM development team is experienced, they may already be performing modeling that adheres to recognized standards and conventions, which may or not be captured in documentation. If they are not documented, then it would be prudent to record those best practices so that they are standardized and available to others for reference. Similarly, accreditation and V&V may be performed informally, but when those procedures are documented they become a recognized set of guidelines that, when followed, ensure an FFM product that conforms to project expectations. Software utilities that aid the modeling process create FFMs that can be easily integrated, verified, and validated. As the FFM development team acquires experience and knowledge, the development of software utilities that customize the modeling process, the analyses performed, and the assessment of results are a typical outcome. Whereas these software utilities may be specific to a particular software package, the general concepts and application are relevant to any FFM development.

2.3.2 FFM CONCEPTUAL DESIGN

The general architecture of the FFM may be composed of subsystems, individual components, other ancillary hardware, and instrumentation; in addition, connectivity between these elements is established and failure modes are defined.

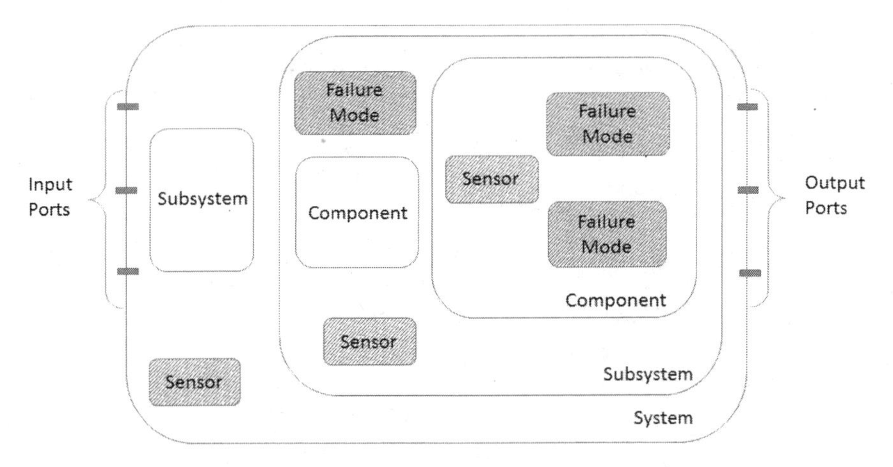

Fig. 2.7 Representative FFM.

Collectively this defines a basic FFM, and a diagram for a representative model is shown in Fig. 2.7.

Here, the representative FFM does not show the interconnections that exist among the system, subsystem, components, sensors, or failure modes because the diagram is only meant to show the basic building blocks of the FFM. In addition, accurate modeling of the flow of information in the FFM can be represented with other advanced modeling elements that are not shown here. More defined and complex FFM diagrams that illustrate the different connections and modeling elements that can be used are shown through the applications in this chapter.

The complexity of a given FFM will depend on the application being modeled, the requirements that need to be satisfied, and the model's intended use by the customer. The customer for the FFM, and the analyses and information it provides, can aid in its construction by interacting with the modeler during all phases of its development, testing, and validation. Likewise, the FFM can be modified and enhanced by the developer based on knowledge obtained through continued collaboration with system designers, the hardware and fabrication processes, and all phases of testing.

The remainder of this section provides a description of these three items: 1) system design and diagnostic requirements traceability, 2) FFM architecture, and 3) modeling conventions and best practices. These three items build the foundation for a well-developed FFM. Effective FFM applications cover the needed system and diagnostic requirements, build a structured architecture that supports the needs of the system designers and the project, and employ appropriate best practices that produce useful models and analyses results for all stakeholders.

2.3.2.1 SYSTEM DESIGN AND DIAGNOSTIC REQUIREMENTS TRACEABILITY

During the knowledge acquisition phase, system design and diagnostic requirements are collected that guide the FFM developer with specifications that must be modeled and analyses that must be performed to ensure that required objectives are achieved. The FFM product may be used for a variety of purposes, such as supporting design development, analytical verification of diagnostic coverage, and real-time monitoring and assessment during operations. These functions are often related to supporting formal project-level reviews and documentation, as well as informal analyses requested from system designers. As such, the FFM provides a systematic way for tracing the propagation of failures in the system design and exposes where requirements are not met or may not be needed. Essentially, the FFM can be used for analyzing the defined system and assessing diagnostic results, and it can also be used as an experimental tool for answering questions and interrogating the system.

FFM requirements should include functional requirements, performance requirements, and interface requirements; these are often defined by the system designers with influence from the project level regarding feasibility and implementation. For example, a requirement from a system designer may not be feasible due to cost or schedule impacts. In that case, the project management and system designers must work together to determine a resolution. Functional requirements define the failure modes the model must detect, the mechanisms available to detect them, and the level of isolation (e.g., component, subsystem, system) required for each failure mode. Functional requirements also include the analyses the model will be required to perform. Under NASA's Constellation and Space Launch System (SLS) Programs, advances in FFMs have resulted in a number of new analytical approaches to support verification of system diagnostic requirements and algorithms. Examples of these new approaches include the verification of fault isolation requirements for proposed line replaceable units, the verification of algorithms to detect launch commit criteria, and sensor sensitivity studies designed to assess the impact of sensor loss to diagnostic capability. Performance requirements describe how well the FFM must function and typically include quantitative bounds for metrics like the time required to detect a failure mode and the accuracy of the diagnoses, the latter being quantified by false positive and false negative rates. Interface requirements may specify the nature of the interface between FFM subsystems that will eventually be integrated into a system-level FFM; or, in the case of real-time assessment, the interface between a data bus and the FFM.

2.3.2.2 FFM ARCHITECTURE

The FFM architecture is primarily based on the model schematic for the system where the subsystems, components, instrumentation, and connectivity are defined. FFM development software usually supports a hierarchical model structure, where a given model element can contain other model elements. In this structure, the lowest model elements represent either failure modes or mapping functions that define the transition of failure effects through nominal components. This hierarchical structure allows the model developer to trace failure modes and other functions through the system. For small FFMs, a flat hierarchical structure—a structure in which the entire system structure can be displayed on a single level—may be sufficient. For larger models, the system may need to be decomposed into subsystems, and then into assemblies, modules, components, and parts, until the required model fidelity is obtained. If required for analysis, system elements that have special designations, such as line replaceable units, can also be easily defined in the FMM.

Component selection criteria should be based on requirements levied on the FFM by the program. In general, the FFM should only include components that are relevant and could, therefore, influence model configuration or contain credible failure modes. Credible failure modes are usually ones that have been categorized as critical and defined as such in formal requirements and system design documentation. It is recommended that the depth of the FFM hierarchy be somewhat consistent across the various subsystems included in the top-level system FFM. The fidelity of data in FMEAs, historically used to provide failure mode data to FFMs, can vary widely from component to component. This has an effect of biasing assessment metrics. FFM developers should be aware of this possibility and strive to balance model fidelity across components and subsystems.

When defining the system elements, it is useful to organize information into a System Breakdown Structure (SBS). This structure ideally captures the breadth and depth of the decomposed system (i.e., hardware and software elements), provides a mapping of other element data that will be used in the FFM (e.g., element and failure modes named to defined conventions), and initiates an understanding of failure-propagation pathways. Here, the breadth of the model is represented by the list of components that must be modeled, while the depth is represented by the various hierarchical levels. Identifying the overall structure of the FFM, both breadth and depth, is an important part of the conceptual design. Using the SBS to explicitly identify components that are to be included in the FFM, to identify components that should be excluded from the FFM, and to define the representation of included components will confirm the expectations of the customers and consumers of the FFM and results of associated diagnostic assessments. For large models, this decomposition of the FFM can facilitate the parsing of model development work to various members of a modeling team.

Standardizing and documenting element names prior to implementation provides for more efficient modeling. This is especially true during the integration of FFM partitions that were previously delegated to different modelers. In addition, making names succinct and consistent aids in readability of both FFM diagrams and reports. One of the challenges to FFM development is that the same failure for a component can be described in numerous, sometimes ambiguous, ways (e.g., a valve can fail open vs fail full open, or fail part open vs fail part closed). And the same name can be interpreted differently by each developer. Providing a failure mode dictionary that consistently names the failures provides continuity among the same elements in different subsystems and catalogs the same failure mode across a large FFM.

The FFM architecture also needs to take into account the mission timeline for the modeled application. The establishment of an operational profile for the system is essential in determining the complexity of the modeling required to represent the various configurations; this is especially important for an FFM that is used for real-time diagnostics. For the FFM, a system mode typically defines a unique model configuration that establishes specific failure mode propagation paths and determines the subset of active failure modes and available tests. In real-time applications, consideration of system mode definition must be made when test thresholds are changed due to the operational profile, even though the model configuration remains the same. Some systems may have very simplistic operational modes (e.g., on or off), whereas other systems may have complex operational modes that are not easily represented.

2.3.2.3 MODELING CONVENTIONS AND BEST PRACTICES

Representing a system's failure space can be done in a number of ways. Each developer will envision that representation differently depending on their domain expertise and modeling experience. Often, a large system model may have multiple developers or be developed over an extended period of time. Inconsistencies in the way model elements are represented or the location of system data within the model can result in modeling errors that are difficult to find and resolve. In addition, inconsistent representation can make interfacing with other supporting software tools and real-time diagnostic applications impossible. The key to minimizing these model development issues is to establish a set of modeling conventions and best practices.

The objectives of the modeling conventions and best practices are to provide a guide for developers to follow that will 1) provide a consistent look and feel to the model, 2) facilitate integration of the model with other functional fault models, 3) enhance diagnostic reporting and model verification by providing traceability to source documents within the model, 4) standardize the interface with external processing tools and real-time diagnostics, 5) allow reusable component libraries to be developed, and 6) accelerate model development and verification through the use of automated or semi-automated tools. In addition, best practices provide modeling solutions to various implementation and development issues. Overall, modeling conventions and practices will reduce the overall model development costs and should result in a model that better supports the intended application.

For the past two to three decades, NASA has been developing and evolving FFM technologies for application to space flight systems. Requirements and expectations for these models have expanded as well. An important advance in FFMs resulted as model development teams identified the need to establish a common and consistent set of modeling conventions. Early drafts of conventions for launch vehicle and ground operations FFM applications [22] appeared in 2008 under the NASA Constellation Program. These conventions and practices were eventually expanded and documented in Report K0000190582-GEN NASA

Ground Systems and Launch Vehicles Testability Engineering and Maintenance System (TEAMS) Modeling Conventions and Practices [14].

This report specifically supports or has supported the model development for the SLS Mission and Fault Management (M&FM) Integrated Vehicle Failure Model (IVFM) team; the Command, Control, Communications & Range (C3R) Project Advanced Ground Systems Maintenance (AGSM) Element–Fault Isolation and Functional Fault Modeling team; and the Autonomous Mission Operations (AMO) EFT-1 ACAWS team. The document covers basic model development concepts, such as the level of model fidelity and how to represent failure modes, as well as specific modeling best practices, such as representing test point elements locally within the model or at a remote location external to the modeled system and the issues to consider with either approach. In some cases, these modeling practices can be directly tied to conventions that can further facilitate and support the development approaches. The conventions address the location of key system information in model element names and labels as well as the elements' appearance, such as color and style.

To illustrate the impact that these conventions and practices can have on a model, we review portions of the Basic FFM described in Section 2.2.2. Fig. 2.8 shows how following the conventional appearance requirements enables immediate distinction of the general component modules that are a blue-no-fill style

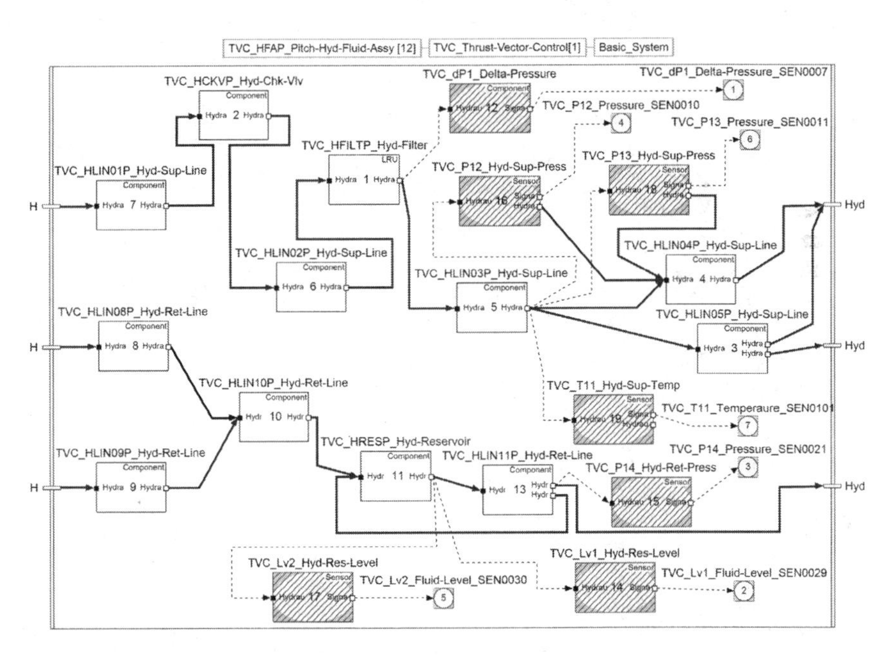

Fig. 2.8 Modeling conventions and best practices [14] applied to example FFM displays distinction between types of components represented.

(shown as no-fill style in Fig. 2.8) with the sensor modules that are a blue-hatched-style (shown as hatched style in Fig. 2.8). Similarly, in Fig. 2.9, failure modes modules in a red-hatched (i.e., modules 1 and 2) style are distinguishable from mapping modules in the green-hatched (i.e., modules 3 and 4) style. Maintaining these appearance conventions throughout the model development creates that common look-and-feel environment that facilitates not only developer peer reviews of the models, but also reviews conducted with domain experts where this common appearance increases the credibility level with the expert.

Defining conventions that require that the location and content of the system design information be specified within the FFM allows modelers and stakeholders to trace model elements to their source documentation. Model element names, property labels, and internal fields can contain key pieces of data that support analysis reporting. Structured, defined content facilitates integration of the models with other independently developed models and interfacing within real-time diagnostic applications. For example, the current FFM convention [14] for the failure mode module name is,

<Subsystem ID>_<Schematic ID>_<Description>_<FMEA ID>

where

<Subsystem ID> is an identifier (ID) of the highest-level subsystem module where the failure mode originates

<Schematic ID> is a unique ID for the module containing the modeled failure mode, taken from the reference schematics used to develop the system model

<Description> is a text description, short and concise, of the failure mode where words are separated by dashes "-"

<FMEA ID> is an ID for this failure mode from the FMEA of the modeled system, if available (if the FMEA ID is not available either because there is no FMEA of the modeled system, or this particular failure mode was not included in the system's FMEA, then "NA" will be entered, indicating the ID is not available)

By formalizing *what* system design information will be included in the model and *where* it will be included, interfaces with supporting software can be more easily created and supported. In addition, the conventions facilitate traceability between the system design and the FFM, which supports overall verification and accreditation. For example, having specific formats for model inputs enables automatic verification and checking of the FFM.

2.3.3 FFM IMPLEMENTATION

Over the past five to seven years, the implementation of FFMs has advanced from being an art form with few documented standards and practices to a reasonably

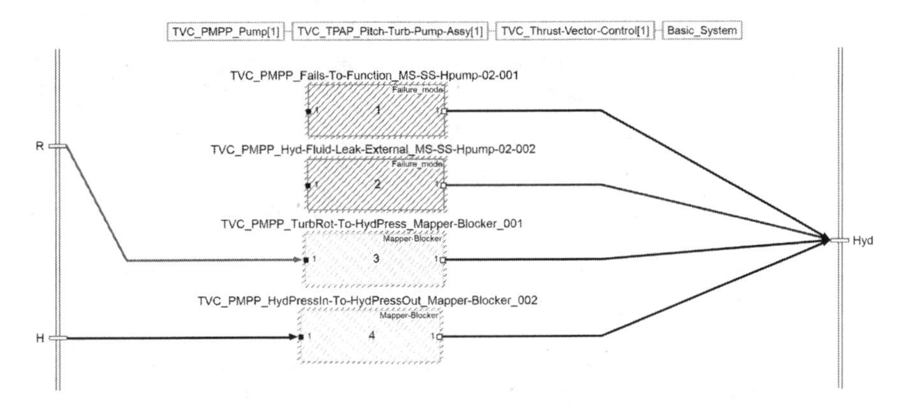

Fig. 2.9 Modeling conventions and best practices [14] applied to example FFM model displays the low-level distinction between failure mode modules and mapping modules.

proficient modeling technology with well-documented guidelines. Still, individual modelers working independently will typically represent the failure space of a given subsystem or component using inconsistent nomenclature and modeling constructs. This can lead to incompatibilities, confusion, and verification issues during the integration of these models into a system-level FFM. Hence, it is important that individual modelers adhere to the modeling conventions and guidelines established during the conceptual design phase. Once the components to be modeled (breadth) and the hierarchical structure (depth) are defined, the FFM development team can begin the process of building the model with FFM COTS development software and in-house software tools.

For example, during much of the time this work was being performed, the COTS software did not provide a user-friendly global find-and-replace capability. Consequently, NASA developed the Batch Editor [15] as a standalone software tool to fill that void. The Batch Editor provides a user with a means to extract data from, and make broad changes to, a model. For example, the user can query an FFM for the names of failure modes, review the resulting CSV (Comma Separated Variable) file to identify and globally correct misspelled or incorrect names, and then use the Batch Editor to impose changes embodied in the revised CSV file on the FFM. The Batch Editor currently supports over 100 editing commands, ranging from simple "get" and "set" commands to complete model creation from a sequence of commands. It is being used by FFM modelers within NASA's SLS and Orion programs.

2.3.3.1 GENERAL MODELING APPROACH

The architecture and construction of the FFM will depend on the purpose of the model, the scope and complexity required, the establishment and organization of the modeling environment, and the preparation and planning to support its intended use. In general, the FFM is expected to provide analyses related to systems health management that pertain to fault detection and isolation. In particular, requirements for analyses, such as those needed to support ground-based diagnostics and maintenance or real-time capabilities, can be specific to the system being modeled. These and other considerations will affect the architecture and how the modeling elements are defined because, if they are not part of the initial model implementation, it may be difficult to include them later.

The scope and complexity required of the FFM will depend on the type and number of physical domains being modeled. For example, an FFM that is being used to model the operation of a battery will be very different than one being used to model an entire electrical power system. In addition, the model may need to include more than one type of physical domain, such as when applied to a main propulsion system that includes systems for the fuel, oxidizer, hydraulic actuation, and electrical power. More complexity must be managed when the application is a space vehicle where there are systems that include the electrical power system, engine, main propulsion system, avionics, and control and actuation systems. Consequently, using the SBS to partition the system into subsystems, which can be modeled independently, should simplify the implementation of the system-level FFM.

The establishment and organization of the modeling environment facilitates FFM best practices and will result in a model that is based on well-defined and agreed upon guidelines. Following the modeling conventions and practices, previously discussed in Section 2.3.2, ensures that the FFM is built with special consideration given to naming components and other ancillary modeling elements, so that it is familiar to users with knowledge of the system design. Furthermore, using a predefined set of rules will ease modifications and automatic updating of the model because it is based on a structured framework. Along with using approved guidelines, the modeling environment can also benefit from using generic components as shown in [19]. Generic components provide a means to create models that have a commonality for like components; these can assist in easing the development of the FFM, especially for inexperienced modelers. In addition, generic components can improve the integration process when smaller models are assembled into a larger one, as similar components are fundamentally set up in the same manner but may be modified for their specific placement in the FFM. For example, this implies that a valve modeled in one subsystem is set up in the same manner when that same type of valve is used in another subsystem, thereby eliminating the need to create two different valve models that may confuse analysis results.

Pertaining to space-based applications, the FFM application ultimately will be used to conduct systems health management analyses as shown in reference [17], and may possibly be used as a tool for trade studies and investigations that might be needed to resolve design changes. For example, the criticality of a failure mode typically needs to be defined in the FFM so that it can be used for identification purposes when performing a testability analysis. Likewise, other analyses might involve knowing caution and warning, redundancy management, or launch commit criteria information because these can specifically identify information related to a failure mode, component, or other modeling element. Instrumentation and test points are used in the FFM much like they are in hardware as a way to obtain information about the system state. Consequently, each test point should be identified by its type (pressure, temperature, etc.), phase of operation (ground, prelaunch, flight, etc.), and objective (operational, ground, experimental, etc.).

2.3.3.2 FFM DEVELOPMENT USING GENERIC COMPONENT MODELS

Even with a core set of conventions and practices and a suite of support tools, FFM development may require significant time and resources. Model elements containing qualitative failure mode and diagnostic testing information that are entered manually are subject to human-touch errors (e.g., typographical errors). The desire to reduce these human-touch errors motivated the advancement of many of the FFM software tools. However, although these tools have been largely successful, there is still significant manual effort involved in establishing the model structure and connecting elements to create the proper propagation paths.

A system schematic may contain hundreds of components, but typically has only a few tens of unique component types that make up the system. If each component is individually crafted by the modeler, it is likely that there will be inconsistencies in the way similar components are represented across the system, such as the textual wording of failure mode data, establishment of propagation flow paths, and transition of functions through a component—or even the simple location and style of modeling elements. Such inconsistencies cause confusion during model reviews and during the transition of an FFM from one modeler to another. To illustrate this, Fig. 2.10 shows the FFM representation of two identical valve components. Although both implementations of the valve provide the same functionality, the structure and details of the models differ. If these two models were included in the same system-level FFM, they would likely frustrate model reviewers, requiring them to unnecessarily spend time assessing the functional equivalence of the models and reconciling the differences.

Looking at these two component models in detail, Fig. 2.11 zooms in to highlight the failure mode naming distinction and illustrate the need for conventions when naming modules within the FFM development software. Both valve models contain failure modes for the valve failing to open when commanded. In the Valve 1 model (upper left frame), Failure mode 1 is named "Valve-Fails-to-Open," while in the Valve 2 model (lower right frame), Failure mode 1 is named "Valve-Remains-Closed." Although this inconsistency appears to be simply a difference in semantics, it will cause unnecessary problems in reporting and interpreting diagnostic assessments generated from the FFM and will add an unnecessary burden during verification and accreditation processes.

In addition, inconsistent FFM module representation can lead to an imbalanced diagnosis across a larger integrated system in which similar components could have varying degrees of model fidelity and detail. This often results in confusion from the overall system diagnostic assessment. For example, similar components in different subsystems within the same FFM could model the same failure mode space in completely different ways, again requiring additional effort on the part of the reviewers and integrators to align them. As an example, refer to Fig. 2.10 where failure mode modules are represented by darker hatched, rectangular boxes. The valve model in part a of Fig. 2.10 represents the failure space with six (6) failure mode modules, while the valve model in part b represents the same failure space with four (4) failure modes.

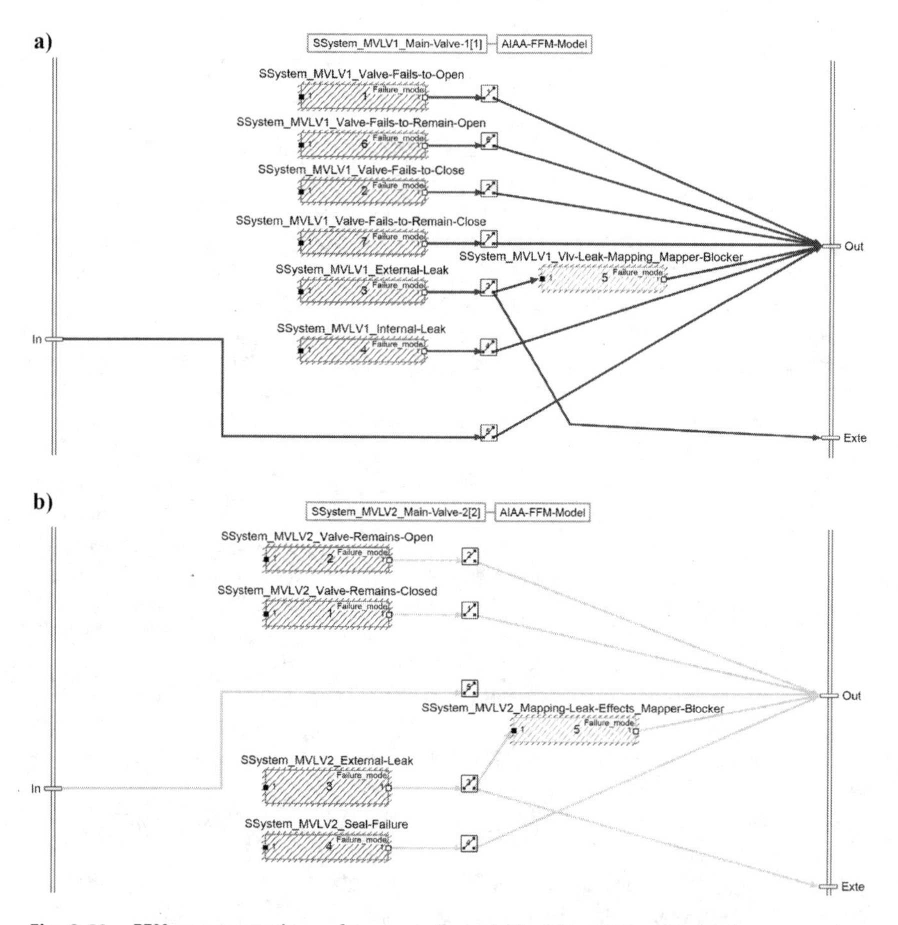

Fig. 2.10 FFM representations of two sets (a and b) of identical valve hardware illustrate the variations that can occur when models of identical components are developed individually.

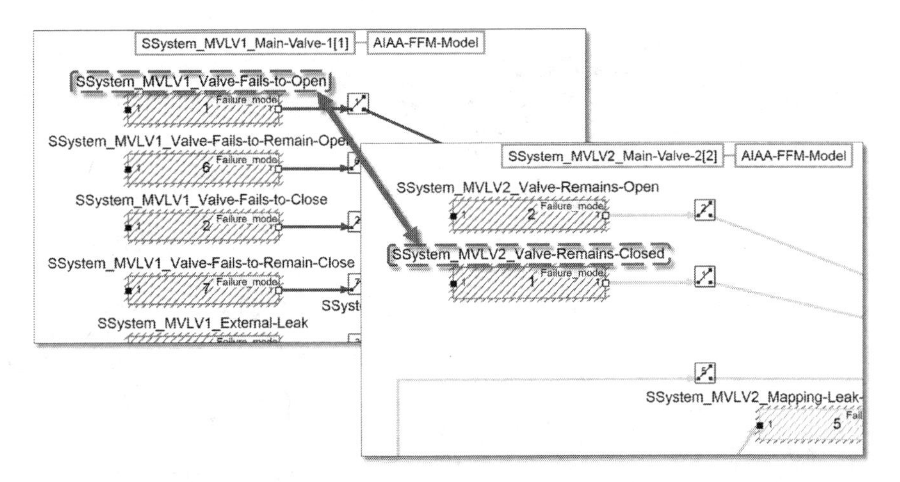

Fig. 2.11 FFM representations of two common valve components zooming in on a specific variation between the models.

Previous diagnostic modeling efforts identified the advantages of establishing a library of component models that could be used as building blocks in constructing future FFM system models [8, 19]. Although the concept of component libraries is not new, current versions of the FFM development software do not natively support reusable component libraries. To work around this limitation and advance FFMs, modelers working under the NASA Ground Systems Development and Operations (GSDO) program's AGSM project developed in-house tools to support the use of FFM component libraries [22].

FFM libraries containing generic component models were advanced as one means of overcoming the previously described issues. Although the idea of generic models is not new, the use of generic models in the FFM domain is recent and still maturing. Figure 2.12 illustrates the FFM development process where Generic Component Models (GCMs) are established initially for candidate component types that are common in the FFM. The GCMs are then "instantiated," meaning that information placeholders are updated to reflect the specific representation of the final component, the Instantiated Component Model (ICM). These ICMs can then be integrated into the general system-level FFM. Under the AGSM application, systematic verifications were performed at the GCM, ICM, general system-level, and final system-level stages of FFM development. Based on experience with the initial application of this process, researchers estimated an 80 percent reduction in component module development costs [19]. The GCM models can be maintained in a library that is then available for future FFM efforts, further reducing overall development cost.

This generic component model development process took advantage of approved modeling conventions [14] and extended them to the generic

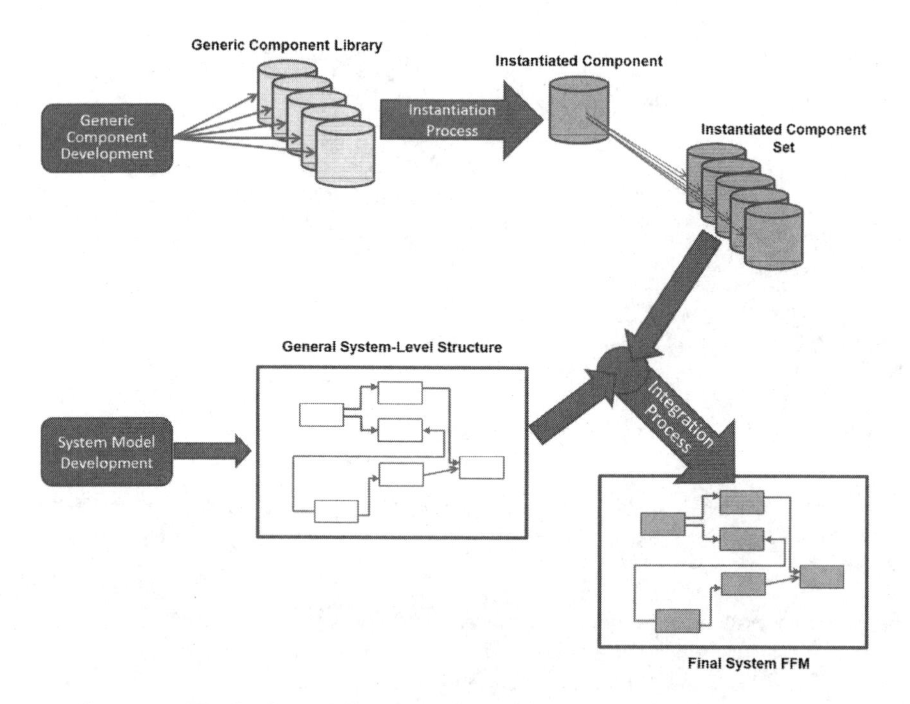

Fig. 2.12 FFM development process with generic component modeling included.

component modeling, which in turn facilitated the development of a software tool called GEMINI. This software tool enforces the modeling conventions during the instantiation process and guides the user through the various places in each GCM that can accept system-specific input.

Generic component model development processes like the one discussed here are essential in areas where similar components are being modeled over and over. Establishing an accepted library of component models that the developer can draw from can result in more rapid model development and models that have a more consistent look and feel. Having a formal development process also enables a structured verification approach to be applied throughout the FFM development cycle, which enhances the process of accrediting FFMs.

2.3.3.3 VERIFICATION AND VALIDATION OF FFMS

Once the FFM is developed and built, it must be evaluated to ensure that the model is constructed properly and that it fulfills its intended purpose. These two assessments make up the verification and validation of the FFM, which can involve both formal and informal methods. In practice, using a combination of formal and informal methods will thoroughly review the model and reveal where issues must be addressed. In addition, the amount of verification and validation required is determined by the model's intended use. An FFM developed for a proof-of-concept application will not need the rigorous assessment that an FFM used for an actual space-flight application may need. Ultimately, verification and validation are important steps that are required if the FFM needs to be certified for its planned function.

Systematic evaluation of the FFM for verification can be performed to ensure that the model is built according to the defined modeling conventions and practices [14]. When this is done in an automated fashion, human-touch errors are eliminated, and the accuracy of the FFM, as measured against the modeling conventions, is quantitatively defined. As a consequence, there is confidence in the ability of the FFM to accurately represent the system design. One automated tool that has significantly advanced the verification of FFMs is the Verification Analysis (VERA) tool [18]. The VERA tool automatically compiles a list of all of the elements in an FFM and programmatically verifies (to the extent possible) that they follow the established conventions and practices [14] that have been developed. This program is able to produce a comprehensive list of model elements and to verify most of the low-level model implementation. The VERA tool is written in Visual Basic for Applications (VBA) and utilizes the extracted FFM information to inspect the model's structure, then produces a comprehensive list of specified elements within the FFM, including all component modules, arcs/links, failure modes, failure effects, tests, and mappers. These elements are then analyzed by the tool to confirm that they follow the approved modeling conventions and practices [14], have the appropriate relationship with their parent and child elements, and contain all available design information specific to the item being modeled.

Specifically, the VERA tool generates a Microsoft Excel workbook that contains all of the components of each verification category. A separate spreadsheet is created in the workbook for each verification category, and every model element from each category is listed along with all pertinent information about that element that the VERA tool is able to collect from the extracted FFM information. Identified errors are divided into categories based on the part of the model that they affect, and scores are assigned to each error based on that error's severity. These errors and the resulting scores are divided into four different groups: technical errors, convention errors, cosmetic errors, and information errors. Each type of error is scored independently so that both the severity and the scope of each error type is readily apparent. Here is a definition of each type of error:

- **Technical errors:** These are identified modeling errors that directly impact the performance of the model. An example would be an unconnected failure mode module. These are the most critical errors and should be addressed immediately. The assigned scores for these errors are one hundred (100).
- **Convention errors:** These are errors that violate the approved modeling conventions and practices [14] and will impact the ability of the model to properly

interact with supporting software or diagnostic assessment output. An example would be a failure mode module with a name containing the incorrect number of fields. These errors should be addressed, but the model will function. The assigned scores for these errors are ten (10).

- **Cosmetic errors:** These errors do violate the approved modeling convention and practices, but their impact is minimal. Often these errors are simply visual representation and meant for human consumption only. An example of this type of error would be that a failure mode module does not have the correct fill color, red. These errors should be corrected as they will enhance the verification reviews of the model. The assigned scores for these errors are one (1).
- **Information errors:** These errors simply detect an omission of design information in the model, such as if no FMEA ID is provided in the failure mode module name. It is possible that the omission is intentional or is because the data are not available yet, in which case it cannot be corrected. Still, this error will indicate what information gaps remain in the model. The assigned scores for these errors are one (1).

Validation of the FFM can begin with informal reviews, which provide a means to capture an expert's knowledge that otherwise might be overlooked. In this case, reviewing the model with domain experts ensures that the model fundamentally represents the application under consideration and is a worthwhile endeavor, because it exposes the model to a level of inspection that cannot necessarily be quantified but is intrinsically beneficial. On the other hand, formal project reviews, such as preliminary and critical design reviews, afford a well-established manner in which the FFM can be methodically and systematically reviewed against the defined requirements and offers an official vetting of the model. This type of FFM validation is typically required when the application is implemented in hardware for an experiment or space-flight project. Stakeholders benefit from both informal and formal validation reviews, because they expose the FFM to what should be a thorough review process.

Further validation of the FFM will most likely be performed with simulated or experimental data to make sure that the model performs as expected and produces accurate results. Because the FFM is used for health management analyses such as fault detection, isolation, and recovery, having data that represent off-nominal and failure scenarios is critical for its validation. For that reason, the numerical data must be representative of the instrumentation data that are expected to be processed by the FFM that is actually put into practice. In addition, software-in-the-loop testing can provide a cost-effective way of validating the FFM that would otherwise be prohibitive in hardware due to the cost or safety concerns; this also provides a controlled environment in which the conditions are repeatable.

The D-Matrix Comparator (DMC) [15] is another software tool that has been developed to advance the validation of FFMs. As described in Section 2.2.1, the

D-Matrix is one critical element of the FFM development software's testability analysis output. The D-Matrix represents the model's failure space where each row is a failure mode within the system diagnostic model and each column a measurement test. For a particular set of conditions, the testability analysis determines whether a measurement test can detect a particular failure mode. If so, the value "1" is placed in the cell where the measurement test column and failure mode row intersect. The set of tests that can detect a particular failure mode is that fault's detection signature.

During the system design process, the FFM will evolve with the system design. To validate the model as it evolves, it is important to identify and understand how changes to the model impact the associated testability analysis output—including the D-Matrix. Differences between D-Matrix files, from various versions of the model, are due to the applied modifications or from errors introduced during modifications. Therefore, a capability to compare D-Matrix output files for multiple versions of the model is needed. When manually comparing D-Matrix files, two problems are encountered. First, unless the model is rather simple, the D-Matrix file will contain an overwhelming number of rows of faults and columns of tests. For example, one Ares I subsystem model had almost 280 tests and over 500 faults. Further, the vehicle-level Ares I model, composed of integrated subsystem models, had over 1300 tests and 8300 failure modes. In addition, when a change has occurred in the model, the number and order of the rows and columns may change, making manual comparison of the two files tedious, time-intensive, and prone to human error. The DMC tool was developed to simplify the verification and validation of model changes. The tool supports model verification and validation by analyzing the various failure modes, tests, and detection signatures in the two D-Matrix files to identify differences. The user can then use this information to determine the need for, and direction of, future modifications to the model.

When utilized, the DMC tool will open an Excel workbook with six worksheets. The first two are the D-Matrices for the specified files and the last four are analysis and comparison reports. The following is a summary of the four analysis and comparison reports:

- **Testability Options Report:** This report compares the options that were selected for each FFM when the Testability Analysis was run. These options include the selected technologies (which are often used to specify the fault criticalities), selected modes (which are used to specify the "switch" positions), the level to which the failure was isolated, and the tests that were selected. This worksheet allows the user to verify that the same testability settings have been selected for each model and, therefore, any difference between the D-Matrices is due to differences between the models, not to different testability conditions.
- **Test Comparison Report:** The second worksheet displays a list of the first model's measurement tests that are not available in the second model

according to the D-Matrix outputs, as well as a corresponding list of measurement tests that are available for the second model, but not for the first. Test names from the first file that have no match in the second file are listed in one column, while test names from the second file that have no corresponding match in the first file are listed in a second column. If all of the test names in a given file are matched with the names in the other file, "None" is inserted as the only entry within the column for that file. Note: at this point the D-Matrix Comparator identifies only differences in test names between the D-Matrices; no attempt is made to directly compare the definition of the test elements.

- **Failure Mode Comparison Report:** The next worksheet displays failure modes found in one model's D-Matrix but not in the other. Similar to the previous Test Comparison Report, failure modes that are in the first D-Matrix file, but not the second file, are listed in column one. The second column contains the failure modes found in the second file, but not the first. If all of the failure modes in a given file are matched with the failure modes in the other file, "None" is inserted as the only entry within the column for that file. Note: at this point the D-Matrix Comparator identifies only differences in failure mode names between the D-Matrices; no attempt is made to directly compare the definition of the failure mode modules.
- **Detection Signature Comparison Report:** The DMC's final report is a comparison between the failure detection signatures from the two D-Matrix files. The worksheet for this report contains four columns. The first column lists all of the common failure modes between the D-Matrices. For each failure mode, the second column lists measurement tests where both D-Matrices indicate the test detects the failure mode. The third column lists measurement tests detecting the failure mode that are indicated only in the first D-Matrix. Similarly, the fourth column lists measurement tests that are only indicated by the second D-Matrix.

2.3.4 FFM APPLICATION

In this section, the application of FFMs to two classes of problems is discussed. First, the use of FFMs in supporting the analytical verification of systems requirements is discussed. Second, the use of FFMs to support operations is discussed. In both cases, implementation approaches, advances, best practices, and potential challenges are addressed.

2.3.4.1 FFM ANALYSES SUPPORT FOR SYSTEM REQUIREMENTS VERIFICATION

FFMs can provide insight into the FM capabilities of the system. Through analyses with the model, system requirements for safety, availability, and maintainability can be evaluated. Developing the ability to verify these requirements early in the design phase enables system engineers to address deficiencies while cost impacts are less severe. The FFMs enable a level of verification early and throughout the design process due to the qualitative aspect of the data contained within the model. FFMs also provide an ability to trace a given failure mode to the sensor(s) available to detect off-nominal effects propagated by the failure mode. This allows the user to assess the overall effectiveness of the monitored system sensor suite to detect and isolate these failures.

The two key concepts defined in this section are "detection signature" and "ambiguity group." For each failure mode, the detection signature is the list of all of the tests that will detect the effects propagated from the failure mode. This signature will include not only tests from local sensors, but from every sensor included in the initial testability analysis. For the failure mode analyses presented here, an ambiguity group is defined as a group of failure modes with identical detection signatures. This indicates that the current set of observations cannot discriminate between the failure modes in the group. If the detection system is required to make a discrimination between failure modes in an ambiguity group, then the system designers may need to consider developing additional tests or adding additional measurements.

Figure 2.13 illustrates a notional FFM that represents a series of propagation paths along which the effects from each failure mode could propagate. Failure effects generated by the failure modes (i.e., boxes labeled FM1, FM2, and FM3) can be transformed along the propagation paths as they pass through various components. Along these paths, connected test points (i.e., circles) represent observation points or sensors that contain tests used to detect failure effects.

The D-Matrix from the FFM testability analysis can be further analyzed in several ways. First, the user can review the tests that detect each failure mode. This is called detectability analysis and is depicted for the notional FFM representation in Fig. 2.14. In the figure, the failure mode, FM1, is under review and the sensors containing tests that detect the effects propagated from FM1 are TP1 and TP3. This detection trace for each failure mode is called the detection signature.

The detectability analysis also identifies failure modes that are not detected by any of the available sensor tests. Failure modes identified as undetectable should be brought to the attention of systems designers, as they may indicate a failure of the system to meet requirements to detect critical failure modes.

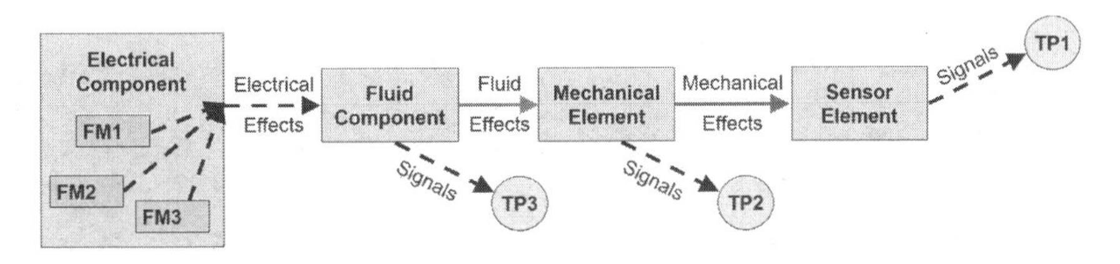

Fig. 2.13 Notional representation of an FFM.

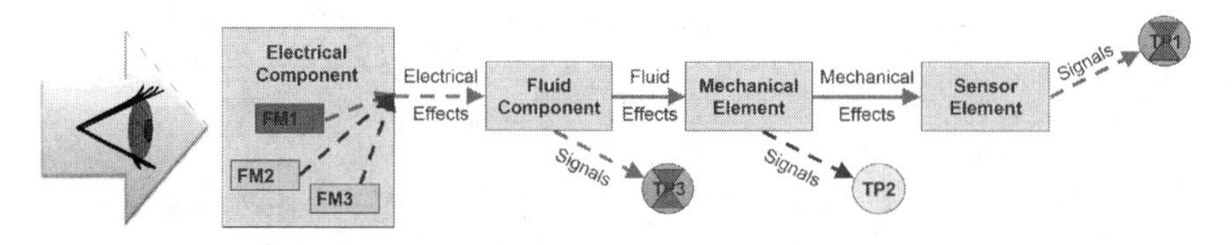

Fig. 2.14 Notional representation of detectability analysis for an FFM.

The ability of the system to detect failures may be quantified by the detection coverage, D_{cov}, as follows:

$$D_{\text{cov}} = \left(\frac{FM_{\text{total}} - FM_{\text{miss}}}{FM_{\text{total}}} \right) \times 100\% \qquad (2.1)$$

Here, FM_{total} is the total number of failure modes available for the testability analysis and FM_{miss} is the number of available failure modes that were not detected from the available test suite for the analysis. If the FFM is developed sufficiently and follows key modeling conventions, then the user can restrict the testability analyses to contain only certain failure modes and tests. For example, the user may want to determine the detection coverage for failure modes resulting in complete loss of the system and restricted to a certain classification of sensor. In space launch vehicle applications, this example could be for criticality 1 failure modes that would result in a "loss of vehicle" and/or a "loss of crew," and restrict the system observability to flight critical sensor measurements. The analysis may also constrain the definition of failure mode detection to require more than one test detection, thus requiring redundancy for failure detection.

From another perspective, the FFM can be reviewed to determine which failure modes are detected by each test. This is called test utilization analysis and is depicted for the notional FFM representation in Fig. 2.15. In this analysis, each individual test is reviewed to determine which failure modes it detects. The results can determine the effectiveness of the test suite and indicate tests that provide no detection for the analyzed conditions. This analysis can be used as a justification for retaining or removing measurements from the sensor suite.

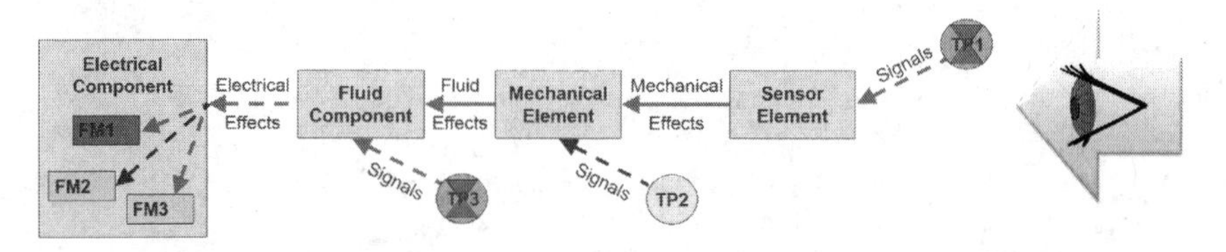

Fig. 2.15 Notional representation of test utilization analysis for an FFM.

This analysis can be further restricted to a small set of tests that represents a specific detection algorithm. The resulting trace of failure modes could be used as part of the algorithm verification process to show that the expected detection of failure modes is possible, that all anticipated failure modes can be detected by the algorithm, and that no unintended failure mode detections occur. Under the SLS M&FM project, this type of approach is being applied to generate failure mode traces for specific monitored conditions. These traces are then used to support quantitative estimates of false positive/false negative values for the algorithms, as well as support probability risk assessments for the monitored conditions.

Extending the detectability analysis, the user can determine the isolation capability of the system. Isolation refers to the system's ability to detect and distinguish between failure modes and groups of failure modes. As previously discussed, each failure mode has a detection signature. If that signature is unique from all other failure modes, then the failure mode is said to be isolatable. Failure modes possessing common detection signatures are ambiguous relative to each other and are reported as an ambiguity group. If that detection signature was observed during actual operation of the system, the diagnostic assessment could only be refined to this grouping.

Because of the hierarchical nature, the FFM failure modes should be contained in modules representing the hardware component where the failures originate. Taking the failure mode isolation analysis one step further, failure mode ambiguity groups can be divided by the originating failed component. Therefore, component isolation analysis is a review of the failure mode ambiguity grouping from the component perspective. This ambiguity grouping helps to determine the associated failure modes that are detected and isolated. In addition, if the component's failure mode is not isolated, a list is provided for the other components that are members of the ambiguity group. This level of isolation is important in establishing maintenance strategies for the system.

These processes have been refined under NASA's Ares I Project and SLS program to provide support for the following analyses:

- Abort Conditions (AC) and Abort Triggers (AT) Failure Mode Traces
- Caution and Warning (C&W) Condition Failure Mode Traces
- Redundancy Management (RM) Detection Coverage Failure Mode Traces
- Launch Commit Criteria (LCC) Failure Mode Traces
- Latent Failure Interaction Analysis
- Prelaunch LRU Analysis
- Post-Flight Observability Analysis

The NASA SLS AC and AT failure mode traces are represented by a list of all failure modes whose effects lead to a particular AC, and support the calculation of AC probability of occurrence and the probability of AT activation for the associated monitored ACs. In addition, these traces are used to indicate potential false alarms (i.e., false positive detections) resulting from noncritical failure modes. The NASA SLS C&W failure mode traces are also represented by a trace of all failure modes that would lead to the violation of a particular C&W condition. This trace is used to identify missed (i.e., false negative) detections for failure modes not traced to the C&W condition that were expected to be detected, as well as false alarms associated with failure modes whose effects trace to the C&W condition but do not warrant a C&W alarm. If this occurs, then the C&W condition detection may need to be reevaluated to ensure proper identification of the condition. This information is also used to support the C&W message storm analysis, which ensures that C&W messages generated from single fault scenarios do not overwhelm the crew with more messages than they can effectively process.

The NASA SLS RM detection coverage failure mode traces are represented by a list of failure modes that could lead to a loss of redundant functionality within the system. This list is used to provide SLS designers with a mapping of the failure modes addressed by the redundancy management strategy implemented. Finally, the LCC failure mode trace is represented by the list of failure modes that would lead to a violation of each LCC condition. For all of the failure mode traces just described, this information is also beneficial in the support of post-test and post-flight operational failure and anomaly investigations. The Latent Failure Interaction Analysis identifies failure modes from redundant elements or safety devices that remain unobservable until some other event makes them visible. The Prelaunch LRU Analysis uses the LRU lists, as supplied by the SLS elements, and determines the effectiveness of the SLS vehicle sensor suite to detect and isolate faults to a LRU during prelaunch operations. The Post-Flight Observability Analysis is performed for individual sensors, or a group of sensors, to identify the set of failure modes that could be detected by the associated sensor tests.

The general approach for generating the failure mode traces associated with the AC/AT, C&W, RM detection coverage, and LCC conditions is to first determine the specific sensor tests associated with the particular condition and align those sensor tests to the ones represented in the SLS IVFM. In addition, the operating phase(s) when the particular condition is active is also determined and aligned to the IVFM operational phases that establish the proper FFM configuration. The SLS IVFM testability analysis is then conducted based on this information, using an advanced automated tool called the Extended Testability Analysis (ETA) tool [17]. The purpose of the ETA tool is to process the testability analysis results from the FFM development software, along with information contained within the associated FFM, to provide the user with detailed documentation of the results. The ETA tool was initially developed under the NASA Constellation Program to support the Functional Fault Analysis (FFA) team for the Ares I Launch Vehicle and has been advanced under the SLS program to support the M&FM IVFM team. The ETA tool extracts information from the testability analysis output and the associated FFM to provide a detailed set of

HTML and XML reports highlighting aspects of the system's ability to detect the effects propagated from failure modes and to isolate system faults or failed components. The following analysis reports may be generated using the ETA tool:

- **Detectability Report:** This report provides the detectability of each failure mode that propagates effects. It lists failure modes detected by the available suite of tests, the tests that detect each failure mode, and failure modes that are not detected by the available suite of tests. Additional information is extracted from the FFM to facilitate the location of the failure modes and test measurements.
- **Test Utilization Report:** This report provides the capability to detect each of the failure modes using the available suite of tests. It lists each test in the available suite of tests, the failure modes detected by each test, and tests that do not detect any of the failure modes. Additional information is extracted from the FFM to facilitate the location of the test measurements and failure modes. This report is used to support the AC/AT, C&W, and LCC failure mode traces.
- **Failure Mode Isolation Report:** This report provides the capability to isolate each of the failure modes using the available testability analysis results. It lists failure mode ambiguity groups that are composed of a unique detection signature, and all failure modes having that unique detection signature. Additional information is extracted from the FFM to facilitate the location of the test measurements and failure modes. This report is used to support the LCC failure mode traces.
- **Component Isolation Report:** This report provides the capability to isolate each of the failure modes to a user-specified component label or to a list of explicitly specified components. The component labels are an internal property in FFM assigned for the module representations of the physical components. Each FFM module can have only one label, but the same label can be assigned to multiple modules. The report lists only failure mode ambiguity groups that contain failure modes from the specified components. Each listed ambiguity group is reported in the same manner as in the Failure Mode Isolation Report. An extended report provides an assessment of each selected component, highlighting undetected failure modes and isolated failure modes.
- **System Effect Mapping Report:** This report provides the failure modes that propagate to special user-created test points that represent system-level effects within the model. The ETA Tool generates five sub-reports. The first lists system-level effects that are not mapped to any failure modes. The second, alternatively, lists failure modes that are not mapped to any system-level effects. The third sub-report provides an Effect-to-Failure Mode mapping similar in format to the Test Utilization report. The fourth and fifth sub-reports provide Effect Isolation results with different levels of detail. These sub-reports are similar to those generated by the Failure Mode

Isolation Report, except that the system-level effects are the detection signatures. This report is used to generate failure mode traces that support RM.

- **Sensor Sensitivity Report:** This report provides the change in diagnostic performance that results from the removal of individual sensors or groups of sensors and the tests mapped to those sensors. The sensors involved in the analysis are specified by the user with a selected input file. Separate reports are generated for the individual sensor removal analysis and each sensor grouping specified (up to two).
- **FMEA Report:** This report provides an FMEA style spreadsheet for the XML output only from the FFM. The report is restricted to the conditions of the testability analysis. Therefore, it only reports failure modes and tests available in the testability analysis.

2.3.4.2 FFM SUPPORT FOR OPERATIONS

The FFM developed for FM analysis can be used for real-time diagnostics when built into an operational environment. The model development tools generate data artifacts that include the D-Matrix described in Section 2.2.1. The artifacts also include a mapping of test names and failure mode names to model indices used by the diagnostic reasoner to associate FFM tests with failure diagnoses. Test results are derived from system sensors and onboard computers using test functions external to the diagnostic model, and these test results are then passed to the diagnostic reasoner to complete the real-time diagnosis. Automated tests are updated regularly, such as every second, to provide continual diagnostic results.

Diagnosis is accomplished using tests that can either implicate or exonerate system failure modes. A failed test can be due to a number of possible causes, and each of these causes is said to be implicated by the test. A different test that passes can prove that one of the possible failure modes implicated by the first test is not responsible. That failure mode is then exonerated, or proven to not be responsible for the failed test, just as the suspect of a crime can be exonerated by an airtight alibi. If all goes well, additional passing tests will exonerate all but one of the failure modes, and the remaining failure mode will be unambiguously diagnosed as the only possible cause for the failed tests. If multiple failure modes are implicated and not exonerated, these failure modes are called an "ambiguity group." In an ambiguity group, one (or more) of the group's members may be the cause of the failure, but the data is insufficient to determine specifically which member or members of the group are at fault.

This section describes several advancements that have been achieved in recent years that mature operational capabilities for system FM using FFMs. These advancements are described in detail in the following paragraphs.

- **Interface definition:** The interfaces required to transfer data between the system, test logic, diagnostic reasoners, failure impact reasoners, decision support reasoners, and displays have been defined and elaborated in both test environments, such as AMPS, and in large-scale systems, such as Orion EFT-1.

- **Data filtering and cleaning:** One of the most difficult aspects of failure diagnostics is to avoid false positives that can result from incorrect or incomplete data. Recent emphasis on data quality used for diagnostics has begun to improve the reliability of diagnostic information available from run-time reasoners.
- **Data framing and synchronization:** Systems often use different data rates that can obscure the true failure sequence as a failure propagates through a system. Techniques have been developed to reduce false failure diagnosis resulting from data timing issues.
- **System mode determination:** A complex system can have vast numbers of possible configurations, such as when equipment is turned on or off or changed between standby and primary status. Improved modeling and data processing methods have reduced the likelihood that a system mode change is mischaracterized as a component failure or that incorrect data interpretation results in a missed diagnosis.
- **Test logic:** Diagnostic reasoning relying on pass or fail results requires highly accurate test logic that is valid over a wide range of configurations and environments.
- **Presentation of results:** The raw output of a diagnostic reasoner can contain information that usually needs to be formatted and filtered for a particular user. Onboard flight crews, mission controllers, or other support personnel, maintenance personnel, and failure analysts may all require different presentations of the outputs of FM reasoners.
- **Failure Impact Determination:** Diagnosing a failure is only a starting point for determining what to do about the failure. Recent work has identified the impact of failures on other parts of a system and how those losses affect the mission itself.

The remainder of this section describes these advances in operational FM based on a system-level FFM that is connected to the system's sensors, processors, and data networks.

The FFM contains the primary information needed to perform fault diagnosis during system operation. The components, their failure modes, the applicable system modes, connectivity between components and failures, and tests that detect the symptoms of failures are all included in the model. Test logic converts system data to pass or fail results that are sent to the diagnostic reasoner. The test logic could be built into the model itself or may be defined in a separate set of data. Within TEAMS-RT (hereafter referred to as the "FFM operational software"), the diagnostic reasoner is segregated from the test logic. However, diagnostic systems could be constructed such that the test logic is built into the fault model. The diagnostic tests that reduce sensor data to the pass/fail test results used by the diagnostic reasoner can be as simple as a limit check on a single data value, such as a temperature exceeding 100°C. They may also be highly complex, such as statistical analysis performed on a number of data elements. The diagnostic reasoner uses the pass/fail results with the D-Matrix and supporting run-time data generated by the model to complete the fault diagnosis.

From the FFM, the D-Matrix, described in Section 2.2.1, is constructed to correlate tests to failure modes. For diagnosis of the root cause of a failure, the essential information is an association, or mapping, of tests to failure modes for a given system mode. The FFM generates run-time data that are loaded at system initialization. These data include the D-Matrix, which is used by the FFM operational software. The diagnostic system obtains system data and uses the test logic to check for conditions that may indicate failures, and the diagnostic reasoner applies the test logic to the stored D-Matrix data to generate the diagnostic results. The relationships between the FFM, the system and its operational data output, and the diagnostic reasoner are shown in Fig. 2.16.

Interfaces for Real-time Diagnostics

Real-time automated diagnosis is performed by establishing interfaces between the system data and the diagnostic system. The operational system interfaces with the diagnostic system to obtain data from the system's sensors and computers. The outputs of the diagnostic system provide system health information to mission operations systems, including both human operators on board or on the ground, and to automated decision reasoners that use the information to reconfigure the system in response to failure or degradation.

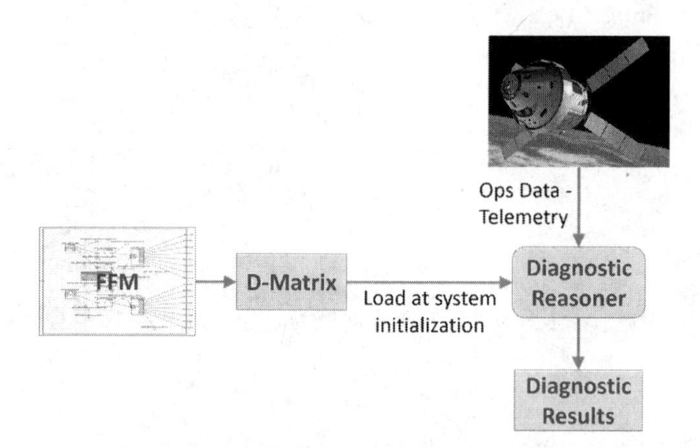

Fig. 2.16 The diagnostic reasoner correlates the FFM with data from the operational system to continually diagnose system health throughout a mission.

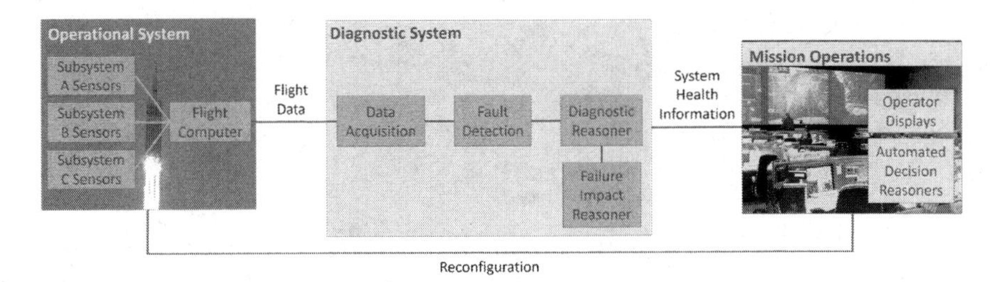

Fig. 2.17 A simplified run-time diagnostic architecture shows the system, its subsystems, and data processors.

As shown in Fig. 2.17, a diagnostic system interfaces with the operational system to obtain data from the system's sensors and flight computers. The diagnostic system data acquisition element cleans and filters the data, then passes them to a fault detection element to reduce the data to a set of pass or fail results. The diagnostic reasoner determines failures from the test results, and the failure impacts associated with diagnoses are determined. The combined failure and failure impact information is then provided to mission operations elements to make decisions about system reconfigurations that will mitigate the impact of failures. Mission operations could include onboard crew, ground-based mission operations, or automated decision systems.

Data Filtering

Real-time data may be messy. It is often necessary to filter and clean the data before performing the fault detection logic to assure quality information and avoid transient, misleading, and distracting outputs. A diagnostic system that generates false results, either a failure when the system is actually behaving nominally (a false positive) or not identifying a failure when there actually is one (a false negative, or missed diagnosis), is not likely to be trusted by the system operators and is a major cause for skepticism about automated reasoners. A primary method for reducing false results is to filter and clean (i.e., pre-process) the data coming into the system. Some common data filters include the following:

Off-scale Data: Sensors are typically calibrated over a range of possible values, but sensor or sensor processing failures can cause data to fall outside practical ranges. A test for a temperature falling below a threshold should not be evaluated as a failed test result if the reading suddenly drops to -459°F, or absolute 0°K. A data filter that prevents performing the temperature threshold test precludes a false positive on the data. A separate test that checks for an off-scale reading can be used to diagnose a failure of a sensor or data processing equipment, but there may be normal modes of operation in which sensors are off scale, such as sensing systems that are turned off when the data aren't needed. If an off-scale temperature reading is detected, qualitative tests on the temperature need to be avoided regardless of the cause of the off-scale readings.

Static or Lost Data: Real-time systems often produce cyclic data that are transmitted at a given rate, even if the actual source of the data has been lost. Various methods may be used to report on the validity of the data, and if available, are a valuable and easy filtering method. If the system does not produce valid values, perhaps due to data bandwidth limitations, the data interface filtering and cleaning logic may need to determine if incoming data is "alive." For example, an Analog to Digital Input/Output (ADIO) data card could lose power and stop generating data signals, but a receiving flight processor may continue to transmit the latest known value from the card. The filtering logic could determine that the value has stopped updating, or that all of the data from the card have stopped updating, and suppress the execution of any test using the data.

Transient Data: Occasionally, a real-time system may produce a temporary erroneous indication that doesn't reflect actual stable conditions. A data value could spike because radiation alters the data, or because vibration affects the electrical output of a sensor independent of the physical measurement. If the test logic generates a failed result based on a transient data value, the diagnostic system could diagnose a failure and then clear that failure during the next cycle. Although the system operators can recognize transient false positives and learn to ignore them, they can be distracting and result in a lack of trust in the diagnostic system. If automated decision reasoners use the faulty diagnostic information, undesired actions could be taken that are unnecessary or detrimental to the mission. One method for avoiding transient conditions is to delay changing the state of a test from pass to fail, or from fail to pass, for a number of consecutive cycles; this is known as persistence. As early as the Apollo program, spacecraft engineers have used a rule of thumb that three readings in a row should be seen before acting on a data change, but other counts can be used, depending on various factors.

One means of addressing the previously described data filtering challenges is Sensor Data Qualification and Consolidation (SDQC) [25, 26]. SDQC is an onboard flight software function that implements algorithmic approaches to determine the validity of sensor data and to reduce valid data from redundant measurements to a single representative measurement for each data frame. The qualification function is designed to mitigate the impact of corrupt data on higher-level decision and control logic; the consolidation function reduces the amount of data that need to be presented to the higher-level logic. The development of SDQC was initiated under NASA's Ares I Upper Stage project [25] and is being implemented in flight software for NASA's SLS [26].

Data Framing and Synchronization

Real-time systems typically execute at a cyclic rate and organize data into various rate groups in a well-defined data frame. The diagnostic system must account for the timing of the data in order to avoid diagnostic errors. If a diagnostic reasoner uses a passing value to exonerate a possible failure, and a failing test to implicate a failure, and some of the data are sampled at 1 Hertz (Hz) and other data are sampled at lower rates, such as every 10 seconds (0.1 Hz), the test results could be inconsistent when a failure occurs. With an FFM and its counterpart diagnostic reasoner, a vector of test values is sent to the reasoner each diagnostic cycle. If the last observed value for a test is sent to the reasoner every data cycle, when a failure occurs, the tests on higher rate synchronous data yield a failed test, but the slower rate data would still indicate a passed test. The reasoner could exonerate the actual fault based on the passed tests from the slower rate data. At the next read cycle for the slower data, the results will become consistent, but if the actual fault has already been exonerated, a reasoner may reach a different conclusion. It could take several additional data cycles to correct itself, while yielding changing and inconsistent results during this critical time. The ISS and Orion Spacecraft both use data rates ranging from 40 Hz to 0.1 Hz.

To avoid such inconsistencies, the diagnostic system needs to account for timing of multi-rate data. A possible simple method to handle multi-rate data may be to execute the diagnostic reasoner's calculation cycle at the slowest rate to sample the data from every data element, and use only the last value of the faster rate data. The test results are collected from all of the data rate groups, and at the completion of the system's major frame (e.g., the 10-second period), all of the test results are transmitted as a complete set to the diagnostic reasoner. One drawback to this approach is that diagnostic results may be needed more quickly than the slowest cyclic data rate. Failures characterized by rapid onset and detected by fast data will not be diagnosed for some time after they occur. Furthermore, because the purpose for slow rate data may be to reduce bandwidth on networks and telemetry downlinks, the 0.1 Hz data probably are not all sampled on the same cycle, so that even slowing the diagnostic cycle may not guarantee consistency of test results. If each 0.1 Hz data element is offset from a 10-second frame boundary in order to balance the data throughput, it may be difficult to always determine a point at which the fault signature is complete.

Another approach, which has been used successfully with the Orion ACAWS Project, avoids the problem of finding a point at which the diagnostic signature is complete. The data and test interface set an UNKNOWN value on tests for all data cycles in which data are not received from the system. It is an accurate, conservative representation of the system, one that avoids assumptions about the persistence of information. If a failure occurs between readings from slow data, any tests based on faster data will be processed and sent to the reasoner. The reasoner might identify a group of possible causes based on the available data, but the slower rate group has not yet reported any results. When the slow data are read and sent to the diagnostic reasoner, tests that pass can exonerate some possible faults, whereas failing tests support the results from earlier failing tests. By using UNKNOWN results for slow tests, a false exoneration of preliminary candidate failures is avoided. Figure 2.18 shows the problem and its resolution. In the

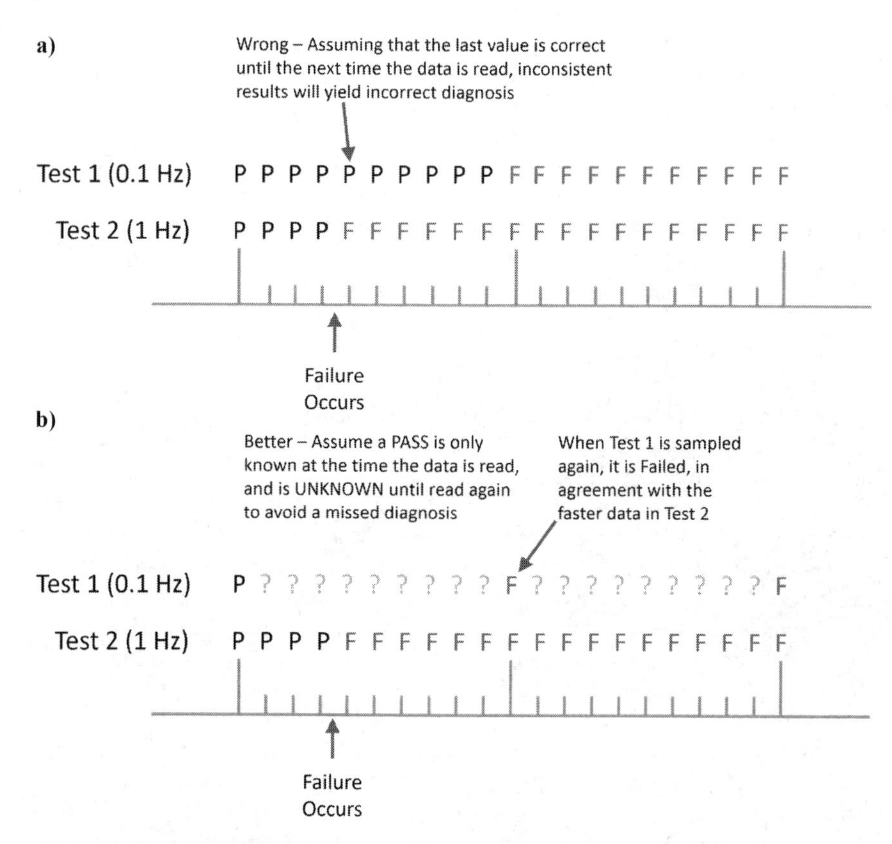

Fig. 2.18 Testing data with multiple rates can yield incorrect conclusions without proper filtering and data framing.

upper diagram, the system assumes that the last observed value remains correct until the next value is sampled, but if a fault occurs between data samples, the actual fault could be exonerated prematurely. A better approach is to treat results as UNKNOWN, except when they are explicitly observed.

System Mode Determination

The failure modes that can occur, the criticality of failure modes, and data available for diagnosis can change significantly over the course of a mission. During spacecraft launch, engine performance, guidance, abort systems, and controls are critical, whereas solar arrays and thermal radiators are not yet deployed. Later on, engines are jettisoned from the vehicle and on-orbit equipment is activated. Throughout the mission, equipment may be added or discarded, turned on or deactivated. The diagnostic system must respond and remain in harmony with the state of the system as it updates in response to its mission plan and the environment in which it operates.

The changes in system state may be planned and known far in advance, but other changes that affect diagnosis may occur with little prior knowledge. The operational phases, mission, and operational modes such as launch, on-orbit, trans-lunar cruise, or Earth return are known early in the operation's concept definition. Decisions about deactivating a payload, switching over to a backup system, or cross-strapping a power bus may be made in response to unforeseen circumstances.

From the perspective of a diagnostic system, the major mission phase and system mode changes that occur during transition from launch through reaching orbit can almost be viewed as using a different fault model. Many failure modes are only applicable in some mission phases, such as the launch vehicle during ascent, and are not possible once orbit is reached. Systems typically change the data that are produced, or change the rates of data as mission configurations are changed. These changes often coincide with, or even define, the mission phases and system mode boundaries. The data available to the diagnostic system contain information that define the mission phases and system configurations. The diagnostic system can use the data in various ways.

One approach to system mode changes is to simply reload a different diagnostic system as the system modes change. For example, when the launch booster segments are jettisoned, the associated diagnostic system and models can be unloaded and discarded. The diagnostic system developed for on-orbit mission phases may not even be loaded into memory until the on-orbit processors are activated.

A diagnostic system and its models that are built and tested for an entire mission will need to adapt to known major mode changes. The diagnostic FFM can be built with awareness of these defined modes. System data that define the mode changes can be passed to the diagnostic reasoner, similar to the passed/failed test results. The mode change can affect which tests are used, as well as which failure modes are applicable, and adjust its reasoning accordingly. The FFM operational software's reasoner can precompute different D-Matrices for each system mode and, when notified, can rapidly swap in the new D-Matrix without interruption.

Every time equipment is turned on or off, the environment changes, or failures occur, the system is in a different configuration. Many of these configuration changes have a relatively minor impact on the diagnostic reasoning, but need to be accommodated nonetheless. The reasoner needs to avoid diagnosing a component as failed simply because the operator decided to turn it off. These somewhat minor, perhaps temporary, system configuration changes can be more difficult to manage than the major, planned mission phase changes because the number of possible combinations of configurations can be very large.

One approach to managing configuration changes is to build all possible configurations into the FFM. Because most equipment can be turned on or off, and many systems allow for rerouting power and other resources, a model can include switches for each one of these configuration options. There is, however, a vast number of possible combinations, with each combination representing an operational mode. The reasoner is capable of receiving data that describe the position of each of these mode switches and updating accordingly. Its approach is to recalculate a D-Matrix in real time, but possibly taking considerable processor time (seconds to minutes, depending on the processor, the model, and the number of switches).

An alternate approach is to manage configuration changes external to the diagnostic FFM, determining which tests do not apply, or require different test logic, for different operational modes. A mode manager external to the diagnostic reasoner can readily determine which tests don't apply, and in certain operating modes can suppress test results by passing UNKNOWN to the diagnostic reasoner. This approach avoids a possible interruption to diagnosis that could result from forcing the reasoner to recompute its diagnostic solution on the fly, and has been used successfully in NASA technology demonstrations.

Test Logic

A discrete diagnostic reasoner, such as that found in the FFM operational software, requires that system data are transformed from digital data to discrete results such as pass/fail, or perhaps multistate discrete values, such as Nominal, Low-Threshold, or High-Threshold. The test logic performs this transformation, obtaining data from the system and transmitting them to the diagnostic reasoner to perform fault diagnosis. The data originate from the system's sensors but may be processed extensively by the system's avionics and software. Diagnostic tests often rely on direct sensor values (e.g., temperature, pressure, voltage) exceeding an operational threshold. Tests often need to combine conditions, perhaps using voltage and current readings to calculate power, and testing the result against a threshold. Modern system avionics and software perform many tests that are used for automated decisions or operator notifications such as Fault Detection, Isolation, and Recovery (FDIR) algorithms commonly used in spacecraft systems.

In addition to data values from sensors and software, the absence or the corruption of data is itself a diagnostic input. For example, suppose there is a failure mode that is characterized by a temperature exceeding 130°F. During operations the temperature sensor jumps from 72°F during one cycle to 732°F during the next cycle, and then remains constant at 732°F. If the data are read once per second, the physics would generally preclude an actual temperature rise of 700°F in two sequential measurements, and the diagnostic reasoner could conclude that the reading is not valid. Further, if the system reports "failed" with the now invalid temperature test, the diagnostic reasoner would attempt to identify a failure that caused the temperature to exceed the threshold and will reach an incorrect conclusion. There is some failure that has caused the data to become invalid; the cause could be a sensor or ADIO device failure, the failure of a flight processor or one of its interfaces, or a failed network component. Because tests on data validity are related to failures that may have little relationship to the failure being checked in the data itself, the data validity tests included in the model should be located in or near the data processing equipment, rather than in the subsystem where the failure effect would manifest itself.

Data validity and data value testing need to be associated. Invalid data need to result in the data value test receiving an unknown result. There are a couple approaches to this. One is to include the data validity testing in the model along with the data value testing, perhaps using a multi-outcome test. A test could be defined as off-scale low, nominal, above-threshold, or off-scale high. The model could then associate the nominal and above-threshold value with a thermal-related failure mode, and associate the off-scale results with other failure modes. An alternative approach that has been used successfully is to test for status external to the FFM and diagnostic reasoner. Any invalid status results in setting the data value test to unknown. The status can then be used as a test input to other FFM tests, such as tests for sensor short or open circuits, ADIO failure, network connectivity failures, or other data errors. If groups of sensors all go invalid concurrently, the system can then isolate the common point of data loss such as an ADIO, connector, network, or processing component. Data validity can include various forms or levels of validity:

- **Off-scale high or low.** Sensor short or open circuits or analog-to-digital conversion equipment can cause sensor values to be reported at the extreme ends of the range. Sensors start with converting a physical attribute, such as temperature or pressure, to a small voltage signal, typically over a small range from 0 to 5 volts. Broken wires or short circuits could result in either no voltage or maximum voltage (or very close to it).
- **Static or missing data.** Depending on how the data architecture is built, a loss of network or data processing capability can result in either no data at all or a continuous transmission of the same value. The Orion spacecraft flight computers produce a constant data frame that is filled with some value, so that if the source of the data is lost due to a power loss of component, or loss of a network element, the last observed value will be transmitted continuously. In other systems, the data may just drop out. If data are not received at all, the data value tests on the data are not performed, and so it is easy to assure that no test value, or an unknown value, is passed to the diagnostic reasoner. The test logic may need to detect that data are not being received, however, and generate failed results for tests associated with the data transmission mechanisms and their failures.

Once it is known that good-quality data have been received by the test logic, the data value tests can be performed. These tests can range from a simple threshold to complex algorithms involving multiple data values and conditions. Modern system software often performs tests for automation, control, or display purposes that can be used as test inputs to a diagnostic reasoner. FDIR algorithms offer a variety of fault conditions at a subsystem level, such as testing for low pressure, overcurrent, and rather than performing the same algorithm in the test logic, the FDIR results can form the basis of the test logic. Orion's FDIR logic often includes testing the validity of data, checks persistence of a condition (five cycles in a row before declaring a test failure, for example), and uses multiple redundant sensors when available. When high-quality processed data are available from the system and convey the necessary failure symptom information, the test logic can be significantly simplified, and the reasoning is more likely to remain consistent with the vehicle's FDIR and automation logic.

For many conditions, the system's sensor data are used directly. If the flight systems perform sensor engineering unit conversion, the test logic can work with understandable data values such as temperatures, pressures, or voltages. Some systems may transmit the raw, unprocessed sensor signals, and the diagnostic test logic will need to do its own engineering unit conversion, or else specify tests based on the unprocessed sensor values. Many sensors, for example, use 12 bits of data scaled over the range of the electrical input that the sensor can generate. These unprocessed values are in "counts," in which 0 volts equates to 0 counts and 5 volts (or whatever the maximum scale is) equates to 4095 counts. A test could be specified as "fail when greater than 1237 counts," which might equate to 130°F. Logic could be added that converts the sensor value to temperature, and then the test could be more readably specified as "fail when greater than 130°F."

Recall that test logic should assure that the incoming data are valid before performing a test on the data. Prior checking for missing, static, or off-scale data will significantly reduce false results and help build operator trust in results generated by the diagnostic system.

When there are redundant or at least similar sensors, the combination of sensors provides additional benefit to the diagnostic logic. Sensor redundancy can be handled in the test logic, the diagnostic FFM, or perhaps in both. Suppose there is a component with two independent sensors measuring the temperature near the component, and a failure mode characterized by temperature over 130°F. The FFM could be built with two tests related to the component, test T1 and test T2. If both T1 and T2 pass, the component is readily recognized as good or not failed, and if both tests fail, the component is implicated with respect to the temperature. If the sensors diverge, T1 may fail while T2 passes, so T1 implicates the component failure mode, while T2 exonerates it. The diagnostic reasoner probably at best identifies the component as suspect, or possibly failed; the human operator may be able to judge which sensor is more likely correct, relying on judgments about data "reasonableness," prior history of a sensor, how rapidly the values diverged, or other intuition. Putting the logic for redundant sensors in the test logic has some advantages. If both sensors agree within a reasonable bound, the test can provide high confidence data to the diagnostic reasoner. If the sensors are right on the edge of a test threshold, but jittering around it, the test logic can suppress a conclusion until the values clearly converge on one side of the threshold or the other. If they diverge significantly from each other, it can be concluded that at least one of them has failed, which is a different diagnostic problem than the component over-temperature condition for which the sensors are testing. Unless there is clear evidence supporting which sensor is in error, a conservative position is to evaluate the temperature test as unknown. Because the sensor validation is dependent on knowledge about the system design and command, and its data handling systems, this level of sensor validation may need to be handled in custom test logic, rather than in the FFM that does not have a capacity for the detailed sensor knowledge involved. If there is a third sensor or a compatible attribute, it may be possible to determine which sensor is in error, but if not, it is advisable to disregard both sensors.

There are often significant timing considerations when designing the diagnostic test logic. Some failure effects propagate rapidly through a system, whereas others may take considerable time to show up in data. Electrical power faults, for example, have an immediate effect when electrical power is lost. Faults in a thermal control system may often propagate slowly and may exhibit variations due to the environment and configuration. Suppose a valve in a coolant loop fails, stuck in the closed position, thus preventing the flow of coolant to a thermal radiator. Tests of the coolant temperature on one side of the valve would be expected to rise as a symptom of the fault, but the temperature rise may take minutes or more. In some conditions, the temperature might remain in normal limits indefinitely, such as if the equipment to be cooled is in a low usage condition, or if the system is operating in a cold environment. Test logic, coupled with the diagnostic FFM, often needs to handle these sorts of timing and variability considerations.

Test logic is one of the more complex aspects of a run-time diagnostics system. Tradeoffs between diagnostic sensitivity and avoidance of false results must often be made. Can the diagnostic system wait until it is certain that a component is failed, or is it more important to provide results quickly, even if this results in incorrect information some of the time? The diagnostic system requirements should be carefully and fully defined in order to properly define the test and diagnostic logic needed for a successful system implementation.

Failure Impacts Determination

Although it is crucial to be able to identify the source, or root cause, of a failure, it is equally important to understand the effects of a system failure on other components and how the loss of function of those components impacts the mission. For example, if a power distribution component fails, such as an AMPS PDU (Fig. 2.5), power to electrical loads is lost, resulting in either a loss function or loss of redundancy. The FFM can be built to contain information that is useful for identifying components and functions affected by a failure. For example, when an AMPS PDU output switch fails to open, there is resulting loss of function of the cabin lights. Critical functions often rely on redundant components, and although loss of one of these redundant components does not result in loss of the function, the loss of redundancy increases the risk. The AMPS system's MBSUs provide redundant paths for providing power to electrical equipment, and loss of one of the MBSUs increases the risk that another failure will result in loss of function of electrical equipment.

To diagnose system faults, the FFM is used to trace back from tests to a specific failure mode. To identify failure impacts, a separate reasoner is used that takes a diagnostic result as its input and follows a directed graph built from the FFM to identify the functional and redundancy impacts. Functional impacts are identified when the model indicates that all required resources for a component's functioning have been lost due to failure. Redundancy impacts occur when a failure impacts one pathway to a component but other pathways remain. The model's AND nodes are used to determine if an impact is loss of function or loss of redundancy. If multiple failures result in loss of all paths to a component, the impact changes from loss of redundancy to loss of function.

Presentation of Results

The outputs from a diagnostic reasoner, such as that found in the FFM operational software, can provide highly useful information to an operator or team that needs to make rapid decisions in the face of a failure and possibly loss of critical functionality. While determining the root cause failure, the reasoner may generate additional information, such as possible additional or alternative failure causes. To provide clear and concise actionable knowledge, system health management engineers, in collaboration with system operators, must judiciously select information to be displayed and devise a means of presentation that efficiently supports the decisions being made by each operator.

Every failure that is diagnosed may not need to be displayed for every type of operator. Flight crew have different information needs from those of ground-based flight controllers of a crewed spacecraft. Backroom or engineering support personnel have different needs from front-room controllers. The diagnostic results typically are filtered and processed for different operators. Diagnostic results can also be used as an input to automated decision reasoners, and the diagnoses presented to these tools must match the reasoners' interfaces and capabilities. Flight crews may only need information about failures requiring immediate action, whereas ground controllers may need information about any failure.

If the FFM and diagnostic reasoner are not able to unambiguously determine a root cause failure, and can only determine a group of possible causes, the health management engineers are challenged to provide useful information without obfuscation. Ambiguity between closely related failure modes may provide useful actionable information. For example, if the reasoner can't distinguish between a short circuit and an open circuit, but either would result in loss of power to a pump, displaying results that indicate that some failure has resulted in loss of particular functionality is probably very useful. On the other hand, if the reasoner can't distinguish between several possible diverse sources, it may be better to report nothing than to report a long list of possible faults. The FFM may include probabilistic information that may be useful in narrowing ambiguity, although it should be used cautiously. A flight crew that needs to make recovery decisions or an engineering support team performing failure analysis have very different needs and expectations. The flight crew need to decide what to do, and if there is uncertainty, procedures typically provide steps to resolve ambiguity before taking actions. It wouldn't be very useful to report "70% probability that Pump A failed, 30% probability that Isolation Valve 1 Stuck Closed," leaving crew to guess which one to work. A procedure that disambiguates possible failures, or a procedure that recovers cooling function regardless of root cause, is needed. In completely unexpected, unanticipated, and unlikely situations, there may not be sufficient information to even begin a diagnostic procedure. Engineering support groups are typically engaged to help in such circumstances, and any information available from the FFM and diagnostic reasoner can be useful.

Ambiguity can also arise from cases in which a single fault could account for the entire fault signature, but multiple faults could also produce the same signature. Suppose a failure of a controller card results in loss of power to six different components. The diagnostic reasoner could ascertain that the card has failed, but the reasoner may also consider it possible that all six components had failed, independently, at the same time as the card failed, and report the electrical loads as possibly faulted along with a controller card fault. A common diagnostic strategy is to report only the single fault when one fault adequately explains the full fault signature, even though it may be feasible that multiple concurrent faults could produce the same fault signature. Figure 2.19 provides an example of a diagnostic display showing the impact of loss of function for an analog-to-digital input-output board. The failed hardware is identified in the PDU-C1 window of Fig. 2.19 as *C1 ADIO*. Loss of redundancy impacts can also be identified and displayed when applicable. In systems with hundreds or thousands of components and failure modes, building graphical displays requires careful management of the display hierarchy. When a failure occurs, the operator needs to rapidly locate the failure on a display and gain situational awareness as quickly as possible. Providing indications of failure at both an overview level and with preciseness requires careful display design. The examples show use of icons to indicate failures, loss of function, and loss of redundancy impacts. Displays may need to indicate that a subsystem or assembly has a failure of a subcomponent, prompting an operator to open another display to retrieve detailed information about the failure. Fault displays can include capabilities to understand what the impact of additional failures might be, so including provisions for what-if analysis can be desirable. For example, if a failure of a coolant pump has occurred, the operator can plan ahead and query the diagnostic system, such as "What will be the condition if I lose a power on a bus?" The what-if capability can help identify worst-case failures and make decisions about continuing a mission or taking precautions to abort the mission.

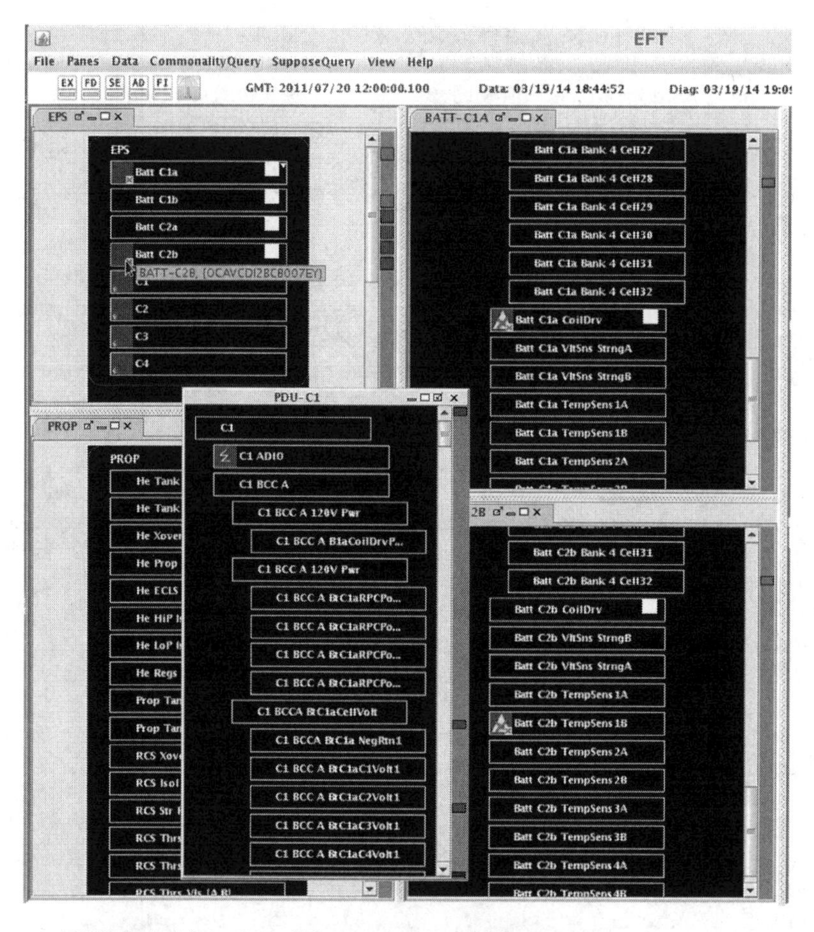

- ACAWS display information to flight control or crew about:
  - The "root-cause" failure behind a C&W event. Cell 19 – Cell short
  - Components affected by the failure.
  - Components at increased risk due to loss of redundancy

- Decision-making assistance with nominal operations
  - "At-a-glance" GO/NOGO Flight Rule Tables
    - E.g., Extended Power-Down Decision at Entry Interface

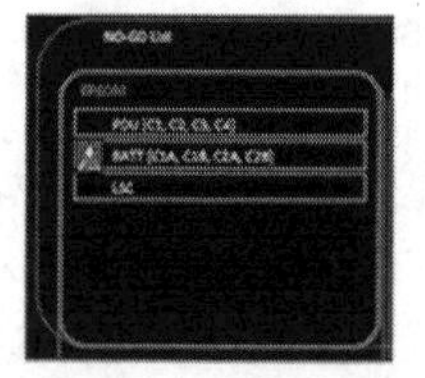

Fig. 2.19 Fault displays showing faults and failure impacts are used to show system health and failure information to operators.

2.4 THE FUTURE OF FFM

Functional models are a key element of conventional systems engineering. Engineers capitalize on functional analysis in the early phases of system design, due to its qualitative nature and flexibility. Functional Fault Modeling and analysis provide the benefits of analyzing the impact of failures on a system's ability to achieve desired goals. Current research efforts to expand the application of FFMs in the early design phase include combining FFMs with simulation to analyze failure scenarios. Other efforts address the knowledge acquisition phase in FFM by providing new methodologies to store and link component-, subsystem-, and system-level information for more efficient reuse.

Development of FFMs is labor-intensive, as described in Section 2.3.3 of this chapter. The recent popularity of Model-based Systems Engineering (MBSE) has increased the understanding of just how powerful an interconnected set of models can be in the design and development of complex systems. NASA is sponsoring research that is aimed at leveraging system models developed in SysML, the modeling language of choice in the MBSE community, to reduce the often tedious and time-consuming manual effort required in the development of FFMs. The ability of SysML to represent failure behavior, and the mapping of modeling elements used in MBSE to those required for FFMs, is being studied. A major goal of this effort is to increase the automation of FFM construction and, simultaneously, to decrease the amount of manual labor and associated human "touch" errors. Current automation efforts capitalize on data sources that are available in digital form, such as FMEAs and channelization (connectivity). Models produced by MBSE efforts hold the possibility of providing similar knowledge, as well as additional information about system behavior, due to the richness of the modeling elements included in SysML. The ability to tap into a robust set of models of the structure and behavior of a complex system would also increase the consistency of system knowledge used in various modeling and analysis efforts.

The impact of having a rich set of functional fault models will never be more evident than in the operational phases of a system's life cycle. As shown in Section 2.3.4.2, FFMs can be utilized to monitor and diagnose system operation in real time. Analysis of FFMs already benefits the design of runtime monitoring systems by providing coverage statistics on the system's ability to detect and diagnose failures, provide false positive and false negative analyses, and analyze latent failures that may exist in the system. Future work on mapping elements of FFMs back to the requirements they satisfy, and testing plans for certifying the implementation of FM algorithms, will increase the benefits being derived from this methodology. For future aerospace systems that have extremely complex interactions among subsystems, large, integrated hardware and software systems, and autonomy requirements, the robust representation of the failure propagation through the system available from FFMs will be essential in the design of the FM for safety- and mission-critical operations.

The theory supporting ISHM is continuing to evolve and coalesce [1–3]. As part of that process, a set of robust FM metrics for state estimation was recently presented with example calculations [5]. Combined, these metrics quantify the performance of FM control loops and, subsequently, the effectiveness of the whole ISHM system. In order for FFMs to support these metrics, qualitative estimates embodied in the FFMs need to be linked to quantitative estimates for the system and relevant analyses applied.

2.5 CONCLUDING REMARKS

This chapter describes key advances in Functional Fault Modeling (FFM) that have occurred under NASA programs over the past five to seven years. These advances are primarily motivated by a need and desire to improve the safety and to reduce the life-cycle cost of crewed and uncrewed space systems. They can benefit intelligent and autonomous systems—systems that are largely recognized as a requirement for future NASA missions—by providing real-time, state self-awareness capabilities, as well as supporting analytical verification of these systems' diagnostic requirements early in the design when they will have the largest impact and benefit. However, they are not limited to this domain and can be used in other FFM applications. These advances have also influenced the FFM paradigm, where the process, analytical, and software tools developed during this period are accepted as the building blocks for standardized FFM by the modeling community.

Initially, background material describing the essence of FFMs, how they are being advanced by NASA, their benefits, and example FFMs are provided as foundational to the discussion and as tutorial information for those unfamiliar with FFMs. An overview of an FFM development process is then used to organize and provide a context for discussing specific advances in FFM capabilities. The development process—which includes phases for knowledge acquisition, conceptual design, implementation and verification, and application—also provides a structured approach for those new to FFM development. The chapter concludes with a discussion of future capabilities and software tools that would further advance the FFM discipline.

It is the authors' objective and sincere hope that, by documenting FFM advances in this way, FM—intelligent and autonomous systems developers—will discover and learn about state awareness tools and techniques previously unknown to them; and that they will be able to use those tools and techniques to advance the capabilities of their applied disciplines. Likewise, as other disciplines use FFMs and the advances presented in this chapter, further acceptance of FFM and the analyses it can provide will be the outcome.

ACKNOWLEDGEMENTS

The advances in Functional Fault Modeling identified in this chapter were supported by the NASA Human Exploration and Operations Mission Directorate under the following programs and projects: Constellation Program, Ares I Project; Space Launch System Program, Mission and Fault Management Project; Ground Systems Development and Operations Program, Advanced Ground Systems Maintenance Project; and the Orion Program, Multi-Purpose Crew Vehicle Project.

REFERENCES

[1] Rasmussen, R. D., "GN&C Fault Protection Fundamentals," *31st American Astronautical Society Guidance, Navigation, and Control Conference*, Breckenridge, CO, Feb. 2008.

[2] Johnson, S., "Introduction to System Health Engineering and Management in Aerospace," *1st Integrated Systems Health Engineering and Management Forum*. Napa, CA, Nov. 2005.

[3] Johnson, S. B., Gormley, T. J., Kessler, S. S., Mott, C., Patterson-Hine, A., Reichard, K. M., and Scandura, P. A., Jr., *System Health Management with Aerospace Applications*, Wiley, Chichester, UK, 2011, pp. 3–27.

[4] Aaseng, G. B., Patterson-Hine, A., and Garcia-Galan, C., "A Review of Systems Health State Determination Methods," AIAA Paper 2005-2528, Jan. 2005.

[5] Johnson, S. B., Ghoshal, S., Haste, D., and Moore, C. M., "Fault Management Metrics," *AIAA Science and Technology Forum*, Grapevine, TX, Jan. 2017.

[6] Poll, S. et al., "Evaluation, Selection, and Application of Model-Based Diagnosis Tools and Approaches," *AIAA Infotech@Aerospace Conference*, Rohnert Park, CA, May 2007.

[7] Maul, W., Melcher, K., Chicatelli, A., and Johnson, S., "Application of Diagnostic Analysis Tools to the Ares I Thrust Vector Control System," *AIAA Infotech@Aerospace Conference*, Atlanta, GA, April 2010.

[8] Ferrell, B., Lewis, M., Perotti, J., Oostdyk, R., and Brown, B., "Functional Fault Modeling of Cryogenic System for Real-Time Fault Detection and Isolation," *AIAA Infotech@Aerospace Conference*, Atlanta, GA, April 2010.

[9] Ferrell, B., Lewis, M., Perotti, J., Oostdyk, R., Spirkovska, L., Hall, D., and Brown, B., "Usage of Fault Detection Isolation & Recovery in Constellation Launch Operations," *AIAA SpaceOps 2010 Conference*, Huntsville, AL, April 2010.

[10] Schwabacher, M., Martin, R., Waterman, R., Oostdyk, R., Ossenfort, J., and Matthews, B., "Ares I-X Ground Diagnostic Prototype," *AIAA Infotech@Aerospace Conference*, Atlanta, GA, April 2010.

[11] Spirkovska, L., Aaseng, G., Iverson, D., McCann, R., Robinson, P., Dittemore, G., Liolios, S., Baskaran, V., Johnson, J., Lee, C., Ossenfort, J., Dalal, M., Fry, C., and Garner, L., "Advanced Caution and Warning System, Final Report–2011," NASA TM-216510, 2013.

[12] Frank, J., Aaseng, G., Dalal, K., Fry, C., Lee, C., McCann, R., Narasimhan, S., Spirkovska, L., Swanson, K., Wang, L., Molin, A., and Garner, L., "Integrating Planning, Execution and Diagnosis to Enable Autonomous Mission Operations," *2013 International Workshop on Planning & Scheduling for Space (IWPSS)*, Moffett, CA, 2013.

[13] Deb, S., and Ghoshal, S., "Remote Diagnosis Server Architecture," *IEEE Systems Readiness Technology Conference (AUTOTESTCON)*, Valley Forge, PA, Aug. 2001.

[14] Maul, W., "Testability Engineering and Maintenance System (TEAMS) Modeling Conventions and Practices," Report K0000190582-GEN, National Aeronautics and Space Administration, 2014.

[15] Barszcz, E., Robinson, P., and Fulton, C., "Tools Supporting Development and Integration of TEAMS Diagnostic Models," *AIAA Infotech@Aerospace Conference*, St. Louis, MO, March 2011.

[16] Maul, W., Fulton, C., and Melcher, K., "Extended Testability Analysis Tool User Guide," *AIAA Infotech@Aerospace Conference*, St. Louis, MO, March 2011.

[17] Maul, W. A., and Fulton, C. E., "Software Users Manual (SUM) Extended Testability Analysis (ETA) Tool," NASA/CR-2011-217240.

[18] Bis, R., and Maul, W., "Verification of Functional Fault Models and the Use of Resource Efficient Verification Tools," *AIAA Infotech@Aerospace Conference*, Kissimmee, FL, Jan. 5C9, 2015.

[19] Maul, W. A., Hemminger, J. A., Oostdyk, R., and Bis, R. A., "A Generic Modeling Process to Support Functional Fault Model Development," *AIAA Infotech@Aerospace Conference*, San Diego, CA, Jan. 4–8, 2016.

[20] NASA Systems Engineering Handbook, NASA/SP-2007-6105 Rev. 1, Dec. 2007.

[21] Aaseng, G. B., Barszcz, E., Valdez, H., and Moses, H., "Scaling Up Model-Based Diagnostic and Fault Effects Reasoning for Spacecraft," *AIAA Space Conference and Exhibition*, Pasadena, CA, Aug. 31–Sept. 2, 2015.

[22] Ferrell, B., Lewis, M., Perotti, J., Oostdyk, R., and Brown, B., "Functional Fault Modeling Conventions and Practices for Real-Time Fault Isolation," *SpaceOps 2010 Conference*, Huntsville, AL, April 2010.

[23] Melcher, K. J., Maul, W. A., and Hemminger, J. A., "Functional Fault Model Development Process to Support Design Analysis and Operational Assessments," *AIAA Space and Astronautics Forum*, Long Beach, CA, Sept. 2016.

[24] Oeftering, R. C., Kimnach, G. L., Fincannon, J., Mckissock, B. I., Loyselle, P. L., and Wong, E., "Advanced Modular Power Approach to Affordable, Supportable Space Systems," *AIAA Space Conference and Exhibition*, Pasadena, CA, Sept. 2012.

[25] Wong, E., Fulton, C. E., Maul, W. A., and Melcher, K. J., "Sensor Data Qualification System (SDQS) Implementation Study," NASA/TM-2009-215442, *International Conference on Prognostics and Health Management*, Denver, CO, Oct. 2008.

[26] Wong, E., "Sensor Data Qualification and Consolidation (SDQC) for Real-Time Operation of Launch Systems," *AIAA SPACE Conferences and Exposition*, Long Beach, CA, Sept. 2016.

Toward Safe Intelligent Unmanned Aircraft Using Formal Methods and Runtime Monitoring

Christoph Torens and Florian-Michael Adolf
German Aerospace Center (DLR), Institute of Flight Systems, Braunschweig, 38108, Germany

Future unmanned aircraft are expected to be autonomous, perform missions automatically, and act intelligently when unforeseen events or degraded situations occur. This results in enormous complexity for modeling and computing the system states, system behavior, and environmental data. Furthermore, the aerospace domain is a safety-critical domain, enforcing specific levels of safety and compliance to extensive standards. Therefore, software has to be of high quality and free of safety-critical errors. But the verification and validation of a complex system, especially the high-level software components, is a critical element. Because of software complexity and the fact that the state-space of theoretically possible executions cannot be covered by testing, a holistic testing concept, utilizing complementary test methodologies, is required. This chapter discusses the high-level autonomous capabilities of the German Aerospace Center (DLR) Autonomous Research Testbed for Intelligent Systems (ARTIS) framework and focuses on the challenges and best practice approach for verification and certification for autonomous unmanned aircraft. One of the first challenges for developing an intelligent unmanned aircraft is the development of a high-quality set of requirements that describes the autonomous behavior of the system. Furthermore, this work proposes the development of a generic set of high-level requirements describing the targeted level of autonomy. To complement traditional verification methodologies, which also play an important role, model checking is also used to proof consistency of behavior and compliance to the requirements. Another way to assure safety, specifically for autonomous behavior, is to utilize runtime monitoring concepts. The idea is to supervise the execution and escalate any error as soon as it occurs to a high-level decision-making unit, such as a pilot. Furthermore, it is commonly understood that self-awareness, maintenance of information about the system status, is necessary to be able to act intelligently.

NOMENCLATURE

ALFURS	Autonomy Levels for Unmanned Rotorcraft Systems
ALFUS	Autonomy Levels for Unmanned Systems
ARTIS	Autonomous Research Testbed for Intelligent Systems
DLR	German Aerospace Center
LTL	Linear Temporal Logic
MiPlEx	Mission Planning and Execution Framework
UAS	Unmanned Aircraft System
G_ϕ	It is ϕ globally true, ϕ is always true
X_ϕ	In the next state, ϕ is true

3.1 INTRODUCTION

New technologies, in particular unmanned aircraft, as well as intelligent software agents, have the potential to change the world. There is a reason why big players like Google, Amazon, Facebook, and others are investing huge efforts to research and develop these technologies. The concept of unmanned aircraft alone, using a remote pilot and without any form of automation, might represent technological progress on its own and would surely have benefits, such as the safety of the pilots during dangerous tasks or missions. However, it is the combination of unmanned aircraft with high levels of automation that enables endless opportunities. The same combination that has this huge potential, however, is difficult to achieve. The general air space is used by manned aircraft, general aviation, and commercial transportation. As a result, there are high requirements on software quality to ensure the overall safety.

To accomplish flight certification, software has to be developed according to industry standards. The standard for safety-critical development of software for the aerospace domain is DO-178. This standard enforces establishment of rigorous development, quality, and verification processes. Specifically, the verification aspects of safety-critical software contribute to the overall costs of the development of aircraft systems today and basically counteract all efforts to incorporate complex software functionality or even any form of autonomous behavior. In its latest version, DO-178C [1], several supplements have been added to the standard. One of these supplements is DO-333 [2], which allows the use of formal methods in a standardized fashion for the first time. Formal methods are promising because they allow an improved level of quality in the verification of systems.

Formal methods have a long history in computer science. Some of the early works on program verification using temporal logic were introduced nearly 40 years ago by Pnueli [3]. The interest in this formalism is growing, but there is still relatively little use of these methods in commercial projects. A study from 2013 that identifies barriers to the introduction of formal methods shows that this is still a problem [4]. The study identifies nine types of barriers and also provides mitigation suggestions for each type: education, tools, industrial environment, engineering, certification, misconceptions, scalability, evidence of benefits, and cost.

The greatest barrier for widespread use of formal methods in research and engineering is proper education. This means prerequisites in skills and training in theoretical concepts, such as formal languages, formal (temporal) logic, and automata theory. Therefore, formal verification is often only considered if the general emphasis on verification is high, which is usually the case for software in safety-critical domains. Even then, however, its use is not common [5, 6], and in general a gap between the academic work and industry use has been identified [7].

Another factor for contributing to the low spread of formal methods, even in safety-critical domains, is uncertainty about the certification credit resulting from the use of these techniques. This uncertainty results from the fact that the software development standard predominantly requested by certification authorities for safety-critical aviation software (e.g., aircraft, airports) did not consider the use of formal methods in DO-178B [8]. It was possible to use these techniques in safety-critical projects then, but details would have to be discussed each time with the certification authority and there was no or little accepted guidance for their use. Since late 2011, the successor standard DO-178C [1] directly supports the use of formal methods with a designated supplement DO-333 [2]. It is therefore important to analyze the use of formal methods according to that new standard.

This work results from efforts to integrate formal methods into our existing test strategy and investigates the use of formal methods specifically in regard to the previously mentioned standards. It is exemplified with a model for the mission manager module of the MiPlEx (mission planning and execution) framework. This chapter shows how the introduction of formal methods to our verification strategy lowers the aforementioned barriers. The described best practices use a combination of traditional verification techniques and show the integration of requirements formalization, model checking, and runtime monitoring. In addition, this chapter discusses the challenges of autonomous unmanned aircraft systems (UAS) and uses existing definitions to address the scope and scalability of autonomy, especially in the context of our autonomous research framework ARTIS. Related work on the concepts of autonomy, verification, and validation, and the growing use of formal methods, specifically runtime monitoring is described in Section 3.2. After the brief introduction of the MiPlEx software and its capabilities in Section 3.3, details on the autonomous capabilities, online replanning, and semantic planning capabilities of ARTIS are shown. The ARTIS showcase is used as a running example for describing our best practices of verification and validation methodologies and to illustrate our efforts toward achieving safe, autonomous UASs. General considerations about existing autonomy frameworks, as well as the modeling and the verification of autonomy, are discussed in Section 3.4. Section 3.5 describes using generic requirements for the specification and modeling of autonomous UASs. This approach supports reuse, improves quality, and facilitates an early validation of the requirements. The relevant ARTIS software verification techniques are described in Section 3.6 and integrated into a general development process for clarification. Next, the concept of runtime monitoring for unmanned aircraft is discussed in Section 3.7, supporting analysis and debugging on one hand, and safety and autonomy on the other. Finally, conclusions are presented in Section 3.8.

3.2 RELATED WORK

The autonomous function capabilities of the ARTIS framework were introduced in previous work [9–11], as has our overall test strategy [12]. The specifics of certification considerations were discussed for small rotorcraft like ARTIS, as well as in general [13]. Analysis of the situation, however, shows that verification is still a stochastic measure. Errors in the code can only be found by a large set of test cases, and coverage criteria like the modified condition/decision coverage (MC/DC) [14] are utilized to assure that these tests are systematically distributed in the global behavior space. However, even the MC/DC design assurance level A criterion may not find common errors, as one study suggests [15]. To complement this stochastic method, we are trying to integrate formal methods into our test concept. With DO-178C [1], the latest successor of DO-178B [8], the use of formal methods is allowed under the supplement DO-333 [2].

The development of complex systems is a challenge; studies show that approximately 70% of faults are introduced during requirement elicitation, architecture development, and design [16]. Specifically, the specification has been identified as the biggest bottleneck in formal methods and autonomy [17]. However, most traditional verification activities address implementation and integration activities. As a result, it is important to find errors early in development. The verification and consistency of requirements has been analyzed using formalization and model checking [18].

The general use of formal methods is already evaluated in several articles. Model-checking approaches are used to verify systems [19], as well as specified mission behavior [20, 21]. Some work is also concentrating on analyzing hybrid systems [22, 23] as well as dividing logic and continuous control [24]. Another interesting approach for the aerospace domain is shown by Webster et al. [25, 26], where behavior is modeled as belief, desire, and intent.

There is also some work on the use and effectiveness of formal methods in regard to certification for safety-critical domains. Before the standardized possibility to use formal methods for certification purposes was introduced, there was previous work, proposing to modify the established certification process for the use of formal methods, for example using Simulink and Scade as tools [27] as well as general guidance to use these methods for certification credit [28, 29]. Now, with the availability of the new standard, the use of formal methods for certification is a very interesting topic. A direct comparison of testing versus formal verification regarding DO-178C is given by Moy [30]. Pires proposes a method to prove certification objectives by formally proving the compliance of low-level requirements to C source code using *Frama-C* [31]. The latest work by Webster is also about generating certification evidence for rational agents using model checking with the aforementioned technique of modeling behavior [32]. An extensive case study was done by NASA in 2014 utilizing three types of formal methods for certification credit [33]. Model checking is among the proposed methods; the described tools are Matlab/Simulink and Kind.

Furthermore, using runtime assertions during and beyond implementation is a well-known technique [34]. Runtime verification and runtime monitoring specifically using temporal logic is a research topic originally founded in theoretical computer science. A detailed explanation and differentiation to model checking is given by Leucker [35]. NASA has utilized runtime monitoring on the Java byte code level with their tool Java PathExplorer [36]. NASA also created a language for runtime verification of safety-critical systems, called Copilot [37]. Recently, however, the concept of runtime monitoring has been used by NASA on a high level, as a means to construct a system health monitor [36].

Finally, NASA published an extensive best-practices survey on a range of techniques for verification of mission-critical software [38]. The discussed methodologies are model checking, constraint solving, monitoring, the need state machine learning, and static code analysis.

3.3 ARTIS SYSTEM DESCRIPTION

As a running example, we introduce our research framework as a basis for discussions of autonomous capabilities and following discussions on verification aspects. The DLR developed the ARTIS framework for the research of intelligent systems and autonomous functions for unmanned fixed wing aircraft and unmanned rotorcraft. The latest addition to this fleet of rotorcraft is SuperARTIS (Fig. 3.1), a rotorcraft with an intermeshing rotor and empty weight of 35 kg, 50 kg of max. payload mass, and a 2.8-m rotor diameter. The autonomous pilot is realized using interacting software components for guidance, navigation, and control. Essentially, this renders the UAS into a flying software product.

One of the key software components for achieving autonomous behavior is the MiPlEx component, a software framework comprising real-time mission plan execution and 3-D world modeling, as well as algorithms for combinatorial motion planning and task scheduling. The control architecture achieves hybrid control by combining the main ideas from the behavior-based paradigm and a three-tier architecture. The behavior-based paradigm reduces the system modeling complexity for composite maneuvers (e.g., land/takeoff) as a behavior module that interfaces with the flight controller.

Fig. 3.1 SuperARTIS, a rotorcraft with an intermeshing rotor and empty weight of 35 kg, 50 kg of max. payload mass, and a 2.8-m rotor diameter.

In the remainder of this chapter, this system will be used to show our approach to developing autonomous UASs. The research goal for ARTIS is to achieve a high level of autonomy. In this chapter, we give two examples of ARTIS functionality that map to the autonomous capabilities of the autonomy levels for unmanned rotorcraft systems (ALFURS) framework. Exploration and online replanning capabilities, as well as semantic planning, are detailed in the following sections.

3.3.1 TERRAIN EXPLORATION AND ONLINE REPLANNING

The three-tier architecture has the advantage of different abstraction layers that can be interfaced directly such that each layer represents a level of system autonomy. The software framework allows our unmanned rotorcraft to navigate through a priori unknown terrain, as shown in Fig. 3.2.

With the onboard control system, it is possible to design and execute mission plans. This planning system automates the translation of user-specified sets of waypoints into a sequence of parametrized behavior commands. The path planner is able to find collision-free paths in an obstacle-constrained three-dimensional space. The task planner determines a near optimal order for a given set of tasks. Moreover, a task planner can solve specialized problems (e.g., specifying the actual waypoints for an object search pattern within an area of interest).

The mission planner is capable of finding unobstructed paths quickly while optimizing task assignments and task orderings for multiple UASs. The scope of the MiPlEx module is targeted toward more than just path planning; obstacle detection, definition and execution of search patterns for a given area, and online exploration of unknown terrain are also supported and thus represent high-level autonomous behavior. Intelligent systems need the ability to plan and assess actions and to be able to choose the most optimal plan. Especially for such high-level decisions, which directly impact the overall behavior of the system and decide the actions which will be executed, verification aspects are important.

3.3.2 SEMANTIC PLANNING

One aspect of intelligent and autonomous behavior is the consideration of semantic data about the known environment. It is not only desirable to be able to fly without collisions with ground structures, but mission planning should also consider additional factors like general risk for the population. To achieve this, the planning needs to incorporate the types of areas that are present in the environment, such as highly populated areas, streets, rural areas, or forests. The type of area influences risk toward the population and other factors, such as noise levels. The system can thus be programmed to optimize the path by

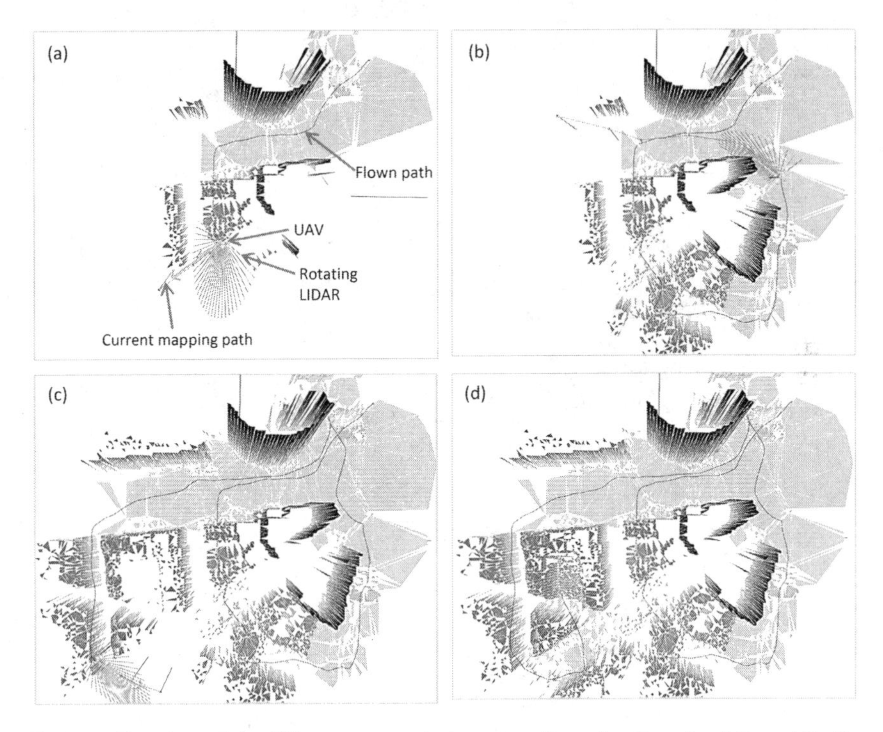

Fig. 3.2 Top view of the different stages during a terrain exploration simulation at Berlin, Potsdamer Platz (from upper left to lower right); the rotorcraft decides for each temporary waypoint based on a cost-modified roadmap.

weighting regions differently or by limiting flight performance according to the risk. On the ALFURS autonomy scale, the MiPlEx capabilities would correspond to autonomy level 4 and above (real-time path planning and replanning; perception capabilities for obstacle, risks, target, and environment changes detection). Path planning for unmanned aircraft often comprises Euclidean cost models that are easy to model and are of low computational cost (Fig. 3.3). Path planning that utilizes more environment information, toward intuitive paths, is often computationally harder to implement and practically harder to adapt to different scenarios. Such path searches can cause complexity and implementation challenges with respect to the path cost model and computational overhead. Hence, MiPlEx comprises an approach that considers location features from an environment map during the path-planning process, as shown in Fig. 3.2. The concept is tailored for irregular free space graph representations and exemplified for low altitude guidance of an unmanned rotorcraft (Fig. 3.4). Path preferences are represented in terms of weighted regions that are implemented into the horizontal prisms. No assumptions about the prism alignment are required so that no assumptions about the regions have to be considered in the graph topology itself. The construction of the graph representing the free space was modified to assure the existing assumptions. The placement of the graph's nodes respects the unknown alignment of the regions and makes sure that the weighting of the graph's edges is feasible. On connecting the nodes to build a graph, the edges are weighted additionally to an admissible underlying cost function. To hold the graph search complete and optimal, an adaptation of the potential heuristic was made. The special construction of the graph adds nodes and edges to the

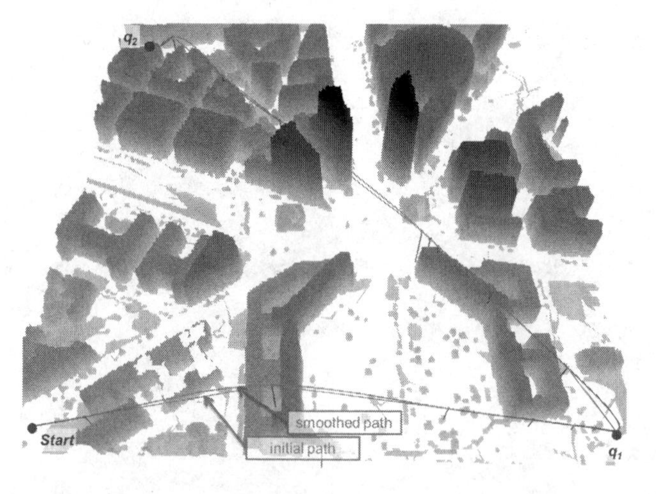

Fig. 3.3 Conventional Euclidean cost-to-go function: no preference for flat ground (white) and allows for flight above terrain objects (gray).

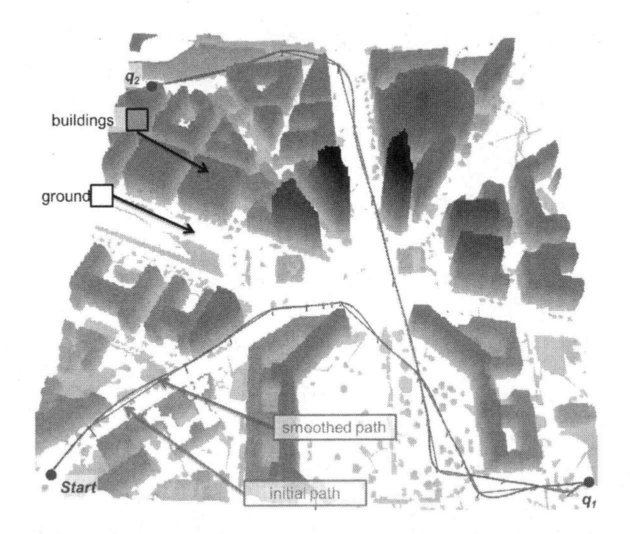

Fig. 3.4 Semantic label planning: Prefer low-level along flat ground like streets (white) over terrain objects (grayscale).

graph. To keep the graph search fast, a number of methods are used to accelerate the graph search phase using a dynamic sampling or a nonadmissible potential heuristic. A geometric optimization for crossing differently weighted region borders was proposed to reduce the path costs even further. The algorithms were implemented for an unmanned rotorcraft in the ARTIS fleet and used for path-planning in obstacle-rich environments. Performance evaluations show that the path can easily be influenced to prefer or avoid regions, but then the planning process becomes slower. As a result, a global graph-based path planner that takes local preferences by weighting regions for finding the shortest path is feasible. It should not be limited to sampling-based roadmaps, but should also fit other sampling-based approaches.

3.4 DEFINITION AND MODELING OF AUTONOMY

Several definitions of *autonomy* exist; often the term is also used as a synonym for automation. Indeed the term autonomy has been discussed thoroughly in scientific publications. For the purpose of this chapter, our definition of autonomy follows the definitions and builds upon the ALFURS concept [39].

3.4.1 CHARACTERISTICS OF AUTONOMOUS BEHAVIOR

The aerospace domain is a safety-critical one. The extreme end of the spectrum of autonomous behavior is that it may not be deterministic, or even predictable, how the system will behave. To achieve flight certification for such a system would be unthinkable today. Deterministic behavior of the aircraft system is key to achieving flight certification. As a result, there are several approaches to reduce efforts in this area. For example, the European Aviation Safety Agency (EASA) has proposed a new category of aircraft operations [40], the so-called specific operation category, where it may be acceptable to compensate certain airworthiness risks by introducing mitigation actions in a defined scenario. Furthermore, the organization ASTM International is working on a standard, named Methods to Safely Bound Flight Behavior of UAS Containing Adaptive Algorithms [41], that proposes allowing parts of the system to be of reduced integrity if the core functionality is supervised and bounded by a high-integrity module that can act as a backup.

3.4.2 AUTONOMY FRAMEWORKS AND AUTONOMY LEVELS

The ALFUS (autonomy levels for unmanned systems) framework [42] and the corresponding term definition [43], established by the National Institute of Standards and Technology (NIST) serve as the basis for common understanding of autonomy itself and the capabilities of autonomous systems. This approach is used and further refined for unmanned rotorcraft systems into the ALFURS framework [44], because the direct application of the ALFUS framework for such rotorcraft systems requires additional effort. Furthermore, a metric to compare different UASs against each other was needed with a normalized scale. These frameworks have been commonly adopted by their corresponding communities and have remarkably helped to establish a common understanding of generically used terms and concepts in the area of autonomy. However, whereas the ALFUS framework seems to be too generic and the ALFURS framework is specific for unmanned rotorcraft, the autonomy framework can also be applied to UASs in general.

For a better understanding of the autonomy concepts used, the autonomy levels from Kendoul [44] are briefly introduced. Autonomy is assessed through the capabilities of the guidance, navigation, and control components of a UAS. For each level from 0 to 10, the ALFURS framework assigns certain capabilities to the guidance, navigation, and control components. Level 0 represents the lowest autonomous mode of operation, with all functionality performed by external systems (usually a human operator), and level 10 represents full autonomy of the system, including human-level decision making and accomplishment of missions without any human intervention. For example, for autonomy level 0, no guidance capability is required onboard a UAS. But for autonomy level 4, real-time replanning of paths is required. For further clarification, the guidance component is detailed for levels 0 to 6 in Table 3.1.

TABLE 3.1 AUTONOMY LEVELS (EXCERPT FOR GUIDANCE COMPONENT OF LEVELS 0 TO 6, ADAPTED FROM KENDOUL [39])

Autonomy Level	ALFURS Level Description	Selected Capabilities
6	Dynamic mission planning	Reasoning, high-level decision making, mission-driven decisions, high adaptation to mission changes, tactical task allocation, execution monitoring. Higher level of perception to recognize and classify detected objects/events and to infer some of their attributes, mid-fidelity situational awareness.
5	Real-time cooperative navigation and path planning	Collision avoidance, cooperative path planning and execution to meet common goals, swarm or group optimization.
4	Real-time obstacle/event detection and path planning	Hazard avoidance, real-time path planning and replanning, event-driven decisions, robust response to mission changes. Perception capabilities for obstacle, risks, target, and environment changes detection, real-time mapping (optional), low-fidelity situational awareness.
3	Fault/event adaptive unmanned aircraft system (UAS)	Health diagnosis, limited adaptation, onboard conservative and low-level decisions, execution of preprogrammed tasks. Most health and status sensing by the UAS, detection of hardware and software faults.
2	External systems independent navigation (e.g., non-GPS)	All sensing and state estimation by the UAS (no external systems such as GPS), all perception and situation awareness by the human operator.

(Continued)

TABLE 3.1 AUTONOMY LEVELS (EXCERPT FOR GUIDANCE COMPONENT OF LEVELS 0 TO 6, ADAPTED FROM KENDOUL [39]) (*Continued*)

Autonomy Level	ALFURS Level Description	Selected Capabilities
1	Automatic flight control	Preprogrammed or uploaded flight plans (waypoints, reference trajectories, etc.); all analyzing, planning and decision making by external systems. Most sensing and state estimation by the UAS, all perception and situational awareness by the human operator.
0	Remote control	All guidance functions are performed by external systems (mainly human pilot or operator).

3.5 INTELLIGENT UAS REQUIREMENTS

The basis for all verification activities is a set of requirements. The standard DO-178 enforces a strict bidirectional traceability of compiled code, software functions, and design, up to low-level and high-level requirements for high levels of criticality. This means, to be able to start verification and certification activities, all software functions including autonomous capabilities need to be defined properly as requirements.

The ALFURS framework gives a rough overview of capabilities for autonomy levels, but there is no direct approach to easily generate requirements from the given set of abstract capabilities. As a consequence, the ALFUS and the ALFURS approaches should be further extended to create a common basis or reference set of UAS autonomy requirements. Such an effort should aim toward generic high-level requirements and reusable software models, which are not specific for a certain aircraft, but rather generic for a whole class of aircraft, and which can be mapped to a certain level of autonomy and corresponding capabilities.

3.5.1 GENERIC REQUIREMENTS

The approach used for the creation of these requirements was to find generic requirements fitting these capabilities and mapping them to the autonomy levels (more details describing this approach can be found in previous work) [44]. The ARTIS requirements were used as a starting point, which then got re-elicited based on our previous efforts with formal modeling of the mission planning and execution component [45]. Originally, these software functions were intended for unmanned rotorcraft only but had to be generalized and separated into generic and specific sets of requirements when Prometheus, a fixed wing aircraft, was added to the ARTIS fleet. This approach is now extended and generalized into an unmanned aircraft generic set of requirements. Some basic requirements and the according autonomy levels are presented in Table 3.2. The field "Activity" is used to describe the action, and the other fields, such as "System" and "Obligation," are positioned to make the resulting row more readable. The terms in capital letters represent keywords with a defined meaning. We say a requirement is *general* if it describes a specification that is common to all UASs. A *generic* requirement describes a specification that has a certain parameter that has to be adapted for specific UASs but can still act as a placeholder until that parameter is further detailed. The rationale to create a common set of UAS requirements is summarized by the following items:

- Requirements can be reused by the UAS community (e.g., safety requirements or to support common autonomy research).
- Discussions and reviews will make requirements more sound and can allow for wider acceptance.
- A working basis for discussion with rule-making bodies and associated certification authorities will be established.
- This starting point can serve as a common basis for verification activites in development standards.

In this example, the requirements shown in Table 3.2 concentrate on the guidance system of a UAS. The proposed requirements are intended to be valid in general, for all UASs, with a guidance component. The presented requirements should therefore represent a basic set of requirements applicable for a UAS. Furthermore, they are generic in the sense that the desired autonomy levels can be picked and the requirements can be filtered according to that level. For example, requirements R1.1 to R1.13 correspond to the level of autonomy that enables pilots to plan waypoint missions while the guidance component of the UAS is able to execute the mission and transform it into actionable commands. The requirements R1.15 to R1.18 allow for a high-level definition of mission tasks, such as mapping an area or searching an area for a specific object. Finally, R1.19 and R1.20 are about the online replanning capability that enables collision avoidance and dynamic tasks, such as following an object or exploring an unknown area.

There are also several requirements that are usually constrained by the missions that have to be performed. For example, speed and endurance fall into this category (e.g., R2.1 and R2.2), see Table 3.3.

TABLE 3.2 GENERAL REQUIREMENTS FOR UASs IN A STRUCTURED, SEMIFORMAL FORMAT

ID	ALFURS Autonomy Mapping	When	Active System/Obligation	Activity
R1.1	0	ALWAYS	the SAFETY-PILOT SHALL	be able to control the UAS
R1.2	1	ALWAYS	the GUIDANCE SYSTEM SHALL	be able to load a list of waypoints/ NEW MISSION
R1.3	1	ALWAYS	the GUIDANCE SYSTEM SHALL	handle automatic takeoff and landing safely
R1.4	1	WHEN a list of waypoints/ NEW MISSION is loaded	the GUIDANCE SYSTEM SHALL	be able to GENERATE MISSION PLAN
R1.5	1	CONTINUOUSLY, during an ACTIVE MISSION	the GUIDANCE SYSTEM SHALL	be able to GENERATE a TRAJECTORY to CONTROL SYSTEM
R1.6	1	ALWAYS	the GUIDANCE SYSTEM SHALL	provide the ability to delete MISSION PLAN
R1.7	1	ALWAYS, during ACTIVE MISSION	the GUIDANCE SYSTEM SHALL	provide the ability to the PILOT to STOP the MISSION and go into STATE STANDBY (hover state or holding pattern)
R1.8	1	CONTINUOUSLY, during an ACTIVE MISSION	the GUIDANCE SYSTEM SHALL	receive inputs from NAVIGATION SYSTEM (system state, IMU/GPS)
R1.9	1	WHEN the PILOT commands start	the GUIDANCE SYSTEM SHALL	execute last uploaded MISSION PLAN
R1.10	1	WHEN the SAFETY-PILOT intervenes	the GUIDANCE SYSTEM SHALL	go into STATE INACTIVE

R1.11	1	WHEN STOP is received during takeoff/landing and UAS is on ground	the GUIDANCE SYSTEM SHALL	go into STATE ONGROUND
R1.12	1	WHEN STOP is received during takeoff/landing and height $\geq$ safe minimum height	the GUIDANCE SYSTEM SHALL	go into STATE STANDBY
R1.13	1	WHEN STOP is received during takeoff/landing and height $<$ safe minimum height	the GUIDANCE SYSTEM SHALL	cancel MISSION, rise to safe minimum height, and go into STATE STANDBY
R1.14	2	ALWAYS	the GUIDANCE SYSTEM SHALL	receive complete sensing information from NAVIGATION SYSTEM (e.g., Height est. using Radar/LIDAR [46])
R1.15	3	ALWAYS	the GUIDANCE SYSTEM SHALL	be able to load a list of TASKS
R1.16	3	WHEN a list of tasks is loaded	the GUIDANCE SYSTEM SHALL	be able to GENERATE MISSION PLAN
R1.17	3	WHEN a task is loaded	the GUIDANCE SYSTEM SHALL	be able to execute TASKS
R1.18	3	CONTINUOUSLY	the GUIDANCE SYSTEM SHALL	monitor HEALTH STATUS and submit data to ground control station
R1.19	4	CONTINUOUSLY, during active mission	the GUIDANCE SYSTEM SHALL	be able to replan current MISSION PLAN
R1.20	4	WHEN an OBSTACLE is detected, blocking the current path	the GUIDANCE SYSTEM SHALL	be able to replan and avoid the OBSTACLE in a safe way

TABLE 3.3 GENERIC REQUIREMENTS FOR UAS, CONSTRAINED BY A MISSION SPECIFIC, OR UAS SPECIFIC PARAMETER

ID	Autonomy Mapping (ALFURS)	When	Active System/Obligation	Activity
R2.1	1	ALWAYS	the GUIDANCE SYSTEM SHALL NOT	command speeds exceeding <max. speed>
R2.2	1	IN ENROUTE FLIGHT	the GUIDANCE SYSTEM SHOULD	command <transit speed>
R2.3	1	IN ENROUTE FLIGHT	the GUIDANCE SYSTEM SHALL	command speeds greater than <min. speed> (e.g., zero if hover is possible)
R2.4	1	IN ENROUTE FLIGHT	the GUIDANCE SYSTEM SHALL NOT	command heights below <min. height>
R2.5	1	ALWAYS	the GUIDANCE SYSTEM SHALL NOT	command heights exceeding <max. height>
R2.6	3	ALWAYS	the GUIDANCE SYSTEM SHALL	ensure via monitoring that flight duration is less than <max. flight duration>
R2.7	3	ALWAYS	the GUIDANCE SYSTEM SHALL	ensure via monitoring that flight distance is less than <safe range without datalink loss>
R2.8	4	ALWAYS	the GUIDANCE SYSTEM SHALL	be able to maintain <safety distance> from known obstacles

3.5.2 SAFETY REQUIREMENTS

The presented generic, pattern-based approach of compiling a set of requirements is especially useful for safety-relevant requirements, as it is essential that no relevant safety requirement is missed during the design and system safety assessment. With current regulations in mind, it is, for example, necessary for all UASs to fulfill R1.1 and R1.10 as a safety requirement because the safety pilot must always be able to take control of the aircraft. On the other hand, a future UAS with a very high autonomy level that accepts missions on an abstract goal basis would possibly omit that feature, because direct control might be too complex and the low-level fallback routine might be waypoint navigation, as shown in requirements R1.2 to R1.9. Additionally, a safe state must be reachable in all situations; this is handled in the requirements R1.3, R1.7, and R1.10. Safety requirements are of particular interest when discussing aspects of runtime monitoring because, by supervising safety requirements, the monitoring can act as an independent safety layer (cf. Section 3.7).

3.5.3 FORMAL REQUIREMENTS SPECIFICATION

The formalization of requirements is important for the later use of these requirements for formal methods, as described later in this chapter with model checking and runtime monitoring. In the previous example a tabular structure was used to write down requirements. The structure of the table is the result of a template approach for eliciting requirements. The template approach made it easy to be able to directly write down requirements, and it was formal enough to facilitate the use of keywords and a fixed structure. This semiformal approach is an adaptation of a *requirements template with conditions* [47]. The disadvantages of using natural language for requirements can thus be minimized. The used structure is a compromise between natural language form and a formal specification. The template was also helpful in formulating the requirements and making sure requirements were considered with all relevant aspects. The building blocks of a requirement were given by a template, as seen in Fig. 3.5.

The formalization procedure was intended to go from natural language descriptions to structured natural language descriptions, and finally to formal

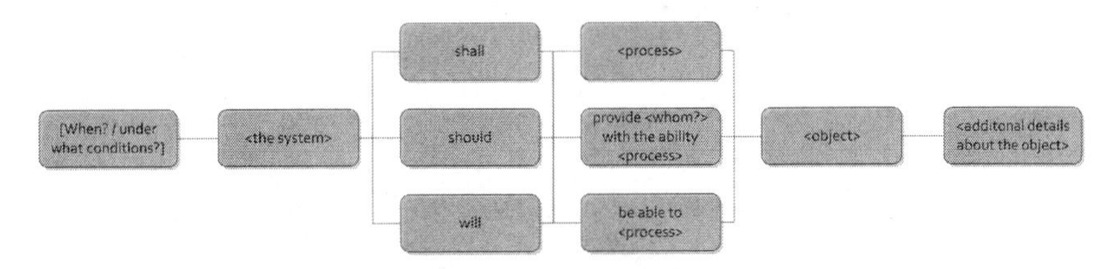

Fig. 3.5 Building blocks of the natural language template for specifying requirements, adapted from [47].

```
1  LTLSPEC G(Status=Landing&Command=Stop&Position=Ground -> X Status=StandbyGround); --R6.1
2  LTLSPEC G(Status=Landing&Command=Stop&Position=Air& MinHeight -> X Status=StandbyAir); --R6.2
3  LTLSPEC G(Status=Landing&Command=Stop&Position=Air&!MinHeight -> X Status=AbortLandTakeoff); --R6.3
```

Fig. 3.6 Functional requirements in linear temporal logic. Each line is concluded with a comment that states the requirements reference (e.g., "–R6.1").

properties. The requirements for an existing software artifact were elicited and then formalized into linear temporal logic (LTL, Fig. 3.6), which is a derivative of temporal logic [3]. The most common LTL operators used were X_ϕ, which means that, in the next state, ϕ is true, and G_ϕ, which states that ϕ is globally always true. A formal model for the software was developed and nuXmv was used as a model-checking tool to analyze the requirements in regard- to the model. NuXmv [48, 49] is popular model-checking software that is frequently used in research projects; this tool is discussed in more detail in a following section. After an initial learning phase, it was relatively easy to transform the tabular requirements into a formal specification in most cases (Fig. 3.7). The figure shows the formalization of a high-level requirement to an LTL formula.

In nuXmv, comments are separated from source code by a "–." The examples show a clear correspondence of keywords on the left side of the property (left of the implication "→") and the "When" column of the tabular requirement specification (e.g., landing, stop, ground, MinHeight). The same observation can be made about correspondence of keywords on the right side of the property and the "Activity" column of the tabular requirement (e.g., StandbyGround, Slowdown, CancelStart). Furthermore, the requirement usually is globally true, and therefore the structure is `LTLSPEC G(lefthand → righthand)`. Not all, but most of the requirements could be handled with this left hand–right hand scheme. The requirements shown in Table 3.2 are transformed with this process into a formal specification, as seen in Fig. 3.6. It was difficult to achieve the abstraction from text to a model. In the above examples, assumptions are already made (e.g., "Landing" is generalized into a "Status" variable, "Stop" is a

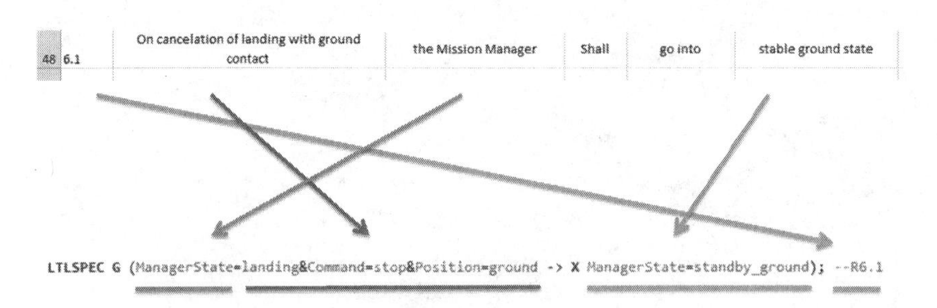

Fig. 3.7 Detailed illustration of the process of requirements formalization from template to linear temporal logic.

value of "Command," and "Position" is either "Air" or "Ground"). To allow this, the requirements have to be specific enough to generalize such keywords to variables and states.

The generic UAS requirements that have been developed and formalized in this section can be further analyzed using model checking. This approach is discussed as one of the verification methodologies of the ARTIS testing concept in the following section.

3.6 ARTIS STATIC AND RUNTIME VERIFICATION APPROACH

The ARTIS software framework encompasses C and C++ Code, Matlab/Simulink models, generated code, and third-party libraries. The overall goal for verification and testing activities is to obtain a full verification of the software against given requirements and to achieve a high-quality, maintainable software. It is important to notice that ARTIS software development is a continuous task. There is consistently new functionality designed, implemented, tested, and either improved or aborted. Testing the software is not only seen as a method for verification, but it has also been shown to increase the overall software quality. For example, refactoring the code structure is less problematic if a set of tests assures the functionality of the code. Furthermore, it is also possible to increase the algorithm quality. For this purpose, benchmarks can be utilized to assess and improve the performance of a given algorithm. However, a thorough verification of a system is a complex task; therefore, there are several layers of testing for the ARTIS system. Details of the development and verification processes can be found in [50]. In this work, the MiPlEx testing efforts and the categories of testing are discussed. In this and following sections, static tests, dynamic tests, software-in-the-loop (SIL) and hardware-in-the-loop (HIL) simulations, flight tests, and model-based tests are elaborated upon as the different testing layers in an integrated test strategy. The presented verification approach utilizes these methodologies in such a way that the tests start from inner, low-level aspects and go to outer, high-level aspects of the system. Different test layers have different test characteristics and different costs. A good test design should combine these techniques to achieve maximum coverage of all test characteristics. The key is to find errors at the earliest possible moment because software errors tend to cause exponentially growing costs the later they are found during software development.

A layered test strategy and automatic tests facilitate this approach. The first identified test dimension is the size of the specific system under test (SUT) that is used by the verification technique (e.g., what is being tested). The possible values range from single lines of code, software functions, a whole software module, interaction of different modules, and software systems up to the complete embedded system. Test effort asks what the costs of a specific test method are. The test or scenario complexity is low if a mathematical function is tested standalone, increases with the combination and interaction of functions, and is highest if an urban scenario is simulated. The coverage assesses the theoretical state-space that can be covered by the method, or the code coverage if that is not applicable. The feedback time describes how much time it takes to get the result from the test back to the developer. A short feedback time is good, as it allows faster development cycles. Finally, the level of automation describes if and to what degree a test can be automatized.

To analyze and assess the ARTIS verification, abilities, and concept, we use the test dimensions to roughly evaluate the approach for each of the used verification methodologies in a qualitative manner in the following sections. These different verification techniques complement each other and each contribute individual aspects to the overall analysis of the system. A visualization of these test dimensions for the ARTIS test concept is shown in Fig. 3.8. In addition to these different dimensions of each verification technique, each has a different applicability in the overall development process of the system (see Fig. 3.9). We use a simple V-model for illustration, as the standards for aircraft development and safety assessment [51, 52] utilize this development model and it is sufficient to show the applicability

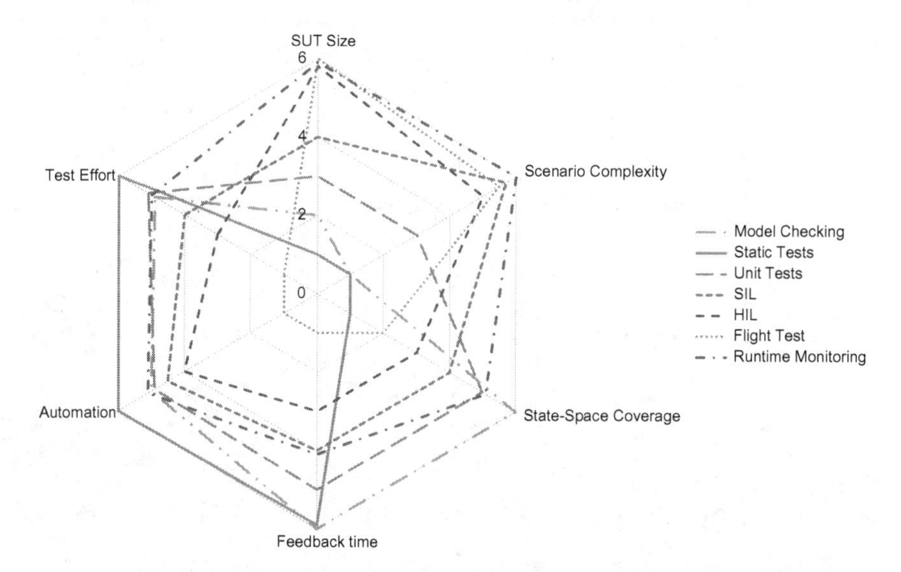

Fig. 3.8 Evaluation of the different ARTIS test dimensions, with various test methods. The axes are: size of the specific system under test (SUT), test effort, scenario complexity, state-space coverage, feedback time, and the level of automation. The different test methodologies are model checking, static tests, unit tests, software-in-the-loop tests (SIL), hardware-in-the-loop tests (HIL), flight tests, and runtime monitoring. Note that a high test effort results in a low score value for test effort in the diagram.

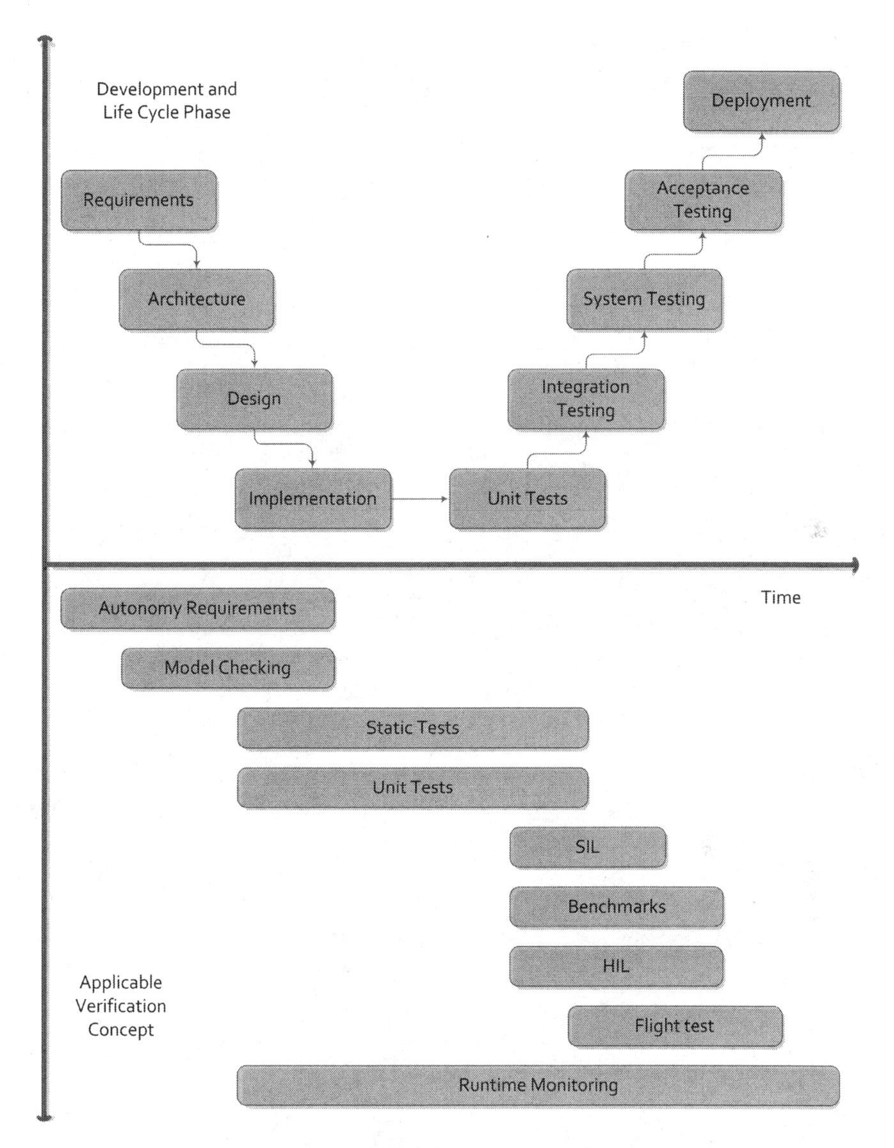

Fig. 3.9 Overview of the proposed verification techniques as part of the standard V-model for software and systems development.

of verification methodologies. However, it should be noted that there are approaches for using more modern development models [16, 53, 54] than the traditional V-model.

3.6.1 MODEL CHECKING

A system engineering process starts with acquiring requirements, which are often formed into models. The models are usually used as a semiformal piece of specification and documentation. If the models can be formalized to a substantial level, then formal approaches can be utilized to analyze that specification. In the previous section, we developed a generic set of requirements, specifying an autonomous UAS with level-zero to level-four capabilities of the referenced ALFURS framework.

For the creation of the formal model, at first a state diagram (Fig. 3.10) was used to visualize thoughts and further clarify requirements as an intermediate step. This model traces back to the described generic requirements, as indicated by requirement annotations (e.g., R1.19). The left diagram shows the operational mode of the mission manager. There are "safe" states: standby_ground, standby_air, and pause_state. The remaining states represent different kinds of movement, such as landing). The right diagram shows a separate task that can receive and validate missions transferred from the ground control station.

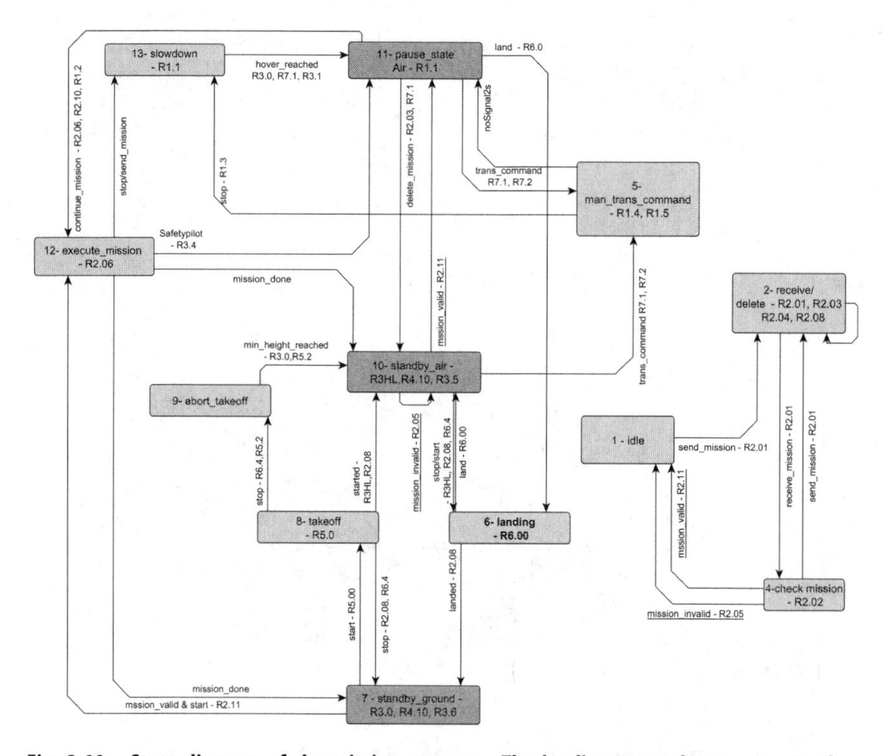

Fig. 3.10 State diagram of the mission manager. The landing state that represents the discussed requirements is highlighted with bold text, standby states are printed in darker shade.

This model was then transformed into a formal nuXmv specification. NuXmv uses a textual input language and allows definition of two separate things: a behavioral model (i.e., a finite state machine (FSM)) and properties that are to be validated on the model in LTL. An FSM can be specified via an `assign` declaration by giving an initial value `init(var)` and state transitions `next(var)`, which are defined by boolean expressions. The different FSMs are then executed concurrently by nuXmv, or, to be precise, the model-checking algorithm tries to find counterexamples for the properties that could occur during parallel execution of the FSMs.

Eclipse was used as an editor, which featured syntax highlighting using an available nuXmv plug-in* and a derivative plug-in distribution.† Also, execution of the model checking could be done directly from within Eclipse using the external tools option.

The two FSMs of Fig. 3.10 can be translated to nuXmv syntax without much difficulty. The transitions are guarded by a logical expression (e.g., check actual state) and can then be executed by assigning the next value the variable gets. To retain general comprehensibleness, the produced expressions were built to mimic a simple syntax: state + event = next-state. But in some cases, additional state variables do further alter the behavior, as in the following example:

```
ManagerState=execute_mission & Command=mission_done &
  Position=air :standby_air; -R3HL
```

This example shows that it is generally possible to translate a previously developed FSM into a formal model.

While building the model, the properties can be used to test it. Therefore, it is most important to have specified requirements. Also, complete traceability greatly supports the incremental building of the model correctly.

3.6.1.1 TEST DIMENSION EVALUATION

The system under test corresponds to the level of abstraction that is being modeled. The correspondence of the model to the system itself is guaranteed by executing the model-generated tests against the implementation. Typically, models give a high-level view of behavior and structure, resulting in a value for the SUT size that corresponds to a module or the interaction of modules on a certain level. The test effort for model-based and formal techniques is very high [55]. The initial costs in particular are high, because formal tools and skilled test engineers with a good theoretical background are needed. The recurring costs for creating and maintaining a model are also an important factor, because the formal models need to be high quality and cannot be automatically derived from existing artifacts in most cases. These costs, however, tend to

*Data available online at http://code.google.com/a/eclipselabs.org/p/nusmv-tools/retrieved 01-03-2018
†Data available online at http://code.google.com/a/eclipselabs.org/p/nuseen/retrieved 01-03-2018

become smaller with more experience. The scenario complexity is abstract, but errors that occur only in combination of several circumstances can be found. Therefore, the complexity is valued as high. Model-based and other formal methodologies have their special strengths in the state-space and code coverage. Using model-based testing, it is often possible to simply generate tests until the desired coverage is achieved. Model-checking techniques can even prove logically the correctness of a given specification. These approaches are especially helpful in finding errors early in development. Thus, the feedback time is optimal. The formal, executable model even allows for finding errors in the model or in the specification itself, before any line of code is implemented. Also, the test automation benefits from the formal model; therefore, tests cannot only be executed but also generated, automatically in most cases.

3.6.2 STATIC TESTS/COMPILE TIME ASSERTIONS

There are several different techniques that are identified by static analysis. For ARTIS, a large number of static assertions were added inside the implementation. This assures a feedback to the programmer directly during the compile time (e.g. platform assumptions). In particular, Boost [56] concept checks are used, which define and test type requirements during compile time. Furthermore, static code checkers are used to assess the code quality. The tools CCCC [57], Cppcheck [58] and CppLint [59] are used for the ARTIS project. Doxygen [60] warnings also are used to ensure code quality by identifying missing comments. Results for these tests are obtained and evaluated on at least a nightly basis via a Jenkins build automation and continuous integration server.

3.6.2.1 TEST DIMENSION EVALUATION

For the covered static analysis tools, the SUT size is small; the checks run over a single or few lines of code within a file. But they have a very low test effort, possibly being the cheapest kind of test. Automatic analysis tools can often be used with minimal adaptation. This however, requires the tools to be used actively during the development and action to be taken based on the tools results. If a tool like Cppcheck is used after the development stage, when the product goes into a dedicated testing phase, the potentially huge number of warnings may lead to a huge effort in fixing these errors or identifying the false alarms. On the other hand, resolving the warnings during the development phase will help to identify possible errors and design flaws in an early stage. The scenario complexity is low, as there is no scenario in static tests and the tests tend to do simple checks. However, the coverage is not an issue, as every file and each line gets tested. The feedback time is excellent; the results of the checks are returned during compile time or are being generated by the continuous integration server. The automation of static tests is good. As described, several tools and frameworks are available. However, false alarms raised by the code-checking tools have to be reviewed and dismissed regularly.

3.6.3 UNIT TESTS/RUNTIME VALIDATION

Dynamic analysis refers to running the executable model or binary, or parts of it, and performing tests. This means unit tests, but also coverage and memory leak tests. In particular, a lot of effort was put into creating an extensive set of unit tests. Units are being tested on various integration levels. The test complexity scales with the integration level of the tested unit component. The first set of tests determines if the used platform is suitable for the execution of the program suite. This is necessary, because various target platforms are being, used and the tests assure the targets' bit-alignment and int-size and that other platform-specific properties meet the requirements. Then, additional basic functionality is tested, like math and vector operations and operations with the relevant data structures. After this, the integrated functionality is tested, such as behavior, UAS classes, trajectory functions, and the world model. As a last step, high-level functionality is tested (e.g., mission manager, mission planner, behavior sequences, the corresponding sequence controller, the flight mechanics, and flight controller). Roadmap features are also tested.

The unit tests go beyond the testing of small isolated software components. Different modules interact with each other, and so the tested functionality gets more and more complex. The test ordering is conceptual but also temporal. This means that, for example, and the world model is tested before the mission planner, which uses the world model. With each test layer, an increasing number of software modules is used for the test. The unit tests integrate a great amount of expert knowledge. For example, it is not only tested whether a path is successfully planned to the correct location, but also how. This means that the software makes sure that hardware constraints are considered (e.g., maximum allowed turn rate limit, maximum speed, and acceleration). Similarly, the generated missions are checked for behavior ordering and parameter restrictions. Some algorithms have been implemented in different variations; for example, graph search is implemented as A* and D*. In such a case, one implementation can be used to test the other, to see if both algorithms arrive at the same result. To further aid in debugging processes, the tests generate a great amount of log data. We implemented an automated test reporting that generates data in the order of hundreds of megabytes, including generated images, graphs, and PDF files, showing the computed paths as trace visualizations and visualizations of imported world models, as shown in Fig. 3.11.

3.6.3.1 TEST DIMENSION EVALUATION

For unit tests the SUT size is primarily targeted to small software functions and modules. For basic isolated functions, the test effort is relatively small, but the effort grows with integrated functions in which a complicated setup may be required. The test complexity is also dependent on the targeted unit level; it is kept to a minimum for the used module functionality. The code coverage is good. Full coverage is not easy to obtain but can be improved by using techniques such as equivalence classes. However, using unit test executions, it is not possible to obtain exhaustive state-space coverage. The feedback time is excellent, because unit tests are used during programming to debug and design the software itself. Unit tests are the key example for test automation, as proposed by test first approaches.

3.6.4 SOFTWARE-IN-THE-LOOP

SIL simulations are used to test the software integration level (i.e., the main parts of the software system as a whole, including the interfaces to other software and hardware components). For this purpose, a simplified software system simulator has been written. The simulation can act in a variable simulation time, where calls are executed synchronously. This enables the MiPlEx framework to perform fast functional tests. Asynchronous execution of tests is also possible to test real-time behavior. The components integrated in this simulation are all the software modules, up to the flight controller and the flight manager. In Fig. 3.12, a test simulation setup is shown that uses a simplified closed-loop model of the ARTIS system to simulate its behavior. It first performs simple obstacle-avoidance maneuvers along linear paths, and then in more complex terrain generated from measurement flights above urban terrain. The scenario test utilizes a virtual world model that a sensor simulation uses to detect obstacles during flight; then the software replans the path accordingly, thus avoiding the obstacles.

The complexity of the simulation can be adapted easily to an arbitrary level (e.g., from abstract scenarios with a single wall in an empty environment up to

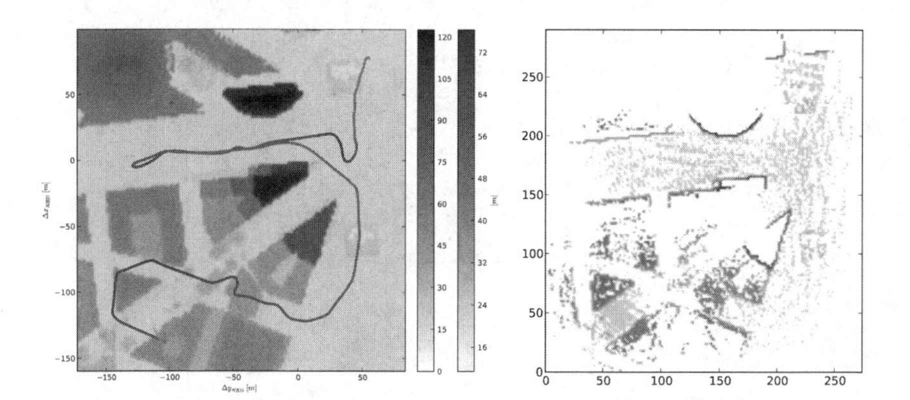

Fig. 3.11 Plot of the fly path in an urban environment, generated by a unit test case (left). Plot of the found obstacles (gray) along the corresponding explored path in unknown environment (right). Color intensity corresponds to height. Plots are being generated by the unit tests for reporting purposes and review by experts.

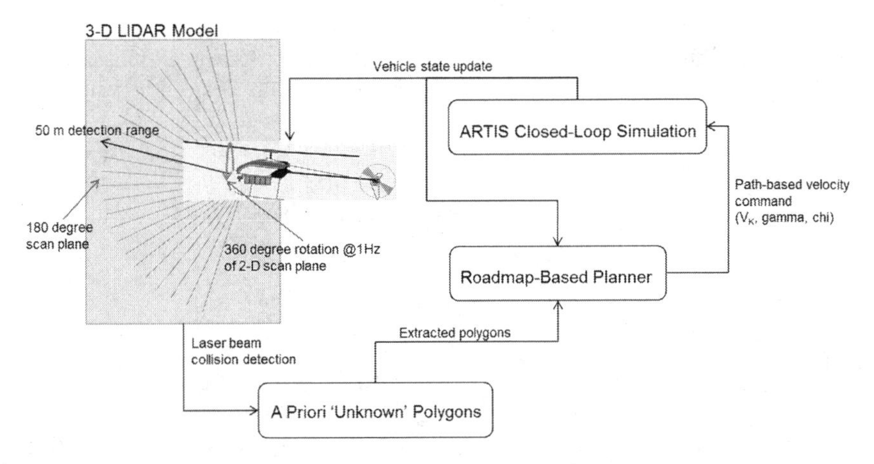

Fig. 3.12 Model of the system simulation for closed-loop flights in unknown terrain: The unknown environment is scanned by a virtual sensor. Explored obstacles are mapped into the world model of the planer. The planer can replan the flight path according to the explored obstacles.

complex urban environments with a dense population of heterogeneous obstacles). A simplified model for testing purposes is also used by NASA in [62]. These scenario tests are a powerful tool for the verification approach, as they can be automatically executed without human interaction. However, real-time tests may take a lot of time. As a result, the test suite executes 3.8 million assertions and 273 test cases, and it has a runtime of more than six hours. After the basic closed-loop system under test is assured with these simpler scenarios, we utilize the benchmark introduced in [62], to assess the obstacle field navigation performance with respect to safety (proximity to obstacles), smoothness (velocity derivatives and turn rates along the path), and performance (minimum time to target), seen in Fig. 3.11. This allows detection of whether any software change reduced the system performance or compromised its safety properties compared to previous revisions. The benchmarking approach is described in more detail in Section 3.6.6.

3.6.4.1 TEST DIMENSION EVALUATION

The SUT size for SIL tests is quite large. On this level, the complete software integration is tested, and basically the whole software or the main parts of the software are used to simulate scenarios. The test effort is greater compared to that of the unit tests because a scenario has to be created and simulated. Moreover, a simplified software system has to be implemented to perform the simulations. From a software perspective, the scenario complexity is considerable. Complete world models with complex environments can be simulated. Scenarios may be even easier to simulate than to be provided for a flight test in reality. But the code as well as state-space coverage is limited. Usually, it is not possible to do exhaustive testing using scenarios. These tests are utilized to test the software integration. The feedback time and test automation are very good because the tests have been automated and are actually executed as part of the unit tests. However, evaluation purposes, the expert still looks at generated outputs and reports to gain more insight into the details.

3.6.5 HARDWARE-IN-THE-LOOP

For the next test step, HIL tests are run on the actual flight hardware. The test setup embeds the target platform, actuators, and sensor fusion into the tests. On this level, system integration is tested, up to the actual hardware. The execution is done asynchronously to enable real-time behavior. The HIL tests can be used to assess the computing time and real-time performance. More details on the simulation framework, which is able to integrate different kinds of systems in various scenarios, can be found in [63].

3.6.5.1 TEST DIMENSION EVALUATION

Here, the SUT size is almost at maximum. In addition to SIL tests, the sensor fusion is also tested. Everything but the environment is tested, and therefore mainly the sensor inputs have to be simulated. Only flight tests utilize a bigger SUT. The test effort is much lower than on a flight test, but it is still quite high. A complete test laboratory has to be available to perform these tests. The scenario complexity and coverage are essentially the same as with SIL tests. For the simulated scenario itself, it is not important whether it is being done in SIL or HIL. As a result of the given test effort and automation, the feedback time is worse than for SIL tests but better than for flight tests. Test automation can be partially achieved because the scenarios can be executed automatically, but the software and hardware integration and the setup for sensor simulation has to be performed manually.

3.6.6 BENCHMARK

Similar to the concept of automatic tests is the concept of a benchmark to assess the score/goodness of an implementation. For example, for finding a path from a start point to a goal, a benchmark would score the length of the path, giving short paths a high score. This sort of benchmarking approach for the path planner is widely accepted. One score system is based on the well-known obstacle field navigation (OFN) benchmarks [62], shown in Fig. 3.13.

In addition to the length of the path, the distance to obstacles, the acceleration, velocity, turn rate, flight time, and computing time can be measured during test simulations. These values must not exceed or go below their respective threshold

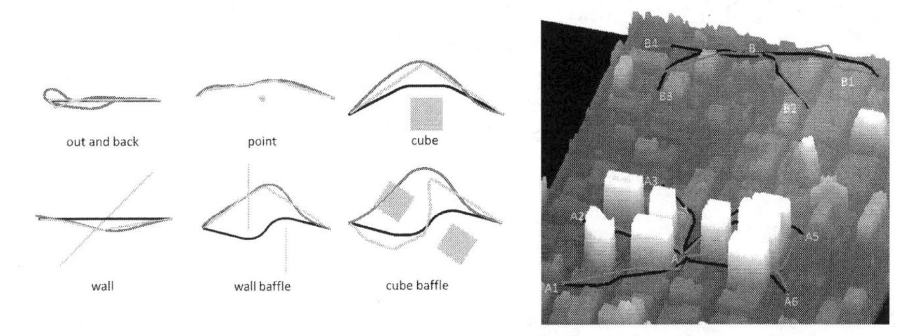

Fig. 3.13 Evaluation of MiPlEx in simple benchmark scenarios (left): Fly around a standardized obstacle (cf. Benchmarks [62]). The referenced baseline solution is shown in black, a simple linear path solution is shown in light gray, and the MiPlEx solution is in dark gray. Evaluation in an urban scenario (right): Fly from a start position towards a center position (either A or B) in an initially unknown location. Only baseline (dark) and MiPlEx solutions (light) are shown to get a clear illustration.

values. These values can also be used to assess the algorithm, for example by giving a score penalty for flying above a certain threshold if the mission is to fly at low altitudes. The aforementioned benchmark criteria are independent of the used planner approach. This approach facilitates automatic validation, for it scores the system and thus assures on a high level that the system does what it should do and assesses the result independently of the implementation. A software engineer can use this score to compare (and thus improve) the implementation. Further metrics have been developed to access *planner-specific* performance values. The most significant one is whether the planner always has the capability to escape into free space from a given position. Thus, we measure the number of collision-free connections of the UAS's node to any connected neighbor node (local visibility). Having a roadmap-based planner supports this measure, as no additional software infrastructure is necessary to measure this aspect of path-planning completeness. This is directly supported by the roadmap.

3.6.6.1 TESTING VERSUS BENCHMARKING

In general, testing and benchmarking are two separate but strongly related subjects. Testing gives a qualitative view of tests, simulations, and implementation, whereas benchmarking gives a quantitative view. However, creating benchmarks is not trivial and incorporates much expert knowledge. Therefore, when starting with a test, the benchmark may be a variation on the test and vice versa. A test can be deferred from a benchmark by setting a threshold as a fail/pass criterion. On the other hand, a benchmark may be used for relative scoring when a fixed test limit cannot be determined.

3.6.6.2 TEST DIMENSION EVALUATION

Benchmarking can be utilized in different stages of the UAS development life cycle. Because of aforementioned strong similarities between testing and benchmarking, the test dimension evaluation for benchmarking will result in the same assessment as the evaluation for unit tests, SIL, HIL, and flight test, respectively. Therefore, the visualization in Fig. 3.8 does not contain separate entries for benchmarking.

3.6.7 UTILIZING FORMAL LANGUAGES FOR DATA EXCHANGE

Missions are defined as sequences of parameterized behavior commands. A formal extended Backus-Naur form (EBNF) grammar is used to formally define missions. This makes it possible to check whether a mission is (syntactically) valid. In addition to this, several consistency checks have been integrated to also check specific semantic aspects of the transmitted mission. Although it is not possible for the mission management to reason whether a mission plan makes "sense," it is possible to implement such plausibility checks to eliminate errors when creating missions and thus increase the trust that only valid missions are allowed to be executed at runtime. These checks are implemented using the EBNF grammar [64] shown in Fig. 3.14. The system simply ignores malformed behavior sequences that cannot be matched against it. EBNF grammars are formal grammars of logical production rules that comprise nonterminal symbols, similar to placeholders or variables, and terminal symbols, which are generated by the grammar. Grammars are widely used to implement compilers for programming languages, and, as such, a behavior sequence is a sequential program. The grammar implemented in this context defines production rules for each behavior

```
ID 20070510
TO -5
HV 0 0 -3 180
avoidance on
WT 10
HV 33.3 3.51 -7 28.9
HV 37.1415 1.63484 -10 90
avoidance off
WT 5
tracker on
HV 37.1415 48.688 -10 90
HV 37.1415 48.688 -10 180
HV 31.81 48.688 -10 180
HV 26.4785 1.63484 -10 180
HV 26.4785 1.63484 -10 90
HV 26.4785 48.688 -10 90
HV 26.4785 48.688 -10 180
HV 21.147 48.688 -10 180
tracker off
LD
WO
```

```
<mission>      ::= "ID" <numbers> <takeoff> [<waypointlist>] <land>;
<numbers_all>  ::= {<nonzero> | "0"};
<nonzero>      ::= "1"|"2"|"3"|"4"|"5"|"6"|"7"|"8"|"9";
<float>        ::= [ {<numbers_all>} "." ] <numbers_all>;
<numbers>      ::= <nonzero> {<nonzero> | "0"};
<waypointlist>::= { <reactive> | <trajectory> | <task> };
<reactive>     ::= {<takeoff> | <land> | <standby >};
<trajectory>   ::= {<hover> | <fastflight> | <pir> | <pir_flight>};
<takeoff>      ::= "TO" [ "-" <float> ];
<hover>        ::= <hover_to> | <hover_turn> | <hover_wait>;
<hover_to>     ::= "HV" ["-"]<float> ["-"]<float> ["-"]<float> ["-"]<float>;
<hover_turn>   ::= "HT" ["-"]<float> ["-"]<float>;
<hover_wait>   ::= "WT" [<number>];
<standby>      ::= "WO";
<fastflight>   ::= <fly_to> <fly_to> <fly_to> { <fly_to> };
<fly_to>       ::= "FT" ["-"]<float> ["-"]<float> ["-"]<float> ["-"]<float>;
<pir>          ::= "FI" ["-"]<float> ["-"]<float> ["-"]<float> ["-"]<float>;
<pir_flight>   ::= "FF" ["-"]<float> ["-"]<float> ["-"]<float> ["-"]<float>;
<task>         ::= <command_src> "on" <waypointlist> <command_src> "off";
<command_src>  ::= "tracker" | "avoidance";
<land>         ::= "LD";
```

Fig. 3.14 The mission (left) is an example for a behavior sequence that schedules a mission plan, including a pattern-tracking section and a reactive obstacle-avoidance section. The extended Backus-Naur form grammar (right) specifies the syntax for plausible behavior sequences.

command or task flag, such as a "hover to" or "tracker off." Its root is defined by "<mission>," which basically enforces every mission to start with a takeoff behavior and end with a land behavior. It implements some basic parameter checks, for instance for the "hover and wait" command where the waiting time must not be a negative number. Furthermore, the grammar allows a check for any necessary prerequisites for specific behaviors. For example, the trajectory planner for the forward flight needs at least three of its behavior commands.

3.6.7.1 TEST DIMENSION EVALUATION

Technically, there is a similarity between checking languages and runtime monitoring, which is described in the next section. As a result, the visualization in Fig. 3.8 does not contain separate entries for EBNF and runtime monitoring.

In the case of EBNF, the SUT size is small because the checks handle a single data item. Runtime monitoring in general may perform checks of the outputs of a subsystem, but it is also capable of joining and merging outputs of the whole aircraft. However, the test effort is considerably higher than for unit tests, as the integration of the monitoring approach into the system may require significant efforts. The scenario complexity is highest because runtime monitoring may be used in flight tests or even during productive use of the system. The coverage on the other hand is low because only the singular run is being verified. The feedback time is real time, but it can be utilized during different stages of the development cycle (e.g., unit tests, SIL, HIL, and flight test). Finally, the automation of runtime monitoring is excellent, as one of the main aspects of this methodology is the generation of automated monitors.

3.7 RUNTIME MONITORING FOR UNMANNED AIRCRAFT

Two of the main challenges for the development of a UAS may first seem to be independent. Initially, with complex systems, there is always the need to have powerful methodologies to debug and analyze the system. For the ARTIS framework, extensive data from all sensors and software modules are logged into files. This logging capability is an important feature for debugging. However, going manually through system log files to analyze system behavior can become a huge undertaking and is prone to errors. Analyzing more complex properties can become infeasible if data have to be derived and set in context to data history or data from a different data source. Therefore, an automation tool for finding, filtering, or tagging specific data with supplementary information can be a huge benefit for data analysis.

Second, future unmanned aircraft are expected to be highly autonomous. Besides functional capabilities, such as obstacle sensing and mission planning, another key issue of autonomy is the concept of health management. Health management enables the aircraft to assess its own capabilities and, in case of degradation, enables it to react in a robust fashion by triggering contingency procedures. The first part of such a health management concept is the monitoring of the system status, to enable a form of self-awareness. Starting from autonomy level 3 of the ALFURS framework, required capabilities include health diagnosis and detection of hardware and software faults. In short, to achieve the concept of autonomy, an awareness of the system itself and its internal states is necessary to cope with abnormal system states, degraded situations, and unforeseen environmental events.

Runtime monitoring is capable of both offline analysis of log files as well as the supervision of system states and violation of specified properties. The use of a formal description language for the specification of properties enables formal specifications and allows the automatic generation of monitors as well as reasoning about the specification itself. As a subsequent step, it would be further possible to trigger a mitigation action if the monitor assesses a system fault. The monitor could either initiate an action by itself, reset and reactivate the faulty system, or deactivate it and activate a backup system. The backup system could have reduced capabilities but would only be able to maintain or ensure an autonomous reaction. As a result, the monitoring of systems and subsystems is the key approach to designing robust autonomous systems. Fig. 3.15 shows our concept of a UAS with an integrated runtime monitoring component as an independent fail-safe device.

3.7.1 RUNTIME MONITORING USING TEMPORAL LOGIC

Runtime monitoring describes a collection of approaches to evaluate formal specifications on traces of systems to verify the correctness of the system. One

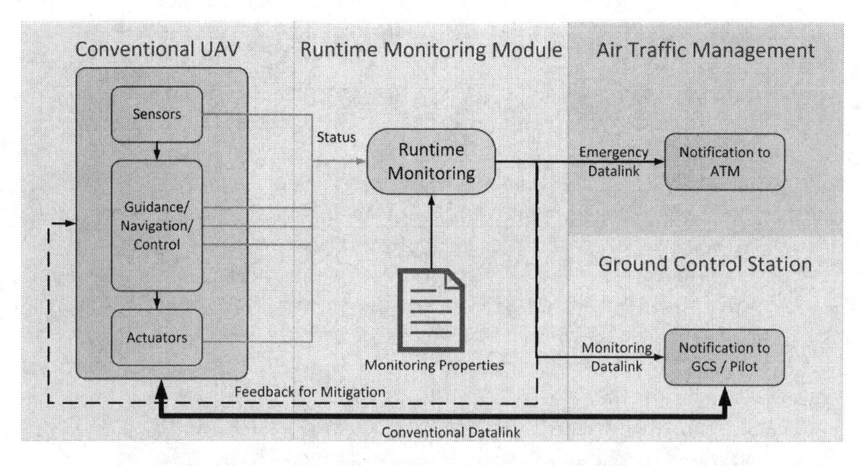

Fig. 3.15 Concept of a high-level module for runtime monitoring to assess the systems status and trigger warnings or mitigation actions.

important benefit of runtime monitoring is that it is possible to capture and formalize observed properties into temporal logic. This gives engineers a powerful tool for verification and debugging. Furthermore, it is possible to make assumptions of system design explicit; therefore, invalid assumptions can be found easily. Not every invalid assumption will cause a failure because it can be masked (dormant) until specific additional conditions are fulfilled. Without monitoring, such inconsistencies would not be noticed. Because these formal requirements can be synthesized into monitors, it is possible to directly verify the implementation against these properties.

Two main modes of operation are distinguished: *online monitoring* and *offline monitoring*. In online monitoring, the interface of a system is observed at runtime one event at a time, and the monitor produces a verdict with respect to the specification according to the trace of observations made up to that point. In contrast, for offline monitoring, we may assume that the trace is immediately fully available and may be traversed in either direction, which allows for more efficient algorithms. Runtime monitoring is a lightweight formal method compared to exhaustive verification methods such as model checking. Model checking analyzes all possible runs of a system and checks for conflicts with the given properties. Because of the state explosion problem of analyzing all possible runs, model checking may be infeasible for complex practical systems. Runtime monitoring does not have this problem because only one system run is analyzed by this methodology. Gradual application of runtime monitoring to existing systems is possible, as there is no need to first formally reengineer the system to a model. Specification languages for monitoring can be more expressive (e.g., types, functions, etc.) because we can evaluate them directly on the given trace. Also, it is possible that verdicts of the monitor can be used in a feedback loop to influence the system behavior itself. Correctness of monitoring algorithms and specification is easier to argue than correctness of system under test. Because the same methodology is used, a single specification can be used for both online and offline monitoring. Additionally, the formal specifications can potentially later be used for the model checking of components, and statistical analyses can be used to evaluate different runs of the system. Model checking uses abstractions of the system; this may result in additional problems because the abstraction might not be correct. With runtime monitoring, it is possible to close this verification gap between the model and the system because assumptions that have been made about the system can be validated at runtime, as seen in Fig. 3.16.

3.7.2 MONITORING FOR AUTONOMY

The same information that can be given to a pilot to increase situational awareness can also be given to the system itself. But beyond increasing the situational awareness of a pilot, a runtime monitor for supporting autonomy and intelligent behavior should be able to not only supervise specific properties, but additionally correlate such information to the known system state. Runtime monitoring is a suitable technique for such tasks because this new data of system inputs must be continuously assessed and correlated to the current system states, as well as its history. Furthermore, it is possible to intelligently change the system state according to specific inputs and thus cause it to react differently to the same situation. For example, when sensory inputs cannot be trusted as fully as they may be under optimal conditions, safety boundaries for velocity and obstacle distance could be increased. Or when an unusual rate of fuel consumption is registered, the mission plan could be optimized for minimal fuel consumption, the mission could be cut short, or a return to base could be initiated.

As stated earlier, the MiPlEx software component takes over active tasks that an onboard pilot should normally perform. To complement this, the runtime monitor takes over the supervisory tasks of the onboard pilot. As such, the runtime monitor not only supervises mere functionality but acts as an intelligent component that assures high-level decisions and actions do not lead to a catastrophic situation and are consistent with known environmental conditions. This functionality can be interpreted as self-awareness of the system.

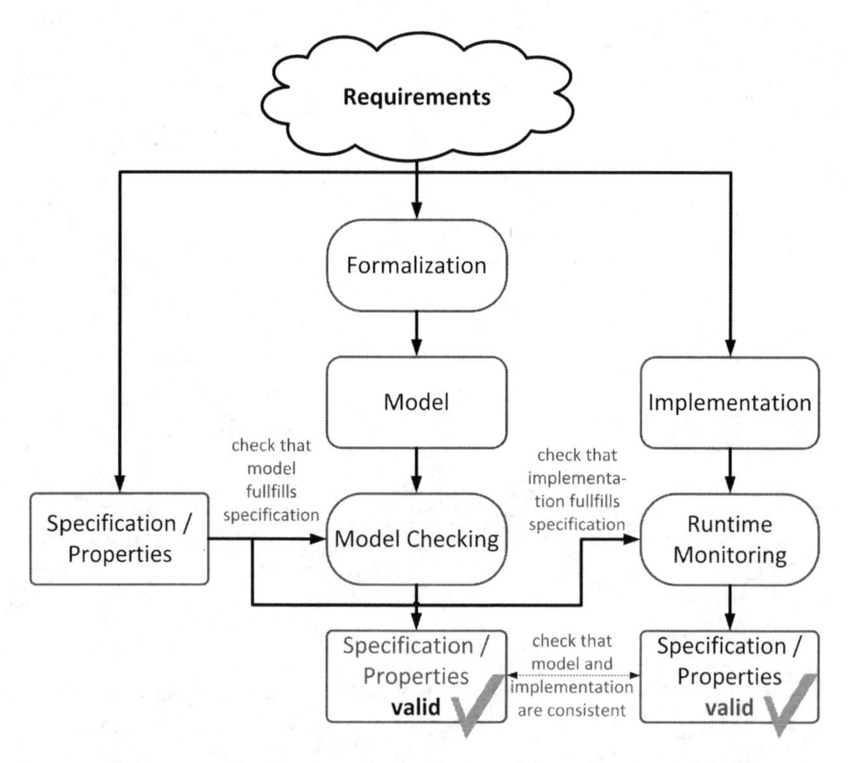

Fig. 3.16 Closing the verification gap, when using model checking of an abstract system model.

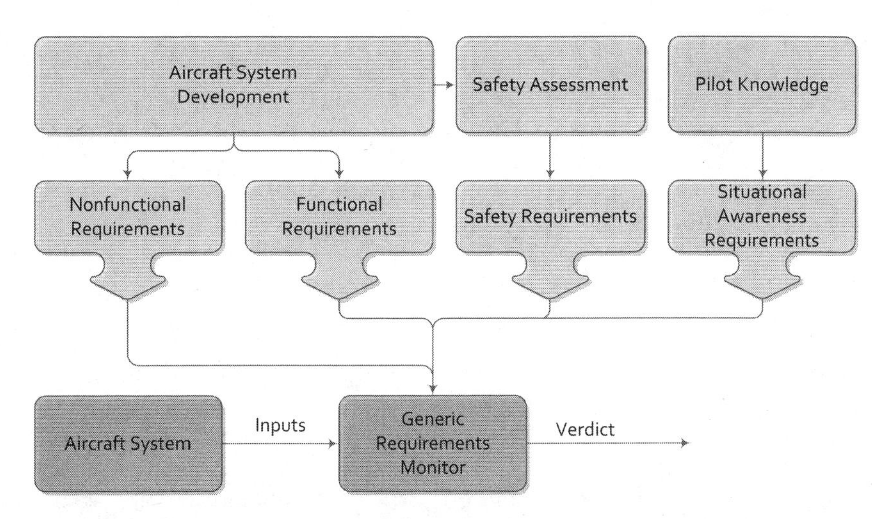

Fig. 3.17 Depiction of the different sources and scopes of inputs for monitoring requirements.

Various kinds of inputs can be used to develop monitoring properties, as shown in Fig. 3.17. Specifically, the goal is to supervise autonomy using the concept of runtime monitoring. In the context of our example capabilities detailed in this chapter, the approach for monitoring the path planning as well as its execution would be to monitor whether the ordering of achieved path waypoints corresponds to the planned sequence. In addition, certain limits of path derivation could be monitored. For the capability of semantic planning, monitoring could ensure that certain areas are not used for planning and that a safety distance to such areas is always maintained.

3.8 SUMMARY AND CONCLUSION

The ARTIS UAS research platform enables the development and test of autonomous software functions. We have used this platform to specify, model, verify, and demonstrate high-level capabilities.

In particular, the specification of autonomy is a challenge. This work proposed the use of generic requirements for autonomous UASs. These generic requirements were based on an established framework of autonomy definitions with a scaled approach to describe capabilities using autonomy levels. As a showcase, high-level requirements for UASs were developed for autonomy levels zero to four. Additionally, parameterized requirements were discussed. As a next step, the formalization to LTL was shown. These high-level requirements could be used to develop a reference model for the ARTIS mission manager, utilizing autonomy capabilities of levels zero to four. Finally, model checking was used to verify the consistency of the requirements, improving both the quality of these requirements as well as the model.

Furthermore, we detailed our best practice approach for the verification of our ARTIS software framework. The utilized verification methodologies are model checking, static testing, unit testing, software-in-the-loop, hardware-in-the-loop, flight testing, EBNF checking, and runtime monitoring. These techniques were discussed and assessed for the ARTIS use case in a qualitative manner in regard to different test characteristics or test dimensions. It could be shown that these verification techniques complement each other and cover a broad spectrum of the test dimensions.

We specifically focused on the concept of runtime monitoring for both offline analysis/debugging and online runtime monitoring. Runtime monitoring is a lightweight formal method that can easily be integrated into existing development processes by analyzing existing log files for debugging purposes. For online runtime monitoring, we have discussed the capabilities and applications toward safe operation of UASs. However, online integration of runtime monitoring requires much effort for the integration of this methodology into the system. Once this integration is achieved, runtime monitoring gives powerful possibilities to advance system awareness, health management, and intelligent behavior. The online runtime monitor supervises high-level behavior, safety, performance, and situational awareness. Therefore, runtime monitoring is a key aspect to increasing autonomy and making software-intense systems more robust and fail safe.

REFERENCES

[1] Radio Technical Commission for Aeronautics, "DO-178C/ED-12C Software Considerations in Airborne Systems and Equipment Certification," RTCA, Washington, D.C., 2011.

[2] Radio Technical Commission for Aeronautics, "DO-333/ED-216 Formal Methods Supplement to DO-178C and DO-278A," RTCA, Washington, D.C., 2011.

[3] Pnueli, A., "The Temporal Logic of Programs," Foundations of Computer Science, 1977, *18th Annual Symposium on, Oct. 1977*, Providence, Rhode Island, pp. 46–57.

[4] Davis, J. A., Clark, M., Cofer, D. D., Fifarek, A., Hinchman, J., Hoffman, J., Hulbert, B., Miller, S. P., and Wagner, L., "Study on the Barriers to the Industrial Adoption of Formal Methods," *Formal Methods for Industrial Critical Systems*, edited by C. Pecheur, and M. Dierkes, Vol. 8187 of *Lecture Notes in Computer Science*, Springer, Berlin, Heidelberg, 2013, pp. 63–77.

[5] Bowen, J., and Hinchey, M., "Ten Commandments of Formal Methods . . . Ten Years Later," *Computer*, Vol. 39, No. 1, Jan. 2006, pp. 40–48.

[6] Bowen, J., and Stavridou, V., "The Industrial Take-Up of Formal Methods in Safety-Critical and Other Areas: A Perspective," *FME '93: Industrial-Strength Formal*

Methods, edited by J. Woodcock, and P. Larsen, Vol. 670 of *Lecture Notes in Computer Science*, Springer Berlin Heidelberg, 1993, pp. 183–195.

[7] Barjaktarovic, M., and Nassiff, M., "The State-of-the-Art in Formal Methods," *AFOSR Summer Research Technical Report for Rome Research Site, Formal Methods Framework-Monthly Status Report, F30602-99-C-0166, WetStone Technologies*, 1998.

[8] Radio Technical Commission for Aeronautics, "DO-178B/ED-12B Software Considerations in Airborne Systems and Equipment Certification," RTCA, Washington, D.C., 1992.

[9] Dittrich, J., Bernatz, A., and Thielecke, F., "Intelligent Systems Research Using a Small Autonomous Rotorcraft Testbed," *Proceedings of the 2nd AIAA Unmanned Unlimited Systems, Technologies, and Operations Aerospace Conference*, San Diego, CA, Sept. 2003.

[10] Adolf, F., and Thielecke, F., "A Sequence Control System for Onboard Mission Management of an Unmanned Helicopter," *AIAA Infotech@Aerospace Conference*, Rohnert Park, California, 2007.

[11] Adolf, F.-M., Andert, F., Lorenz, S., Goormann, L., and Dittrich, J., "An Unmanned Helicopter for Autonomous Flights in Urban Terrain," *Advances in Robotics Research: Theory, Implementation, Application*, edited by T. Kröger, and F. Wahl, Vol. 9, Springer Berlin/Heidelberg, June 2009, pp. 275–285.

[12] Torens, C., and Adolf, F.-M., "Automated Verification and Validation of an Onboard Mission Planning and Execution System for UAVs," *AIAA Infotech@Aerospace Conference*, Boston, MA, Aug. 19–22, 2013.

[13] Torens, C., and Adolf, F., "Software Verification Considerations for the ARTIS Unmanned Rotorcraft," *51st AIAA Aerospace Sciences Meeting including the New Horizons Forum and Aerospace Exposition*, Jan. 2013.

[14] Hayhurst, K., Maddalon, J., Miner, P., Szatkowski, G., Ulrey, M., DeWalt, M., and Spitzer, C., "Preliminary Considerations for Classifying Hazards of Unmanned Aircraft Systems," NASA/TM-2007-214539, 2007.

[15] Bhansali, P. V., "The MCDC Paradox," *SIGSOFT Software Engineering Notes*, Vol. 32, No. 3, May 2007, pp. 1–4.

[16] Hoffman, J., "V&V of Autonomy: UxV Challenge Problem (UCP)," *Safe and Secure Systems and Software Symposium (S5)*, Dayton, Ohio, 2016.

[17] Rozier, K. Y., "Specification: The Biggest Bottleneck in Formal Methods and Autonomy," *Proceedings of the 8th Working Conference on Verified Software: Theories, Tools, and Experiments (VSTTE)*, Volume 9971 of *Lecture Notes in Computer Science*, Springer-Verlag, Toronto, Canada, July, 2016, pp. 1–19.

[18] Miller, S. P., Tribble, A. C., Whalen, M. W., and Heimdahl, M. P. E., "Proving the Shalls: Early Validation of Requirements Through Formal Methods," *International Journal on Software Tools for Technology Transfer*, Vol. 8, No. 4, 2006, pp. 303–319.

[19] Mehlitz, P., "Trust Your Model-Verifying Aerospace System Models with Java Pathfinder," *IEEE Aerospace Conference, Big Sky*, MT, March 1–8, 2008.

[20] Sirigineedi, G., "Towards Verifiable Approach to Mission Planning for Multiple UAVs," *AIAA Infotech@Aerospace Conference and AIAA Unmanned...Unlimited Conference AIAA 2009-1853*, April 6–9, 2009, Seattle, Washington, 2009.

[21] Sirigineedi, G., Tsourdos, A., Zbikowski, R., and White, B. A., "Modelling and Verification of Multiple UAV Mission Using SMV," *Electronic Proceedings in Theoretical Computer Science*, Vol. 20, March 2010, pp. 22–33.

[22] Platzer, A., and Quesel, J.-D., "KeYmaera: A Hybrid Theorem Prover for Hybrid Systems (System Description)," *Automated Reasoning*, edited by A. Armando, P. Baumgartner, and G. Dowek, Vol. 5195 of *Lecture Notes in Computer Science*, Springer Berlin Heidelberg, 2008, pp. 171–178.

[23] Platzer, A., "Verification of Cyberphysical Transportation Systems," *Intelligent Systems, IEEE*, Vol. 24, No. 4, July 2009, pp. 10–13.

[24] Dennis, L. A., Fisher, M., Lincoln, N. K., Lisitsa, A., and Veres, S. M., "Practical Verification of Decision-Making in Agent-Based Autonomous Systems," *ArXiv e-prints*, Oct. 2013.

[25] Webster, M., Fisher, M., Cameron, N., and Jump, M., "Model Checking and the Certification of Autonomous Unmanned Aircraft Systems," *Computer Safety, Reliability, and Security*, edited by F. Flammini, S. Bologna, and V. Vittorini, Vol. 6894 of *Lecture Notes in Computer Science*, Springer, Berlin, Heidelberg, 2011.

[26] Webster, M., Fisher, M., Cameron, N., and Jump, M., "Formal Methods for the Certification of Autonomous Unmanned Aircraft Systems," *Computer Safety, Reliability and Security*, edited by F. Flammini, S. Bologna, and V. Vittorini, Vol. 6894 of *Lecture Notes in Computer Science*, Springer, Berlin, Heidelberg, 2011, pp. 228–242.

[27] Whalen, M., Cofer, D., Miller, S., Krogh, B. H., and Storm, W., "Integration of Formal Analysis into a Model-Based Software Development Process," *Formal Methods for Industrial Critical Systems*, edited by S. Leue, and P. Merino, Volume 4916 of *Lecture Notes in Computer Science*, Springer, Berlin, Heidelberg, 2008, pp. 68–84.

[28] Habli, I., and Kelly, T., "A Generic Goal-Based Certification Argument for the Justification of Formal Analysis," *Electronic Notes in Theoretical Computer Science*, Vol. 238, No. 4, 2009, pp. 27–39.

[29] Brown, D., Delseny, H., Hayhurst, K., and Wiels, V., "Guidance for Using Formal Methods in a Certification Context," *Proc. Embedded Real Time Software and Systems*, Toulouse, France, 2010.

[30] Moy, Y., Ledinot, E., Delseny, H., Wiels, V., and Monate, B., "Testing or Formal Verification: DO-178C Alternatives and Industrial Experience," *Software, IEEE*, Vol. 30, No. 3, 2013, pp. 50–57.

[31] Pires, A. F., Polacsek, T., Wiels, V., and Duprat, S., "Use of Formal Methods in Embedded Software Development: Stakes, Constraints and Proposal," *Embedded Real Time Software and Systems*, 2014.

[32] Webster, M., Cameron, N., Fisher, M., and Jump, M., "Generating Certification Evidence for Autonomous Unmanned Aircraft Using Model Checking and Simulation," *Journal of Aerospace Information Systems*, Vol. 11, No. 5, 2014, pp. 258–279.

[33] Cofer, D., and Miller, S. P., "Formal Methods Case Studies for DO-333," NASA/CR-2014-218244, NF1676L-18435, 2014.

[34] Rosenblum, D. S., "A Practical Approach to Programming with Assertions," *IEEE Transactions on Software Engineering*, Vol. 21, No. 1, Jan. 1995, pp. 19–31.

[35] Leucker, M., and Schallhart, C., "A Brief Account of Runtime Verification," *The Journal of Logic and Algebraic Programming*, Vol. 78, No. 5, 2009, pp. 293–303.

[36] Reinbacher, T., Rozier, K. Y., and Schumann, J., "Temporal-Logic Based Runtime Observer Pairs for System Health Management of Real-Time Systems," *Tools and*

Algorithms for the Construction and Analysis of Systems, Springer, Berlin, 2014, pp. 357–372.

[37] Pike, L., Goodloe, A., Morisset, R., and Niller, S., "Copilot: A Hard Real-Time Runtime Monitor," *Runtime Verification*, edited by H. Barringer, Y. Falcone, B. Finkbeiner, K. Havelund, I. Lee, G. Pace, G. Roşu, O. Sokolsky, and N. Tillmann, Vol. 6418 of *Lecture Notes in Computer Science*, Springer, Berlin, Heidelberg, 2010, pp. 345–359.

[38] Groce, A., Havelund, K., Holzmann, G., Joshi, R., and Xu, R.-G., "Establishing Flight Software Reliability: Testing, Model Checking, Constraint-Solving, Monitoring and Learning," *Annals of Mathematics and Artificial Intelligence*, Vol. 70, No. 4, 2014, pp. 315–349.

[39] Kendoul, F., "Survey of Advances in Guidance, Navigation, and Control of Unmanned Rotorcraft Systems," *Journal of Field Robotics*, Vol. 29, No. 2, 2012, pp. 315–378.

[40] European Aviation Safety Agency, "Concept of Operations for Drones, A Risk Based Approach to Regulation of Unmanned Aircraft," 2015.

[41] ASTM Standard Committee F38.01 Workgroup WK53403, "Standard Practice for Methods to Safely Bound Flight Behavior of Unmanned Aircraft Systems Containing Complex Functions," 2016.

[42] Huang, H.-M., Messina, E., and Albus, J., "Autonomy Levels for Unmanned Systems (ALFUS) Framework-Volume II: Framework Models Version 1.0," NIST Special Publication 1011-II-1.0, National Institute of Standards and Technology, 2007.

[43] Huang, H.-M., "Autonomy Levels for Unmanned Systems (ALFUS) Framework-Volume I: Terminology," NIST Special Publication 1011-I-2.0, National Institute of Standards and Technology, Oct. 2008.

[44] Torens, C., Adolf, F.-M., Patil, G., and Vernekar, G. K., "Towards Generic Requirements, Models for Automated Mission Tasks with RPAS," *AIAA Infotech @ Aerospace, AIAA SciTech Forum. 3rd Software Challenges in Aerospace Workshop/ SciTech*, Jan. 4–8, 2016, San Diego, CA, 2016.

[45] Torens, C., and Adolf, F., "Using Formal Requirements and Model-Checking for Verification and Validation of an Unmanned Rotorcraft," *American Institute of Aeronautics and Astronautics, AIAA Infotech @ Aerospace, AIAA SciTech*, 05–09 January 2015.

[46] Kendoul, F., "Towards a Unified Framework for UAS Autonomy and Technology Readiness Assessment (ATRA)," *Autonomous Control Systems and Vehicles*, edited by K. Nonami, M. Kartidjo, K.-J. Yoon, and A. Budiyono, *Intelligent Systems, Control and Automation: Science and Engineering*, Vol. 65, Springer Japan, 2013, pp. 55–71.

[47] Pohl, K., *Requirements Engineering: Fundamentals, Principles, and Techniques*, Springer, 2010.

[48] Cimatti, A., Clarke, E., Giunchiglia, E., Giunchiglia, F., Pistore, M., Roveri, M., Sebastiani, R., and Tacchella, A., "Nusmv 2: An Opensource Tool for Symbolic Model Checking," *Computer Aided Verification*, Springer, 2002, pp. 359–364.

[49] Cavada, R., Cimatti, A., Dorigatti, M., Griggio, A., Mariotti, A., Micheli, A., Mover, S., Roveri, M., and Tonetta, S., "The nuXmv Symbolic Model Checker," *CAV*, 2014.

[50] Torens, C., Adolf, F.-M., and Goormann, L., "Certification and Software Verification Considerations for Autonomous Unmanned Aircraft," *Journal of Aerospace Information Systems*, Vol. 11, No. 10, 2014, pp. 649–664.

[51] The Engineering Society for Advancing Mobility Land Sea Air and Space, "4754A Guidelines for Development of Civil Aircraft and Systems," 2010.

[52] The Engineering Society for Advancing Mobility Land Sea Air and Space, "ARP4761 Guidelines and Methods for Conducting the Safety Assessment Process on Civil Airborne Systems and Equipment," 1996.

[53] Gross, K., Fifarek, A., and Hoffman, J., "Incremental Formal Methods Based Design Approach Demonstrated on a Coupled Tanks Control System," *HASE, IEEE*, 2016, pp. 181–188.

[54] Hoffman, J., "Utilizing Assume Guarantee Contracts to Construct Verifiable Simulink Model Blocks," *Safe and Secure Systems and Software Symposium (S5)*, 2015.

[55] Menzies, T., and Pecheur, C., "Verification and Validation and Artificial Intelligence," *Advances in Computers*, Vol. 65, 2005, pp. 153–201.

[56] *Boost Unit Test Suite*, Aug. 2012, http://www.boost.org/doc/libs/1_50_0/libs/test/, [last accessed 24 June 2018].

[57] "CCCC - C and C++ Code Counter," http://cccc.sourceforge.net/, [last accessed 24 June 2018].

[58] Marjamäki, D., "Cppcheck - A Tool for Static C/C++ Code Analysis," http://cppcheck.sourceforge.net/, [last accessed 24 June 2018].

[59] Google, "google-styleguide-Style Guides for Google-Originated Open-Source Projects," https://code.google.com/p/google-styleguide/, [last accessed 24 June 2018].

[60] van Heesch, D., "Doxygen—Generate Documentation from Source Code," http://www.doxygen.org/, [last accessed 24 June 2018].

[61] Gundy-Burlet, K., "Validation and Verification of LADEE Models and Software," *51st AIAA Aerospace Sciences Meeting including the New Horizons Forum and Aerospace Exposition*, American Institute of Aeronautics and Astronautics, Jan. 2013.

[62] Mettler, B., Kong, Z., Goerzen, C., and Whalley, M., "Benchmarking of Obstacle Field Navigation Algorithms for Autonomous Helicopters," *Journal of Intelligent and Robotic Systems*, Vol. 57, No. 1–4, 2010, pp. 65–100.

[63] Dauer, J. C., and Lorenz, S., "Modular Simulation Framework for Unmanned Aircraft Systems," *AIAA Modeling and Simulation Technologies Conference*, Boston, MA, Aug. 19–22, 2013.

[64] International Organization for Standardization, ISO/IEC 14977, "Information Technology Syntactic Metalanguage Extended BNF," 2001.

CHAPTER 4

Artificial Immune System for Comprehensive and Integrated Aircraft Abnormal Conditions Management

Mario G. Perhinschi
West Virginia University

Hever Moncayo
Embry-Riddle Aeronautical University

4.1 INTRODUCTION

Failures, malfunctions, and damage affecting aircraft subsystems, as well as general environmental and dynamic upset conditions, have been consistently identified as the primary sources or aggravating circumstances of the majority of aviation accidents and incidents [1–3]. It is important to properly address safety under normal and abnormal operational conditions throughout the entire life cycle of aerospace systems, including design, production, maintenance, and operation [4], within a thoroughly conducted aircraft health management process [5–8]. Toward this objective, a new computational paradigm, mimicking the biological immune system, has been extended and implemented for aerospace applications in recent years. The formulation of an immunity-inspired framework for comprehensive and integrated system monitoring and control under normal and abnormal operation, specific methods and algorithms, and example implementations are presented in this chapter.

Increasing safety of aircraft operation has been identified as a critical objective [9–12] and significant research efforts have been recently focused on developing sophisticated onboard hardware and software aimed at avoiding unrecoverable flight conditions, maintaining control, and continuing the mission in the presence of abnormal or upset conditions [13]. Although a variety of promising solutions to these issues have been constantly investigated and demonstrated throughout the years [14, 15], the dominant approach was to isolate individual causes and effects and solve specifically and narrowly defined problems. Many of these methods have reached maturity for successful practical implementation; however, when exposed to the overall complexity and multidimensionality of the aircraft health management problem, they do not reach on their own the desired level of generality and effectiveness.

The search for new technologies aimed at ensuring safe mission completion at post-failure conditions has primarily focused on the development of fault-tolerant adaptive control laws [16-19]. Although promising results have been obtained in addressing specific subsystem failures, such as actuator malfunctions or wing damage, a more general applicability and reliability is still to be achieved. A similar challenge is faced by the development of failure detection and identification schemes. Aircraft actuator malfunctions have been widely approached using various state estimation or observer-based schemes [20–22], system models [23, 24], artificial neural networks [25–27], or adaptive thresholds [28]. Although aircraft sensor multiple redundancy is common practice, sensor failure detection and identification schemes have been proposed, beyond typical voting approaches, based on neural network estimation of sensor outputs [29], analytical [30] and fuzzy logic models [31], or Kalman filtering of aircraft dynamics [32]. The use of large networks of sensors has been investigated for aircraft structural health monitoring [33, 34] in conjunction with data fusion algorithms and model-based prediction [35]. Indirect assessment of damage to main structural components via parameter identification was used to develop fault tolerant control laws [36]. Available methodologies have been extensively investigated for health monitoring of aircraft propulsion subsystems [37], including a variety of model-based approaches [38–40] and data-driven approaches [41–43]. Although the need for considering the pilot as a critical component of the pilot + aircraft system is more and more widely acknowledged, especially due to the progress in adaptive/intelligent control systems, the potential of monitoring and detecting in-flight pilot-related abnormal conditions has received limited attention [44]. Evaluating the type, severity, and dynamic implications of subsystem abnormal conditions has not yet been addressed systematically. Some detection and identification approaches implicitly handle some of its aspects. Efforts to predict and protect flight envelope under subsystem damage or failure conditions have led to promising solutions to specific problems [45–51]. However, they are generally targeting limited and highly constrained scenarios and have yet to be completely integrated with all phases of the process, such as detection, identification, and accommodation.

The need for a holistic methodology that can address the problem of ensuring safety of aircraft operation throughout and outside the design flight envelope with desirable robustness and survivability capability has been recognized [52–57]. Such a methodology is expected to handle the complexity of the problem in a highly effective, integrated, and comprehensive manner. The high potential of data-driven approaches has been emphasized [58, 59]. A new strategy was formulated [52] involving three main components. First, adequate characterization of the dynamic fingerprint of off-nominal conditions must be accomplished. Second, integrated onboard systems must be developed that can predict, detect, identify, evaluate, and counteract off-nominal conditions. Finally, these technologies must rely on an epistemological background solid enough to allow for the level of predictability and reliability required by current and future certification processes. Considering the multitude of factors (aircraft subsystem abnormalities, external hazards, pilot abnormal conditions, aircraft upset conditions, and the list may continue), their variability, versatility, and uncertainty, the resulting multidimensionality is enormous. Therefore, the development of a holistic methodology providing a general and complete solution to the off-nominal or abnormal condition (AC) detection, identification, evaluation, and accommodation (ACDIEA) problem for aerospace systems requires suitable tools and strong theoretical background.

Research in biomedical sciences has revealed that the functionality of the biological immune system exhibits excellent robustness, adaptiveness, and cognition capability [60, 61]. Relying on a highly distributed structure and simple-in-principle yet powerful mechanisms, it can collect, store, process, and transfer an enormous amount of information within an extremely complex and multidimensional environment. All of these characteristics led to the emergence of the relatively new computational paradigm in artificial intelligence, the artificial immune system (AIS) [62–67]. Different aspects of immune system components have provided the source of inspiration for promising approaches in a variety of applications [66] including anomaly detection [68, 69], optimization [70, 71], data mining [72, 73], scheduling [74, 75], computer network protection [76], pattern recognition [77], cooperative control [78, 79], and adaptive control [80, 81].

The potential of the immunity metaphor for aerospace applications has been acknowledged quite early [82, 83] and pioneer steps in the area have addressed the detection of control surfaces failures [84], detection of wing and tail damage [85], structural monitoring [86, 87], product life health management [88], engine subsystem fault diagnosis [89], selecting and constructing air combat maneuvers [90], flight path generation [91, 92], adaptive control augmentation [93], and unmanned aerial system coordination [94]. In recent years, these efforts have been continued and extended toward the development of an all-inclusive immunity-based methodology for a complete and general solution of the aircraft ACDIEA problem [95–97].

This chapter presents the development of the immunity-based methodology for addressing aircraft ACDIEA. The chapter is structured as follows. The general outline of the immunity-inspired aircraft AC management (ACM) process is presented in Section 4.2. Some of the basic elements of biological immunity serving as source of inspiration are described in Section 4.3. The concepts within the AIS paradigm and how they can be used for a comprehensive and integrated solution to the general aircraft ACDIEA problem are presented in Section 4.4. Approaches for the generation of self and non-self, as critical components of the AIS, are introduced in Section 4.5. The four phases of the ACDIEA process (detection, identification, evaluation, and accommodation) are discussed in Sections 4.6 through 4.9, respectively.

4.2 AIRCRAFT ABNORMAL CONDITION MANAGEMENT

The extended aircraft system is understood to include the vehicle, all equipment, pilot and crew, environment, and any other relevant factor. Assume that the operation of the aircraft under normal conditions is represented by points in a hyperspace defined by relevant variables or features. These points correspond to simultaneously reachable values of the features. Within the AIS paradigm, the regions of the feature hyperspace under normal conditions will be referred to as "the self." All other points are simply unreachable or may potentially be reached under ACs and form the "non-self." Generally, ACs may include any situation that raises concerns relative to safety or performance, such as faults and failures of aircraft hardware, human pilot–related abnormal situations, conditions associated to upset dynamics, and severe environmental phenomena.

A holistic aircraft ACM process is envisioned as consisting of four major components: AC detection, identification, evaluation, and accommodation. From this perspective, performing aircraft ACM becomes an extremely challenging, complex, and multidimensional task. The AIS paradigm allows that all four processes be performed within a consistent framework gravitating around the AIS as a comprehensive information depository, which is constructed using measurements and/or simulation data and does not require sophisticated explicit models [97].

The *detection* process consists of acknowledging that an AC has occurred affecting one or several of the extended aircraft subsystems and components. The outcome of the detection process $\mathcal{O}_D$ is binary:

$$\mathcal{O}_D = \begin{cases} 0 & \text{operation under normal conditions} \\ 1 & \text{operation under AC} \end{cases} \tag{4.1}$$

The *identification or isolation* process is expected to determine which subsystem is directly affected by the AC. Depending on the structure and the complexity of the targeted system, it may be more effective to perform the AC identification in multiple subsequent phases. For instance, a first identification phase could discriminate between major subsystems such as actuators, sensors, structural components, or human pilot. If a structural damage is identified, a second phase could determine whether the damage affects the fuselage, the wing, or the horizontal tail. A next phase could further locate the damage, for example, along the span of the wing. The outcome of the identification process, $\mathcal{O}_I$, can be expressed as an N_S-dimensional vector with binary components, where N_S is the total number of subsystems considered:

$$\mathcal{O}_I = \begin{bmatrix} o_{I1} & o_{I2} & \dots & o_{IN_S} \end{bmatrix}, \quad o_{Ij} = \begin{cases} 0 & \text{for subsystem } j \text{ under normal conditions} \\ 1 & \text{for subsystem } j \text{ under AC} \end{cases} \tag{4.2}$$

Alternatively, $\mathcal{O}_I$ can only include the list of values j, for which $o_{Ij} = 1$. In general, the identification process is ulterior to or simultaneous with the AC detection. The possibly multiple identification phases may occur simultaneously or in succession.

The AC *evaluation* process is expected to assess the AC qualitatively and quantitatively, in a direct and indirect manner depending on how the evaluation outcomes characterize the AC itself or its consequences. Three major aspects must be addressed. One is of a direct qualitative nature and consists of establishing the type of the AC. For example, if a failure affecting the left elevator is detected and isolated, the qualitative evaluation is expected to determine if the aerodynamic control surface is locked, or freely moving, or constrained to move in a limited range, or its aerodynamic effectiveness is reduced. The other two major aspects of the evaluation process are quantitative in nature and can be characterized as direct and indirect. The outcome of the direct AC evaluation represents an estimate of the severity of the AC using properly defined metrics. For example, right elevator is locked at +12 deg, or left engine can only produce 70 percent of nominal thrust. The indirect AC evaluation is envisioned as a flight envelope reassessment process. It is expected to predict physical and/or recommended limits on relevant system states with consequences on aircraft performance and handling qualities. In general, the outcome of the evaluation process has three components corresponding to the three main aspects:

$$\mathcal{O}_E = \{o_{ET} \quad o_{ES} \quad o_{EFE}\} \tag{4.3}$$

The first component is categorical and refers to the type of the AC. It may be a scalar if only one type of failure affects the system or a vector if multiple ACs affect the same or different subsystems. The second component addresses the AC severity and is typically a numerical scalar or vector. However, a categorical metric for the severity of some AC may also be defined (e.g., "low," "medium," and "high"). The third component typically represents the set of new ranges for relevant system states that are affected by the detected AC. The number of these relevant states depends on the nature of the AC. The size of o_{EFE} depends on the number of relevant states and the number of detected ACs.

The AC accommodation can be achieved indirectly through warnings and information provided to the pilots to increase their situational awareness (passive accommodation) and directly by triggering compensation within the control laws (active accommodation). The passive accommodation consists of simply reprocessing the outcomes of the detection, identification, and evaluation processes and delivering them in an ergonomic way to the pilot. Within the AIS paradigm, the direct accommodation can be achieved based on three different concepts. One limited approach uses the outcomes from the detection, identification, and evaluation processes to trigger the activation of pre-existing control laws in a similar manner to the classical gain scheduling approach. Another approach uses immunity-inspired mechanisms to develop adaptive control laws. Finally, the information structuring capabilities of the AIS can be extended to extract directly from the self/non-self construction adequate control compensation in the presence of ACs.

The immunity-inspired aircraft ACM process consists of four main components functionally connected and closely interacting as illustrated in Fig. 4.1:

- ACM system design and implementation
- Online ACDIEA
- Post-processing of flight data and ACDIEA outcomes
- AIS updating

The objective of the offline ACM system design and implementation process is to develop an integrated and comprehensive ACDIEA scheme with high flexibility and adaptability. The following design elements must be clearly defined: extended aircraft subsystems that are targeted, the types of AC (including known and unknown ACs), the AC severity scale, the flight envelope relevant variables, and the nature and level of passive and active accommodation. A critical element of the immunity-based ACM is the generation and structuring of the self/non-self. For a comprehensive solution to the ACDIEA problem, large amounts of measurement data are necessary, at least under nominal conditions, over the entire operational envelope. It should be noted that all of these data do not have to be available ab initio and may be collected during regular operation of the system as part of the post-processing and updating processes. Alternative approaches would typically require the development of extensive accurate models and sophisticated design tools. It is a major advantage of the AIS approach

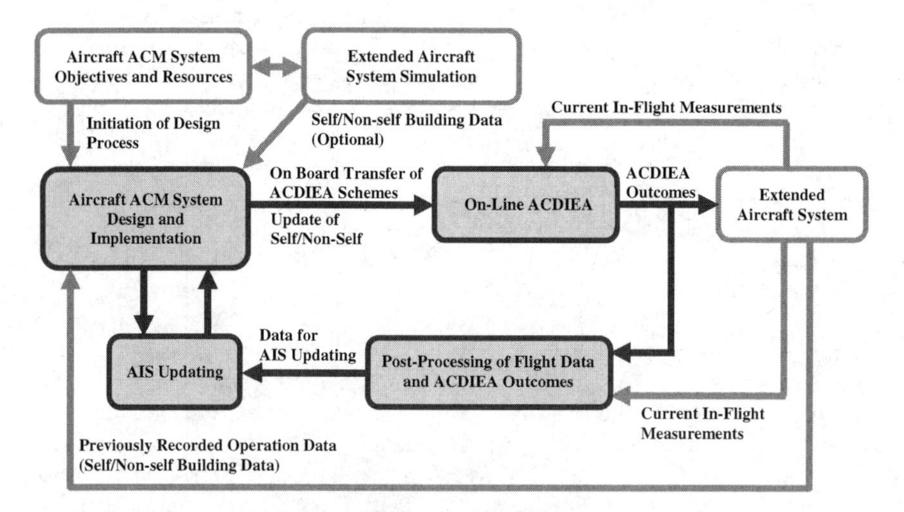

Fig. 4.1 AIS-based aircraft ACM process.

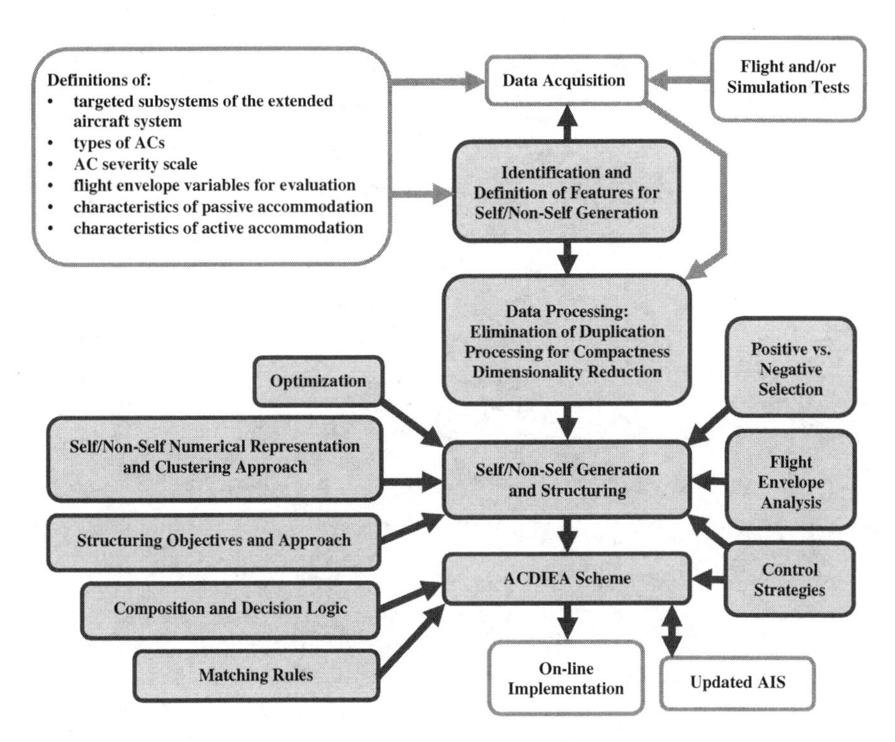

Fig. 4.2 AIS-based ACM system design and implementation.

that it relies only on acquiring and processing experimental data, which is considerably less difficult and expensive. The main elements and interactions within the ACM system design and implementation component are presented in Fig. 4.2. Further description and details are provided in following sections.

The online ACDIEA module represents the real-time operation of the ACDIEA scheme. Instantaneous measurements of features at a certain sampling rate are compared against the self/non-self representation, which acts as a data-driven de facto system model. The outcomes of the ACDIEA are produced through interactions with elements of the AIS, typically referred to as detectors, identifiers, evaluators, and compensators. These outcomes are transferred to the pilot, the onboard monitoring and recording system, and the automatic fault tolerant control laws. Note that the AIS framework allows for continuous updating of the ACDIEA scheme. The functionality of the online ACDIEA component is described in Fig. 4.3.

The post-processing of flight data and ACDIEA outcomes is expected to facilitate the updating of the ACDIEA scheme. Within this process, collected system data and scheme outcomes are analyzed to identify the need for two types of updating. One is referred to as evolutionary updating and it is triggered by modifications of the system due to aging, wear off, or retrofits. The other is referred to

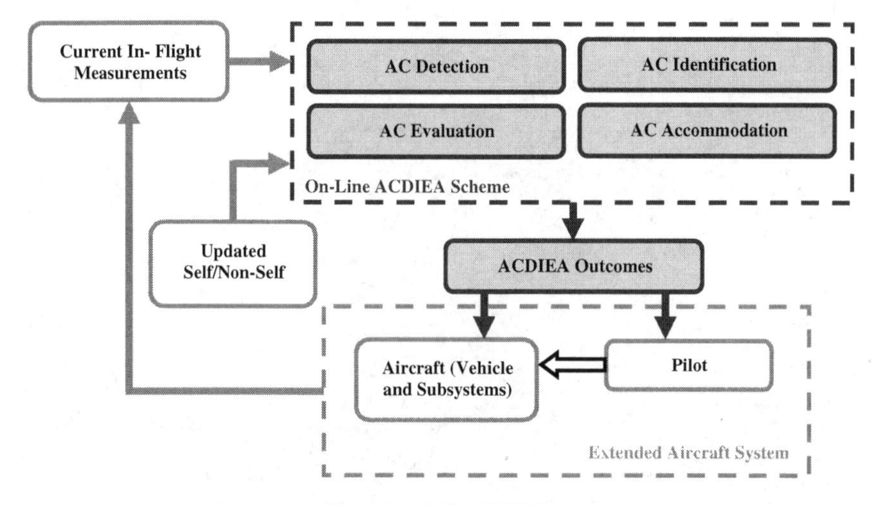

Fig. 4.3 Online ACDIEA.

as non-evolutionary updating and it is triggered by the exploration of new regions of system operational envelope and the collection of data that were not initially available for ACM system design. This category also includes adjustments of the self/non-self representation in areas where the outcomes of ACDIEA prove not to be adequate, presumably due to imperfections of self/non-self generation data and algorithms. Figure 4.4 presents the general diagram of the post-processing module.

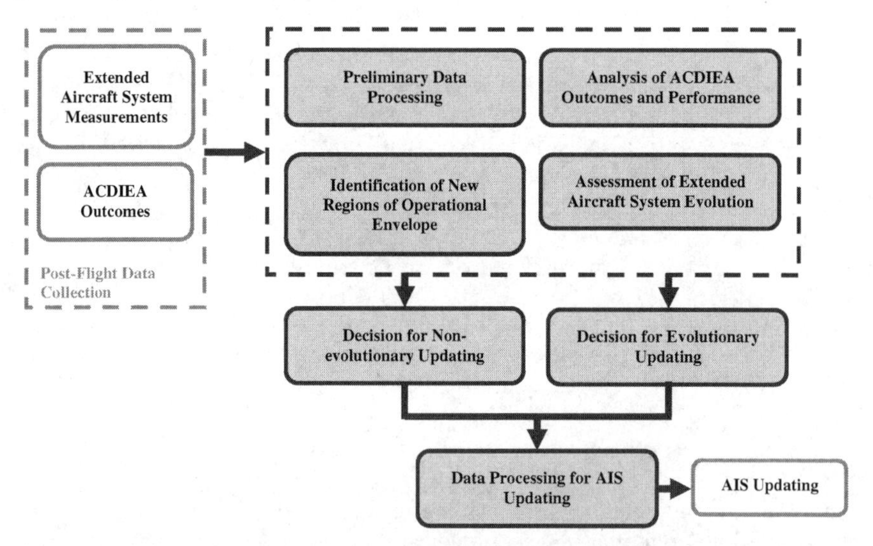

Fig. 4.4 Post-processing of flight data and ACDIEA outcomes.

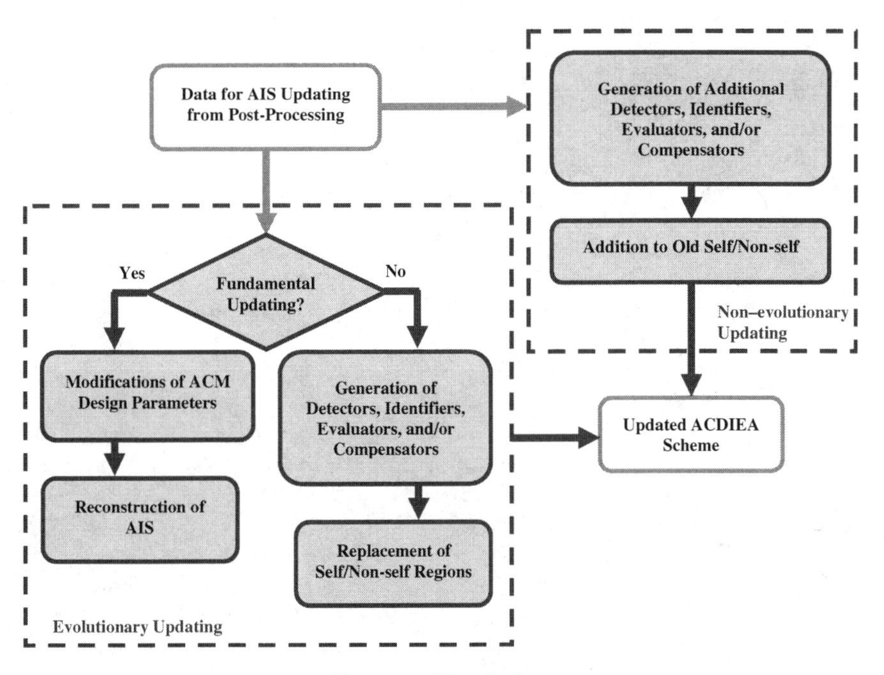

Fig. 4.5 AIS updating.

The AIS updating component uses newly acquired information to update the ACDIEA scheme. Non-evolutionary updating involves the generation of new AIS constituents and the extension of the old self representation into areas for which adequate data were missing. The evolutionary updating may involve modifications of ACM system design parameters, case in which a more fundamental reconstruction of the AIS is necessary, or may represent a simpler reassessment of self/non-self within the same general framework of ACM system design. The functionality of this module is presented in Fig. 4.5.

4.3 BIOLOGICAL IMMUNITY AS SOURCE OF INSPIRATION

The biological immune system is a rich source of inspiration that has shown promising potential for developing intelligent systems for a variety of applications [64–66]. It exhibits a highly sophisticated, complex, adaptive, and self-organized architecture and functionality for protecting the organisms from invading antigens (also called pathogens) such as bacteria, viruses, parasites, and any hazardous external substances that can prevent the organism from operating within nominal ranges in a balanced and stable manner. The immune system is characterized by powerful properties such as robustness, adaptability, memory, reactivity, and response specificity, which allow it to maintain or recover overall equilibrium and correct functionality of the living organism.

Invading antigens trigger specific immunity responses as a sophisticated line of defense consisting of different types of specialized organs, cells, organic compounds, and functional mechanisms. This biological line of defense includes two main components: the innate and the adaptive immune systems [61]. The first component is genetically inherited and responds to intruders immediately and in a rather indiscriminate manner. The second one is evolved through previous interactions with external intruders. It relies on current and old information acquired by the innate immune system to provide a secondary response that is specific, more effective, and more efficient.

Lymphocytes are specialized cells instrumental in producing antibodies, which are in charge of binding to antigens and marking them for destruction by phagocytes. Lymphocytes can be classified in two categories, B-cells and T-cells. These two types of cells are important for triggering the production of specific antibodies and maintaining the balance between antibodies and antigens. B-cells are produced by the bone marrow and are directly controlling the generation of antibodies. T-cells produced in the thymus gland are in charge of controlling the number of B-cells in the system. A dynamic balance in the production of antibodies with respect to the number and virulence of the antigens is achieved through the interaction between helper T-cells (T_h-cells), which accelerate cell generation, and suppressor T-cells (T_s-cells), which inhibit cell generation. Both of these sets of T-cells are typically referred to as effector T-cells to differentiate them from cytotoxic T-cells (T_c-cells)—which destroy the infected own cells—and memory T-cells—which store antigen information and can provide a faster and more aggressive response in future encounters with the same antigen [60]. It is also known that B-cells can be differentiated into memory cells and play a similar role. The proper amount of helper and suppressing T-cells is modulated by the activity of dendritic cells (DC) through the production of specific chemical markers: interleukin-12 (IL12), which favors the generation of T_h-cells, and interleukin-10 (IL10), which favors the generation of T_s-cells.

DCs are responsible for establishing a vital link between the innate and adaptive components of the immune systems. They are typically considered to be part of the innate immune system and are responsible for collecting antigen information and presenting it to other immunity constituents, thus playing a critical role in the development and activation of the adaptive immune response [98]. DCs are generated in the bone marrow through a negative selection process that allows them to only react to antigen and not to their own organism cells. When they detect pathogens, the DCs engulf them, break them down into their constituent chemical compounds, display these markers on their surface, and produce IL12. If during this maturation process the DC does not encounter antigens, IL10 is produced instead. Mature DCs migrate to lymph nodes where the balance between IL12 and IL10 triggers the generation of T-cells. The interaction between the innate and adaptive immune systems is illustrated in Fig. 4.6.

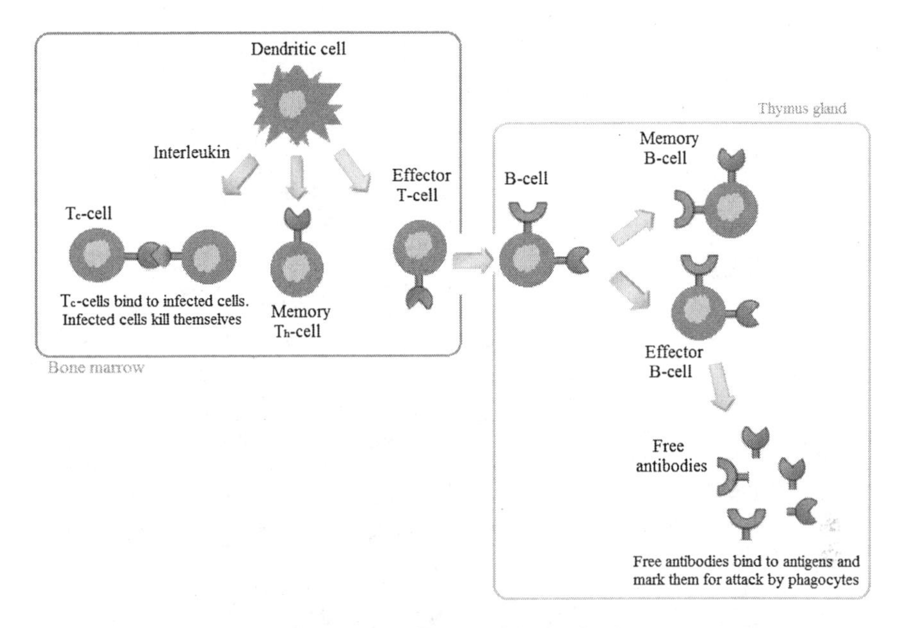

Fig. 4.6 Interaction between innate and adaptive immune systems.

Antibodies are characterized by their shaped molecular structure and the regions on the surface, called *paratope*, that host chemical markers compatible with antigens only. Antigens possess similar regions on their surface, called *epitope*, where their own specific chemical markers reside [60], as presented in Fig. 4.7. An antibody attaches to an antigen at the level of paratope/epitope, if chemical compatibility exists. If an antibody paratope matches with an antigen epitope, the antigen is classified by the immune system as fully recognized and an elimination process is started.

Due to its complexity and intricate functionality, the immune system is still only partially understood and continues to be a matter of intense study in biological and medical sciences. However, several mechanisms and processes have been identified as critical to the immune response and as valuable sources of inspiration for tackling a variety of system design and control problems.

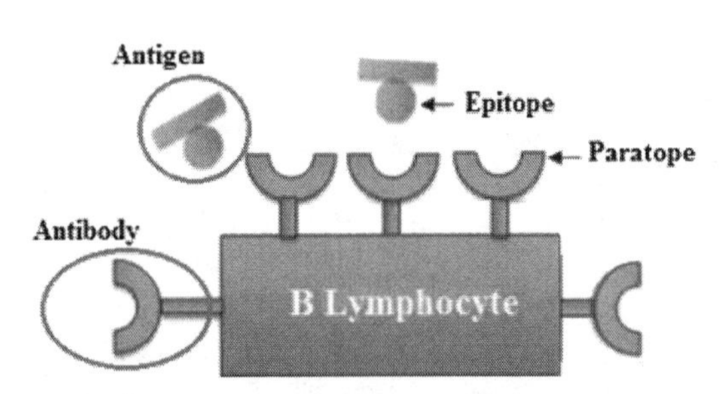

Fig. 4.7 Antigen discrimination based on paratope/epitope matching.

They include primarily negative and positive selection, humoral feedback, and clonal selection principle.

One of the main capabilities of the immune system is its ability to detect, bind to, and mark for destruction unknown exogenous entities (referred to as non-self), while not reacting to cells of the host organism (referred to as self). This is achieved by generating specialized immunity cells through a process called negative selection (NS) [60, 61] that allows the survival of those cells only that do not recognize self-patterns. However, the property of recognizing self-cells may also prove valuable for effective immune system operation. A similar process, positive selection (PS) [60, 61], may lead to the production of immunity cells that are capable of matching chemical markers of the self-cells. For example, to ensure high variability of the new T-cells in terms of biological features (typically proteins or polysaccharides), they are first generated through a pseudo-random genetic rearrangement mechanism. In the thymus, these cells are subjected to a two-phase process that involves both PS and NS mechanisms [99]. First, through a PS mechanism, those T-cells that attach to the self's major histo-compatibility complex molecules are allowed to live while the ones that do not are destroyed. Second, during an NS phase, T-cells that bind to both the self major histo-compatibility molecules and self-peptides are eliminated because they can potentially cause a detrimental autoimmune response. T-cells that do not bind to self-peptides are allowed to leave the thymus and are released through the living organism for immune functionality. Both concepts are illustrated in Fig. 4.8 [100], where the different chemical markers are represented by different geometrical shapes.

In the event of an infection, the immune system exhibits both a humoral and cellular type of response by producing specialized cells to detect and destroy antigens, while achieving a proper balance between the number and virulence of antigens on one hand and the number and effectiveness of antibodies on the other. This regulatory action drives an optimization process aimed at achieving maximum defense effectiveness with minimum resource expenditures. When an antigen invasion is first detected, helper T-cells stimulate B-cells to increase the production of antibodies. The information acquired by DCs is instrumental in establishing the proper balance between the number of antibodies and the

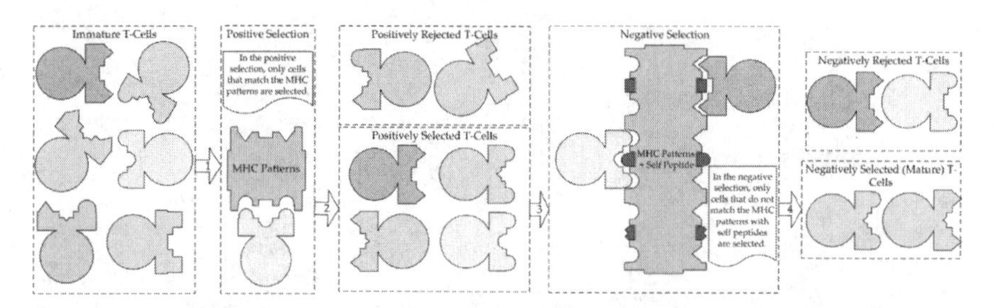

Fig. 4.8 Negative and positive selection mechanisms [100].

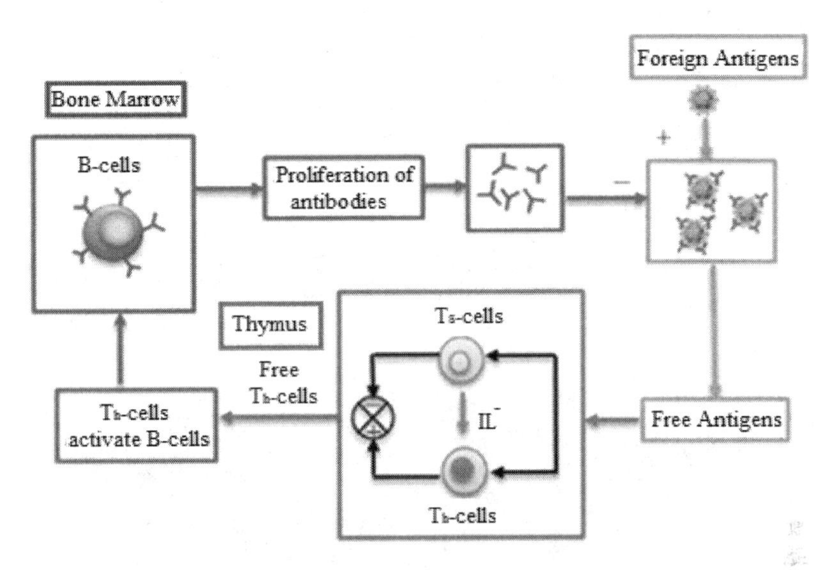

Fig. 4.9 Humoral feedback immune mechanism [101].

current level of threat, as they may evolve in time. This balance is achieved through the regulation of T_h-cells, which have a stimulating effect on antibodies production, and of T_s-cells, which have an inhibiting effect. The interaction between these cells allows a rapid immune response to antigen invasion, while ensuring optimal effectiveness and stability of this dynamic response [61]. The process is illustrated in Fig. 4.9 [101].

The production of specialized immunity cells, such as B-cells, through NS mechanisms is further enhanced by allowing with priority the proliferation of those cells that recognize particular antigens. In other words, the production of cells susceptible to matching non-self receives more stimulation once matching antigens are present. This *clonal selection* process takes place in phases and depends on the affinity or potential suitability of a candidate immune-specialized cell to match antigens. Selected specialized cell candidates are first cloned rapidly and the new cells are subjected to a high-rate mutation process. New lymphocytes that contain self-reactive receptors are eliminated and mature cells are induced to proliferate rapidly. Those specialized cells that are most suitable to match targeted antigens are stimulated more strongly to multiply [102]. The clone selection mechanism is illustrated in Fig. 4.10.

After an initial response to a specific antigen, the information related to the T-cells and B-cells produced during the encounter is stored in specialized cells that act as an immunological memory [103]. This capability to recognize and "remember" previously encountered pathogens is instrumental for long-term specific immunity that allows for a faster and more robust antibody generation and optimization in response to subsequent similar invasions.

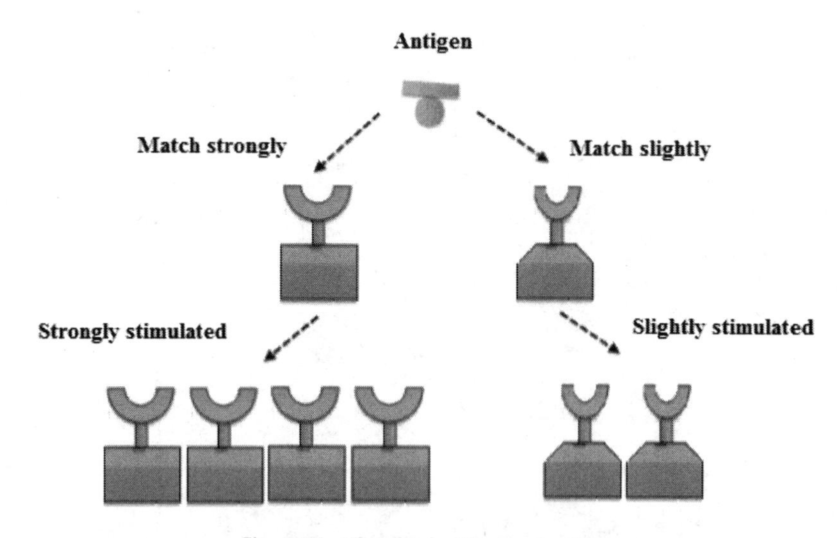

Fig. 4.10 Clonal selection mechanism.

4.4 THE ARTIFICIAL IMMUNE SYSTEM PARADIGM

In recent years, the biological immune system has inspired a number of new algorithms that have been implemented with promising results, primarily in the area of immunity-inspired fault detection [68, 69]. The general self/non-self discrimination principle is at the core of these approaches. The basic idea supporting the AIS paradigm for subsystem fault detection is that an AC can be declared when a current configuration of "features" does not match with any configuration from a predetermined set known to correspond to normal situations. Therefore, a failure affecting one of the aircraft subsystems is assimilated to an invasion by antigens, while the feature values are similar to the biological chemical markers, such as those located on antibody paratopes and represent the primary means for encoding the self/non-self. The set of features may include various sensor outputs, state estimates, filtered data, statistical parameters, or any other information expected to be relevant to the behavior of the system and able to capture the dynamic signature of abnormal situations, either known or unknown. Features can also represent computational results from other types of algorithms such as observer-based residuals, expert system outcomes, or artificial neural network evaluations, to give just a few examples [97, 104].

Extensive experimental data are necessary to determine the self or the feature hyperspace under normal conditions. This places the AIS paradigm among data-driven methodologies. Adequate numerical representations of the self/non-self must be used and the data processed such that they are manageable given the computational and storage limitations of the available hardware. If built properly, the self is sufficient for successful AC detection. The artificial counterpart of the antibodies—the detectors—can be generated and optimized. This process typically attempts to mimic the variation followed by selection of the T-cells and can be performed using either NS-type or PS-type algorithms. NS-type algorithms use the non-self region of the feature hyperspace to generate detectors, through clustering or partitioning. An AC is declared if the explored current feature point matches any of the detectors. If the PS strategy is adopted, the detectors are generated to coincide with the self and the process is equivalent to clustering or partitioning the self data. In this case, an abnormal situation is declared if the explored current configuration does not match any of the detectors. For detection purposes, NS and PS methods are equivalent; however, using PS within a detection scheme is typically more computationally intensive than using NS because it is necessary to test the complete set of positive antibodies before classifying a sample as abnormal. With the NS approach, the activation of a single negative antibody is enough to declare the presence of an abnormal situation.

For a comprehensive approach, the computational handling of the large feature hyperspace as a whole is impractical, if at all possible, in many instances. Significant computational issues related to distances, thresholds, and geometry of hyper-volumes occur when the dimensionality increases [105, 106]. Alternatively, lower order projections of the self can be used as part of the hierarchical multi-self (HMS) strategy [107]. These lower order projections or sub-selves may directly represent subsystems or components, thus providing valuable support for AC identification. In particular, using all 2D projections produced by the set of features allows intuitive analytical and graphical representations of distances, thresholds, and boundaries. All of the sub-selves produce their own self/non-self discrimination outcomes, which must be properly processed for global monitoring decisions. The functionality of DCs provides inspiration for an information fusion approach [108, 109] capable not only of reliable detection outcomes, but also AC identification and evaluation. The production of IL10 and IL12 by the biological DC is converted into a counting mechanism for processing discrimination outcomes from multiple self/non-self projections.

The specific sub-selves or projections that produce positive AC detection in conjunction with the location of the triggered detectors within the projection provide information regarding the affected subsystem or component, the AC type, and the AC severity. Direct extraction of this information can be achieved if the non-self is structured using data available under AC and processed through PS-type of mechanisms [110, 111]. This approach will attach specific labels with AC identification and evaluation information to the non-self detectors, thus converting them into identifiers and evaluators. AC severity may also be inferred based on the distance between the feature point and the closest self region. Alternatively, the artificial DC mechanism can be used to provide different feature patterns that are useful in identifying the subsystem directly affected by the AC and qualitatively evaluating it [109].

Typically, AC detection and identification must be successfully completed prior to AC evaluation. Methods used for AC identification can be extended to address AC qualitative and quantitative direct evaluation. However, the quantitative indirect AC evaluation necessitates customized approaches depending on the affected subsystem, the nature and type of AC, and the evaluation outcomes of interest. It also requires information from qualitative and quantitative direct evaluation phases. Customized algorithms for analyzing the effects of specific classes of ACs on the reduction of flight envelope have been developed and demonstrated successfully [112].

Passive accommodation simply relies on the capability of the pilot to handle the AC given relevant information from prior ACM processes. The capability of immunity cells to "memorize" past antigen encounters and accelerate subsequent responses can be mimicked to develop active AC accommodation control laws [113]. The humoral feedback response responsible for proper balance between activation and suppression of antibodies generation provides the supporting structure for the design of adaptive immunity-based control mechanisms [101].

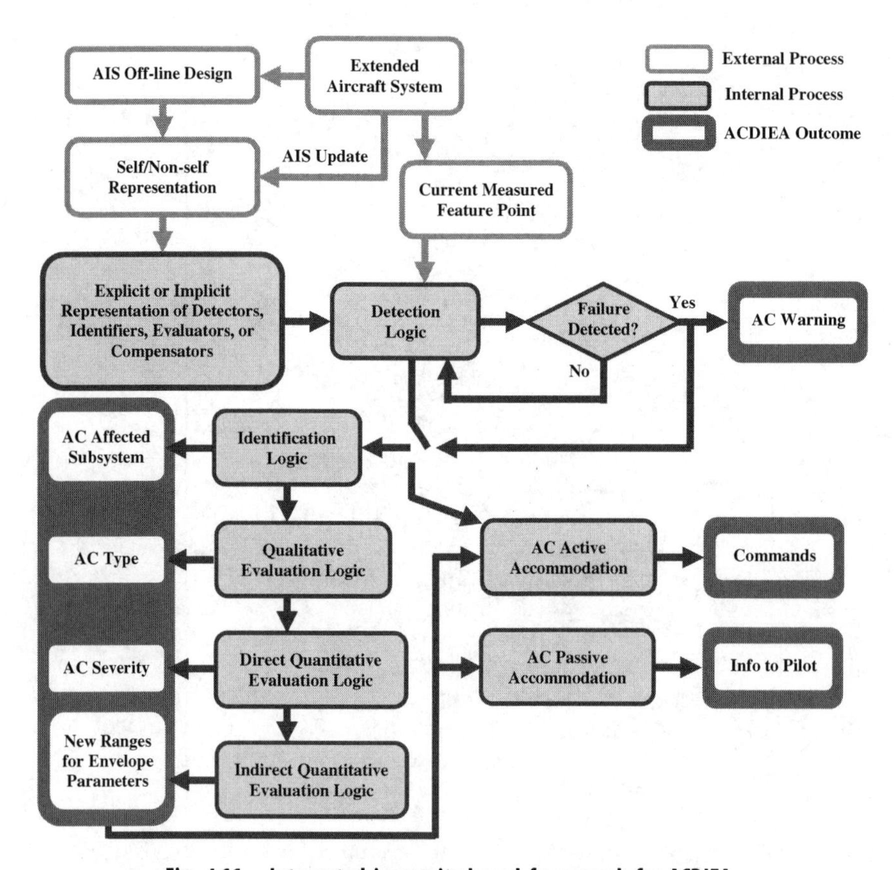

Fig. 4.11 Integrated immunity-based framework for ACDIEA.

The interaction of all of the processes just described, within an integrated immunity-based framework for ACDIEA, is presented in Fig. 4.11. A summary of main components of the AIS paradigm and their biological sources is presented in Table 4.1. It should be noted that the AIS is not intended to be a model of the biological immune system. Therefore, the parallels and similarities may only be approximate and attempted only to the extent that they offer valuable tools for solving aspects of the ACDIEA problem. The main similarities between biology and the AIS paradigm are also illustrated in Fig. 4.12.

4.5 GENERATION OF SELF AND NON-SELF

The generation of the self/non-self is a critical element of the AIS paradigm and must address several important issues such as feature selection, adequate numerical representation, data collection and processing, covering of the self and of the non-self, and optimization [95, 97].

4.5.1 SELF/NON-SELF REPRESENTATION

One of the primary steps in the development of an AIS as the core of aircraft ACM process is the proper selection and definition of features. The *features* are those variables (in general, functions of time t) that completely define the targeted system and are expected to discriminate the fingerprints of all AC considered, in terms of occurrence, presence, type, severity, and consequences. They can be (sub)system states, inputs, control system variables, estimated values, and all of those relevant variables that may be able to capture the characteristics of normal and abnormal system operation. The implication is that the definition of features requires the pre-definition of targeted AC categories. However, it should be noted that the framework provided by the AIS paradigm allows for the consideration of "unknown" categories as well.

The set $\mathcal{F}$ of all features φ_i defines an N-dimensional real hyperspace $\mathcal{U}$ (referred to as *Universe*):

$$\mathcal{F} = \{\varphi_i \quad | \quad i = 1, 2, \ldots, N\}, \quad P \in \mathcal{U}, \;\; \mathcal{U} \subset R^N \tag{4.4}$$

where P (*feature point*) is the N-dimensional vector of all simultaneous values at any given instant $\bar{t}$ of features φ_i:

$$P = [\, \varphi_{1P} = \varphi_1(t = \bar{t}) \quad \varphi_{2P} = \varphi_2(t = \bar{t}) \quad \ldots \quad \varphi_{NP} = \varphi_N(t = \bar{t})\,] \tag{4.5}$$

Features are typically normalized based on known or estimated reference values under AC to take values between 0 and 1:

$$\varphi_i \in [0, \quad 1] \tag{4.6}$$

The point O with coordinates $[\varphi_1 = 0 \quad \varphi_2 = 0 \ldots \varphi_N = 0]$ is considered the origin of an orthogonal coordinate system $(\overline{\mathcal{U}})$ associated to the hyperspace $\mathcal{U}$.

TABLE 4.1 MAIN BIOLOGICAL TERMS AND THEIR AIS PARADIGM COUNTERPARTS

Biological Term	AIS Paradigm Counterpart
Host organism or self	System operation under normal conditions represented by regions of the feature hyperspace reachable under normal conditions Sets of feature clusters and/or their projections under normal operating conditions
Alien entities or non-self	Regions of the feature hyperspace that are unreachable or reachable under abnormal conditions Sets of complementary feature clusters and/or their projections outside of normal operating conditions
Organic markers (proteins and other compounds)	System features or characteristic variables and their values
Antigen	Set of current features values (feature point) at abnormal conditions
Antibody	Data cluster in the non-self feature hyperspace (detector)
Paratope/epitope binding	Matching algorithms
Innate immune system	Self/non-self discrimination outcome from lower-order self projections
Adaptive immune system	AIS updating process
Memory lymphocytes	Labeled clusters in the non-self hyperspace (identifiers or evaluators)
Clonal selection	Mechanism for increased exploitation within an evolutionary optimization algorithm
Negative selection	Selection or characterization logic based on non-matching
Positive selection	Selection or characterization logic based on matching
Dendritic cell (DC)	Artificial DC, a computational unit that processes the outcomes of the self/non-self discrimination process
DC life	Time allowed for the computational unit to be active for processing inputs and maturation
Interleukin-10	Counter of negative self/non-self discrimination outcomes (feature point belongs to self)

(*Continued*)

TABLE 4.1 MAIN BIOLOGICAL TERMS AND THEIR AIS PARADIGM COUNTERPARTS (*Continued*)

Biological Term	AIS Paradigm Counterpart
Interleukin-12	Counter of positive self/non-self discrimination outcomes (feature point belongs to non-self)
Regulatory T-cells	Computational units (artificial DCs) voting a normal condition over a moving time window
Stimulatory T-cells	Computational units (artificial DCs) voting an abnormal condition over a moving time window

Therefore, the Universe of interest becomes a hyper-cube of unit side and the feature point P can be represented by its position vector with respect to O, $\vec{r}^{OP}$, whose coordinates with respect to $\overline{\mathcal{U}}$ are denoted as:

$$\left[\vec{r}^{OP}\right]_{\overline{\mathcal{U}}} = [\varphi_{1P} \quad \varphi_{2P} \quad \cdots \quad \varphi_{NP}]_{\overline{\mathcal{U}}} \tag{4.7}$$

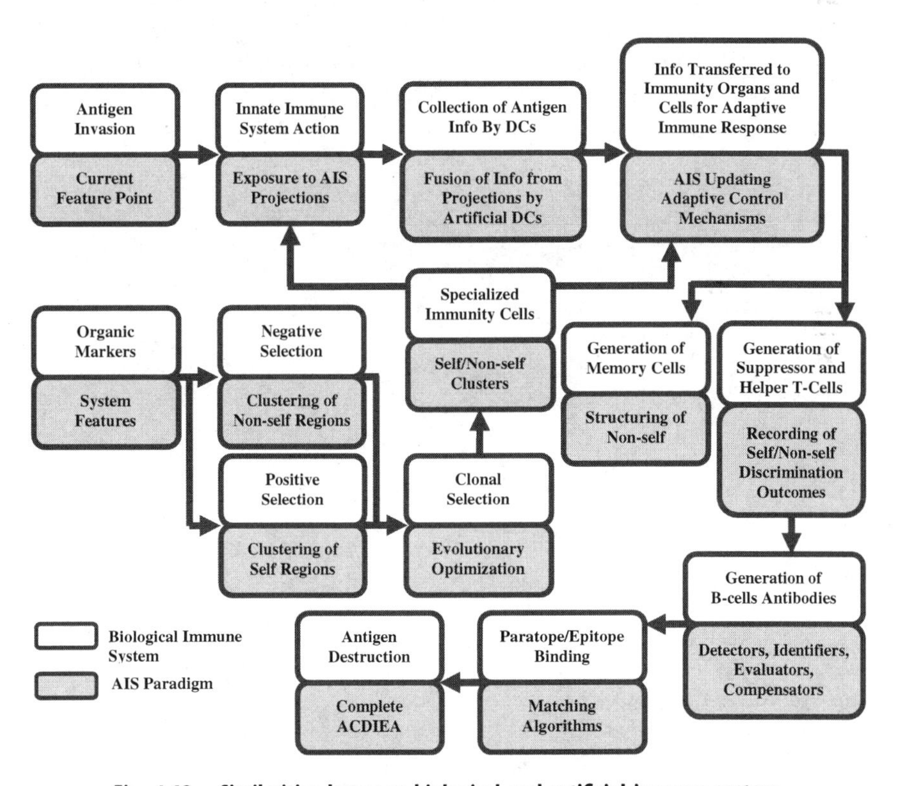

Fig. 4.12 Similarities between biological and artificial immune system.

The self $\mathcal{S}$ is defined as the hyper-sub-space of all possible feature points at normal conditions and all other points in $\mathcal{U}$ form the non-self $\overline{\mathcal{S}}$. Therefore:

$$\mathcal{S} \cup \overline{\mathcal{S}} = \mathcal{U} \quad \text{and} \quad \mathcal{S} \cap \overline{\mathcal{S}} = \emptyset \tag{4.8}$$

For computational tractability and practical reasons, the self points are clustered and the self $\mathcal{S}$ is represented as a set S of hyper-bodies including these clusters. The non-self $\overline{\mathcal{S}}$ is covered with similar hyper-bodies to produce a set of non-self clusters $\overline{\mathrm{S}}$. The geometry of these hyper-bodies can potentially have an impact on the computational efficiency and performance of the AIS generation and the overall ACDIEA process [114]. The following shapes for the self/non-self representation can typically be considered:

- Hyper-cubes: determined by an N-dimensional center and one scalar value for the side
- Hyper-rectangles: determined by an N-dimensional center and N values for the sides
- Hyper-spheres: determined by an N-dimensional center and one value for the radius
- Hyper-ellipsoid of rotation: determined by an N-dimensional center and two values for the axes
- Generalized hyper-ellipsoid: determined by an N-dimensional center and N values for the axes

For all shapes (except hyper-spheres) variable orientation can be considered as determined by an additional N-dimensional vector. For example, for the hyper-spherical representation with N_c clusters c_i, the self and the self clusters can be expressed as:

$$S = \{c_1 \quad c_2 \quad \ldots \quad c_{Nc}\}, \quad c_i = [C_i \quad R_{ci}] = [\varphi_{1i} \quad \varphi_{2i} \quad \cdots \quad \varphi_{Ni} \quad R_{ci}] \tag{4.9}$$

where C_i is the center and R_{ci} is the radius of the self cluster i. For the same hyper-spherical representation with $N_{\bar{c}}$ non-self clusters, the non-self and the non-self clusters can be expressed as:

$$\overline{S} = \{\bar{c}_1 \quad \bar{c}_2 \quad \ldots \quad \bar{c}_{N\bar{c}}\}, \quad \bar{c}_j = [\overline{C}_j \quad R_{\bar{c}j}] = [\varphi_{1j} \quad \varphi_{2j} \quad \cdots \quad \varphi_{Nj} \quad R_{\bar{c}j}] \tag{4.10}$$

where $\overline{C}_j$ is the center and $R_{\bar{c}j}$ is the radius of the non-self cluster j. The self clusters c_i or the non-self clusters $\bar{c}_j$ will be referred to as detectors, if either a positive selection-type of detection algorithm or a negative selection-type of algorithm is used, respectively. Structuring the non-self by adding information to non-self clusters converts them into identifiers usable for AC identification purposes. Clusters of the self or non-self that are processed to be used for evaluation or accommodation will be referred to as evaluators and compensators, respectively. The structuring process will add sets of tags with appropriate information to the cluster representations. For example, assuming that N_S subsystems or components are considered, the clusters of a non-self structured for affected subsystem identification will have an additional tag k representing the affected subsystem:

$$\bar{c}_j = \left[\overline{C}_j \quad R_{\bar{c}j} \quad k\right] = \left[\varphi_{1j} \ \varphi_{2j} \dots \varphi_{Nj} \ R_{\bar{c}j} \ k\right], \quad k \in \{1, 2, \dots N_S\} \tag{4.11}$$

The frontier or threshold between "normal" and "abnormal" operation (i.e., between self and non-self) is represented by an N-dimensional surface:

$$\Sigma(\varphi_1, \quad \varphi_2, \quad \dots, \quad \varphi_N) = 0 \tag{4.12}$$

Similar multidimensional threshold surfaces may be defined to discriminate between different AC categories. Note that this formulation incorporates the concept of "variable threshold" and the surface Σ can be regarded as such a variable or multidimensional threshold. It can be conveniently augmented with fuzzy logic to account for uncertainties and gradual phenomena.

The number of system features for any system of aircraft complexity is typically very large. Handling the resulting hyperspace as a whole may produce critical numerical and conceptual issues [105]. However, by considering lower order self/non-self projections, referred to as sub-selves, these issues may be mitigated or eliminated with little or no penalty in performance, by adopting the HMS strategy [107]. The representation of the self/non-self may then consist of a set of projections or sub-selves π_i:

$$S = \{\pi_1 \quad \pi_2 \quad \dots \quad \pi_{N\pi}\} \tag{4.13}$$

while the representation of each sub-self may consist of a set of clusters. For example:

$$\begin{aligned} \pi_i &= \{c_{i1} \quad c_{i2} \quad \dots \quad c_{iNic}\}, \\ c_{ij} &= [C_{ij} \ \ R_{cij}] = [\varphi_{1ij} \ \ \varphi_{2ij} \ \dots \ \varphi_{Nij} \ \ R_{cij}] \end{aligned} \tag{4.14}$$

A graphical representation of a 3D self projection for a supersonic fighter aircraft is presented in Fig. 4.13. The features are roll, pitch, and yaw angular rates. A pre-imposed number of 500 clusters with variable radii was used within the clustering algorithm.

4.5.2 GENERAL STRATEGIES FOR SELF/NON-SELF GENERATION

The self/non-self generation process, as shown in Fig. 4.14, requires several steps. It is important that the data acquisition is performed properly and data sanity is ensured. The clustering methodology must guarantee that there is no overlapping among self and non-self clusters; there is minimum empty space inside the self clusters; there is minimum uncovered area by the non-self clusters; there is minimum overlapping among self clusters; there is minimum overlapping among non-self clusters; and there is a minimum number of detectors for a

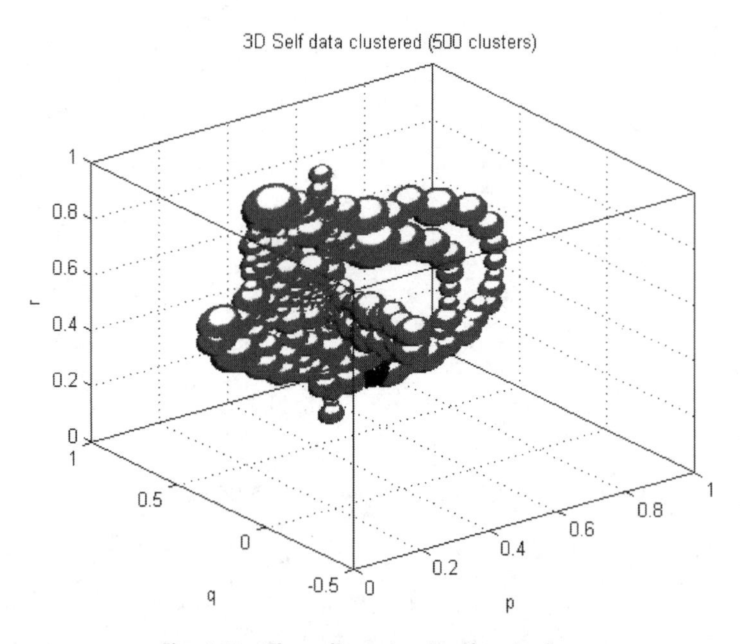

Fig. 4.13 Three-dimensional self projection.

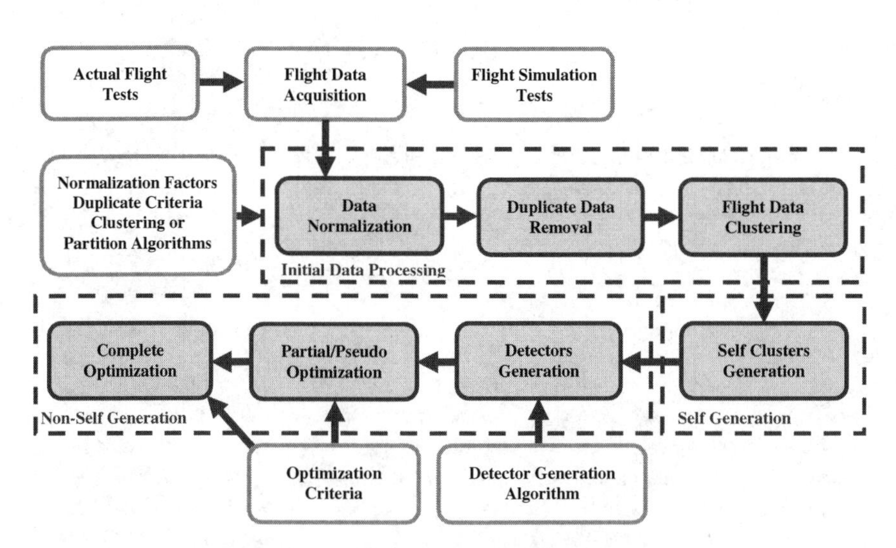

Fig. 4.14 Self/non-self generation process.

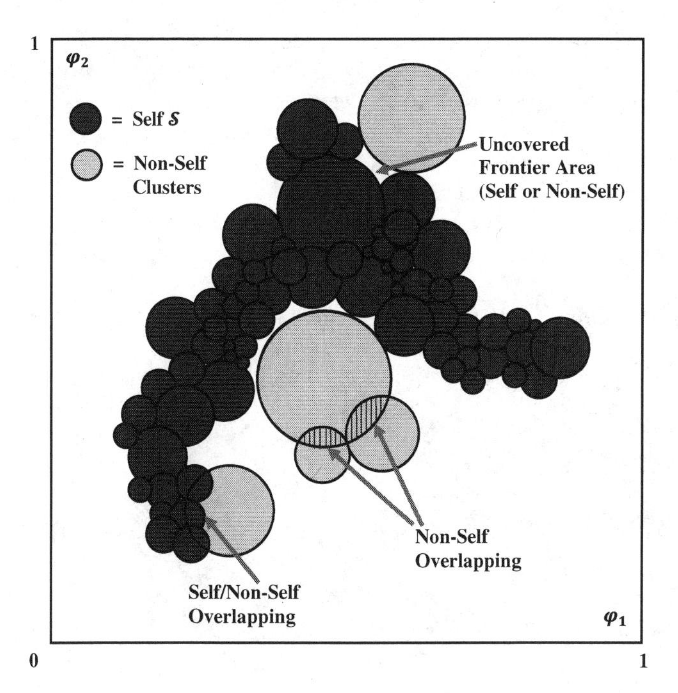

Fig. 4.15 Self/non-self cluster characteristics.

desirable resolution. Figure 4.15 illustrates the definitions of these concepts. Some clustering algorithms include partial optimization properties for some of these criteria. Evolutionary techniques must be used to achieve a complete optimization [115, 116].

For a comprehensive and integrated solution to the aircraft ACDIEA problem using the AIS paradigm, the number of system features will be very large. Using the feature hyperspace as one single entity produces critical issues related to computational resources and effort and to the characteristics of hyperspaces relative to distances and thresholds [105].

The issues related to computational resources and effort are obvious considering that in order to achieve similar high resolution of the feature space—a critical element especially for AC identification, evaluation, and accommodation—exponentially larger computational effort is required with unitary increase of the space dimensionality.

The implications of the characteristics of hyperspaces relative to distances and thresholds are subtler. It should be noted that when the dimensionality of the hyperspace goes to infinity the volume of the unit hyper-cube (the "Universe") remains equal to one, while the volume of the inscribed hyper-sphere goes to zero. As a result, distances to border hyper-planes of the hyper-cube may be extremely small, whereas distances to hyper-cube corners may be extremely large. This means that the intuition in establishing thresholds and assessing distances, which is built in the 3D physical space, becomes inoperational. This counterintuitive effect starts to be significant once the number of features goes beyond 10 [105].

To eliminate or at least mitigate these effects, the HMS strategy was proposed [107] based on the observation that, although the entire set of features is necessary to capture the dynamic fingerprint of "all" ACs, only limited subsets may be necessary to capture the dynamic fingerprint of any individual AC. With proper integration and ACDIEA logic, subsets of features may be used to build projections of the self that eventually yield similar ACDIEA performance as when considering the entire multidimensional self as a whole. The full set of features must first be divided into subsets, which define projections or sub-selves. The capabilities of each projection for capturing specific ACs must be evaluated and considered in a ranking process that will eventually drive the outcome decision logic. This outcome decision logic is based on individual outcomes from each projection, their ranking, and a composition logic and can be typically

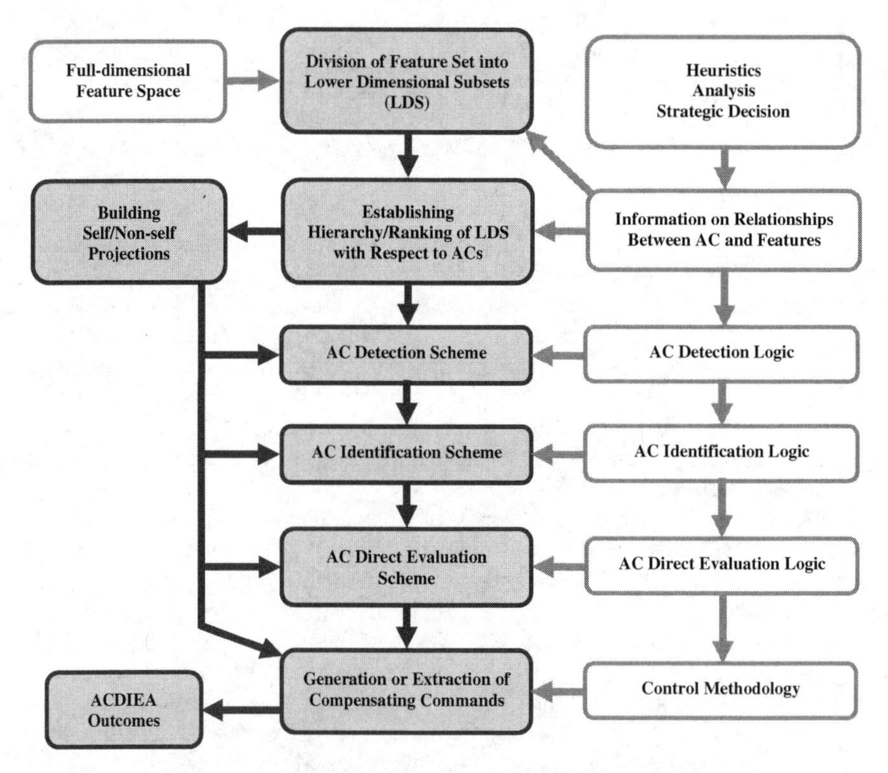

Fig. 4.16 The hierarchical multi-self strategy for ACDIEA.

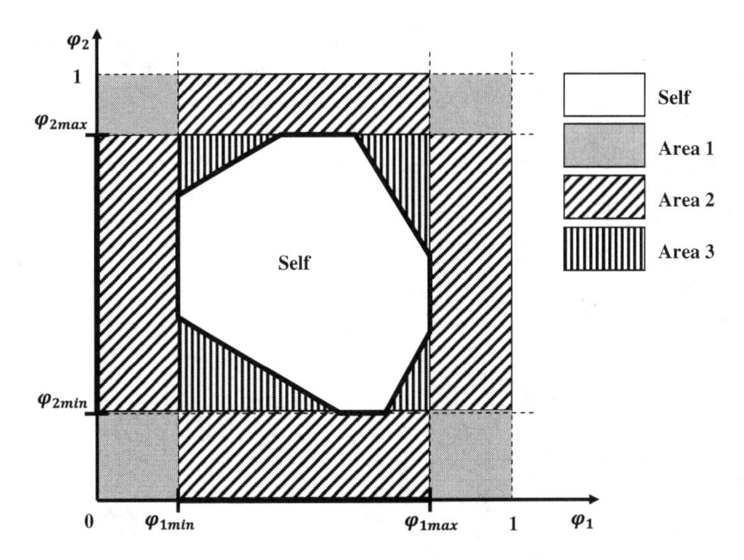

Fig. 4.17 Two-dimensional self/non-self and 1-dimensional projections.

formulated separately for each main process: detection, identification, and evaluation. The general concept of the HMS strategy is illustrated in Fig. 4.16.

The process of establishing what low dimensional projections to use and how to integrate them may be difficult [114]. A conservative approach in which all possible projections of a certain size are considered has been developed successfully [109]. Because each projection produces a discrimination outcome, the integration of the large number of outcomes for a global decision must be properly addressed with the help of information fusion algorithms.

Using lower dimensional self projections, instead of the full-dimensional self, may adversely affect the discrimination performance if regions of the non-self are "hidden" during the process of projecting the self. The specific structure characteristics that lead to hidden non-self areas are illustrated in Fig. 4.17. Let us assume that a system is completely determined by two features (φ_1, φ_2) and, using the full 2D self, perfect self/non-self discrimination is achieved. The complete set of lower dimensional projections/sub-selves consists, in this example, of two 1-dimensional projections, the segments between ($\varphi_{1\min}$, $\varphi_{1\max}$) and ($\varphi_{2\min}$, $\varphi_{2\max}$). The current feature point may belong to any of the four areas depicted in Fig. 4.17. If the feature point belongs to Area 1, then both projections correctly detect the AC. If the feature point falls inside Area 2, then one projection will correctly detect the AC while the other will not. In this situation the detection logic must decide to either declare an AC or not. Finally, if the feature point belongs to Area 3, neither of the two projections has the capability to detect the AC. The regions of the non-self corresponding to Area 3 are completely hidden in the lower dimensional projections.

Let us define the *order* of an AC as the minimum number of features necessary to completely detect that AC. A complete detection is said to be possible if all feature points produced under that particular AC fall within the non-self. For example, if all feature points under a particular AC belong to Area 1, then a complete detection is possible with only 1 feature (either φ_1 or φ_2) and the order of AC is 1. All feature points under the AC fall outside any of the two segments, that is, inside the non-self as defined by the 1-dimensional projections. If all feature points under another particular AC belong to Area 2, some points will fall inside the 1-dimensional sub-selves; therefore, the order of this AC is 2. However, note that if the AC is such that feature points fall only in the top and/or the bottom band of Area 2, then the order is 1. In conclusion, the performance penalty when using lower dimensional projections may be reduced down to zero, if the dimensionality of the projections is larger or equal to the order of the ACs. In practice, even if hidden regions of the non-self exist, the overall discrimination performance may still be good, if those regions are reached with low probability.

4.5.3 ANTIBODIES GENERATION APPROACHES

4.5.3.1 CLUSTERING APPROACHES

Two alternative clustering approaches have been proposed and implemented [117]. Data for each projection or sub-self are processed separately to produce a set of antibodies by covering the respective lower dimensional non-self. In the first approach, referred to as the raw data set union (RDSU) method, all raw test data available are collected in one file before a set of antibodies is generated. In the second approach, referred to as cluster set union (CSU) method, the processing of smaller individual sets of data is performed by clustering and then combining the clusters in a single set for detector generation. The flowchart of these AIS antibodies generation methods is presented in Fig. 4.18 [117] and described as follows.

The RDSU method processes experimental data at normal conditions by combining different flight test samples in one single data file to be then normalized based on the span of the flight data plus a percentage margin. Elimination of duplicated points within the generated hyperspace is performed to decrease the amount of storage and computing resources needed, while preserving the information content of the data. An optimized "*k-means*" algorithm [96] may be used to cluster the reduced data using geometric hyper-bodies specified by the user. Finally, the generated self-clusters are used to generate antibodies through a negative selection-type process by covering the non-self hyperspace. The antibodies generation process can be stopped after a prescribed number of iterations, when a preset maximum number of acceptable detectors is reached, or when a desired coverage of the non-self is achieved.

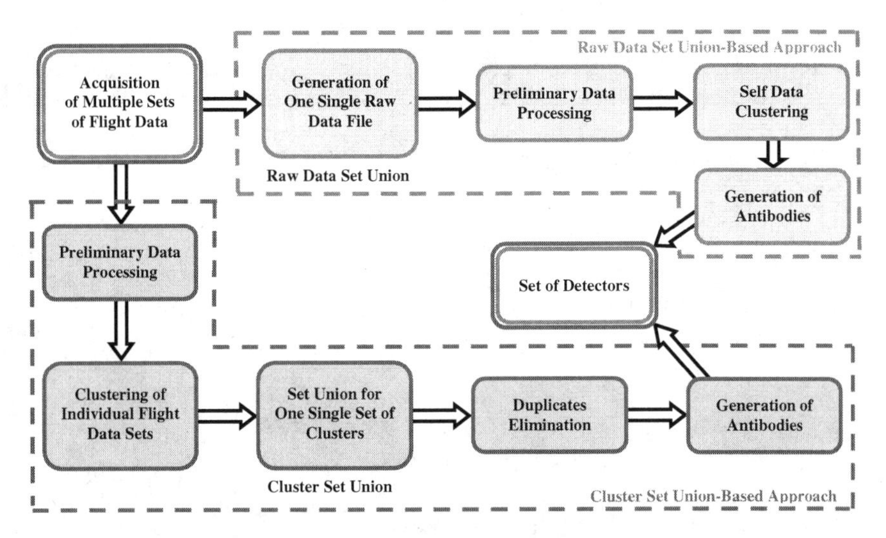

Fig. 4.18 Two clustering methods for AIS antibodies generation [117].

The CSU method uses an optimized fusion algorithm to combine different sets of clusters generated from single sets of experimental flight test data. Within this approach, the data are first split in subsets and normalized as preparation for clustering. The normalization factor for each dimension of the hyperspace is determined as the span of the flight data plus a percentage margin. Note that when multiple sets of experimental data are used for antibodies generation, the same normalization factors must be used for all data. The process continues with clustering of each individual set using a parallel computation approach to improve performance. A fusion phase is then performed that consists of set union accompanied by overlapping elimination among the hyper-clusters. To guarantee minimum overlapping among detectors, the distance between cluster centers must be greater than or equal to the sum of the radii minus the permitted overlapping threshold. The approach allows the update of the AIS when new flight tests are available by clustering only the newly acquired data and then putting old and new clusters together and eliminating any duplication. Finally, an enhanced negative selection algorithm for real-valued representation with variable detector radius is applied to the cluster union set. The algorithm ensures no overlapping between detectors and self clusters and minimizes the uncovered areas in the non-self [117].

It has been found through simulations that the computational time needed by the RDSU method is consistently lower than the one needed by the CSU method [117]. The difference varies quite largely depending on the self. The lower the amount of data, the lower the computational time RDSU needs for clustering. Typically, the raw data file reduces size significantly after the elimination of duplicate points. The clustering module is invoked only one time per (sub-)self for RDSU. The CSU method clusters the same number of flight data points 11 times (the number of flight tests used for the analysis in [117]) without any duplicate point removal. Only then, duplicate clusters are eliminated. For some of the selves generated using RDSU, the number of unique data points was around 60,000 (which is comparable to the number of records in each of the 11 flight files). It should be noted that the single file, including all flight data used for detector generation with RDSU, has about 600,000 records. In most cases, this results in a still large data set, even after duplicates are eliminated. Typically, this amount of data cannot be handled by the *k-means* clustering method on computers with less than 8 GB RAM.

The detection performance in terms of percentage detection rate and false alarms of the two sets of detectors obtained with the different methods has been previously compared for a subset of relevant 2D projections or sub-selves under several types of failure [117]. The performance is similar and the two methods can be considered equivalent from this point of view. The RDSU method requires large computer memory, but the total computation time is lower. The CSU method can be implemented on lower memory computers; however, the overall computation time increases, unless parallel computation is used.

4.5.3.2 PARTITION OF THE UNIVERSE APPROACH

The partition of the universe approach consists of dividing the universe into uniformly distributed partition clusters with predefined shape and resolution [100]. The raw self data points are tested against partition clusters, and self clusters are identified and labeled. Proper balance must be achieved between partition resolution and number of triggering data points such that low empty space within the self is ensured. As a result, the self is represented by strings of integers identifying the self partition clusters. The non-self is implicitly defined and clustering is not necessary. However, the size of the non-self representation may become prohibitive. AC detection can be performed through a positive selection–type of logic using the entire full-dimensional self, thus preventing non-self region hiding. The matching between current feature points and detectors can be assessed using computationally inexpensive integer string matching. As compared to the clustering methods, the approach seems to be better suited when the subsystems or components are highly decoupled and their mutual interactions are dominated by few features. The approach has not yet been investigated for aerospace applications; however, it shows promise especially if hybrid methods with clustering methods are developed.

4.5.4 EXAMPLE APPLICATION OF SELF/NON-SELF GENERATION METHODS

4.5.4.1 SUPERSONIC FIGHTER AIRCRAFT

The extensive immunity-based aircraft ACDIEA has been investigated using a supersonic fighter aircraft model implemented within West Virginia University

(WVU) six-degrees-of-freedom motion-based flight simulator [118]. In this example, several ACs were tested including actuator failures, sensor faults, structural damage, and engine malfunction. Different flight scenarios were considered over a wide range of the flight envelope for Mach numbers between 0.6 and 0.9 and altitudes between 9,000 ft and 31,000 ft. Each flight test lasted between 15 and 20 minutes. The data acquisition rate was 50 Hz. Given the nature and type of the ACs targeted, a total of 32 features have been considered for self/non-self definition including aerodynamic and attitude angles, Mach number, altitude, angular rates, translational and angular accelerations, pilot stick and pedal displacements, commanded angular rates, quadratic and decentralized estimation errors, and neural network estimates of angular accelerations [110]. A two-phase approach was used in which lower dimensional projections were selected based on their discrimination performance with a detection rate equal to 70 percent or larger. This threshold was selected as representation of adequate ability to capture the dynamic fingerprint of the subsystem failures considered. The generated projections/sub-selves were processed to obtain detection, identification, and evaluation outcomes. Alternatively, a complete set of 496 2D projections for the 32 features was considered within the HMS strategy. The outcomes of individual projections were fused using an artificial DC algorithm [108]. A graphical representation of a 2D self/non-self projection for the supersonic fighter aircraft is presented in Fig. 4.19. The features are sideslip angle and roll rate. High resolution uniform radius was used for clustering the self, while variable radius detectors were covering the non-self.

4.5.4.2 SPACECRAFT VEHICLE TEST-BED

The process of generating antibodies has been applied to a spacecraft vehicle test-bed [119]. Figure 4.20 shows an Asteroid Free-Flyer (AFF) cold gas thruster prototype vehicle designed and built at NASA Kennedy Space Center for supporting asteroid exploration missions. The vehicle, with a 2-m diameter, is mounted within a three-degrees-of-freedom gimbaled platform that allows free motion in the roll, pitch, and yaw axes. Future space missions will require high levels of autonomy to increase operational performance and mission success. These goals can only be achieved if the system is enhanced with onboard intelligence for system health monitoring, self-recovery, and autonomous decision making under dynamically changing and extreme environments. The AIS paradigm has been applied to this test-bed to demonstrate the capabilities of bio-inspired algorithms to perform such tasks.

The AFF is actuated using 12 cold gas nitrogen thrusters that are designed to control the vehicle attitude in all three axes. The system is stabilized by a nonlinear, quaternions-based controller that allows attitude tracking and angular rate regulation. Figure 4.21 depicts the main components of the test-bed.

Feature selection yielded a set of 20 variables including actual and reference attitude quaternions, actual and commanded attitude angles, angular rates, and

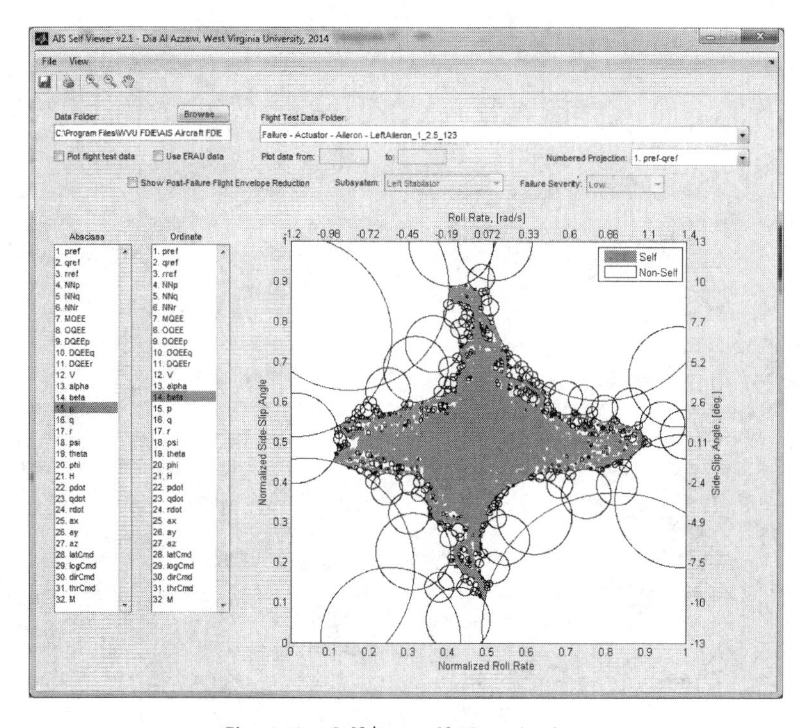

Fig. 4.19 Self/non-self 2D projection.

Fig. 4.20 AFF prototype.

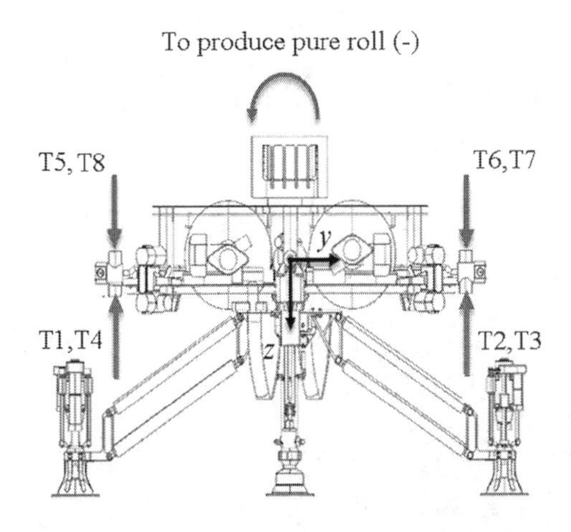

Fig. 4.21 Schematic of test-bed and hardware.

translational accelerations. Data under nominal conditions were collected for a variety of roll-pitch-yaw maneuver sequences. The operation of the system has also been simulated under ACs to gather information for non-self structuring for AC identification purposes. Two solenoid valves, T1 and T2, were artificially blocked one at a time at a certain instant during the test. Figure 4.22 shows typical normalized fingerprint of the system dynamics under nominal conditions. As an example, 2D projections of quaternion errors 2, 3, and 4 versus actual roll rate are presented in Fig. 4.22.

In Fig. 4.23, an example of the clusters and generated antibodies is presented for a 2D self projection defined by nondimensional quaternion error #1 and yaw

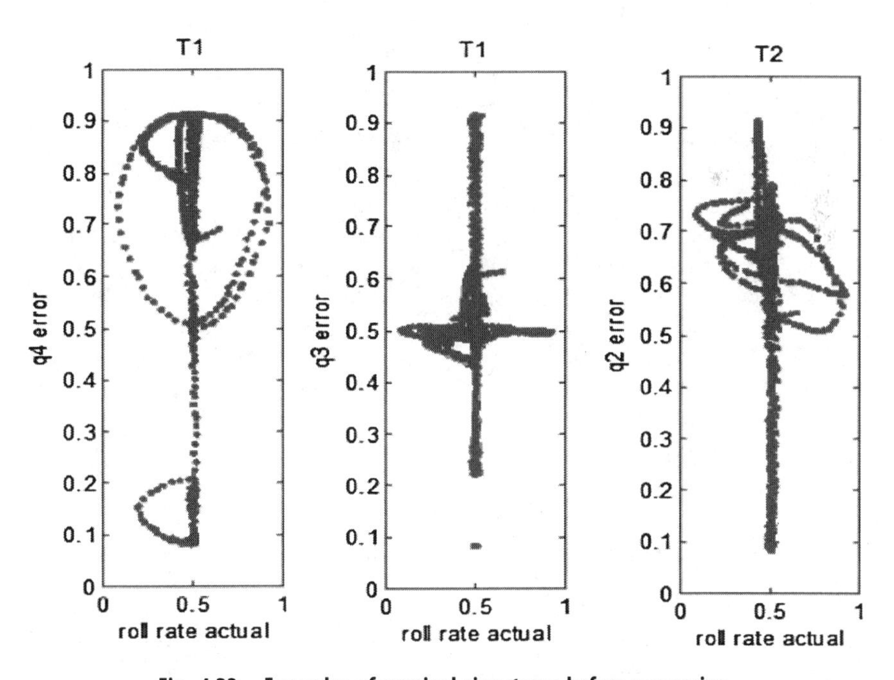

Fig. 4.22 Examples of nominal signatures before processing.

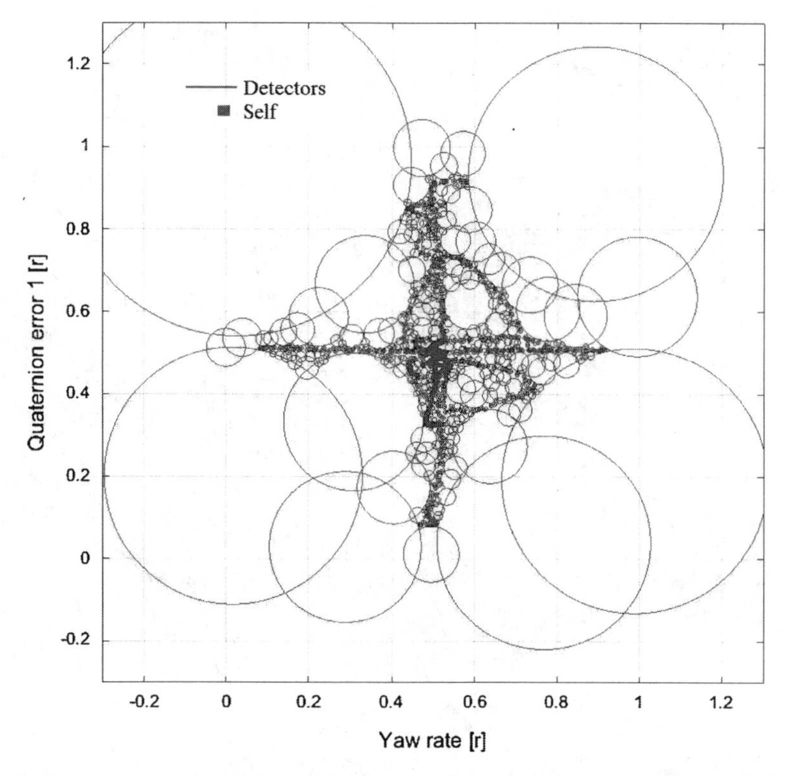

Fig. 4.23 Example of 2D self/non-self projection for the AFF AIS.

rate features using the CSU method. Variable radius detectors are covering the non-self, while higher resolution clusters form the self. The total number of detectors generated for this projection was 600. It is important to note that an allowed overlap value has to be selected in such a way that the balance between the holes and the covered space is adequate within the non-self region. The detector generation algorithm guarantees minimum overlapping among detectors and self clusters, minimum uncovered areas inside the non-self, minimum overlapping among detectors, and minimum number of detectors for a desirable resolution and online implementation capabilities.

4.6 IMMUNITY-BASED AIRCRAFT ABNORMAL CONDITION DETECTION

4.6.1 SELF/NON-SELF DISCRIMINATION

The two most powerful mechanisms that the natural immune system uses to produce specialized cells that are able to detect and recognize alien invading entities are NS and PS. NS allows the production of those immunity specialized cells that do not recognize self-patterns, while PS leads to the generation of cells that are capable of matching chemical markers of the self-cells. These two mechanisms as complementary processes provide inspiration for self/non-self discrimination (SNSD) algorithms relying on the so-called detectors, which are critical for the aircraft AC detection process. It is important to note that the intensive and lengthy procedure that is required for adequate detectors generation is performed offline and does not affect the real-time operation of onboard detection schemes.

By recalling the definitions of Sections 4.2 and 4.5, let the current feature point at time $t = \bar{t}$ be defined by the coordinates of its position vector with respect to origin O:

$$P(t = \bar{t}) = \begin{bmatrix} \varphi_{1P} & \varphi_{2P} & \cdots & \varphi_{NP} \end{bmatrix} \tag{4.15}$$

If one N-dimensional self/non-self is used, then an NS-type of detection algorithm is formulated for each time sample $\bar{t}$ such that an AC is declared if the feature point falls inside any detector:

$$\mathcal{O}_D(\bar{t}) = \begin{cases} 0 & \text{if } \left|\vec{r}^{OP} - \vec{r}^{O\bar{C}_j}\right| > R_{\bar{C}_j} \quad \text{for all } j = 1, 2, \ldots, N_{\bar{C}} \\ 1 & \text{otherwise} \end{cases} \tag{4.16}$$

A PS-type of algorithm will declare an AC if the current feature point is outside all of the self clusters:

$$\mathcal{O}_D(\bar{t}) = \begin{cases} 1 & \text{if } \left|\vec{r}^{OP} - \vec{r}^{OC_i}\right| > R_{C_i} \quad \text{for all } i = 1, 2, \ldots, N_C \\ 0 & \text{otherwise} \end{cases} \tag{4.17}$$

To obtain a "normal condition" outcome with the NS-type of approach, all detectors must be checked at each sample $\bar{t}$. With the PS approach, the testing of the current feature point is stopped as soon as a matching self cluster is found. Therefore, at normal conditions, the PS-based detection is less computationally intensive. On the other hand, an AC will be declared as soon as a matching detector has been found with the NS approach instead of testing the entire set of self clusters, as is required by the PS approach. Therefore, an AC may be detected faster with the NS approach. Depending on the number of self clusters and detectors in conjunction with the system update rate, these differences may be relevant or not. However, note that building and structuring the detectors is critical for the failure identification process, if the structured non-self approach is used.

If N_σ sub-selves are used within the HMS strategy, then Eq. (4.16) or (4.17) is applied for each sub-self to obtain individual discrimination outcomes at instant $\bar{t}$, $\mathcal{O}_{Dk}(\bar{t})$, where $k = 1, 2, \ldots, N_\sigma$. All individual outcomes must be composed to produce a global instantaneous detection outcome by using a composition rule, which may be as simple as:

$$\mathcal{O}_D(\bar{t}) = \max_k [\mathcal{O}_{Dk}(\bar{t})] \tag{4.18}$$

The number of activated detectors at each sample time can also be used to monitor in real time the health status of the system. In order to improve false alarm performance, the global instantaneous detection outcome may be produced by using a detection logic that incorporates current and past outcomes over fixed or variable time windows.

Typically, the performance of the discrimination process is assessed in terms of sample percentage of false alarms (FA) and sample percentage detection rate (DR). FA is defined as the percentage ratio between the number of time samples when an AC was declared and the total number of samples for a test under normal conditions. DR is defined as the percentage ratio between the number of time samples when an AC was declared and the total number of samples for a test under AC.

A detection logic using directly the SNSD has been proposed [110], in which several sets of self-projections can be organized based on weighting their potential and capability to capture the dynamic fingerprint of targeted ACs. A reduced set of projections is selected for the detection process and different potential weights are assigned to each projection based on their discrimination capabilities. The weighted average of all individual detection outcomes over a past time window of preset size is used to calculate a detection parameter ζ. If ζ is between two preset thresholds, a failure warning is issued. If ζ exceeds the larger threshold, then an AC is declared and the identification process starts.

4.6.2 ARTIFICIAL DENDRITIC CELLS FOR AC DETECTION

The functionality of the DCs inspired an alternative detection logic that acts as an information fusion algorithm capable of consolidating individual self/non-self discrimination outcomes from numerous projections within the HMS strategy.

Let us consider N_σ sub-selves or projections σ_i. At each sample time j, $j = 1, 2, \ldots, w$ over a moving time window, the feature point projections p_{ij} produce instantaneous individual discrimination outcomes:

$$\mathcal{O}_{Dij} = \begin{cases} 0 & \text{if } p_{ij} \in \sigma_i \\ 1 & \text{if } p_{ij} \notin \sigma_i \end{cases} \tag{4.19}$$

that are used to build the detection matrix at each sample time k, D_k as:

$$D_k = \begin{array}{c} \begin{matrix} t_1 & t_2 & \cdots & t_w \end{matrix} \\ \begin{bmatrix} \mathcal{O}_{D11} & \mathcal{O}_{D12} & \cdots & \mathcal{O}_{D1w} \\ \mathcal{O}_{D21} & \mathcal{O}_{D22} & \cdots & \mathcal{O}_{D2w} \\ \vdots & \vdots & \ddots & \vdots \\ \mathcal{O}_{DN_\sigma 1} & \mathcal{O}_{DN_\sigma 2} & \cdots & \mathcal{O}_{DN_\sigma w} \end{bmatrix} \end{array} \begin{matrix} \sigma_1 \\ \sigma_2 \\ \vdots \\ \sigma_{N_\sigma} \end{matrix} \tag{4.20}$$

Let the complement of matrix D_k be defined as:

$$\overline{D}_k = \left\{ \overline{\mathcal{O}}_{Dij} \;\middle|\; \overline{\mathcal{O}}_{Dij} = \text{non}\left(\mathcal{O}_{Dij}\right) \right\} \tag{4.21}$$

The artificial DCs are computational units that fuse the information from the detection matrix to support the generation of the global detection outcome [120]. They are randomly selected to collect information from the current detection matrix. Once a maturation threshold is exceeded, they become part of the mature DCs set and are used in the decision process. For the AC detection process, critical artificial DC parameters are IL10 (indicating normal conditions) and IL12 (indicating ACs). They mimic the production of the corresponding chemical compounds by the biological DCs, thus activating suppressor T-cells (by regulatory DCs) or cytotoxic T-cells (by stimulatory DCs), respectively [60]. The accumulation of IL10 and IL12 within the DC is modeled using these two parameters initialized as 0 and updated each time the DC is selected for exposure to the detection matrix:

$$IL10_k = \begin{cases} IL10_{k-1} & \text{if artificial DC is } \underline{\text{not}} \text{ selected} \\ IL10_{k-1} + \Gamma_{10}\left(W_0, W_{t0}, \overline{D}_k\right) & \text{if artificial DC is selected} \end{cases} \tag{4.22}$$

$$IL12_k = \begin{cases} IL12_{k-1} & \text{if artificial DC is } \underline{\text{not}} \text{ selected} \\ IL12_{k-1} + \Gamma_{12}(W_1, W_{t1}, D_k) & \text{if artificial DC is selected} \end{cases} \tag{4.23}$$

W_0 and W_1 are sub-self performance weights for accurately capturing the self and non-self, respectively. They reflect the sensitivity of the features involved in each projection of discriminating between self/non-self and may also be regarded as a measure of how trustworthy the outcome is from each projection. Establishing these weights must consider the relationship between features and ACs, the existence of hidden non-self, and any test results, if available. W_{t0} and W_{t1} are time sample weights allowing for more recent outcomes to have more impact on the decision process than older ones. Functions Γ_{10} and Γ_{12} provide the flexibility necessary to handle the various characteristics of the targeted system in conjunction with the specifics of the AIS design. For instance, for systems for which the AIS is built as a set of uniform dimensionality projections, with reduced hidden non-self regions, and with balanced distribution of projection discrimination capability [108, 121], the interleukin accumulation functions can be defined as:

$$\Gamma_{10} = W_0 \cdot \overline{D}_k \cdot W_{t0} \tag{4.24}$$

$$\Gamma_{12} = W_1 \cdot D_k \cdot W_{t1} \tag{4.25}$$

Two other critical components of the artificial DC structure are the "triggered features matrix" F_1 and the "non-triggered features matrix" F_0. They are instrumental for both detection and identification. The two matrices represent quantification tools for feature involvement with projections that yield positive detection outcomes and projections that do not, respectively. The structure of both matrices is the same:

$$F_* = \begin{array}{c} \begin{matrix} \sigma_1 & \sigma_2 & \cdots & \sigma_{N_\sigma} \end{matrix} \\ \begin{bmatrix} F_{11}^* & F_{12}^* & \cdots & F_{1N_\sigma}^* \\ F_{21}^* & F_{22}^* & \cdots & F_{2N_\sigma}^* \\ \vdots & \vdots & \ddots & \vdots \\ F_{N1}^* & F_{N2}^* & \cdots & F_{NN_\sigma}^* \end{bmatrix} \end{array} \begin{matrix} \varphi_1 \\ \varphi_2 \\ \vdots \\ \varphi_N \end{matrix} \tag{4.26}$$

or:

$$F_* = \{F_{qi}^* \mid q = 1, 2, \ldots, N, \quad i = 1, 2, \ldots, N_\sigma\} \tag{4.27}$$

where the symbol * stands for either 0 or 1. The two matrices are initialized at 0 and each time an artificial DC is randomly selected to process the discrimination outcomes, their elements are updated as follows:

$$F_{qi}^1 = \begin{cases} F_{qi}^1 + 1 & \text{if } \varphi_q \in \sigma_i \quad \text{and} \quad \mathcal{O}_{Diw} = 1 \\ F_{qi}^1 & \text{otherwise} \end{cases} \tag{4.28}$$

$$F_{qi}^0 = \begin{cases} F_{qi}^0 + 1 & \text{if } \varphi_q \in \sigma_i \quad \text{and} \quad \mathcal{O}_{Diw} = 0 \\ F_{qi}^0 & \text{otherwise} \end{cases} \tag{4.29}$$

Let the total number of mature DCs at instant k be denoted by N_{MDC}. The set of mature DCs is the union of two disjoint subsets, the stimulatory DCs, for which $IL12 \geq IL10$ and the regulatory DCs, for which $IL12 < IL10$. Let the set of activated cytotoxic T-cells be represented by:

$$K = \{K_q \mid q = 1, 2, \ldots, N\} \tag{4.30}$$

where K_q is the number of activated cytotoxic T-cells corresponding to feature φ_q. Then:

$$K_q = \sum_{m=1}^{N_{MDC}} \sum_{i=1}^{N_\sigma} F_{qi}^1 \tag{4.31}$$

for all m for which $IL12 \geq IL10$. Let the set of activated suppressor T-cells be represented by:

$$R = \{R_q \mid q = 1, 2, \ldots, N\} \tag{4.32}$$

where R_q is the number of activated suppressor T-cells corresponding to feature φ_q. Then:

$$R_q = \sum_{m=1}^{N_{MDC}} \sum_{i=1}^{N_\sigma} F_{qi}^0 \tag{4.33}$$

for all m for which $IL12 < IL10$. The role of the suppressor T-cells is to regulate the adaptive immune response by counter-acting a corresponding number of activated cytotoxic T-cells. This results in a set of residual cytotoxic T-cells given by:

$$\widetilde{K} = \{\widetilde{K}_q = K_q - R_q \quad | \quad q = 1, 2, \ldots, N\} \tag{4.34}$$

The global detection outcome at sample k is finally obtained as:

$$\mathcal{O}_D = \begin{cases} 0 & \text{if } \sum_{q=1}^{N} \widetilde{K}_q \leq 0 \\ 1 & \text{otherwise} \end{cases} \tag{4.35}$$

4.6.3 EXAMPLE APPLICATIONS

4.6.3.1 SUPERSONIC FIGHTER AIRCRAFT

The investigation of the direct SNSD approach was conducted on the WVU supersonic fighter aircraft [122], for which the self/non-self was generated as previously described. A total of 496 2D projections were tested against 26 different ACs varying in affected subsystem, type, and severity. Actuator failures consisted of control surface lockage at different deflections affecting stabilators, ailerons, or rudders. Angular rate sensors on all three channels were subjected to different amounts of bias. Damage of wings was simulated to produce different percentage alterations of aerodynamic characteristics. Individual engine shut down was also considered. This investigation confirmed that, for any given AC, some projections exhibit better discrimination capabilities than others. It was also revealed that any given projection may, typically, capture the fingerprint of few ACs and be insensitive to others.

With the online direct SNSD detection logic, the recorded FA is zero. The average DR for all subsystems considered is 92 percent, with the engine and rudder achieving slightly lower averages than the other subsystems [122].

The approach was also successfully tested for detection of AC affecting engine internal actuators and sensor [123]. Seven different internal actuators and eight different sensors used in the engine control system were considered in this investigation. The actuator failures consisted of lockage, while the sensor failures consisted of output bias or constant output. A limited set of only five features was used for building the self. The sample percentage detection rate varied between 31 percent and 100 percent with an average of 71 percent with no or minimal false alarms. It should be noted that even with low sample percentage detection rates, due to the absence of false alarms and the fast output of positive detection outcomes, high detection performance can be achieved with proper data processing logic.

The same simulated flight data were used for implementing and testing the artificial DC mechanism approach [120]. All 2D projections of the self/non-self were used with equal weights. The algorithm was initialized with a pool of 100 DCs and a moving time window of 1 s at 50 Hz. This investigation has demonstrated that the residual cytotoxic T-cells parameter has excellent discrimination capabilities. The parameter is negative and remains in a narrow band for normal conditions. When an AC occurs, this parameter typically experiences an abrupt increase and remains positive. The actual values vary and depend on the AC; however, the separation is clear capturing the occurrence of the AC and providing a reliable detection criterion, as shown by the example in Fig. 4.24. The artificial DC algorithm achieved zero FA as well. The average DR for all subsystems considered is 94 percent. Similar to the SNSD approach, the detection of AC affecting engine and rudder achieved slightly lower averages than the other subsystems.

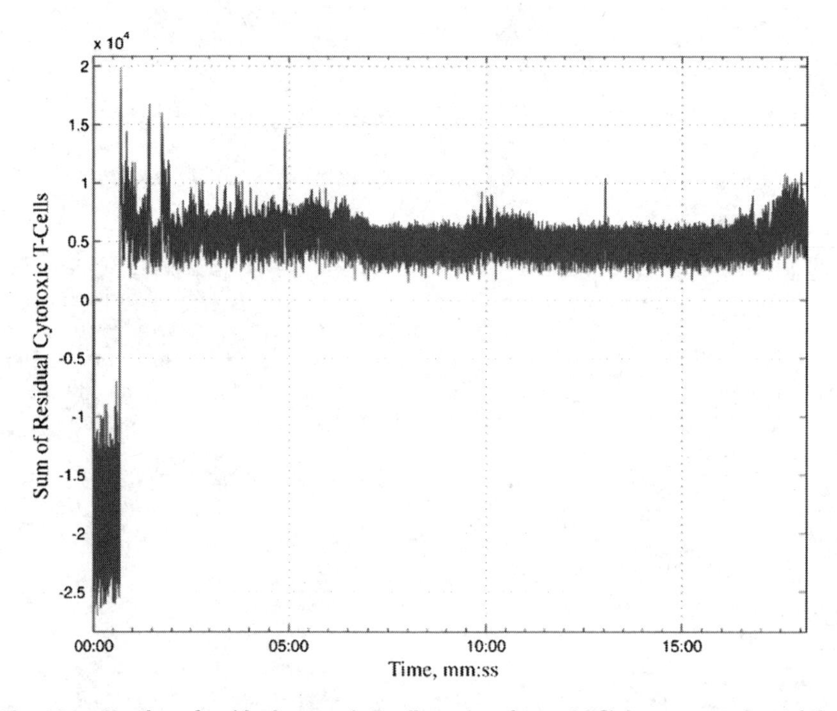

Fig. 4.24 Number of residual cytotoxic T-cells vs time for an AC flight test: 15% loss of the left wing; actual failure time = 40 s; detection time = 40.52 s.

These investigations demonstrated that the two approaches have the potential to produce similar excellent AC detection performance. The direct SNSD approach requires additional investigation for selecting the reduced number of most significant projections, which results in less computational effort during the operation of the detection scheme. The artificial DC approach may perform well with the entire set of low-dimensional projections; however, if tuning of sub-self performance weights is needed, it must be supported by additional information.

4.6.3.2 QUADROTOR UNMANNED AERIAL SYSTEM

The capabilities of the AIS methodology to detect subsystem failures during operation of a quadrotor system have been successfully tested [124]. The direct SNSD approach has been implemented to monitor the health status of the vehicle, while performing specific mission maneuvers. The aircraft used is a low-cost platform featuring an onboard microcontroller that includes embedded sensors such as an inertial measurement unit, barometer, gyroscopes, and accelerometers. It has logging capabilities and it is equipped with a global positioning system, magnetometer, and radio telemetry for ground station monitoring.

A total of 23 features were used to define self/non-self including attitude angles, angular rates, translational accelerations, velocity components, commanded attitude angles and angular rates, throttle command, and the four inputs to quadrotor motors. Ten flight tests were conducted under nominal conditions with an average flight time of 45 s. Moderate roll and pitch maneuvers of ± 10 degrees were performed. Flight tests under failure conditions were also recorded. They correspond to efficiency reduction of 2.5 percent in one of the motors. After some post-processing, 24 2D projections have been identified as possessing the maximum potential for AC detection.

At every sample time, the detection performance can be measured in terms of the number of detectors that are activated, which determines the capability of each set of detectors to capture the dynamic fingerprint at failure conditions within the HMS strategy. An online ACM system can then be designed by adding the detection performance of all of the most effective selected selves. Figure 4.25 shows results of a flight test under failure injected at 8 s. Therefore, the first 8 s correspond to nominal condition, for which the AIS is not expected to produce any activation.

It is important to mention that the time history for activated detectors under failures shows that, once a failure is injected, the number of activated detectors does not remain constant or oscillate around a constant value. This could be explained by the fact that the quadrotor system is already being stabilized by an onboard control system. In this case, a nonlinear dynamic inversion controller will in fact minimize the failure effects and drive the system toward a pseudo-nominal behavior [124]. However, this phenomenon does not prevent the online ACM system from detecting the presence of a failure, even if the robustness

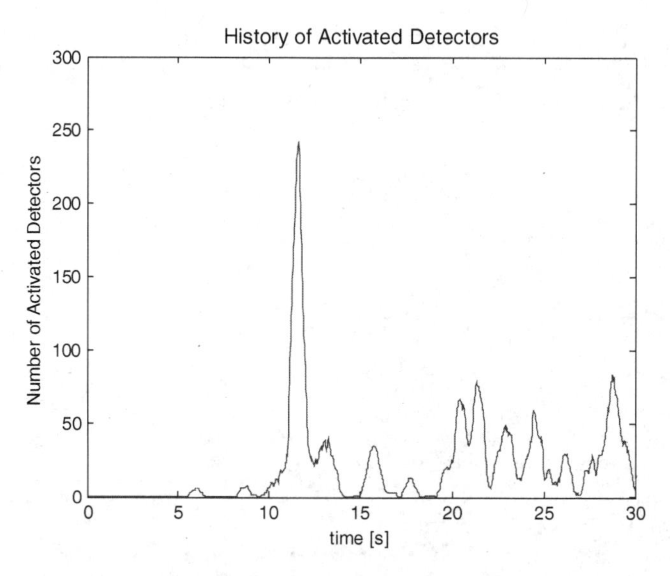

Fig. 4.25 Detectors activation for reduced efficiency in a motor.

of the controller allows its partial rejection. Finally, a validation flight, as presented in Fig. 4.26, illustrates the performance under nominal conditions, for which low or no activation of detectors is expected. Figure 4.26a shows the time history of a nominal flight test that included roll and pitch maneuvers of approximately ± 10 degrees aircraft attitude angle. Figure 4.26b shows the activation for that specific flight, which demonstrates the acceptable performance of the AIS biomimetic system.

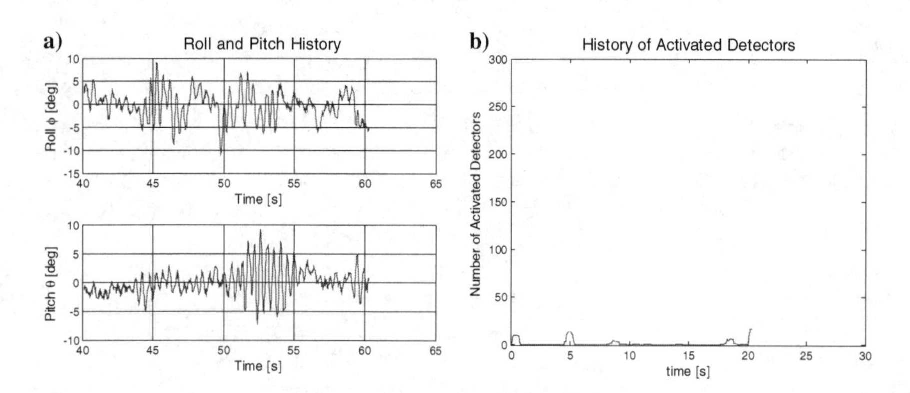

Fig. 4.26 Nominal validation flights.

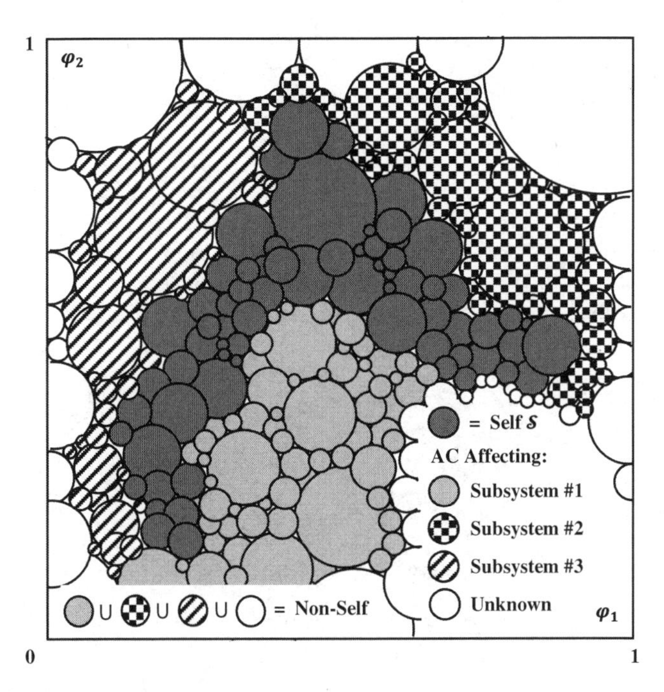

Fig. 4.27 Failed subsystem identification using the structured non-self approach.

4.7 IMMUNITY-BASED AIRCRAFT ABNORMAL CONDITION IDENTIFICATION

Two immunity-based approaches to AC identification (subsequent to or simultaneous with AC detection) are described in this section: structuring the non-self and artificial DC mechanism. The first approach relies on the heuristic ranking of lower order self/non-self projections and the generation of selective immunity identifiers through structuring of the non-self. The second approach is based on an information-processing algorithm inspired by the functionality of the DCs. The artificial DC is defined as a computational unit that centralizes, fuses, and interprets information from multiple selves (or self projections) to produce a unique detection and identification outcome.

4.7.1 STRUCTURED NON-SELF APPROACH

Structuring the non-self for AC identification and converting the detectors into identifiers consists of attaching to each non-self cluster a tag representing the subsystem(s) affected by the specific AC triggering that non-self cluster. If the full-dimensional non-self is used with ideal resolution, then the non-self clusters converted into identifiers are represented by Eq. (4.11), and the tag k is an integer scalar labeling the affected subsystem. The concept is illustrated in Fig. 4.27 [97] for the 2D case. Otherwise, as is the case when using the HMS strategy, the identification tag may be a vector containing labels of several subsystems affected. Therefore, a structured non-self cluster will be:

$$\bar{c}_j = \begin{bmatrix} \overline{C}_j & R_{\bar{c}j} & \mathcal{K} \end{bmatrix} = \begin{bmatrix} \varphi_{1j} & \varphi_{2j} & \cdots & \varphi_{Nj} & R_{\bar{c}j} & \mathcal{K} \end{bmatrix},$$
$$\mathcal{K} = \begin{bmatrix} k_1 & k_2 & \cdots & k_{N_{id}} \end{bmatrix}, \quad k_i \in \{1, 2, \ldots N_S\} \tag{4.36}$$

The structuring of the non-self is performed through a PS-type of algorithm, which implies the prior generation of the non-self with adequate resolution and knowledge of AC characteristics. Such information can be obtained from tests, simulation, or analysis [110].

If the instantaneous detection outcome is positive ($\mathcal{O}_D(\bar{t}) = 1$), then the triggering detector is checked for structural parameters $\mathcal{K}$, and an instantaneous or current identification outcome can be formulated as:

$$\mathcal{O}_I(\bar{t}) = \mathcal{K} \tag{4.37}$$

If full-dimensional non-self is used with perfect feature definition, then the identification process outcome can be simply defined as:

$$\mathcal{O}_I(t) = \mathcal{O}_I(\bar{t}) = k \tag{4.38}$$

If the HMS strategy is used, then Eq. (4.36) is applied to each sub-self i, and a composition logic $\mathbb{C}_I$ must be designed to obtain the current identification outcome as:

$$\mathcal{O}_I(t) = \mathbb{C}_I(\mathcal{O}_{Ii}(\bar{t})), \quad i = 1, 2, \ldots, N_\sigma \tag{4.39}$$

However, if only information from the current instant is used, a large number of incorrect identifications may be produced. This can be mitigated by reprocessing the current outcomes $\mathcal{O}_I(t)$ over a moving time window. The composition logic $\mathbb{C}_I$ can also rely on the artificial DC mechanism.

A novel structured non-self approach was developed [110] consisting of a dual-phase algorithm where lower-dimensional self/non-self projections, previously generated using an NS-type mechanism and tested in simulation under several ACs, are selected according to their ability to detect ACs at a predefined percentage detection rate. By using a PS-type mechanism, the selected projections are processed in order to generate identifiers capable of differentiating similar dynamic fingerprints among several ACs and declaring, not only affected subsystems, but also correct failure types and magnitudes. A more robust combined AC identification and direct evaluation is thus achieved. The identification logic is summarized in Fig. 4.28 [110].

Within the two phases of the structured non-self approach, the process of generating identifiers follows a multi-step optimization based on a positive selection mechanism. Flight test data under AC, previously recorded throughout the entire flight envelope, are used to expose the dynamic fingerprint of the AC. Flight raw

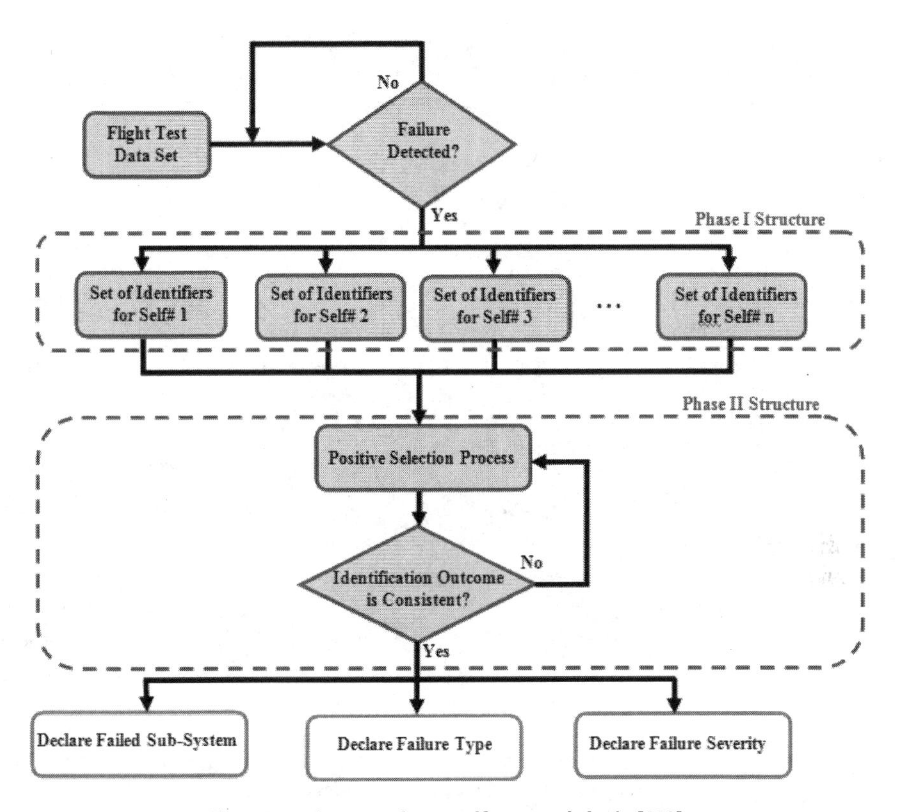

Fig. 4.28 Structured non-self approach logic [110].

data are normalized between 0 and 1 with the same normalization factor used during the antibodies generation. The unit hyper-cube determined during the normalization process delimits the hyperspace of the nominal conditions. High magnitude failures may contain data points that lay far away from the unit hyper-cube of the initial self/non-self. Therefore, outward concentric hyper-cubes are defined in order to determine the distance of the AC point from the self, which subsequently allows the algorithm to determine the severity of the AC as part of the direct quantitative evaluation. Note that these phases may be performed consecutively or simultaneously. The radii of the identifiers may depend on their distance to the self. Therefore, the radius of an identifier increases as the position of its center lies within an outward hyber-cube. Finally, because the number of initial identifiers depends on the number of data points obtained from the flight tests, this may yield an enormous number of identifiers that will produce a degradation of the computer processing capability. A simple elimination algorithm is then implemented in order to reduce the number of identifiers. Identifiers that lay inside the radius of another identifier plus a tolerance are eliminated. A fusion process is then performed to reduce the overlapping and hence the number of identifiers.

4.7.2 ARTIFICIAL DC MECHANISM FOR AC IDENTIFICATION

The artificial DC-based approach for aircraft AC identification consists of determining which one of the N_S subsystems has been affected by the AC, and it starts once an AC has been detected. The affected subsystem is inferred based on the topography of the detection outcomes from self/non-self projections, within the HMS strategy. Therefore, with the DC approach, the AC identification is formulated as a pattern recognition problem [108]. Patterns for AC identification purposes may be defined based on the F_1 matrix to form a reference library. Note that the F_0 matrix may also be introduced into the process. N_S different patterns must be established, one associated with each subsystem. The structure of the patterns depends on the adopted matching algorithm. The flowchart of the process of establishing the library of patterns in conjunction with the Naïve Bayes classifier [125, 126] is presented in Fig. 4.29, where category γ refers to affected subsystem ($\gamma = k$, $k = 1, 2, \ldots, N_S$). There are three possible alternatives for defining the patterns: the feature-pattern (FP) approach, the projection-pattern (PP) approach, and the matrix-pattern (MP) approach [121]. Only the FP approach has been implemented and tested to date [109].

The FP approach for AC identification relies on establishing for each subsystem k and for each AC targeted (AC_j, $j = 1, 2, \ldots, N_{ftk}$) the vector of membership

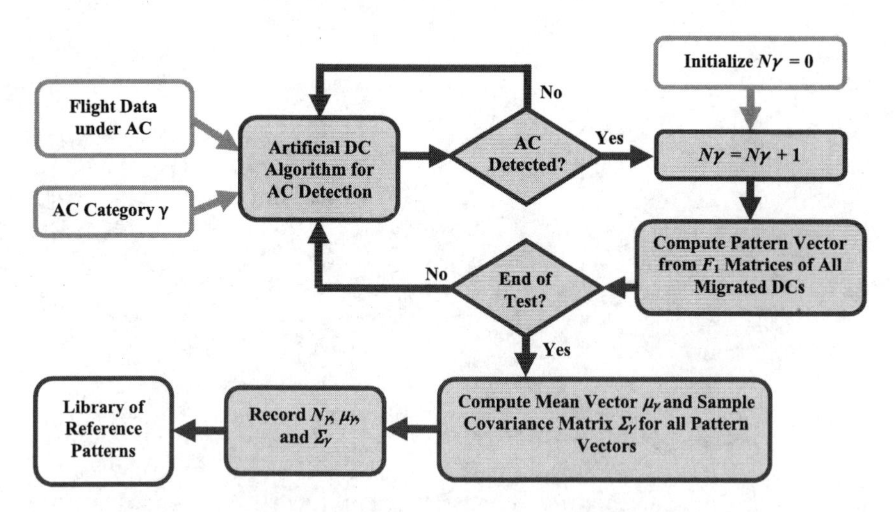

Fig. 4.29 Generation of library of reference patterns for identification using the DC mechanism and the Naïve Bayes classifier.

values of each feature to the subset of features that capture the dynamic fingerprint of the AC. These features will be later referred to as equivalent directly involved variables. Therefore:

$$M_{AC_j} = [\mu_1 \quad \mu_2 \quad \dots \quad \mu_N]^T, \quad \mu_i \in [0, \ 1] \tag{4.40}$$

For each subsystem k, define the reference FP vector as:

$$FP_k = \frac{1}{\max_i \left(\sum_{j=1}^{N_{ftk}} M_{AC_j} \right)} \sum_{j=1}^{N_{ftk}} M_{AC_j} = [\overline{\mu}_1 \quad \overline{\mu}_2 \quad \dots \quad \overline{\mu}_N]^T, \quad \overline{\mu}_i \in [0, \ 1] \tag{4.41}$$

All of these membership values can be determined based on flight data, simulation, heuristics, or analysis. Note that binary logic can be used instead of fuzzy logic. In this case, $\mu_i \in \{0 \quad 1\}$ and $\overline{\mu}_i \in \{0 \quad 1\}$. At each sample time, after an AC is detected, F_1 matrices of the mature DCs are used to compute the current AC FP vector as:

$$F_{1\varphi} = \left[\sum_{m=1}^{N_{MDC}} F_{1m} \right] \cdot I_{N_S}, \quad I_{N_S} \in \{1\}^{N_S \times 1} \tag{4.42}$$

The outcome of the AC identification process will be established by using a matching algorithm ($\mathcal{A}[\,\cdot\,]$) to determine which one of the N_S reference FP vectors best matches $F_{1\varphi}$.

$$\mathcal{O}_I(t) = k, \quad k = \mathcal{A}[F_{1\varphi}] \tag{4.43}$$

The flowchart of the aircraft AC identification process using the artificial DC approach and the Naïve Bayes classifier is presented in Fig. 4.30. The main parameter required by the classifier is the discriminant:

$$\Delta_\gamma(F_{1\varphi}) = \ln(N_\gamma) - \frac{1}{2}\ln|\Sigma_\gamma| - \frac{1}{2}(F_{1\varphi} - \mu_\gamma)^{\mathrm{T}} \Sigma_\gamma^{-1} (F_{1\varphi} - \mu_\gamma) \tag{4.44}$$

where N_γ is the number of samples, μ_γ is the mean vector, and Σ_γ is the sample covariance matrix.

4.7.3 APPLICATION EXAMPLE (SUPERSONIC FIGHTER AIRCRAFT)

Data collected from the WVU six-degrees-of-freedom flight simulator for the supersonic fighter aircraft were used to evaluate the performance of the structured non-self and artificial DC approaches for identification of subsystem failures.

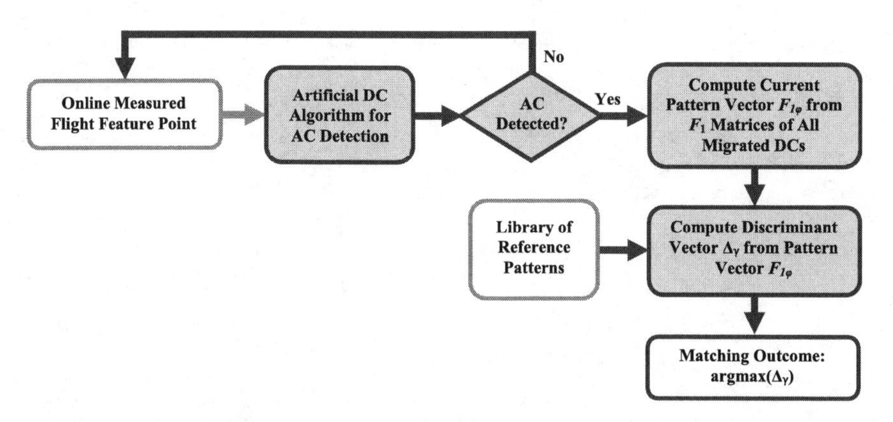

Fig. 4.30 AC detection and identification using the artificial DC mechanism and the Naïve Bayes classifier.

4.7.3.1 STRUCTURED NON-SELF APPROACH FOR AC IDENTIFICATION

The first phase of the structured non-self approach [110] is the result of the AC detection within the HMS strategy and consists of selecting lower dimensional projections with high specific detection potential. A total of 496 2D self/non-self projections were generated and tested against over 20 different failures. As a result of this phase, referred to as Phase I Non-Self Structuring, a total of 183 projections were identified as reaching a sample percentage detection rate of at least 70 percent, for one or more ACs. Various projections may present the ability to capture the dynamic fingerprint of several ACs, while others can only capture the dynamics of a small set or just one single AC. The negative selection-type of logic behind the Phase I Structuring resulted in the reduction of the number of the relevant projections, thus reducing the complexity and the hardware requirements for its implementation.

Phase II Non-Self Structuring includes a PS-type of process where flight test data under ACs are used to generate higher resolution identifiers covering non-self, which are capable of differentiating dynamic fingerprints among ACs. The identifiers are organized in a single array such that the index of each identifier corresponds to a specific AC type and magnitude, thus merging AC identification and direct evaluation. The arrangement of the identifiers is inspired by a mapping-based algorithm, which simplifies the selection scheme. The positive selection process is performed in parallel by all of the projections included in the identification algorithm. Each projection outputs one single identification index. The outputs of all projections are fused to generate the identification outcome.

The structured non-self approach was successfully tested against several different failures affecting actuators, sensors, wing integrity, and engines at different levels of severity [110]. Sample percentage success rate ranged between 86.8 percent and 99.9 percent, with an average of 94.5 percent, while only 93 of the most well-performing 2D projections were used.

4.7.3.2 ARTIFICIAL DC APPROACH FOR AC IDENTIFICATION

Among the three pattern approaches possible within the artificial DC method, the FP approach requires less computational resources, and it is the only one implemented and tested to date [108, 109]. The Naïve Bayes classifier was trained with the current FP vectors $F_{1\varphi}$ from a set of training tests under ACs for each subsystem k to implicitly define the reference features-pattern FP_k of that subsystem. As an example, Fig. 4.31 [109] illustrates the variation of the current FP vectors $F_{1\varphi}$ over the entire test time of a failed right wing subsystem from one of the validation tests.

The artificial DC mechanism was successfully tested for identification of different failures affecting the three axes aerodynamic control surfaces, integrity of wing, and angular rate sensors used in the control augmentation system [108]. The sample identification percentage rate was very high, between 99 and 100 percent. The approach used all 2D projections possible with 28 features.

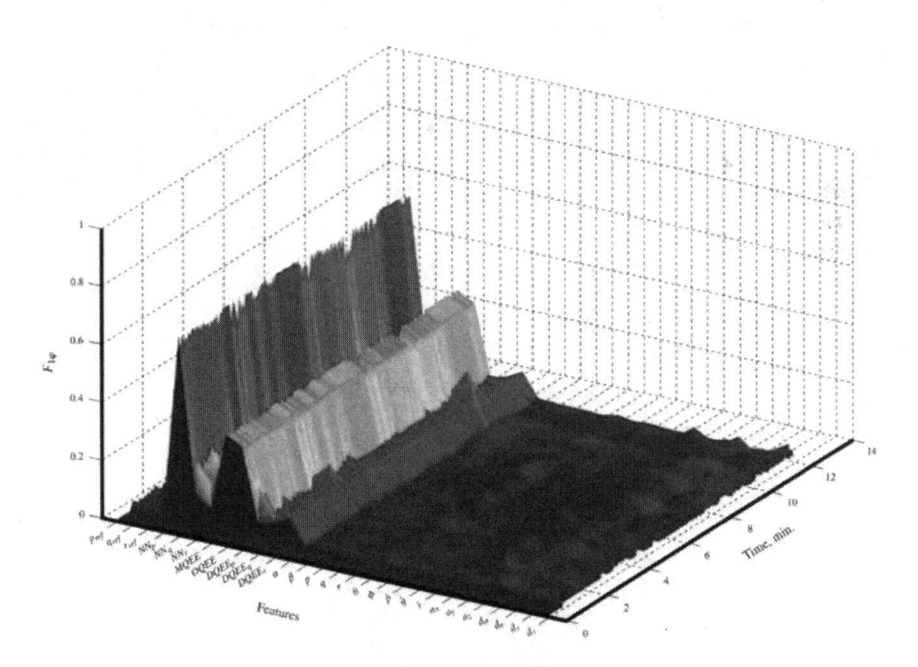

Fig. 4.31 Variation of the feature-pattern vector with time of a damaged right wing [109].

4.8 IMMUNITY-BASED AIRCRAFT ABNORMAL CONDITION EVALUATION

4.8.1 AIRCRAFT AC DIRECT EVALUATION

Aircraft AC direct evaluation within the AIS paradigm addresses two main aspects: determination of AC type or nature (referred to as qualitative direct evaluation) and determination of AC severity or magnitude (referred to as quantitative direct evaluation). Both aspects can be approached using similar tools as for AC identification, and AC direct evaluation may also be viewed as a phase of the AC identification process.

4.8.1.1 STRUCTURED NON-SELF APPROACH FOR AC DIRECT EVALUATION

The structured non-self approach for AC identification relies on the availability of information regarding the regions of the non-self hyperspace to which the feature points are expected to migrate when ACs affecting specific subsystems occur. If such information exists, it is likely that information on the AC type and severity is available for those subsystems as well. This allows for the development of a mapping-based, PS-type of approach that would consider AC identification and direct evaluation simultaneously. When detectors (clusters in the non-self) are generated, they must be tagged not only for the affected subsystem, but also for the type and severity of the AC. For example, the structure of a hyper-spherical detector (thus converted into an identifier and evaluator) becomes:

$$d = \begin{bmatrix} c & R & T_{id} & T_q & T_{dq} \end{bmatrix} \tag{4.45}$$

where c is an N-dimensional vector representing the center of the hyper-sphere within the feature hyperspace, R is the scalar radius of the cluster, T_{id} is the subsystem identifier, T_q is the AC type identifier, and T_{dq} is the AC severity evaluation on a predefined scale. Note that T_{dq} can also be inferred based on the distance from the current feature point to the self [95]. This distance can actually be T_{dq}, which then has to be processed to provide AC evaluation outcomes. An illustration of this concept for AC direct quantitative evaluation is presented in Fig. 4.32 [97].

Assuming that a full N-dimensional self/non-self representation is used, once the detection outcome is positive ($\mathcal{O}_D = 1$), the most recent triggered detector is checked for structural parameters. A current AC direct evaluation outcome is defined as:

$$o_{ET} = T_q, \quad o_{ES} = T_{dq} \tag{4.46}$$

If the HMS strategy is used, then the triggered detectors from all N_π projections must be considered and a composition logic $\mathbb{C}$ must be designed to obtain the current AC direct evaluation outcome as:

$$o_{ET} = \mathbb{C}(T_{q_i}), \quad o_{ES} = \mathbb{C}(T_{dq_i}), \quad i = 1, 2, \ldots, N_\pi \tag{4.47}$$

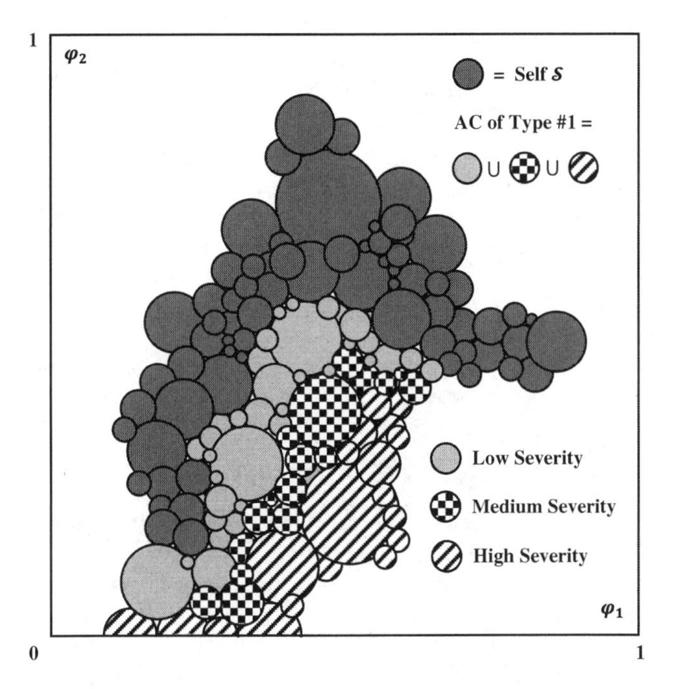

Fig. 4.32 AC direct quantitative evaluation using the structured non-self approach.

Note that, within the HMS strategy, explicit structuring of the non-self can be avoided if an adequate composition logic $\mathbb{C}$ is formulated [107].

The structured non-self approach performing simultaneous AC detection, subsystem identification, and AC severity evaluation has been tested for a complex AC scenario using data from the WVU motion-based flight simulator of a supersonic fighter aircraft [110], which has been introduced in previous sections. The self/non-self was created covering large regions of the flight envelope. Nine different failures were used for verification including lockage of aerodynamic control surface actuators, wing damage, propulsion system malfunction, and angular rate sensor bias. Each AC was tested at low and high severity levels.

The performance evaluation of the approach was based on similar metrics as for the AC detection. A correct identification/evaluation percentage and a missed-identification/evaluation percentage were calculated depending on an accurate declaration of subsystem failure for every time step in which an AC is present. The identifier/evaluator generation algorithm was implemented for all nine different ACs considered to be high magnitude using a total of 183 selected projections. After a heuristic optimization, it was established that the number of relevant projections can be further reduced to 93, with significant computational savings and no performance penalty. Based on the assumption that lower magnitude ACs of the same type generate similar dynamic fingerprints with a close proximity to the self, the set of identifiers was subdivided into two groups: a high magnitude set and a low magnitude set. Therefore, the total number of AC categories was 18. Each set of identifiers/evaluators generated for each AC contained on average 36 identifiers. Considering that every set of identifiers for all failures were integrated into each projection, an average total of 648 identifiers/evaluators per projection are used within this PS-type of process.

The structured non-self algorithm for AC identification and direct evaluation was tested under 16 different failures at low and high severity levels affecting actuators, sensors, structural elements, and engines. The rates of correct detection, identification, and evaluation for all ACs considered range between 86.8 percent (for a right stabilator locked at 2 degrees) and 99.9 percent (right engine malfunction), with an average of 94.5 percent.

4.8.1.2 ARTIFICIAL DC APPROACH FOR AC DIRECT EVALUATION

With the artificial DC approach, the AC direct evaluation is converted into a specific identification problem, using the same algorithms. While the AC identification presented previously is a pattern recognition problem, where one must distinguish between N_S different patterns each corresponding to one of the N_S subsystems considered, the AC direct evaluation is equally a pattern recognition problem, where one must distinguish between $N_{\gamma k}$ different categories of ACs corresponding to subsystem k. Note that AC categories will be considered to be "type" ($\gamma = \tau$) or "severity" ($\gamma = \sigma$).

In a similar manner to the structured non-self approach, the AC direct evaluation based on the DC mechanism may be viewed as an identification process where the target is the "AC," which has the type or the severity of the failure as a defining element in association with the affected subsystem. Patterns for direct evaluation are produced in a similar manner to the ones defined for identification and rely on the F_1 matrix. The flowchart of the process of establishing direct evaluation patterns in conjunction with the Naïve Bayes classifier is similar to the one presented in Fig. 4.29, with proper assignment of γ. For each subsystem, $N_{\gamma k}$ different patterns must be established, one associated to each AC category considered. As is the case with AC identification, three possible alternatives exist for the definition of the patterns: FP, PP, and MP approaches. The online AC direct evaluation, once a library of reference patterns is available, follows the outline presented in Fig. 4.30.

The artificial DC approach for AC direct evaluation has been implemented for the WVU supersonic aircraft fighter [108, 109]. An example of a reference FP vector is presented in Fig. 4.33 for 28 features for a stabilator high-severity failure. The approach was tested for ACs affecting four main categories of subsystems: the aerodynamic control surfaces, wing and tails, angular rate sensors, and engines. Three levels of severity were tested for the first two categories and two levels for the last two categories. The success rate of direct evaluation after

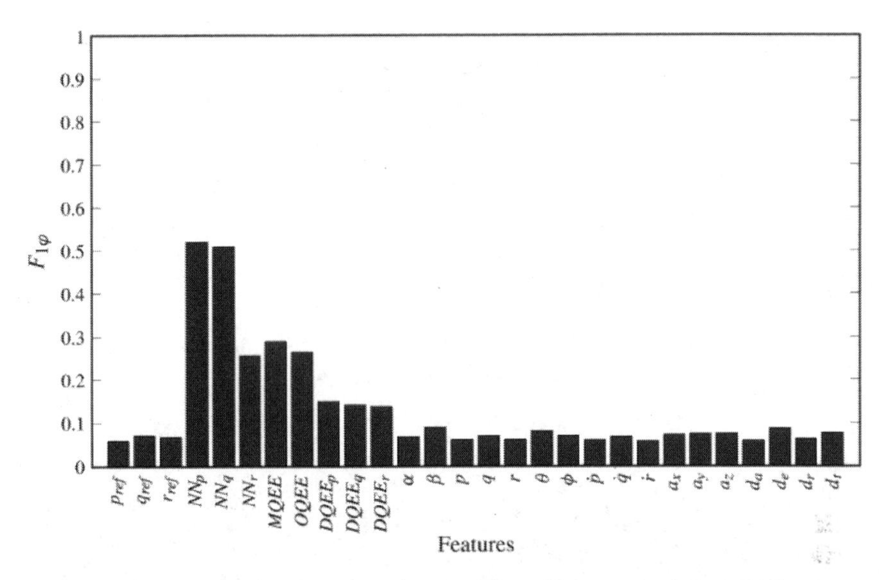

Fig. 4.33 Reference FP vector for a stabilator high-severity failure [109].

successful identification varied between 95.5 percent and 100 percent, with an average of 99.8 percent.

4.8.2 AIRCRAFT AC INDIRECT EVALUATION

The AC indirect evaluation requires that the direct evaluation is successful. It is a highly specific process and requires individualized approaches due to the synergistic interaction between aircraft subsystems, aircraft states, the nature of the ACs, and the multidimensionality of the space of variables subjected to analysis, namely the variables defining the generalized flight envelope [112, 127]. A comprehensive indirect evaluation or altered flight envelope prediction must rely on a combined strategy based on analytical flight envelope assessment and AIS-based approaches for parameter space alteration assessment. In most situations, the analytical methods require accurate modeling of the failures and significant online computational capabilities. The AIS method implies that all pertinent parameters to the flight envelope—considering its generalized meaning—are part of the feature sets that define the "self."

For a targeted class of ACs, specific algorithms for immunity-based AC indirect evaluation have been proposed and successfully demonstrated through simulation for actuators of aerodynamic control surfaces [128], angular rate sensors [129], and main aerodynamic surfaces [130]. All of these efforts have been focused on classes of ACs that alter subsystem aerodynamics in a manner consistent with and predictable based on aircraft operation at normal condition, referred to as "self" within the AIS paradigm. While the approach covers a large area of aircraft operation under AC, the extension of the AIS paradigm to the evaluation of other classes of ACs is still to be investigated.

Let a directly involved variable (DIV) v_δ in the AC be a variable whose alteration or abnormal variation is directly and significantly the result of the AC. Typically, DIVs are used to define/characterize the AC. They may be part of the feature set or not. If they are not, then a relationship between the DIV and some other variable(s) in the feature set must be established. This process will define equivalent directly involved variables (EDIV) v_ε, which are part of the feature set. For example, consider the case of the left elevator locked failure. The DIV can be defined as the left elevator deflection δ_{eL}. It obviously defines the AC; however, let us assume that it is not part of the feature set. A relationship can be formulated between the left elevator deflection δ_{eL} and the longitudinal stick displacement d_e—which presumably is a feature—of the form $d_e = f(\delta_{eL})$, where $f(x)$ is a known function. Therefore, the EDIV is, in this case, d_e. For each AC considered, a set of variables v_E must be determined that are affected by the failure, that are part of the feature set, and that are relevant from the point of view of aircraft operation (they are relevant variables with respect to the generalized flight envelope). These variables may be determined through analytical means but also through the analysis of the 2D selves within the AIS paradigm.

The self can be viewed as a generalized flight envelope based on features φ_i. Let us assume that each failure $f_l \in F, l = 1, 2, \ldots, N_F$, produces a set of N_{Γ_l} constraints Γ_l on a set of known variables X_l where:

$$\Gamma_l = \{\gamma_{l1} \quad \gamma_{l2} \quad \cdots \quad \gamma_{lN_{\Gamma l}}\} \tag{4.48}$$

$$\gamma_{lm} = \gamma_{lm}(X_l) = 0, \quad l = 1, 2, \ldots, N_F \quad \text{and} \quad m = 1, 2, \ldots, N_{\Gamma_i} \tag{4.49}$$

The variables X_l must be part of the feature set, that is:

$$X_l = \{\, x_{l1} \quad x_{l2} \cdots \quad x_{lN_X} \,\} \subset \mathcal{F} \tag{4.50}$$

Let us assume that the alteration of the flight envelope is assessed in terms of a set $Y \subset \mathcal{F}$ of N_Y variables y_n, that is:

$$Y = \{\, y_1 \quad y_2 \quad \cdots \quad y_{N_Y} \,\} \subset \mathcal{F} \tag{4.51}$$

If the constraints Γ_l are known, then they specify any alteration of the variables y_n, and the problem would be solved. This is generally not the case. For this class of ACs, a new self under AC may be obtained [127] based on a set of features. If $Y = \mathcal{F}$, structuring of the non-self is necessary for a solution. In general, $N_Y < N$, and an assessment of flight envelope reduction is equivalent to projecting the non-self region corresponding to failure f_l onto the N_Y-dimensional sub-space. If proper structuring is not possible, one can still obtain the N_Y-dimensional projection of the self, and, for specific types of ACs, this makes it possible to infer AC effects on the flight envelope. For example, let us consider a stuck aileron failure (the left or right aileron is locked at a constant deflection $\delta_a = \delta_{a0}$). The effects on aircraft roll rate p must be assessed. Therefore, $x_1 = \delta_a$, $y_1 = p$, and $\gamma_1 \equiv \delta_a - \delta_{a0} = 0$. It can be stated that the values of the roll rate at post failure conditions will be in the range of normal conditions corresponding to δ_{a0}. Therefore, there will be an overlap between the non-self and self on the 2D plane (x_1, y_1). Then, a "new" self $\hat{S}$ or reduced envelope under AC can be defined as:

$$\hat{S} = \{c_i \mid \gamma_{lm} = 0\} \tag{4.52}$$

The AIS-based concept for the indirect evaluation of this class of ACs is illustrated in Fig. 4.34 [97] for the 2D case.

The AC detection, identification, and direct evaluation must be correctly completed prior to indirect evaluation. Therefore, the sets of DIV (v_δ), EDIV (v_ε), and envelope relevant variables (v_E) are determined. Note that, in general:

$$\begin{aligned} v_\delta &= \begin{bmatrix} v_{\delta 1} & v_{\delta 2} & \dots & v_{\delta N_\delta} \end{bmatrix} \\ v_\varepsilon &= \begin{bmatrix} v_{\varepsilon 1} & v_{\varepsilon 2} & \dots & v_{\delta N_\varepsilon} \end{bmatrix}, \quad v_\varepsilon \subset \mathcal{F} \\ v_E &= \begin{bmatrix} v_{E1} & v_{E2} & \dots & v_{EN_E} \end{bmatrix}, \quad v_E \subset \mathcal{F} \end{aligned} \tag{4.53}$$

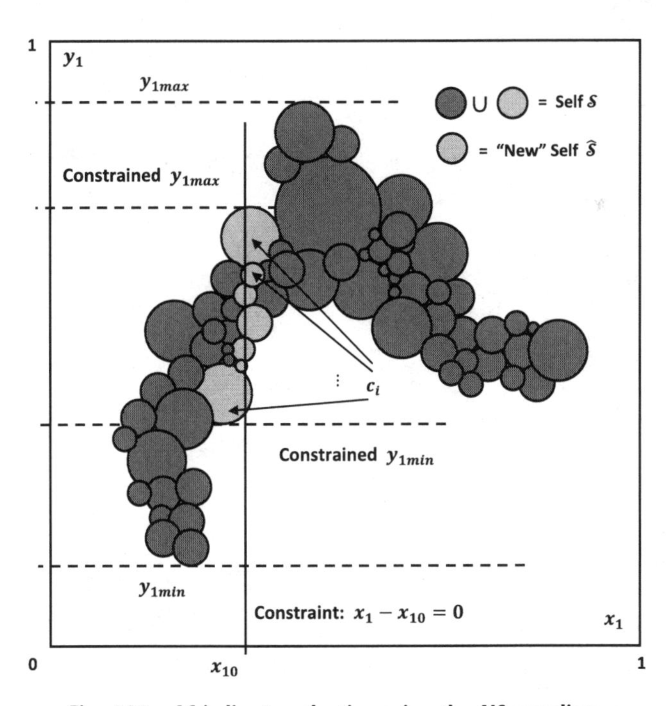

Fig. 4.34 AC indirect evaluation using the AIS paradigm.

and the relationship between v_ε and v_δ must be known:

$$v_\varepsilon = f_{\varepsilon\delta}(v_\delta) \tag{4.54}$$

where

$$f_{\varepsilon\delta} = \begin{bmatrix} f_{\varepsilon\delta 1} & f_{\varepsilon\delta 2} & \cdots & f_{\varepsilon\delta N_\varepsilon} \end{bmatrix} \tag{4.55}$$

such that

$$v_{\varepsilon i} = f_{\varepsilon\delta i}(v_\delta), \quad i = 1, 2, \ldots, N_\varepsilon \tag{4.56}$$

Specific algorithms for immunity-based AC indirect evaluation or flight envelope prediction must be developed for all targeted variables v_E in conjunction with the nature and type of the failure. For example, let as assume that we need to assess the impact on the flight envelope of a left elevator deflection (δ_{eL}) lockage. The approach is applied similarly to failures affecting the right aerodynamic control surface. The following parameters can be specified [127, 128]:

$$\begin{aligned} v_\delta &= [v_{\delta 1}] = [\delta_{eL}] \\ v_\varepsilon &= [v_{\varepsilon 1}] = [d_e], \quad d_e \subset \mathcal{F} \\ v_E &= \begin{bmatrix} a_x & a_z & H & M & p & q & \dot{q} & \alpha & \theta \end{bmatrix}, \quad v_E \subset \mathcal{F} \end{aligned} \tag{4.57}$$

where d_e is the longitudinal stick displacement and the envelope-relevant variables include translational acceleration, altitude, Mach, roll rate, pitch rate, pitch acceleration, angle of attack, and pitch attitude. The relationship between DIV and EDIV is:

$$d_e = k_e \delta_e = k_e \frac{\delta_{eL} + \delta_{eR}}{2} \tag{4.58}$$

It is assumed that the left elevator is locked at a value determined through direct quantitative evaluation:

$$\delta_{eL}(t) = \delta_{eLF} \tag{4.59}$$

The nominal range for d_e is:

$$d_{e\max} = k_e \frac{\delta_{eL\max} + \delta_{eR\max}}{2}, \quad d_{e\min} = k_e \frac{\delta_{eL\min} + \delta_{eR\min}}{2} \tag{4.60}$$

After failure, the range for d_e becomes:

$$d_{e\max F} = k_e \frac{\delta_{eLF} + \delta_{eR\max}}{2}, \quad d_{e\min F} = k_e \frac{\delta_{eLF} + \delta_{eR\min}}{2} \tag{4.61}$$

Note that this is an equivalent or virtual constraint range that reflects the constraint on surface deflection. The controls in the cockpit are still able to move over full range. New ranges for all v_E may be obtained from the self corresponding to the equivalent range of v_ε at post-failure conditions ($d_{e\min F}$, $d_{e\max F}$).

A computationally convenient approach is to use 2D projections (v_ε, v_E) for those pairs of variables that are mutually constrained, which is similar to a cause/effect relationship. If the two variables are not mutually constrained, then each may reach all full-range values irrespective of the other, and the 2D self projection is a rectangle. In this case, further analysis is needed. For example, a locked elevator will limit the control authority on the roll channel, because it is producing undesirable roll that must be compensated with aileron deflection. The projection (d_e, p) does not capture this effect. The limitations of producing roll rate may be converted into an aileron pseudo-failure, which consists of a reduced deflection range that is available after the elevator failure induced roll rate has been compensated. The reduced d_a range after failure may be used with the (d_a, p) projection to obtain the new range of controllable roll rate [128].

The WVU supersonic fighter motion-based flight simulator was used to develop and test AC indirect evaluation schemes. The performance of the post-failure envelope prediction methodology is considered to be adequate, if the data points collected under ACs are within the predicted ranges and cover them up to the proximity of range borders. Therefore, the metrics defined for the performance assessment of AC indirect evaluation schemes must capture the level of prediction confidence, the level of possible range exceedance, and the level of predicted range conservativeness [127]. The metric for the first criterion is a prediction rate (*PR*) defined as the percentage of all verification data points for a targeted relevant variable v_{Ei} that fall inside the predicted range $[v_{Ei\min}, v_{Ei\max}]$. The prediction rate is calculated as:

$$PR = \frac{N_R}{N_V} \cdot 100 \tag{4.62}$$

where N_V is the total number of validation points, N_R is the number of points inside, and $N_{\bar{R}}$ outside the predicted range, such that $N_V = N_R + N_{\bar{R}}$. The index of the predicted range exceedance (*REI*) is calculated as:

$$REI = \frac{\max_{N_{\bar{R}}}[\min(|v_{Ei} - v_{Ei\max}|, |v_{Ei} - v_{Ei\min}|)]}{|v_{Ei\max} - v_{Ei\min}|} \cdot 100 \tag{4.63}$$

and it represents the maximum relative distance between verification points and predicted range limits. Obviously, it is desirable to have large values of *PR* and low values of *REI*. However, it should be noted that, if the predicted range is exceedingly/unrealistically large, then *PR* = 100% and *REI* = 0% without representing good performance. Therefore, the values of the previous two evaluation metrics must be considered in conjunction with a margin index (*MI*) that is expected to capture the level of predicted range conservativeness. *MI* is defined as:

$$MI = \frac{\max_{N_{\bar{R}}}[\min(|v_{Ei} - v_{Ei\max}|, |v_{Ei} - v_{Ei\min}|)]}{|v_{Ei\max} - v_{Ei\min}|} \cdot 100 \tag{4.64}$$

For proper performance assessment, the verification tests must cover the range of the tests used for self-generation. Otherwise, test data may be located far from predicted range limits and *MI* may reach large values, which would not necessarily reflect inaccurate range prediction.

Implementation of the approach with the WVU supersonic fighter aircraft model [127–130] has demonstrated its capability to predict accurately post-failure ranges for relevant state variables such as angular rates and accelerations. A wide variety of ACs have been considered, including lockage of aerodynamic control surfaces (stabilator, aileron, and rudder) [128], bias in angular rate sensors output used in augmentation control laws (pitch, roll, and yaw) [129], and structural damage (wing and horizontal tail) [130].

Figure 4.35 shows an example of a 2D projection (pitch rate q versus longitudinal stick d_e) of self/non-self. The vertical lines limit the equivalent range of the longitudinal stick displacement, while the horizontal lines limit the range of achievable pitch rate at post-failure conditions. In this example, the right

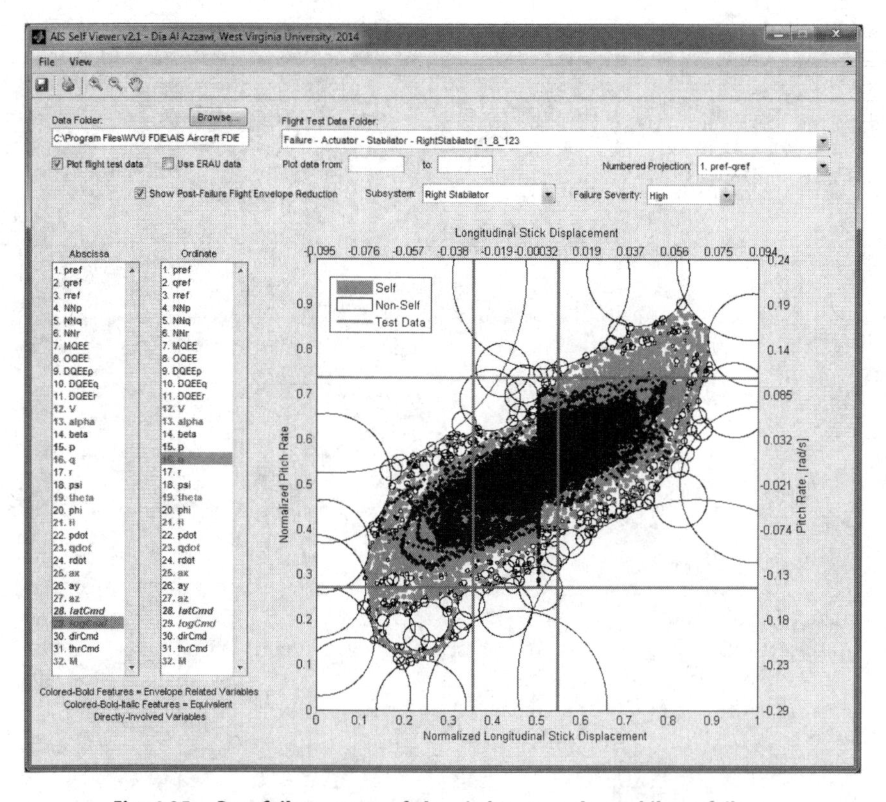

Fig. 4.35 Post-failure range of the pitch rate under stabilator failure.

stabilator was locked at 8 degrees. Verification test points are also presented. They yielded a *PR* of almost 100 percent with only a few points exceeding the upper limit of the predicted pitch rate range with 0 margin index.

4.9 IMMUNITY-BASED AIRCRAFT ABNORMAL CONDITION ACCOMMODATION

The aircraft AC accommodation within the AIS paradigm can be approached along two directions. One consists of developing adaptive control laws based on the antibody activation/suppression mechanism [80]. The other relies on extending the classification capabilities of the AIS and using them not only to detect and diagnose the problem but also to select or find the adequate compensation [93].

4.9.1 IMMUNITY-INSPIRED ADAPTIVE CONTROL MECHANISM

The biological feedback that establishes a balance between the activation and suppression of antibodies generation can be converted into an adaptive mechanism augmenting a baseline controller [97, 112, 131]. For the biological system, the antigens α_a active in triggering the immunity reaction are the result of the antiseptic action of the antibodies α_d on the invading antigens α. Therefore:

$$\alpha_a(t) = \alpha(t) - \alpha_d(t) \tag{4.65}$$

The active antigens α_a trigger the excitation τ_e of mechanisms that produce helper T-cell τ_H, such that:

$$\tau_e(t) = \Phi_1(\alpha_a(t)) \tag{4.66}$$

and

$$\tau_H(t) = \Phi_2(\tau_e(t)) \tag{4.67}$$

The number of helper T-cells (τ_H) reflects the number and virulence of the antigens in the organism; hence, helper T-cells favor the generation of B-cells, which in turn accelerate the production of antibodies. Suppressor T-cells (τ_S) reflect the level of success of the immune system in counteracting the antigens and have an inhibitory effect on the generation of B-cells. Therefore, B-cell generation is stimulated by:

$$\tau_B(t) = \tau_H(t) - \tau_S(t) \tag{4.68}$$

The production and activation of B-cell β is regulated by the balance between helper and suppressor T-cells, such that the production of antibodies is at the necessary level to defend the organisms without being excessive:

$$\beta(t) = \Phi_3(\tau_B(t)) \tag{4.69}$$

Furthermore, the generation of antibodies u depends on the activity of B-cells:

$$u(t) = \Phi_4(\beta(t)) \tag{4.70}$$

The antibodies may undergo a maturation process, typically modeled as a delay function, such that:

$$u_m(t) = \Phi_5(u(t)) \tag{4.71}$$

Suppressor T-cells are produced depending on the current number of mature antibodies, their rate, and the current amount and virulence of antigens:

$$\tau_S = \Phi_6(u, \dot{u}, \tau_e) \tag{4.72}$$

The mature antibodies are released in the bloodstream and may take an active role in locating and destroying antigens, such that:

$$u_a = \Phi_7(u) \quad \text{and} \quad \alpha_d = \Phi_8(u_a) \tag{4.73}$$

The immune system mechanisms are not completely understood; therefore, the functions Φ_1 through Φ_8, which model them, must be determined. The block diagram of the immune system feedback model is presented in Fig. 4.36.

An immunity-inspired adaptive control mechanism has been investigated based on this architecture for an autonomous aerial vehicle [131] and the WVU supersonic fighter [101]. The proposed adaptive element was used to augment a baseline controller and tested for its capability to provide enhanced ACs accommodation. The baseline controller relied primarily on a model reference architecture that uses the pilot stick inputs to generate desired angular rate and angular acceleration commands that are then used to produce compensating

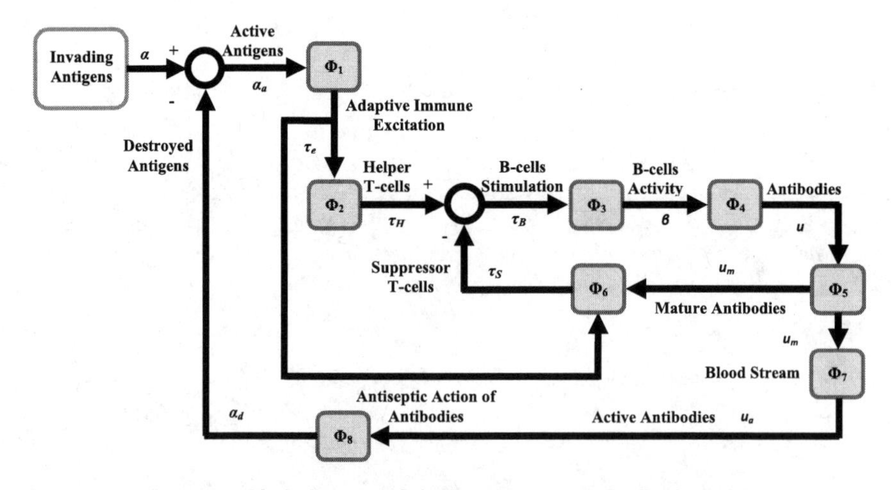

Fig. 4.36 Block diagram of the immune system feedback model.

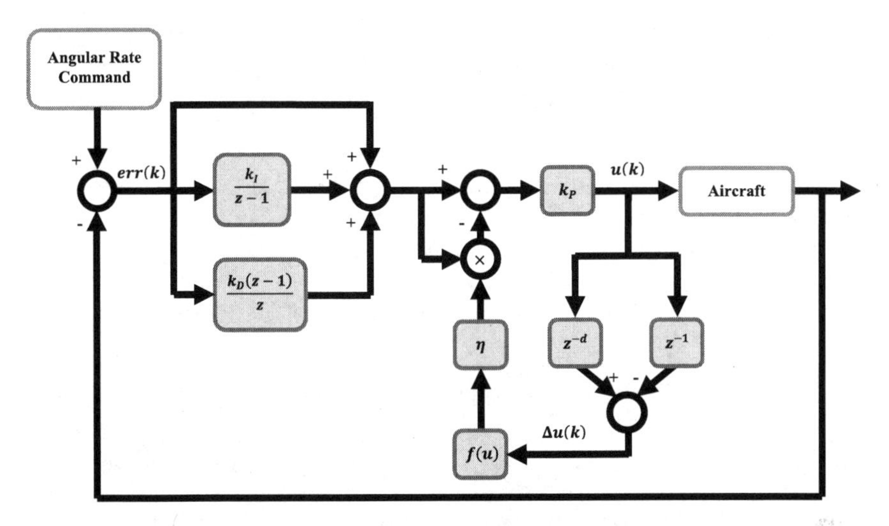

Fig. 4.37 Block diagram of the PID-AIS-based adaptive mechanism.

commands based on a proportional-integral-derivative (PID) approach in conjunction with approximate nonlinear dynamic inversion (NLDI). The immunity-based adaptive mechanism augments the PID part of the control laws as presented in Fig. 4.37, which yields the following discrete expression:

$$u(k) = k_P\left(1 + \frac{k_I}{z-1} + k_D\frac{z-1}{z}\right)[1 - \eta f(\Delta u(k))]err(k) \tag{4.74}$$

where

$$f(\Delta u(k)) = 1 - \frac{2}{e^{[\Delta u(k)^2]a} + e^{-[\Delta u(k)^2]a}} \tag{4.75}$$

The performance of the immunity-augmented architecture was assessed based on specific metrics defined in terms of total pilot input activity, angular rate tracking errors, and total amount of work required by each of the control surfaces [101]. Extensive simulation tests have shown the capability of the immunity-inspired adaptive mechanism to improve failure effects rejection by reducing tracking errors, while maintaining similar or lower levels of pilot and control surface activity.

4.9.2 STRUCTURED SELF/NON-SELF APPROACH FOR AIRCRAFT AC ACCOMMODATION

The second conceptual approach for using the AIS paradigm for control purposes relies on the assumption that the classification capabilities of the AIS can be extended [93] and used not only to detect the problem (AC) but also to select or find the solution for mitigating the AC effects. The control action is considered as a mapping between control and controlled variables under specific performance requirements or constraints. Therefore, if the AIS is to be used for control purposes, both control and controlled variables must be part of the feature set. Then the AIS framework provides the relationships between the control and controlled variables for the normal conditions and for those ACs for which detailed structuring of the non-self is available. Alternatively, the control variables associated with the adaptive immune feedback system and with the best compensation performance under specific upset conditions may be differentiated into memory cells, which can mount a faster and more aggressive secondary response in future encounters with the same AC.

Assume that there is a cause-and-effect relationship between a control variable x and the controlled variable y. If both variables are part of the feature set, then this relationship is intrinsically present in the self. If structuring of the non-self under AC is available, then similar information is present in the non-self as well. Therefore, either under normal or failure conditions, clusters c_j can be extracted for any desired value of the controlled variable y_d, as illustrated in Fig. 4.38 [97]. Compensatory commands x_{cmd} can be generated through further processing of clusters c_j. With proper dimensionality of y and proper self/non-self

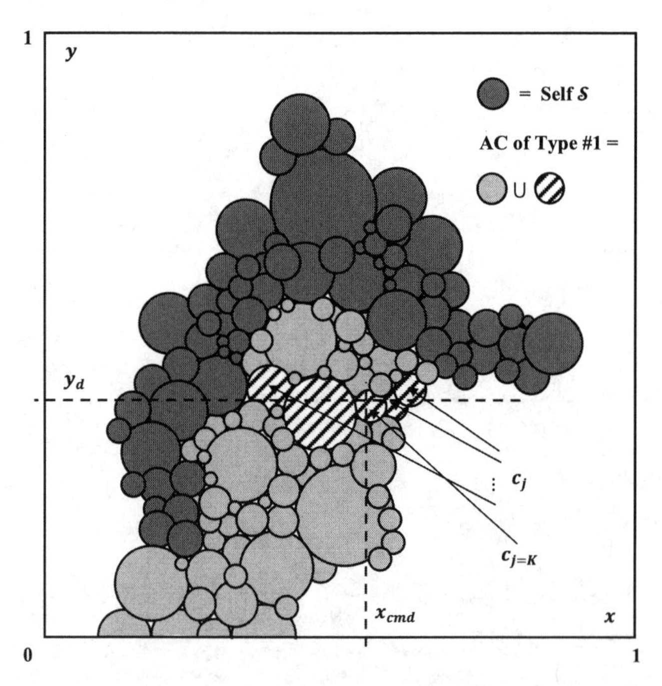

Fig. 4.38 Failure accommodation based on structured non-self [97].

resolution, one command detector or compensator, $c_{j=K}$, can be identified. Converting the compensator into a numerical value x_{cmd} may still require an adequate operator, which can generally be formulated as:

$$x_{cmd} = O(c_K) \tag{4.76}$$

The operator $O(*)$ may be as simple as selecting the center of the compensator.

With the structured non-self approach for AC accommodation, the AIS already built for previous ACM processes can be used for control. If substantial information about the AC and its compensation is available and already stored in the AIS, then algorithms are needed to extract that information and use it for generating commands. This can be achieved by memorizing the dynamic fingerprint of the AC during first encounter situations or simulations and by generating and recording adequate pilot compensation. A second scenario involves an unknown failure that requires a specific new compensation. In this case, algorithms must be developed capable of generating compensatory commands by properly composing bits of information within the self and non-self.

For the first scenario, a preliminary investigation on an aircraft has been performed [113, 132] with the objective of assessing the possibility of extracting useful control/compensation information out of properly structured AIS. This general feasibility study was performed using data collected from the WVU motion-based flight simulator for the supersonic fighter aircraft at normal and failure conditions for two simple maneuvers: uniform symmetric climb/descent and coordinated turn. The ACs considered were actuator locked (stabilator, aileron, and rudder) and pitch rate sensor bias. The strategy was to build through simulation a structured self/non-self that would allow for the extraction of compensatory commands if an AC is detected, identified, and evaluated.

The feature set must be extended to include a description of the mission/task to be performed. The pilot commands necessary to accomplish it must be available and separated from the rest of the features for convenience. Therefore, the extended feature set $\mathcal{F}_E$, is expressed as:

$$\mathcal{F}_E = \{\mathcal{F} \quad \mathcal{F}_{REF} \quad \mathcal{F}_{PC}\} \tag{4.77}$$

$$\mathcal{F} = \{\varphi_i \quad | \quad i = 1, 2, \ldots, N\} \tag{4.78}$$

$$\mathcal{F}_{REF} = \{\varphi_{REFi} \quad | \quad i = 1, 2, \ldots, N\} \tag{4.79}$$

$$\mathcal{F}_{PC} = \{\varphi_{PCi} \quad | \quad i = 1, 2, \ldots, N_{PC}\} \tag{4.80}$$

where $\mathcal{F}_{REF}$ represents the input to the pilot defining the task, and each variable φ_{REFi} represents the desirable values of corresponding φ_i. Pilot commands are defined by N_{PC} variables φ_{PCi}.

A reference task matrix $\mathcal{M}_{REF}$ is predefined based on required values $\mathcal{F}_{REF}$ at each time instant k, which form one row of $\mathcal{M}_{REF}$, denoted as $\mathcal{M}_{REF}(k, :)$. Piloted tests in the WVU flight simulator were performed with the pilot attempting to execute as accurately as possible the designed tasks. The values of features $\mathcal{F}$ are stored as a feature point matrix for pilot performance $\mathcal{M}_{\mathcal{F}P}$. The values of pilot produced commands are stored as a pilot command matrix $\mathcal{M}_{PC}$. Within the AIS framework, for AC accommodation purposes, the aircraft following the desirable flight path was assimilated to a healthy organism. The residual between current feature point $P(k)$ and the reference feature values were defined as invading entities or antigen. The commands provided by the pilot that are trying to bring the aircraft to the desirable state were regarded as immune system antibodies. The generation of antibodies for subsequent invasions is governed by memory T-cells and B-cells. Because pilot compensation is triggered by tracking errors, the residuals between desired feature values and actual features are stored as an artificial immunity memory cell matrix $\mathcal{M}_M$. This process is an analogy to the immune system being exposed to a disease for the first time. The adaptive immune system fights back and generates antibodies, which eventually eliminate the disease and are ready to do the same in the future should similar antigens invade again. Therefore, the set of antigens is expressed as:

$$\mathcal{M}_{AG} = \mathcal{M}_{REF}(k, :) - P(k) \tag{4.81}$$

A set of immunity memory cells is defined as:

$$\mathcal{M}_M = \mathcal{M}_{REF} - \mathcal{M}_{\mathcal{F}P} \tag{4.82}$$

Finally, antibodies are simply:

$$\mathcal{M}_{AB} = \mathcal{M}_{PC} \tag{4.83}$$

At each time sample k, an incoming antigen $\mathcal{M}_{AG}$ is compared to the sets of artificial immunity memory cells from matrix $\mathcal{M}_M$. Once the "best match" is found, a corresponding set of antibodies or compensatory commands is retrieved from matrix $\mathcal{M}_{AB}$.

The study performed with the WVU supersonic fighter aircraft [132] included short duration flights consisting of steady state flight conditions and moderate maneuvers. These flight scenarios were simulated under both nominal and ACs for design/development and validation purposes. Only one failure at a time was injected. Two simplified maneuvers were considered: symmetric climb and coordinated turn. Two types of ACs were considered: left stabilator locked at +4 degrees and roll rate sensor step bias. Sensor noise and mild atmospheric turbulence were included in the simulation. For example, for a steady climbing maneuver, the following set of flight features FF was selected: aircraft velocity V, roll rate p, pitch rate q, yaw rate r, altitude H, and pitch angle θ:

$$\mathcal{F} = \{V, \quad p, \quad q, \quad r, \quad H, \quad \theta\} \tag{4.84}$$

The pilot generated commands were: longitudinal channel stick displacement d_e, lateral channel stick displacement d_a, pedals displacement d_r, and throttle displacement d_T. Therefore:

$$\mathcal{F}_{PC} = \{ d_e \quad d_a \quad d_r \quad d_T \} \tag{4.85}$$

The antigen $\mathcal{M}_{AG}$ would typically consist of elements that represent the differences between the desired and actual values of corresponding features, as defined in Eq. (4.81). However, the investigation in [132] revealed that augmenting the antigen vector with actual values of a subset of features may increase command extraction performance.

For all flight conditions investigated, it was demonstrated that the proposed approach is capable of extracting compensatory commands under nominal and failure conditions out of a properly structured self/non-self. Simple maneuvers were performed automatically with good accuracy based on AIS-extracted commands. These results represent preliminary steps; however, they create the premises for further investigation into closing the loop on a comprehensive immunity-based aircraft monitoring and control process that integrates, within one consistent methodology, aircraft ACDIEA.

REFERENCES

[1] "Aviation Accident Database and Synopses," National Transportation Safety Board, 1962–2016, http://www.ntsb.gov/_layouts/ntsb.aviation/index.aspx [retrieved Aug. 2016].

[2] "ASN Aviation Safety Database 1919–2016," Flight Safety Foundation, http://aviation-safety.net/database/ [retrieved Aug. 2016].

[3] "Accident and Incident Data," Federal Aviation Administration, 1978–2007, https://www.faa.gov/data_research/accident_incident/ [retrieved Aug. 2010].

[4] Oster, C. V., Jr., Strong, J. S., and Zorn, C. K., "Analyzing Aviation Safety: Problems, Challenges, Opportunities," *Research in Transportation Economics*, Vol. 43, No. 1, July 2013, pp. 148–164.

[5] Aaseng, G. B., "Blueprint for an Integrated Vehicle Health Management System," *Proceedings of the AIAA 20th Digital Avionics Systems Conference*, Vol. 1, 2001, pp. 3C1/1–3C1/11.

[6] Scandura, P. A., Jr., "Integrated Vehicle Health Management as a System Engineering Discipline," *Proceedings of the AIAA 24th Digital Avionics Systems Conference*, Vol. 2, 2005, pp. 7.D.1–1.

[7] Tipaldi, M., and Bruenjes, B., "Spacecraft Health Monitoring and Management Systems," *MetroAeroSpace Conference*, IEEE, 29–30 May, 2014, ISBN 978-1-4799-2069-3.

[8] Avram, R., Zhang, X., and Khalili, M., "Quadrotor Actuator Fault Diagnosis with Real Time Experimental Results," *Annual Conference of the Prognostics and Health Management Society*, 2016.

[9] Stolzer, A. J., Halford, C. D., and Goglia, J. J., "Safety Management Systems in Aviation," Ashgate Publishing, Surrey, United Kingdom, 2008.

[10] “Concept of Operations for the Next Generation Air Transportation System”, Joint Planning and Development Office, Version 3.2, Washington, DC, 2011.

[11] “Advancing Aeronautical Safety: A Review of NASA’s Aviation Safety-Related Research Programs,” Committee for the Review of NASA’s Aviation Safety-Related Programs, Aeronautics and Space Engineering Board, Transportation Research Board, Division on Engineering and Physical Sciences, National Research Council, Publisher: National Academies Press, Washington, DC, 2010.

[12] “Global Aviation Safety Plan: 2014 – 2016,” International Civil Aviation Organization, Montréal, Canada, 2013.

[13] Jennions, I. K. (editor), *Integrated Vehicle Heath Management: The Technology*, SAE International, Warrendale, PA, 2013.

[14] Fudge, M., Stagliano, T., and Tsiao, S., “Non-Traditional Flight Safety Systems & Integrated Vehicle Health Management Systems—Descriptions of Proposed & Existing Systems and Enabling Technologies & Verification Methods,” Final Report, The Office of the Associate Administrator for Commercial Space Transportation, Federal Aviation Administration, 2003.

[15] Benedettini, O., Baines, T. S., Lightfoot, H. W., and Greenough, R. M., “State-of-the-Art in Integrated Vehicle Health Management,” *Proceedings of the Institute of Mechanical Engineering, Part G: Journal of Aerospace Engineering*, Vol. 223, No. 2, 2009, pp. 157 – 170.

[16] Nguyen, N., and Krishnakumar, K., “Hybrid Intelligent Flight Control with Adaptive Learning Parameter Estimation,” *Journal of Aerospace Computing, Information, and Communication*, Vol. 6, No. 3, March 2009, pp. 171 – 186, doi: 10.2514/1.35929.

[17] Campbell, S., Kaneshige, J., Nguyen, N., and Krishnakumar, K., “Implementation and Evaluation of Multiple Adaptive Control Technologies for a Generic Transport Aircraft Simulation,” *AIAAInfotech@Aerospace 2010*, AIAA Paper 2010-3322, April 2010.

[18] Marwaha, M., and Valasek, J., “Fault Tolerant Control Allocation for Mars Entry Vehicle Using Adaptive Control,” *International Journal of Adaptive Control and Signal Processing*, Vol. 25, No. 2, Feb. 2011, pp. 95 – 113.

[19] Lakshmikanth, G. S., Padhi, R., Watkins, J. M., and Steck, J. E., “Adaptive Flight Control Using Single Network Adaptive Critic Aided Nonlinear Dynamic Inversion,” *AIAA Journal of Aerospace Information Systems*, Vol. 11, pp. 785 – 806, doi: 10.2514/1.I010165, 2014.

[20] Shin, J. Y., Wu, N. E., and Belcastro, C., “Linear Parameter Varying Control Synthesis for Actuator Failure, Based on Estimated Parameter,” *AIAA Guidance, Navigation and Control Conference*, AIAA Paper 02-4546, Aug. 2002.

[21] Azam, M., Pattipati, K., Allanach, J., and Patterson-Hine, A., “In-Flight Fault Detection and Isolation in Aircraft Flight Control Systems,” *Aerospace Conference, 2005 IEEE*, 5 – 12 March 2005, pp. 3555 – 3565, doi: 10.1109/AERO.2005.1559659.

[22] Boskovic, J., Redding, J., and Knoebel, N., “An Adaptive Fault Management (AFM) System for Resilient Flight Control,” *Proceedings of the AIAA Guidance, Navigation, and Control Conference*, Chicago, IL, 2009.

[23] Byington, C., Watson, M., Edwards, D., and Stoelting, P., "A Model-Based Approach to Prognostics and Health Management for Flight Control Actuators," in *Proceedings of IEEE Aerospace Conference*, 2004.

[24] Ducard, J. G., "SMAC-FDI: A Single Model Active Fault Detection and Isolation System for Unmanned Aircraft," *International Journal of Applied Mathematics and Computer Science*, Vol. 21, No. 1, 2015, pp. 189–201.

[25] Napolitano, M. R., Younghawn, A., and Seanor, B., "A Fault Tolerant Flight Control System for Sensor and Actuator Failures Using Neural Networks," *Aircraft Design*, Vol. 3, No 2, 2000.

[26] Balaban, E., Bansal, P., Stoelting, P., Saxena, A., Goebel, K. F., and Curran, S., "A Diagnostic Approach for Electro-Mechanical Actuators in Aerospace Systems," *IEEE Aerospace Conference*, Paper #1345, 2009.

[27] Oonk, S., Maldonado, F. J., Figueroa, F., and Lin, C. F., "Predictive Fault Diagnosis System for Intelligent and Robust Health Monitoring," *Journal of Aerospace Computing, Information, and Communication*, Vol. 9, 2012, pp. 125–143, doi: 10.2514/1.54961.

[28] Perhinschi, M. G., Napolitano, M. R., Campa, G., Seanor, B., Burken, J., and Larson, R., "An Adaptive Threshold Approach for the Design of an Actuator Failure Detection and Identification Scheme," *IEEE Transactions on Control Systems Technology*, Vol. 14, No. 3, May 2006, pp. 519–525.

[29] Perhinschi, M. G., Napolitano, M. R., Campa, G., Fravolini, M. L., and Seanor, B., "Integration of Sensor and Actuator Failure Detection, Identification, and Accommodation Schemes within Fault Tolerant Control Laws," *Control and Intelligent Systems*, Vol. 35, No. 4, pp. 309–318, Actapress, Dec. 2007.

[30] Bharadwaj, R., Kim, K., Kulkarni, C. S., and Biswas, G., "Model-Based Avionics Systems Fault Simulation and Detection," *AIAA Infotech Aerospace*, AIAA-2010-3328, Atlanta, GA, 2010.

[31] Oosterom, M., and Babuska, R., "Virtual Sensor for Fault Detection and Isolation in Flight Control Systems—Fuzzy Modeling Approach," *Proceedings of the 39th IEEE Conference on Decision and Control*, Sydney, Australia, Dec., 2000, pp. 2645–2650.

[32] Hajiyev, C., and Caliksan, F., "Sensor and Control Surface/Actuator Failure Detection and Isolation Applied to F-16 Flight Dynamic," *Aircraft Engineering and Aerospace Technology*, Vol. 72, No. 2, 2005, pp. 152–160.

[33] Tessler, A., "Structural Analysis Methods for Structural Health Management of Future Aerospace Vehicles," *NASA/TM-2007-214871*, Langley Research Center, Hampton, Virginia, 2007.

[34] Yuan, F. G., (editor), "Structural Health Monitoring (SHM) in Aerospace Structures," *Woodhead Publishing Series in Composites Science and Engineering*, No. 68, Elsevier, 2016.

[35] Roemer, M. J., Ge, J., Liberson, A., and Kim, R. Y., "Autonomous Impact Damage Detection and Isolation Prediction for Aerospace Structures," *Proceedings of the 2005 IEEE Aerospace Conference*, Big Sky, MT, 2005, pp. 3592–3600.

[36] Nguyen, N., KrishnaKumar, K., Kaneshige, J., and Nespeca, P., "Dynamics and Adaptive Control for Stability Recovery of Damaged Asymmetric Aircraft," *Proceedings of the AIAA Guidance, Navigation, and Control Conference and Exhibit*, AIAA, Keystone, CO, Aug. 2006.

[37] Litt, J. S., Simon, D. L., Garg, S., Guo, T.-H., Mercer, C., Millar, R., Behbahani, A., Bajwa, A., and Jensen, D. T., "A Survey of Intelligent Control and Health Management Technologies for Aircraft Propulsion Systems," *NASA/TM—2005-213622*, NASA Glenn Research Center, May 2005.

[38] Kobayashi, T., and Simon, D. L., "Application of a Bank of Kalman Filters for Aircraft Engine Fault Diagnostics," *Proceedings of the ASME Turbo Expo 2003, Power for Land, Sea, and Air*, Atlanta, GA, June 16–19, 2003.

[39] Yedavalli, R. K., Shankar, P., Siddiqi, M., and Behbahani, A., "Modeling, Diagnostics and Prognostics of a Two-Spool Turbofan Engine," *Proceedings of the 41st AIAA/ASME/SAE/ASEE Joint Propulsion Conference and Exhibit*, AIAA 2005-4344, Tucson, AZ, July10–13, 2005.

[40] Melcher, K. J., Sowers, T. S., and Maul, W. A., "Meeting the Challenges of Exploration Systems: Health Management Technologies for Aerospace Systems with Emphasis on Propulsion," *NASA TM-2005-214026*, 2005.

[41] Romessis, C., and Mathioudakis, K., "Setting up of a Probabilistic Neural Network for Sensor Fault Detection Including Operation with Component Faults," *Journal of Engineering for Gas Turbines and Power*, Vol. 125, No. 3, 2003, pp. 634–641.

[42] Volponi, A., Brotherton, T., Luppold, R., and Simon, D. L., "Development of an Information Fusion System for Engine Diagnostics and Health Management," *NASA TM-2004-212924*, Feb. 2004.

[43] Romessis, C., and Mathioudakis, K., "Bayesian Network Approach for Gas Path Fault Diagnosis," *Journal of Engineering for Gas Turbines and Power*, Vol. 128, No. 1, 2006, pp. 64–72.

[44] Perhinschi, M. G., Smith, B., and Betoney, P., "Fuzzy Logic-Based Detection Scheme for Pilot Fatigue," *Aircraft Engineering and Aerospace Technology: An International Journal*, Vol. 82, No. 1, Jan.–Feb. 2010, pp. 39–47.

[45] Urnes, J. M., Reichenbach, E. Y, and Smith, T. A., "Dynamic Flight Envelope Assessment and Prediction," *Proceedings of the AIAA Guidance, Navigation and Control Conference and Exhibit*, Honolulu, HI, Aug. 2008.

[46] Yavrucuk, I., Unnikrishnan, S., and Prasad, J. V. R., "Envelope Protection for Autonomous Unmanned Aerial Vehicles," *Journal of Guidance, Control, and Dynamics*, Vol. 32, 2009, pp. 248–261.

[47] Tang, L., Roemer, M., Ge, J., Crassidis, A., Prasad, J. V. R., and Belcastro, C., "Methodologies for Adaptive Flight Envelope Estimation and Protection," *Proceedings of the AIAA Guidance, Navigation, and Control Conference*, Chicago, IL, Aug. 2009.

[48] Keller, J., McKillip, R., and Kim, S., "Aircraft Flight Envelope Determination using Upset Detection and Physical Modeling Methods," *Proceedings of the AIAA Guidance, Navigation, and Control Conference*, Chicago, IL, Aug. 2009.

[49] Lombaerts, T., Schuet, S., Wheeler, K., Acosta, D., and Kaneshige, J., "Safe Maneuvering Envelope Estimation Based on a Physical Approach," *Proceedings of AIAA Guidance, Navigation and Control Conference*, AIAA 2013-4618, Aug. 2013.

[50] Menon, P. K., Sengupta, P., Vaddi, S., Yang, B., and Kwan, J., "Impaired Aircraft Performance Envelope Estimation," *Journal of Aircraft*, Vol. 50, No. 2, March–April 2013, pp. 410–424.

[51] Zhang, Y., de Visser, C. C., and Chu, Q.-P., "Online Safe Flight Envelope Prediction for Damaged Aircraft: A Database-Driven Approach," *Proceedings of AIAA*

Modeling and Simulation Technologies Conference, AIAA SciTech, AIAA 2016-1189, doi:10.2514/6.2016-1189, 2016.

[52] Tang, L., Roemer, M., Bharadwaj, S., and Belcastro, C., "An Integrated Aircraft Health Assessment and Fault Contingency Management System," *Proceedings of the AIAA Guidance, Navigation and Control Conference and Exhibit*, Honolulu, HI, Aug. 2008.

[53] Belcastro, C. M., and Jacobson, S. R., "Future Integrated Systems Concept for Preventing Aircraft Loss-of-Control Accidents," *Proceedings of the AIAA Guidance, Navigation, and Control Conference*, Toronto, Canada, Aug. 2010.

[54] Belcastro, C. M., "Validation and Verification of Future Integrated Safety-Critical Systems Operating under Off-Nominal Conditions," *Proceedings of the AIAA Guidance, Navigation, and Control Conference*, Toronto, Canada, Aug. 2010.

[55] KrishnaKumar, K., Nguyen, N. T., and Kaneshige, J. T., *Integrated Resilient Aircraft Control*, Encyclopedia of Aerospace Engineering, Wiley, 2010, doi: 10.1002/9780470686652.eae510.

[56] Figueroa, F., and Melcher, K., "Integrated Systems Health Management for Intelligent Systems," *Advances in Intelligent and Autonomous Aerospace Systems, Progress in Astronautics and Aeronautics*, edited by J. Valasek, AIAA, New York, 2012, pp. 173–200.

[57] NASA. (2012) Fault Management Handbook NASA-HDBK-1002. [Online]. http://www.nasa.gov/sites/default/files/636372main_NASA-HDBK-1002_Draft.pdf [retrieved September 2016].

[58] Uhlig, D., Neogi, N., and Selig, M., "Health Monitoring via Neural Networks," *AIAA Infotech@Aerospace* 2010, Infotech@Aerospace Conferences, doi:10.2514/6.2010-3419.

[59] Bowman, C., and Tschan, C., "Condition-Based Health Management (CBHM) Architectures," *Infotech@Aerospace 2011*, AIAA 2011-1431, St. Louis, MO, 2011.

[60] Benjamini, E., *Immunology, A Short Course*, Wiley-Liss, New York, 1992.

[61] Janeway, C. A., Travers, P., Walport, M., and Shlomchik, M. J., *Immunobiology: The Immune System in Health and Disease*, 6th ed., Garland Science, New York, 2005.

[62] Farmer, J. D., Parkard, N. H., and Perelson, A. S., "The Immune System, Adaptation, and Machine Learning," *Physica D: Nonlinear Phenomena*, Vol. 22, No. 1–3, 1986, pp. 187–204, doi: 10.1016/0167-2789(86)90240-X.

[63] Dasgupta, D., and Attoh-Okine, N., "Immunity-Based Systems: A Survey," *IEEE International Conference on Systems, Man, and Cybernetics*, Vol. 1, Orlando, FL, 12–15 Oct. 1997, pp. 396–374.

[64] Dasgupta, D. (ed.), *Artificial Immune Systems and Their Applications*, Springer-Verlag, New York, 1998.

[65] Dasgupta, D., and Forrest, S., "Artificial Immune Systems in Industrial Applications," *Proceedings of the Second International Conference on Intelligent Processing and Manufacturing of Materials*, Vol. 1, Honolulu, HI, July 1999, pp. 257–267.

[66] DeCastro, L. N., and Timmis, J., *Artificial Immune Systems: A New Computational Intelligence Approach*, Springer-Verlag, New York, 2002.

[67] Dasgupta, D., and Nino, L. F., *Immunological Computation: Theory and Applications*, CRC Press, Boca Raton, 2008.

[68] González, F., *A Study of Artificial Immune Systems Applied to Anomaly Detection*, Ph.D. Dissertation, University of Memphis, Memphis, TN, 2003.

[69] Yeom, K.-W., *Immune-Inspired Algorithm for Anomaly Detection*, Computational Intelligence in Information Assurance and Security, Vol. 57, Springer-Verlag, Berlin, Germany, doi: 10.1007/978-3-540-71078-3_5, 2007, pp. 129–154.

[70] Coello, C. A. C., and Cortés, N. C., "Solving Multiobjective Optimization Problems Using an Artificial Immune System," *Genetic Programming and Evolvable Machines*, Vol. 6, No. 2, 2006, pp. 163–190, doi: 10.1007/s10710-005-6164-x.

[71] Bernardino, H. S., and Barbosa, H. J. C., "Artificial Immune Systems for Optimization," *Nature-Inspired Algorithms for Optimization*, Studies in Computational Intelligence, Vol. 193, Springer Verlag, Berlin, Heidelberg, 2009, pp. 389–411.

[72] Castro, L. N., and Von Zuben, F. J., "aiNet: An Artificial Immune Network for Data Analysis," *Data Mining: A Heuristic Approach*, edited by H. A. Abbas, R. A. S. Charles, and S. Newton, Idea Group Publishing, Hershey, PA, 2001, pp. 231–259, doi: 10.4018/978-1-930708-25-9.ch012.

[73] Freitas, A. A., and Timmis, J., "Revisiting the Foundations of Artificial Immune Systems for Data Mining," *IEEE Transactions on Evolutionary Computation*, Vol. 11, No. 4, 2007.

[74] Coello, C. A. C., Rivera, D. C., and Cortés, N. C., "Use of an Artificial Immune System for Job Shop Scheduling," *Lecture Notes in Computer Science, Artificial Immune Systems*, Vol. 2787, 2003, pp. 1–10.

[75] Kroemer, P., Plato, J., and Snel, V., "Independent Task Scheduling by Artificial Immune Systems, Differential Evolution, and Genetic Algorithms," *4th International Conference on Intelligent Networking and Collaborative Systems (INCoS)*, Bucharest, Romania, 2012.

[76] Forrest, S., Hofmeyr, S. A., Somayaji, A., and Longstaff, T. A., "A Sense of Self for UnixProcesses," *Proceedings of the IEEE Symposium on Computer Security and Privacy*, Oakland, CA, July 1996, pp. 120–128, doi: 10.1109/SECPRI.1996.502675.

[77] De Castro, L., and Timmis, J., "Artificial Immune Systems: A Novel Paradigm to Pattern Recognition," *Artificial Neural Networks in Pattern Recognition*, edited by J. M. Corchado, L. Alonso, and C. Fyfe, University of Paisley, Paisley, England, UK, 2001, pp. 67–84.

[78] Khan, M. T., and de Silva, C. W., "Autonomous Fault Tolerant Multi-Robot Cooperation Using Artificial Immune System," *Proceedings IEEE International Conference on Automation and Logistics*, ICAL 2008, Qingdao, China, Sept. 2008, pp. 623–628.

[79] Singh, S., and Thayer, S. M., "Immunology Directed Methods for Distributed Robotics: A Novel, Immunity-Based Architecture for Robust Control and Coordination," *Proceedings of SPIE: Mobile Robots XVI*, Vol. 4573, Boston, MA, Oct. 2001, doi: 10.1117/12.457453.

[80] Takahashi, K., and Yamada, T., "Application of an Immune Feedback Mechanism to Control Systems," *JSME International Journal, Series C: Mechanical Systems, Machine Elements and Manufacturing*, Vol. 41, No. 2, 1998, pp. 184–191.

[81] Ko, A., Lau, H., and Lau, T., "An Immuno Control Framework for Decentralized Mechatronic Control," *Proceedings of the Third International Conference on*

Artificial Immune Systems (ICARIS 2004), Lecture Notes in Computer Science, Vol. 3239, Springer, Berlin, 2004, pp. 91–105.

[82] KrishnaKumar, K., Satyadas, A., and Neidhoefer, J., "An Immune System Framework for Integrating Computational Intelligence Paradigms with Applications to Adaptive Control," *Computational Intelligence: A Dynamic System Perspective*, edited by M. Palaniswami, Y. Attikiouzel, R. J. Marks, II, D. Fogel, and T. Fukuda, IEEE Press, 1995, pp. 32–45.

[83] KrishnaKumar, K., "Artificial Immune System Approaches for Aerospace Applications," AIAA-2003-0457, *Proceedings of the 41st Aerospace Sciences Meeting & Exhibit*, Reno, NV, 2003.

[84] Dasgupta, D., KrishnaKumar, K., Wong, D., and Berry, M., " Immunity-Based Aircraft Fault Detection System," AIAA 2004-6277, *Proceedings of the AIAA 1st Intelligent Systems Technical Conference*, Chicago, IL, Sept. 2004.

[85] Dasgupta, D., KrishnaKumar, K., Wong, D., and Berry, M., "Negative Selection Algorithm for Aircraft Fault Detection," edited by G. Nicosia, V. Cutello, P. J. Bentley, and J. Timmis, ICARIS 2004, LNCS 3239, 2004, pp. 1–13.

[86] Parra dos Anjos Lima, F., Chavarette, F. R., dos Santoe e Souza, A., Silva Frutuoso de Souza, S., and Martins Lopes, M. L., "Artificial Immune Systems with Negative Selection Applied to Health Monitoring of Aeronautical Structures," *Advanced Materials Research*, Vol. 871, 2014, pp. 283–289.

[87] Anaya, M., Tibaduiza, D., and Pozo, F., "Data Driven Methodology Based on Artificial Immune Systems for Damage Detection," *7th European Workshop on Structural Health Monitoring*, La Cité, Nantes, France, July 8–11, 2014.

[88] Pelham, J. G., Fan, I.-S., Jennions, J., and McFeat, J., "Application of an AIS to the Problem of Through Life Health Management of Remotely Piloted Aircraft," *AIAA Infotech @ Aerospace, AIAA SciTech*, AIAA 2015-1797, doi: 10.2514/6.2015-1797.

[89] Wang, L., Zhang, L., Xu, M., and Shi, X., "Research on Fault Diagnosis Method of Civil Aviation Engine Variable Bleed Valve System Based on Artificial Immune Algorithm," *International Journal of Pattern Recognition and Artificial Intelligence*, Vol. 30, No. 07, 2016.

[90] Kaneshige, J., and KrishnaKumar, K., "Artificial Immune System Approach for Air Combat Maneuvering," *Proceedings of SPIE—The International Society for Optical Engineering*, May 2007, doi: 10.1117/12.718892.

[91] KrishnaKumar, K., and Neidhoefer, J., "Immunized Adaptive Critic for an Autonomous Aircraft Control Application," *Artificial Immune System and Their Applications*, edited by D. Dasgupta, Springer-Verlag, Berlin, 1999, pp. 221–241.

[92] Das, P. K., Pradhan, S. K., Patro, S. N., and Balabantaray, B. K., "Artificial Immune System Based Path Planning of Mobile Robot," *Soft Computing Techniques in Vision Science*, SCI 395, Springer-Verlag Berlin Heidelberg, Elsevier, 2012, pp. 195–207.

[93] Karr, C. L., Nishita, K., and Graham, K. S., "Adaptive Aircraft Flight Control Simulation Based on an Artificial Immune System," *Applied Intelligence: The International Journal of Artificial Intelligence, Neural Networks, and Complex Problem-Solving Technologies*, Vol. 23, No. 3, Dec. 2005, pp. 295–308, doi: 10.1007/s10489-005-4614-z.

[94] Weng, L., Liu, Q., Xia, M., and Song, Y. D., "Immune Network-Based Swarm Intelligence and its Application to Unmanned Aerial Vehicle (UAV) Swarm Coordination," *Neurocomputing*, 124, 2014, pp. 134–141.

[95] Perhinschi, M. G., Moncayo, H., and Davis, J., "Integrated Framework for Artificial Immunity-Based Aircraft Failure Detection, Identification, and Evaluation," *AIAA Journal of Aircraft*, Vol. 47, No. 6, Nov.–Dec. 2010, pp. 1847–1859, doi: 10.2514/1.45718.

[96] Moncayo, H., and Perhinschi, M. G., *Aircraft Fault Tolerance: A Biologically Inspired Immune Framework for Sub-System Failures*, VDM Verlag Dr. Muller GmbH & Co. KG, VDM Publishing House, Saarbruecken, Germany, 2011.

[97] Perhinschi, M. G., Moncayo, H., and Al Azzawi, D., "Integrated Immunity-Based Framework for Aircraft Abnormal Conditions Management," *AIAA Journal of Aircraft*, Vol. 51, No. 6, Nov.–Dec., 2014, pp. 1726–1739, doi: 10.2514/1.C032381.

[98] Banchereau, J., and Steinman, R. M., "Dendritic Cells and the Control of Immunity," *Nature*, Vol. 392, March 1998, doi: 10.1038/32588.

[99] William, C. S., Nelson, C. A., Newberry, R. D., Kranz, D. M., Russell, J. H., and Loh, D. Y., "Positive and Negative Selection of an Antigen Receptor on T Cells in Transgenic Mice," *Nature*, Vol. 336(6194), 1988, pp. 73–76, doi: 10.1038/336073a0.

[100] Al-Sinbol, G., and Perhinschi, M. G., "Generation of Power Plant Artificial Immune System Using the Partition of the Universe Approach," *International Review of Automatic Control*, Vol. 9, No. 1, Jan. 2016, doi: 10.15866/ireaco.v9i1.8170.

[101] Perez, A., Moncayo, H., Perhinschi, M. G., Al Azzawi, D., and Togayev, A., "A Bio-Inspired Adaptive Control Compensation System for an Aircraft Outside Bounds of Nominal Design," *Journal of Dynamic Systems, Measurement and Control*, ASME, Vol. 137, Sept. 2015.

[102] Rajewsky, K., "Clonal Selection and Learning in the Antibody System," *Nature*, Vol. 381(6585), No. 7, 1996, pp. 51–758, doi: 10.1038/381751a.

[103] Hongwei, M., *Handbook of Research on Artificial Immune Systems and Natural Computing: Applying Complex Adaptive Technologies*, 1st Edition, IGI Global, Hershey, PA, December 2008.

[104] Perhinschi, M. G., Moncayo, H., Wilburn, B., Wilburn, J., Karas, O., and Bartlett, A., "Neurally Augmented Immunity-Based Detection and Identification of Aircraft Sub-System Failures," *The Aeronautical Journal*, Vol. 118, No. 1205, July 2014, pp. 775–796.

[105] Stibor, T., Timmis, J., and Eckert, C., "On the Appropriateness of Negative Selection Defined Over Hamming Shape-Space as a Network Intrusion Detection System," *Proceedings of the Fourth International Conference on Artificial Immune Systems*, 2006.

[106] Ji, Z., and Dasgupta, D., "Applicability Issues of the Real-Valued Negative Selection Algorithms," *Proceedings for the Genetics and Evolutionary Computation Conference*, 2006, Seattle, WA, 8–12 July 2006, pp. 111–118.

[107] Moncayo, H., Perhinschi, M. G., and Davis, J., "Aircraft Failure Detection and Identification Using an Immunological Hierarchical Multi-Self Strategy," *Journal of Guidance, Control, and Dynamics*, Vol. 33, No. 4, July–Aug. 2010, pp. 1105–1114, doi: 10.2514/1.47445.

[108] Al Azzawi, D., Perhinschi, M. G., and Moncayo, H., "Artificial Dendritic Cell Mechanism for Aircraft Immunity-Based Failure Detection and Identification," *AIAA Journal of Aerospace Information Systems,* Vol. 11, No. 7, July 2014, pp. 467–481, doi: 10.2514/1.I010214.

[109] Al Azzawi, D., Perhinschi, M. G., Moncayo, H., and Perez, A. E., "A Dendritic Cell Mechanism for Detection, Identification, and Evaluation of Aircraft Failures," *Journal of Control Engineering Practice*, Vol. 41, Aug. 2015, pp. 134–148, doi: 10.1016/j.conengprac.2015.04.010.

[110] Moncayo, H., Moguel, I., Perhinschi, I., Al Azzawi, D., Togayev, A., and Perez, A., "Structured Non-Self Approach for Aircraft Failure Identification within an Immunity-Based Fault Tolerance Architecture," *The Aeronautical Journal*, Vol. 120, No. 1225, March 2016, pp. 415–434, doi: 10.1017/aer.2016.15.

[111] Moncayo, H., Perhinschi, M. G., and Davis, J., "Artificial Immune System-Based Aircraft Failure Detection and Identification Over an Extended Flight Envelope," *The Aeronautical Journal*, Vol. 115, No. 1163, Jan. 2011, pp. 43–55.

[112] Moncayo, H., Perhinschi, M. G., and Davis, J., "Artificial-Immune-System-Based Aircraft Failure Evaluation over Extended Flight Envelope," *Journal of Guidance, Control, and Dynamics*, Vol. 34, No. 4, July–Aug. 2011, pp. 989–1001, doi: 10.2514/1.52748.

[113] Togayev, A., Perhinschi, M. G., Al Azzawi, D., Moncayo, H., and Perez, A., "Immunity-Based Accommodation of Aircraft Failures," *Aircraft Engineering and Aerospace Technology*, Vol. 89, No. 1, Jan. 2017, pp. 164–175, doi: 10.1108/AEAT-08-2014-0124.

[114] Moncayo, H., "Immunity-Based Detection, Identification, and Evaluation of Aircraft Sub-System Failures," Ph.D. Dissertation, West Virginia Univ., Dec. 2009.

[115] Davis, J., Perhinschi, M. G., and Moncayo, H., "Evolutionary Algorithm for Artificial Immune System-Based Failure Detector Generation and Optimization," *Journal of Guidance, Control, and Dynamics*, Vol. 33, No. 2, March–April 2010, pp. 305–320, doi: 10.2514/1.46126.

[116] Davis, J., "The Design of an Evolutionary Algorithm for Artificial Immune System Based Failure Detector Generation and Optimization," M.S. Thesis, West Virginia Univ., Aug. 2010.

[117] Perhinschi, M. G., Moncayo, H., Al Azzawi, D., and Moguel, I., "Generation of Artificial Immune System Antibodies Using Raw Data and Cluster Set Union," *International Journal of Immune Computation*, Vol. 2, No. 1, March 2014, pp. 1–15.

[118] Moncayo, H., Perhinschi, M. G., and Davis, J., "Simulation Environment for the Development and Testing of Immunity-Based Aircraft Failure Detection Schemes," *Proceedings of the AIAA Modeling and Simulation Technologies Conference*, Portland, OR, Aug. 2011.

[119] Garcia, D., Moncayo, H., Perez, A., Dupuis, M., and Mueller, R., "Spacecraft Health Monitoring Using a Biomimetic Fault Diagnosis Scheme," *AIAA Science and Technology Forum and Exposition, Integrated Systems Health Management (ISHM)*, Texas, 2017.

[120] Al Azzawi, D., "Aircraft Abnormal Conditions Detection, Identification, and Evaluation Using Innate and Adaptive Immune Systems Interaction," Ph.D. Dissertation, West Virginia Univ., Aug. 2014.

[121] Al Azzawi, D., Moncayo, H., Perhinschi, M. G., Perez, A., and Togayev, A., "Comparison of Immunity-Based Schemes for Aircraft Failure Detection and Identification," *Engineering Applications of Artificial Intelligence*, Vol. 52, pp. 181–193, June 2016, doi: 10.1016/j.engappai.2016.02.017.

[122] Moguel, I., "Bio-Inspired Mechanism for Aircraft Assessment under Upset Conditions," Master Thesis, Embry-Riddle Aeronautical Univ., 2014.

[123] Perhinschi, M. G., Porter, J., Moncayo, H., Davis, J., and Wayne, W. S., "Artificial Immune System-Based Detection Scheme for Aircraft Engine Failures," *Journal of Guidance, Control, and Dynamics*, Vol. 34, No. 5, Sept.–Oct. 2011, pp. 1423–1440, doi: 10.2514/1.52746.

[124] Garcia, D., Moncayo, H., Perez, A., and Jain, C., "Low Cost Implementation of a Biomimetic Approach for UAV Health Management," *American Control Conference*, Boston, MA, July 2016.

[125] Bishop, C. M., *Pattern Recognition and Machine Learning*, Springer, Singapore, 2006.

[126] Murphy, K. P., *Machine Learning: A Probabilistic Perspective*, MIT Press, Cambridge, MA, 2006.

[127] Perhinschi, M. G., Al Azzawi, D., Moncayo, H., Togayev, A., and Perez, A., "Immunity-Based Flight Envelope Prediction at Post-Failure Conditions", *Aerospace Science and Technology*, Vol. 46, pp. 264–272, Oct.–Nov., 2015, doi: 10.1016/j.ast.2015.07.014.

[128] Perhinschi, M. G., Al Azzawi, D., Moncayo, H., Perez, A., and Togayev, A., "Immunity-Based Actuator Failure Evaluation", *Aircraft Engineering and Aerospace Technology,* Vol. 88, No. 6, 2016, doi: 10.1108/AEAT-07-2014-0117.

[129] Perhinschi, M. G., Al Azzawi, D., Moncayo, H., and Perez, A., "Evaluation of Aircraft Sensor Failures Effects Using the Artificial Immune System Paradigm," *International Review of Aerospace Engineering*, Vol. 8, No. 2, April 2015, pp. 71–80.

[130] Perhinschi, M. G., Al Azzawi, D., and Moncayo, H., "Simplified Estimation Algorithms for Aircraft Structural Damage Effects Using an Artificial Immune System," *International Review of Aerospace Engineering*, Vol. 8, No. 4, Aug. 2015, doi: 10.15866/irease.v8i4.7461.

[131] Wilburn, B., Perhinschi, M. G., Moncayo, H., Karas, O., and Wilburn, J., "Unmanned Aerial Vehicle Trajectory Tracking Algorithm Comparison," *International Journal of Intelligent Unmanned Systems*, Vol. 1, No. 3, 2013, pp. 276–302.

[132] Togayev, A., "Immunity-Based Accommodation of Aircraft Subsystem Failures," M.S. Thesis, West Virginia Univ., Dec. 2014.

CHAPTER 5

Prognostics and Health Monitoring: Application to an Unmanned Electric Aircraft

Chetan Kulkarni
SGT, Inc., NASA Ames Research Center, Moffett Field, CA, USA

Matthew Daigle
NIO USA, San Jose, CA, USA

Indranil Roychoudhury and Shankar Sankararaman
SGT, Inc., NASA Ames Research Center, Moffett Field, CA, USA

Kai Goebel
NASA Ames Research Center, Moffett Field, CA, USA

5.1 INTRODUCTION

Prognostics and health management of engineered systems has recently emerged as a potentially game-changing technology to provide safety assurance and optimize cost and mission objectives. Such technology would facilitate operational decision making based on the condition of the system of interest. This predictive technology is critically important for the operation of electric aircraft where there is a fundamental need to operate under available power resources and maintain safe operation while meeting mission objectives [1]. In particular, an *autonomous* aircraft needs to be aware of its own state of operation, compute its own capabilities, and identify how best to complete flight operations and missions. The topic of prognostics and the prediction of remaining flying time are relevant emerging challenges in this context. The computation of remaining flying time is also safety-critical: if the aircraft is powered entirely by battery packs, and if it runs out of available charge while in the air it may lose control. This chapter focuses on a computational methodology for real-time prediction of the remaining flying time of an unmanned electric battery-powered aircraft while the aircraft is airborne during a waypoint-based mission.

Prior to predicting the future behavior of the aircraft, it is necessary to be aware of the current state of the system and the health of its components. In addition it is important to understand that the aforementioned prediction is closely related to the future operation of the aircraft. This approach is, therefore, condition-based and thus, can make use of knowledge regarding the current operational state, as well as future operations, in order to meaningfully predict the future states of the system. Using available information and models regarding the current and future system behavior, this chapter employs a generic model-based prognostics methodology [2–5] to facilitate prediction of remaining flying time for electric aircraft.

For an electric aircraft, the propulsion system is controlled using power drawn from a battery charge. Thus, the prediction of remaining flying time is, at its core, equivalent to predicting the end-of-discharge (EOD) of the constituent battery packs. Therefore, it is critical to monitor battery charge and to estimate the ability of the battery to support future flight activities while being discharged. The ability of the vehicle to complete its given mission depends on the current state of charge for respective battery packs and the future required charge, which is based on the planned route, maneuvers, weather conditions, and battery health. Consider a scenario where an unmanned electric aircraft needs to fly through a preplanned sequence of waypoints and return to its base. Typically, a certain amount of time (chosen to be two minutes, for an experimental aircraft discussed later in this chapter) is required to safely land the aircraft. Thus, the point in time at which the aircraft must begin to head back to the runway for safe landing needs to be calculated. The aircraft cannot power its propellers when the voltage supplied by battery packs falls below a specific threshold voltage; therefore, the goal is to predict when this threshold voltage will be reached.

Because the central theme of this work is based on predicting the future operation of the unmanned aircraft, it is essential to understand that it is almost impossible to predict future behavior with complete precision. Particularly from the point of view of the batteries that power the aircraft, the future power requirements may not be known precisely. Although it may be possible to establish a relationship between the power requirement and the planned maneuvers of the aircraft, it is not possible to accurately estimate such future power requirements with 100 percent confidence. For example, future wind conditions that partially determine the required power will be uncertain to some extent. As a result, it is necessary to treat these quantities as uncertain variables and develop a probabilistic framework that can include these sources of uncertainty. In addition to uncertainty regarding the future, there are additional sources of uncertainty that affect future predictions. For example, (1) the state of health (equivalent of state of charge, in the context of batteries) cannot be measured directly and is usually estimated through filtering approaches (these approaches usually account for measurement errors). As result, the estimate of the state of health is uncertain, (2) the state-space model that predicts the evolution of state of health until EOD (EOD is assumed to occur when the voltage supplied by the batteries falls below a predetermined critical limit) may not be an accurate representation of reality and it is necessary to account for modeling errors. A systematic prediction framework should identify all possible sources of uncertainty, quantify each of them individually, and mathematically estimate their combined effect on the system-level quantity of interest, which in this case is the remaining flying time of the unmanned aircraft [6]. An understanding of the level of uncertainty in the resulting predictions is critical to making robust decisions based on those predictions.

The computational methodology presented in this chapter first estimates the current state of the batteries, and predicts the time of EOD and remaining time until discharge (RTD), in the presence of uncertainty in both the current system state and future system operation. For an electric aircraft powered by batteries, EOD is a direct indicator of the latest time at which the aircraft must land, and RTD is a direct indicator of remaining flying time (RFT).

In order to demonstrate and validate the approach, a sub-scale electric aircraft, the Edge 540T, is adopted as a case study. From the high-level requirements listed by the operator, prognostics-level requirements are derived that can be quantitatively assessed, and these are used to verify that the battery health management algorithm fulfills them. The approach is validated on a set of example flights. As mentioned earlier, this particular aircraft needs a two-minute lead time to landing, and therefore this constraint is treated as a requirement.

The rest of this chapter is organized as follows. Section 5.2 introduces the Edge electric aircraft, including hardware configuration, relevant sensors, batteries, and mission details. Section 5.3 describes the prognostics methodology, as applied to the vehicle. This methodology includes steps for modeling, state estimation using the Unscented Kalman Filter (UKF), and prediction. In addition, it also discusses the approach for deriving requirements that are used for verification. Section 5.4 presents numerical results from the case study. Finally, Section 5.5 concludes the chapter by providing a brief summary of the methodology and identifying technological gaps that exist to facilitate the application of such methods to fielded aircrafts.

5.2 SYSTEM DESCRIPTION

The focus of this chapter is on the Edge 540T [7] electric aircraft used for testing and validating developed prognostic algorithms. This section describes the vehicle hardware systems and its operations and requirements.

5.2.1 VEHICLE HARDWARE

The 33 percent-scaled Edge 540T is shown in Fig. 5.1. It is 2.49 m. long with a 2.54 m. wing span, weighs 21.5 kgs and has 1.21 m^2 of wing area with an average wing load of 0.172 kPa.

The aircraft features exclusively electric propulsion. Two outrunner brushless DC electric motors are mounted in tandem to drive a single 0.67-m propeller. The motor assembly turns the propeller up to 6000 RPM to develop about 16.8 kgs of thrust. Its airspeed ranges from a stall of 12 m/s to a dash of about 40 m/s (23–77 knots).

Fig. 5.1 The Edge 540T during landing.

The electric powertrain is illustrated in Fig. 5.2. A set of four lithium-polymer (LiPo) batteries provides power to the motors, M1 and M2, through two Jeti 90 Pro Opto electric speed controllers (ESCs). The ESCs send synchronized voltages to the motors at a duty cycle that is determined by a throttle input signal sent either by remote control from a pilot or by an onboard autopilot.

Each battery consists of 10 distinct pouch cells, with two parallel-connected sets of five series-connected cells. Each cell has a voltage of 4.2 V at full charge and a nominal capacity of 3900 mAh, so each battery as a whole provides 21 V and a nominal capacity of 7800 mAh. A 50-C max burst discharge is supported, allowing a maximum takeoff current draw of up to 390 A. On average, takeoff current peaks at 140 A. A single motor is supplied power by two series-connected batteries, receiving 42 V at full charge.

Inductive loop current sensors are mounted on the positive lead feeding each

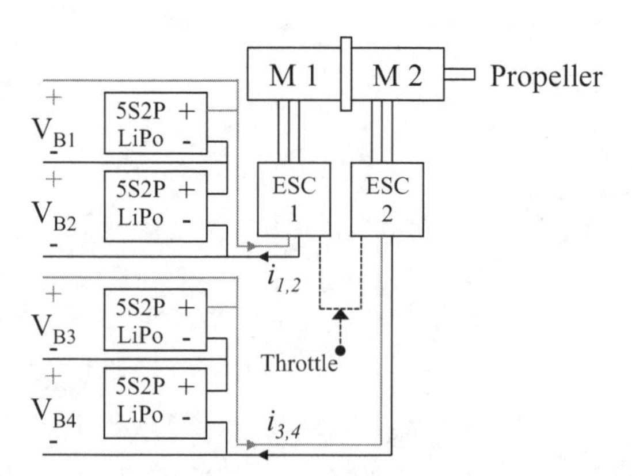

Fig. 5.2 Aircraft electric powertrain.

ESC. Additional current sensors are also mounted on the positive feed from each of the four batteries. The positive lead of each battery is tapped to provide the data system withbattery voltage measurements.

5.2.2 VEHICLE OPERATIONS

The aircraft is operated either by a ground-based pilot or an onboard autopilot that has the capability to navigate using a stored flight plan. Figure 5.3 shows the ground track for a typical flight. Flight activities typically occur at 2000 m above ground level. Flights last about 15 minutes, with the flight duration largely depending on throttle management, wind conditions, and battery health.

A typical flight consists of essentially four activities. These include a takeoff followed by a piloted flight or an autopilot executing a flight plan at constant cruise flight speed of 21 m/s. The throttle is increased to attain a high speed of 23 m/s, which could be during altitude change. The throttle is then decreased to maintain a cruise speed of 21 m/s until the aircraft is brought in for landing.

The aircraft must land before the batteries can no longer provide the required voltage to power the motors. The remaining flying time is defined as the time remaining until the first battery reaches its EOD condition, as defined by the minimum voltage needed to supply the motor system. In order to give the pilot/autopilot ample time to land the aircraft before this occurs, a two-minute warning is required (i.e., the prognostic system must be able to determine when this time point is going to occur at least two minutes prior to it occurring). This buffer is meant to ensure that there is enough time to land, given the potential impact of uncertainty, such that the aircraft does not lose power at any point before landing.

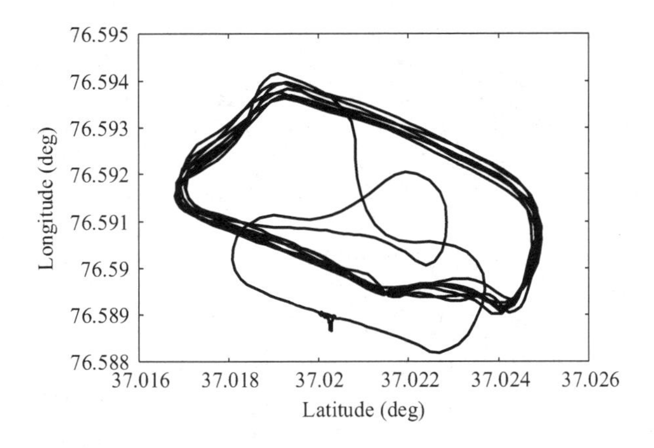

Fig. 5.3 Ground track of a typical flight [7].

5.3 PROGNOSTICS METHODOLOGY

In the context of this chapter, prognostics is focused on predicting a future system event or state and determining how long it will take until this event EOD or state (i.e., degradation/aging threshold) is reached. It is an online process, in that predictions are made regularly, conditioned on the current system state and its planned future operation.

In order to make useful predictions, it is necessary to use an accurate model of how the system will behave in the future and carefully incorporate the different sources of uncertainty regarding such future behavior. This section presents a generic computational architecture that will account for sources of uncertainty and system modeling, facilitating efficient prognostics methodologies. Note that the presented architecture is general enough to be applicable to a variety of engineering systems, and the application of this architecture to the specific problem of predicting remaining flying time for the electric Edge 540T is discussed in Section 5.4.

5.3.1 UNCERTAINTY IN PROGNOSTICS

It is practically impossible to make accurate, precise predictions regarding future behavior of such systems due to the inherent uncertainty associated with the future evolution of the system. The presence of uncertainty poses an important challenge in prognostics and needs to be addressed in a systematic manner using sound fundamental mathematics and statistics.

Uncertainty in prognostics stems from several sources, including incomplete knowledge regarding the system under consideration, lack of accurate sensing, and, in general, the inability to accurately assess the future operating and loading conditions that in turn affect the performance of the system under consideration. In fact, because the future behavior of the system is invariably clouded with uncertainty, it is not even useful to provide results of prognostics without uncertainty estimates. So, it is necessary to identify the various sources of uncertainty that affect predictions and compute the overall impact of uncertainty on prognostics. In order to achieve this goal, it is important to assess the combined effect of the different sources of uncertainty and estimate the overall uncertainty in prognostics.

In order to efficiently address the impact of uncertainty on prognostics, it is necessary to formulate and solve formal mathematical problems. These formulations, as enumerated below, need to deal with how uncertainty can be mathematically represented and its impact on prognostics quantified.

1. **Uncertainty Representation:** The formal problem of uncertainty representation deals with identifying and leveraging upon a suitable, existing mathematical framework that can be used to represent, measure, and express uncertainty in a meaningful manner. This chapter uses a probabilistic methodology for representing uncertainty, although alternate representations

using Dempster-Shafer theory, fuzzy probabilities, etc., are also available in the literature. Using tools of probability, it is straightforward to represent discrete random variables using probability mass functions and continuous random variables using probability density functions. One challenge in prognostics is to represent uncertain trajectories. Specifically, consider a generic-quantity a that varies continuously from the time of prediction (denoted by k_p, in discretized time-index notation) until a predetermined time horizon (denoted by k_h). Let this trajectory be denoted by $\mathbf{A}_{k_p,k_h}$ and the density function of this trajectory be denoted by $p(\mathbf{A}_{k_p,k_h})$. Although the value of a at every time instant is indeed stochastic, it is impossible to represent each of these values as a probability distribution because such an approach would significantly increase the dimensionality. In order to overcome this issue, Daigle and Sankararaman [8] proposed that a trajectory can be expressed as a parameterized function of time and that the parameters can be chosen as random variables; these parameters can equivalently represent the uncertainty in the original trajectory and, therefore, are referred to as surrogate variables. This idea is discussed later in the chapter as a way to account for the uncertainty in future input trajectory for the battery being used in this case study.

2. **Uncertainty Interpretation:** Probability can be interpreted in two ways: (1) the classical, frequentist approach of dealing with true randomness, and (2) the subjective, Bayesian approach of dealing with randomness due to lack of knowledge. Based on the development of these approaches, it is observed that a subjective, Bayesian approach is more suitable for interpreting uncertainty in the context of prognostics, because the focus is on predicting the behavior of one particular component or system.

3. **Uncertainty Quantification:** The formal problem of uncertainty quantification deals with identifying all quantities (independent quantities) that affect prognostics (dependent quantities) and individually expressing/quantifying the uncertainty in those quantities using probability theory. While probability mass functions are useful for discretely varying quantities, probability density functions are more useful for continuously varying quantities.

4. **Uncertainty Propagation:** The formal problem of uncertainty propagation deals with measuring the joint impact of the uncertainty in all of the independent quantities on the predicted (dependent) quantity of interest. In the context of probability tools, this means computing the probability distribution of a dependent quantity $Y = \mathrm{G}(\boldsymbol{X})$, given the probability distributions of the quantities in the vector $\boldsymbol{X}$.

5. **Uncertainty Management:** The formal problem of uncertainty management deals with a variety of post-prognostics activities that aid in managing the uncertainty in prognostics and support operational decision-making activities. This activity is concerned with quantifying the amount of existing uncertainty or its effects, but is focused on dealing with existent uncertainty in a systematic manner.

Whereas the approach for uncertainty representation and interpretation is discussed above within each of the bullets, aspects of uncertainty quantification and propagation are discussed in detail in this chapter. Though uncertainty management is a very important topic, a detailed discussion on this issue is out of the scope of this chapter.

Having explained the importance of uncertainty, the next section formally defines the problem of prognostics, considering that uncertainty is expressed using tools of probability.

5.3.2 PROBLEM DEFINITION

Consider the following model-based prognostics paradigm [9], using a dynamic model of the system under consideration. A general time-invariant state-space formulation of this model can be described as follows:

$$\mathbf{x}(k+1) = \mathbf{f}(\mathbf{x}(k),\ \mathbf{u}(k),\ \mathbf{v}(k)), \tag{5.1}$$

$$\mathbf{y}(k) = \mathbf{h}(\mathbf{x}(k),\ \mathbf{u}(k),\ \mathbf{n}(k)), \tag{5.2}$$

where $k \in \mathbb{N}$ is the discrete time variable, $\mathbf{x}(k) \in \mathbb{R}^{n_x}$ is the state vector, $\mathbf{u}(k) \in \mathbb{R}^{n_u}$ is the input vector, $\mathbf{v}(k) \in \mathbb{R}^{n_v}$ is the process noise vector, $\mathbf{y}(k) \in \mathbb{R}^{n_y}$ is the output vector (corresponding to the system sensors), $\mathbf{n}(k) \in \mathbb{R}^{n_n}$ is the sensor noise vector, $\mathbf{f}$ is the state update function, and $\mathbf{h}$ is the output function.

Without loss of generality, it is of interest to understand how the *state* of the system, $\mathbf{x}$, will evolve in time. Specifically, it is necessary to determine whether some subset of the state space, $\mathcal{X}_l \subseteq \mathcal{X}$, will be reached in some finite time, and, if so, *when* it will be reached [10]. This subset $\mathcal{X}_l$ can represent component failure states, goal states, etc. In the case of the electric aircraft, the subset of the state space we are interested in is those states in which at least one of the batteries is considered to be discharged. Consider the labeling labeling function, $l_l : \mathbb{R}^{n_x} \rightarrow \{true,\ false\}$, that maps a given state to true if the label l applies to a given state $\mathbf{x}$, and false otherwise. Then, $\mathcal{X}_l = \{\mathbf{x} : \mathbf{l}_l(\mathbf{x}) = true\}$.

The inputs to the prediction problem are the following [11]:

1. a time of prediction, k_p;
2. a time horizon of prediction, $[k_p,\ k_h]$;
3. the initial state probability distribution, $p(\mathbf{x}_P(k_p))$;
4. the future input trajectory distribution, $p(\mathbf{U}_{k_p,k_h})$, where $\mathbf{U}_{k_p,k_h} = [\mathbf{u}(k_p),\ \mathbf{u}(k_p+1), \ldots, \mathbf{u}(k_h)]$;[1] and
5. the future process noise trajectory distribution, $p(\mathbf{V}_{k_p,k_h})$, where $\mathbf{V}_{k_p,k_h} = [\mathbf{v}(k_p),\ \mathbf{v}(k_p+1), \ldots, \mathbf{v}(k_h)]$.

---

[1]Note that the following notation is adopted. For a vector $\mathbf{a}$, a trajectory of that vector over a time interval $[k_p,\ k_h]$ is denoted by $\mathbf{A}_{k_p,k_h}$ and $\mathbf{A}_{k_p,k_h}(k) = \mathbf{a}(k)$, for $k \in [k_p,\ k_h]$.

These inputs may change for every new time of prediction.

Specifically, it is necessary to predict the time at which a state with label l would be reached. For a given time k, this is defined as:

$$k_l(k) = \min\{k' : k' \geq k \quad \textbf{and} \quad \mathbf{x}(k) \in \mathcal{X}_l\}, \tag{5.3}$$

i.e., k_l is the earliest time point at which the system state is assigned the label l. For the electric aircraft, k_l will represent the battery discharge and, hence, the end of powered flying time. The remaining flight time is then $k_l(k_p) - k_p$.

Formally, then, the computational problem is defined as:

Problem 1 Given a label l, a time interval $[k_p, k_h]$, an initial state $p(\mathbf{x}(k_p))$, process noise $p(\mathbf{V}_{k_p,k_h})$, and future inputs $p(\mathbf{U}_{k_p,k_h})$, determine $p(k_l)$ and/or $p(k_l - k_p)$.

5.3.3 COMPUTATIONAL ARCHITECTURE FOR PROGNOSTICS

This section presents the computational architecture for prognostics, used to solve the aforementioned predication problem. This architecture, as shown in Fig. 5.4, consists of two main steps:

1. **Estimation:** This step consists of estimating the system state (includes battery state-of-charge and battery state-of-health) at any generic desired time of prediction.
2. **Prediction:** This step consists of predicting the future system state (based on knowledge regarding predicted future system usage) and determining when a predetermined set of conditions is violated, specified by using the label l, as discussed earlier in this chapter (this corresponds to the two-minute threshold for electric UAV application).

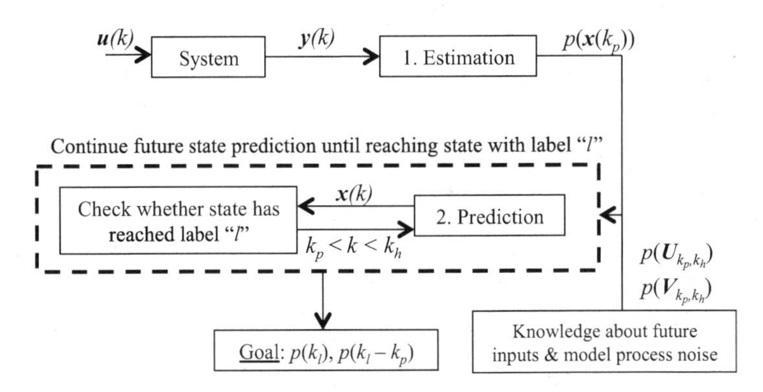

Fig. 5.4 Architecture for prognostics.

5.3.3.1 ESTIMATION

In order to accurately predict the future behavior of any system, it is necessary to first estimate its state. For this purpose, it is common to use Bayesian filtering techniques such as the Kalman filter, the particle filter, and their extensions. In particular, the use of the Unscented Kalman Filter (UKF) [12, 13] is recommended because (1) it can account for nonlinear models and non-Gaussian variables, unlike Kalman filtering, (2) it generally has better accuracy than the extended Kalman filter, and (3) it avoids the high computational cost of particle filters [14]. While this chapter summarizes the UKF methodology, additional details may be found in [12, 13].

The UKF approximates a distribution using the unscented transform (UT). The UT takes a random variable $\mathbf{x} \in \mathbb{R}^{n_x}$, with mean $\bar{\mathbf{x}}$ and covariance $\mathbf{P}_{xx}$, which is related to a second random variable $\mathbf{y}$ by some nonlinear function $\mathbf{y} = \mathbf{g}(\mathbf{x})$ and computes the mean $\bar{\mathbf{y}}$ and covariance $\mathbf{P}_{yy}$ using a set of *deterministically* selected weighted samples, called *sigma points* [12]. $\mathcal{X}^i$ denotes the ith sigma point from $\mathbf{x}$ and w^i denotes its weight. The sigma points are always chosen such that the mean and covariance match those of the original distribution, $\bar{\mathbf{x}}$ and $\mathbf{P}_{xx}$. Each sigma point is passed through $\mathbf{g}$ to obtain new sigma points $\mathcal{Y}$, that is,

$$\mathcal{Y}^i = \mathbf{g}(\mathcal{X}^i) \tag{5.4}$$

with mean and covariance

$$\bar{y} = \sum_i w^i \mathcal{Y}^i \tag{5.5}$$

$$\mathbf{P}_{yy} = \sum_i w^i (\mathcal{Y}^i - \bar{\mathbf{y}})(\mathcal{Y}^i - \bar{\mathbf{y}})^T. \tag{5.6}$$

The symmetric unscented transform selects $2n_x + 1$ sigma points symmetrically about the mean [13], as:

$$w^i = \begin{cases} \dfrac{\kappa}{(n_x + \kappa)}, & i = 0 \\ \dfrac{1}{2(n_x + \kappa)}, & i = 1, \ldots, 2n_x \end{cases} \tag{5.7}$$

$$\mathcal{X}^i = \begin{cases} \bar{\mathbf{x}}, & i = 0 \\ \bar{\mathbf{x}} + \left(\sqrt{(n_x + \kappa)\mathbf{P}_{xx}}\right)^i, & i = 1, \ldots, n_x \\ \bar{\mathbf{x}} - \left(\sqrt{(n_x + \kappa)\mathbf{P}_{xx}}\right)^i, & i = n_x + 1, \ldots, 2n_x \end{cases}, \tag{5.8}$$

where $\left(\sqrt{(n_x + \kappa)\mathbf{P}_{xx}}\right)^i$ refers to the ith column of the matrix square root of $(n_x + \kappa)\mathbf{P}_{xx}$ (e.g., computed using the Cholesky decomposition). The number κ is a free parameter that can be used to tune the higher order moments of the distribution, and if x is assumed Gaussian, then selecting $\kappa = 3 - n_x$ is recommended [12].

The UKF assumes the general nonlinear form of the state and output equations, but is restricted to additive Gaussian noise. First, n_s sigma points $\hat{\mathcal{X}}_{k-1|k-1}$ are derived from the current mean $\hat{\mathbf{x}}_{k-1|k-1}$ and covariance estimates $\mathbf{P}_{k-1|k-1}$. The prediction step is:

$$\hat{\mathcal{X}}^i_{k|k-1} = \mathbf{f}\left(\hat{\mathcal{X}}^i_{k-1|k-1}, \mathbf{u}_{k-1}\right),\ i = 1, \ldots, n_s \tag{5.9}$$

$$\hat{\mathcal{Y}}^i_{k|k-1} = \mathbf{h}\left(\hat{\mathcal{X}}^i_{k|k-1}\right),\ i = 1, \ldots, n_s \tag{5.10}$$

$$\hat{\mathbf{x}}_{k|k-1} = \sum_i^{n_s} w^i \mathcal{X}^i_{k|k-1} \tag{5.11}$$

$$\hat{\mathbf{y}}_{k|k-1} = \sum_i^{n_s} w^i \mathcal{Y}^i_{k|k-1} \tag{5.12}$$

$$\mathbf{P}_{k|k-1} = \mathbf{Q} + \sum_i^{n_s} w^i \left(\mathcal{X}^i_{k|k-1} - \hat{\mathbf{x}}_{k|k-1}\right)\left(\mathcal{X}^i_{k|k-1} - \hat{\mathbf{x}}_{k|k-1}\right)^T, \tag{5.13}$$

where $\mathbf{Q}$ is the process noise covariance matrix.

The update step is:

$$\mathbf{P}_{yy} = \mathbf{R} + \sum_i^{n_s} w^i \left(\mathcal{Y}^i_{k|k-1} - \hat{\mathbf{y}}_{k|k-1}\right)\left(\mathcal{Y}^i_{k|k-1} - \hat{\mathbf{y}}_{k|k-1}\right)^T \tag{5.14}$$

$$\mathbf{P}_{xy} = \sum_i^{n_s} w^i \left(\mathcal{X}^i_{k|k-1} - \hat{\mathbf{x}}_{k|k-1}\right)\left(\mathcal{Y}^i_{k|k-1} - \hat{\mathbf{y}}_{k|k-1}\right)^T \tag{5.15}$$

$$\mathbf{K}_k = \mathbf{P}_{xy}\mathbf{P}_{yy}^{-1} \tag{5.16}$$

$$\hat{\mathbf{x}}_{k|k} = \hat{\mathbf{x}}_{k|k-1} + \mathbf{K}_k(\mathbf{y}_k - \hat{\mathbf{y}}_{k|k-1}) \tag{5.17}$$

$$\mathbf{P}_{k|k} = \mathbf{P}_{k|k-1} - \mathbf{K}_k\mathbf{P}_{yy}\mathbf{K}_k^T, \tag{5.18}$$

where $\mathbf{R}$ is the sensor noise covariance matrix.

In the most general scenario, it may be necessary to estimate the model parameters simultaneously along with the state variables. This leads to a joint state-parameter estimate $p(\mathbf{x}(k_p), \boldsymbol{\theta}(k_p))$, which will be used for prediction. However, in the case of the Edge battery, the parameters of the model are assumed to be deterministic and not estimated; therefore, it is not necessary to calculate the joint-parameter estimate. It is only necessary to estimate the seven state variables that are used within the battery model, the details of which are explained later in Section 5.4. Therefore, the rest of the prognostics methodology is presented using the state estimate $p(\mathbf{x}(k_p))$ only and not using the joint state-parameter estimate.

5.3.3.2 PREDICTION

The goal of prediction is to compute $p(k_l(k_p))$ using the following quantities:

1. State estimate, $p(\mathbf{x}(k_p))$ at the time of prediction.
2. Information regarding future input values, $\mathbf{U}_{k_p,k_h}$.
3. Information regarding future model parameter values, $\mathbf{\Theta}_{k_p,k_h}$.
4. Information regarding model process noise values, $\mathbf{V}_{k_p,k_h}$.
5. Predict till time T_E which is the EOD threshold.

Consider one realization of each of the uncertain quantities at prediction time k_p: the state $\mathbf{x}(k_p)$, the parameter trajectory $\mathbf{\Theta}_{k_p,k_h}$, the input trajectory $\mathbf{U}_{k_p,k_h}$, and the process noise trajectory $\mathbf{V}_{k_p,k_h}$. For this combination of realizations of these four quantities, the corresponding realization of k_l can be computed with the system model, as shown in Algorithm 5.1 [8]. In Algorithm 5.1, the function **P** simulates the system model until the state reaches the predetermined label "l," as explained earlier in this section.

Algorithm 5.1 $k_l(k_p) \leftarrow \mathbf{P}(\mathbf{x}(k_p), \mathbf{\Theta}_{k_p,k_h}, \mathbf{U}_{k_p,k_h}, \mathbf{V}_{k_p,k_h})$

1: $k \leftarrow k_p$
2: $\mathbf{x}(k) \leftarrow \mathbf{x}(k_p)$
3: **while** $T_E(\mathbf{x}(k), \mathbf{\Theta}_{k_p}(k), \mathbf{U}_{k_p}(k)) = 0$ **do**
4: $\quad \mathbf{x}(k+1) \leftarrow \mathbf{f}(k, \mathbf{x}(k), \mathbf{\Theta}_{k_p}(k), \mathbf{U}_{k_p}(k), \mathbf{V}_{k_p}(k))$
5: $\quad k \leftarrow k+1$
6: $\quad \mathbf{x}(k) \leftarrow \mathbf{x}(k+1)$
7: **end while**
8: $k_E(k_p) \leftarrow k$

Algorithm 5.1 requires computing first realizations of the system state, the parameter trajectory, the input trajectory, and the process noise trajectory. These realizations need to be generated outside of Algorithm 5.1, and various methods of prediction generate these realizations in different ways. The distribution for the state comes from the estimation step, and the distributions for the parameter, input, and process noise trajectories are defined indirectly by the set of surrogate variables (which are, fundamentally, the parameters of the trajectory function). If the trajectory represents a constant function, then there is only one surrogate variable; a linear function needs two surrogate variables, and so on. Therefore, sampling a realization of a random trajectory would require two steps: (1) select a random sample of each of the surrogate variables, and (2) use the resultant realization of surrogate variables to construct a trajectory. Then, this trajectory is passed to Algorithm 5.1, in order to compute k_l.

Fundamentally, prediction methods need to solve an uncertainty propagation problem, that is, propagate the uncertainty in $\mathbf{x}(k_p)$, $\mathbf{U}_{k_p,k_h}$, $\mathbf{\Theta}_{k_p,k_h}$, and $\mathbf{V}_{k_p,k_h}$

(where the uncertainty is represented using corresponding probability distributions) through **P** to obtain the uncertainty in k_l represented using the probability density function $p(k_l)$. Prediction methods can be broadly classified into two categories:

1. **Sampling methods:** In these methods, (1) samples are drawn from the distributions of uncertain quantities to obtain multiple sets of realizations for the uncertain quantities, (2) each set of realization is used to call **P**, and the corresponding value of k_l is calculated, and (3) the ensemble of values of k_l are used to estimate the statistics (and in fact, the entire probability distribution) of k_l. Popular sampling methods include Monte Carlo sampling, Latin hypercube sampling, unscented transform sampling, etc.
2. **Analytical methods:** In these methods calculation of the statistics of k_l is based on analytical uncertainty propagation methods such as the first-order reliability method (FORM), inverse first-order reliability method, second-order reliability method, etc.

Both sampling methods and analytical methods have been investigated and successfully applied to prognostics of a variety of applications [6, 8, 11]. It is important to understand that different methods may be efficient/useful in answering different questions regarding the probability distribution of k_l.

For example, sampling-based approaches are usually more useful to obtain accurate estimates of the central behavior (mean, standard deviation, etc.) of the probability distribution of k_l, while analytical approaches like FORM may be more useful to calculate the tail behavior of the probability distribution of k_l (for example, what is the probability that the remaining flight-time is greater than two minutes?). Depending upon the application under consideration and information available, it would be necessary to choose an appropriate prediction method. The choice of the prediction method and the complete numerical details are provided in Section 5.4.

5.3.4 DERIVING REQUIREMENTS

Several metrics have been designed in literature in order to characterize the performance of prognostics algorithms [15]. These metrics enable offline evaluation, assuming that the ground truth information is available. In the context of a system in which prognostics is implemented, high-level requirements about safety can be used to derive explicit requirements on the performance of the prognostics system. Toward this end, prognostics metrics for accuracy, precision, and timeliness are used.

5.3.4.1 PREDICTION ACCURACY

Predictions may either under- or overestimate the actual event time, and both kinds of predictions are attributed with penalties resulting in actions taken by an operator using the predictions. An underestimate (i.e., the event is predicted to occur *before* it actually occurs) results in underutilization of the asset. An overestimate (i.e., the event is predicted to occur *after* it actually occurs) is even more costly, as it may result in the event occurring before an action is taken in response to the prediction.

For a given prediction time λ, the accuracy bound is defined as $\pm\alpha$. Typically, for a prediction of a time of event occurrence, α is defined as a percentage relative to the true value. For example, at $\lambda = 50\%$ to the true event time, a requirement could be that the prediction is within $\alpha = 10\%$ of the true value.

5.3.4.2 PREDICTION CONFIDENCE

Prognostic algorithms inherently contain uncertainties that arise from a variety of factors and, therefore, it is expected for a prediction algorithm to account for such uncertainties and provide an estimate for them. Consequently, prognostic algorithms often estimate the uncertainties in the predicted quantity and express them as probability distributions, sampled outputs, histograms, etc. These estimates can be used to infer the variability (spread) in predictions. In order to make decisions based on prognostics, point estimates are first computed from these PDFs to identify the most probable time for expected event time. Choosing a mean of the PDF has been the most favored choice in most applications; however, choosing a median is better in practice [16]. Furthermore, it was suggested that a comparable result is obtained by guaranteeing that the total probability between the two α bounds is at least 50 percent represented by the parameter β. The higher the value of β, the higher the confidence in a prediction that it will remain within acceptable α bounds. If the sum total of probability between the α bounds is less than 50 percent for any PDF, it should be interpreted as a low confidence prediction and should not be used for decision making.

5.3.4.3 ACTIONABLE LEAD TIME

With respect to timeliness, an operator must be given sufficient prior warning to the event that is to be predicted, so that the operator has ample time to act based on that prediction. This warning must also be sufficiently accurate, such that the operator can extend flight time as much as possible, within acceptable risk tolerances.

The prognostic horizon (PH) metric is used to specify when a prognostic algorithm has sufficiently accurate predictions. In the worst case, this should be achieved when the operator has just enough time to act before the event that is being predicted occurs.

5.4 APPLICATION: EDGE 540T

This section focuses on the application of the prognostics methodology to the prediction of remaining flying time for the Edge 540T aircraft. As described earlier, end-of-flight corresponds directly to EOD of the batteries that power the aircraft. Thus, to predict end-of-flight, EOD must be predicted, and this requires a model for the battery discharging process. First, this section describes the modeling of the batteries, then discusses the various sources of uncertainty, and finally presents the results from applying the prognostics methodology for predicting the remaining flying time for the Edge 540T aircraft.

5.4.1 BATTERY MODELING

Each lithium-polymer prism cell produces around 4.2 V, leading to a total battery output of 21 V. As discussed earlier, two of these are in series to give a total output voltage of 42 V in parallel with another set that power motors M1 and M2, respectively. The discharging of the batteries is represented using an electrochemistry battery model originally developed in [17], which describes how the internal charge moves within the battery, given current as an input, and how voltage is computed based on the internal charge. The details of this model are briefly summarized as follows.

The battery is divided into two electrodes, the positive (subscript p) and the negative (subscript n). Each electrode is split into two control volumes, a bulk volume (subscript b) and a surface layer (subscript s). As the battery discharges, Li ions move from the surface layer at the negative electrode, through the bulk, and to the surface layer at the positive electrode, in order to match the flow of electrons. So, there are four states representing charge (q), described by

$$\dot{q}_{s,p} = i_{app} + \dot{q}_{bs,p}, \tag{5.19}$$

$$\dot{q}_{b,p} = -\dot{q}_{bs,p} + i_{app} - i_{app}, \tag{5.20}$$

$$\dot{q}_{b,n} = -\dot{q}_{bs,n} + i_{app} - i_{app}, \tag{5.21}$$

$$\dot{q}_{s,n} = i_{app} + -\dot{q}_{bs,n}, \tag{5.22}$$

where i_{app} is the applied electric current. The flow of charge between the surface and bulk volumes is driven by diffusion based on Li ion concentration. The concentrations are computed as

$$c_{b,i} = \frac{q_{b,i}}{v_{b,i}}, \tag{5.23}$$

$$c_{s,i} = \frac{q_{s,i}}{v_{s,i}}, \tag{5.24}$$

where, for control volume C in electrode i, $c_{C,i}$ is the concentration and $v_{C,i}$ is the volume. The diffusion rate from the bulk to the surface in electrode i is expressed as

$$\dot{q}_{bs,i} = \frac{1}{D}(c_{b,i} - c_{s,i}), \tag{5.25}$$

where D is the diffusion constant.

The battery voltage, which comprises several electrochemical potentials, is computed as a function of the charge variables. These potentials include the equilibrium potential, concentration overpotential, surface overpotential, and ohmic overpotential [18]. The equilibrium potential is captured using the Nernst equation:

$$V_{U,i} = U_{0,i} + \frac{RT}{nF}\ln\left(\frac{1-x_i}{x_i}\right) + V_{\mathrm{INT},i}, \tag{5.26}$$

where for electrode i, $U_{0,i}$ is a reference potential, R is the universal gas constant, T is the electrode temperature (in K), n is the number of electrons transferred in the reaction ($n = 1$ for Li-ion), F is Faraday's constant and x is the mole fraction of lithium ions in the lithium-intercalated host material [19]. Mole fraction x_i is related to charge using

$$x_i = \frac{q_i}{q^{\max}}, \tag{5.27}$$

where $q^{\max} = q_p + q_n$ refers to the total amount of available Li ions. It follows that $x_p + x_n = 1$. In the Nernst equation, $V_{\mathrm{INT},i}$ is the activity correction term for which the Redlich-Kister expansion [19] is used, as follows:

$$V_{\mathrm{INT},i} = \frac{1}{nF}\left(\sum_{k=0}^{N_i} A_{i,k}\left((2x_i-1)^{k+1} - \frac{2x_i k(1-x_i)}{(2x_i-1)^{1-k}}\right)\right). \tag{5.28}$$

The concentration overpotential is the difference in voltage between the surface and bulk control volumes due to the difference in concentration and is captured by using $x_{s,i}$ in the expression for equilibrium potential.

The ohmic resistance is described using the constant R_o, producing the ohmic overpotential:

$$V_o = i_{app}R_o. \tag{5.29}$$

The surface overpotentials are described by the Butler-Volmer equation, which, for Li ion batteries, reduces to

$$V_{\eta,i} = \frac{RT}{F\alpha}\operatorname{arcsinh}\left(\frac{J_i}{2J_{i0}}\right), \tag{5.30}$$

where J_i is the current density, J_{i0} is the exchange current density, and α is the symmetry factor (0.5 for Li ion). The current densities are defined as

$$J_i = \frac{i}{S_i}, \tag{5.31}$$

$$J_{i0} = k_i(1-x_{s,i})^{\alpha}(x_{s,i})^{1-\alpha}, \tag{5.32}$$

where k_i is a lumped parameter of several constants including a rate coefficient, electrolyte concentration, and maximum ion concentration.

Battery voltage can now be expressed as follows:

$$V = V_{U,p} - V_{U,n} - V'_o - V'_{\eta,p} - V'_{\eta,n}, \tag{5.33}$$

where

$$\dot{V}'_o = (V_o - V'_o)/\tau_o, \tag{5.34}$$

$$\dot{V}'_{\eta,p} = \left(V_{\eta,p} - V'_{\eta,p}\right)/\tau_{\eta,p}, \tag{5.35}$$

$$\dot{V}'_{\eta,n} = \left(V_{\eta,n} - V'_{\eta,n}\right)/\tau_{\eta,n}, \tag{5.36}$$

and the τ parameters are empirical time constants (used because the voltages do not change instantaneously).

5.4.2 MODELING UNCERTAINTY

Recall from Section 5.3.2 that for prediction, there are three main sources of uncertainty to contend with: the system state at the time of prediction, $p(\mathbf{x}_P(k_p))$; the future input trajectory, $p(\mathbf{U}_{k_p,k_h})$; and the future process noise trajectory, $p(\mathbf{V}_{k_p,k_h})$.

The state uncertainty is determined by the estimation algorithm, in this case, the UKF. The UKF represents the uncertainty through sigma points, which capture the mean and covariance of the system state. From this distribution, many different realizations of the system state can be sampled at the time of prediction. For the process noise, it is assumed that the noise for each state at each time instant is independent; hence, the process noise trajectory can be constructed by sampling a new process noise vector at each future time instant, with the same mean and covariance used for the UKF.

Future input trajectories are divided into five different segments where each segment corresponds to different future power demands. Figure 5.5 shows the battery power, for a single battery, as a function of time over a single flight. It first-experiences a sharp increase due to the initial takeoff (about the first 10 s), then the aircraft reaches a constant speed, and as the aircraft rises in altitude, the required power starts to reduce as the air density reduces. Once altitude is reached (around 120 s), different maneuvers are performed, requiring different amounts of power. The first set is at a higher speed (120–350 s), followed by those at a lower speed (350–775 s), followed by landing.

These five different segments are each represented by two parameters that represent the entire trajectory. For the first two segments, the slope and duration are used, and for the remaining, the magnitude and duration are used. So, given samples of these defining trajectory values, a future input trajectory is defined. From a past set of sample flights, the mean values and ranges for these values

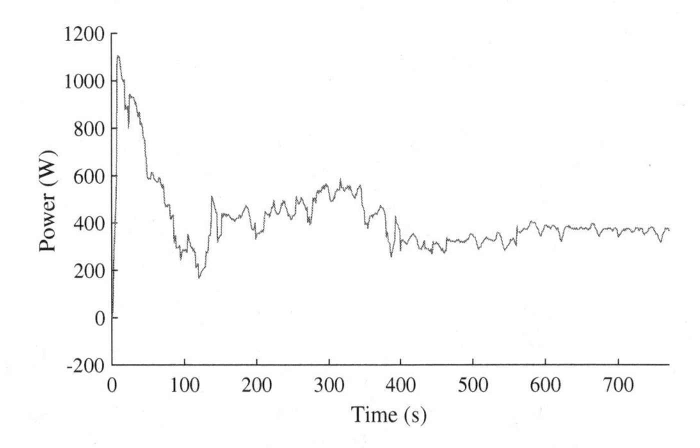

Fig. 5.5 Battery power over a single flight.

have been determined. Uniform distributions with the determined means and ranges are used and sampled from to obtain different realizations of the future input trajectories. Sample trajectories are shown in Fig. 5.6.

5.4.3 REQUIREMENTS FOR REMAINING FLYING TIME PREDICTION

High-level mission requirements for the Edge 540T can be flowed down to low-level requirements, written in terms of the offline evaluation metrics mentioned earlier. If an algorithm conforms to the specified low-level requirements, it will, in turn, also fulfill the high-level operational requirements with respect to remaining flying time prediction.

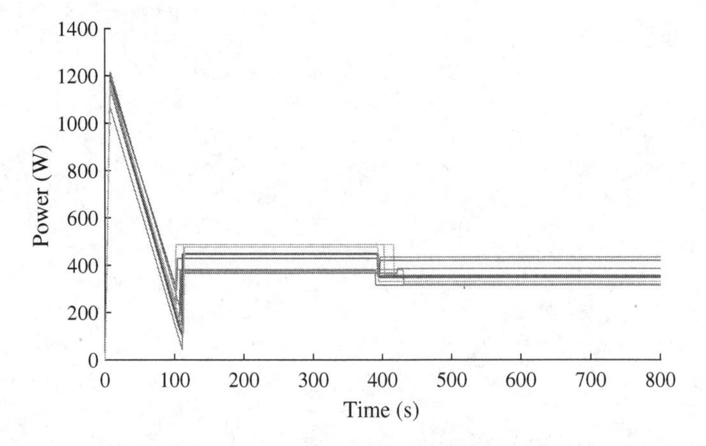

Fig. 5.6 Different realizations of future required battery power over a single flight.

Figure 5.7 shows the requirements flowdown for the electric aircraft. It goes from high-level requirements to performance requirements for the prognostic system. Ultimately, the high-level requirements are transformed into the prognostics metrics for accuracy, precision, and timeliness.

Upon consulting the aircraft operators, it was determined that, contingent upon wind conditions and flight range, the aircraft operator can take at most two minutes to land the aircraft. Based on this information, the corresponding metrics requirements for the prognostics algorithm can be derived:

1. *The prognostic algorithm shall predict end-of-flight no later than two minutes before the end-of-flight condition is reached.*
2. *The prognostic algorithm shall raise an alarm when the remaining flying time reduces to two minutes.*
3. *The prognostic algorithm performance shall meet desired prediction confidence levels at least two minutes from the end-of-flight condition.*
4. *The aircraft shall not land more than 0.4 minutes too early before the end-of-flight condition more than 10 percent of the time.*

Here, the two-minute alarm is biased to occur early rather than late because the landing becomes unsafe when the EOD of the batteries is reached. The early

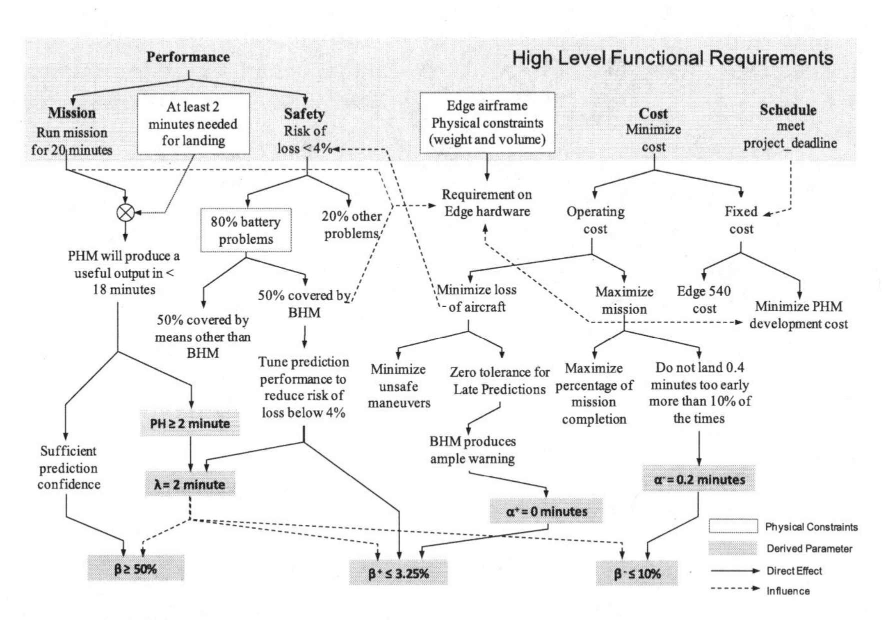

Fig. 5.7 Requirements flowdown tree for the electric aircraft scenario [16].

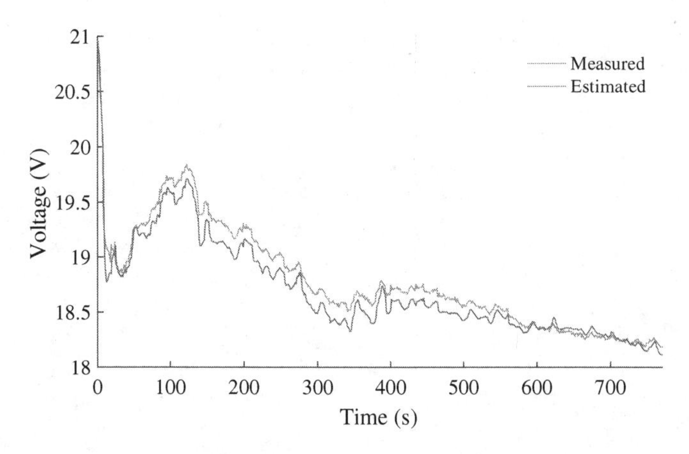

Fig. 5.8 Measured and estimated battery voltage for Flight 1.

alarm prediction bound limits the opportunity cost of unnecessarily denied flying time.

5.4.4 ESTIMATION AND PREDICTION RESULTS

This subsection presents the results of both estimation and prediction steps. The prognostic framework is demonstrated over two representative flights. Figures 5.8 and 5.9 show the measured and estimated battery voltages for the same battery over the two different flights. Clearly, the voltage can be tracked with little error, as the underlying battery model is fairly accurate. A symmetric unscented

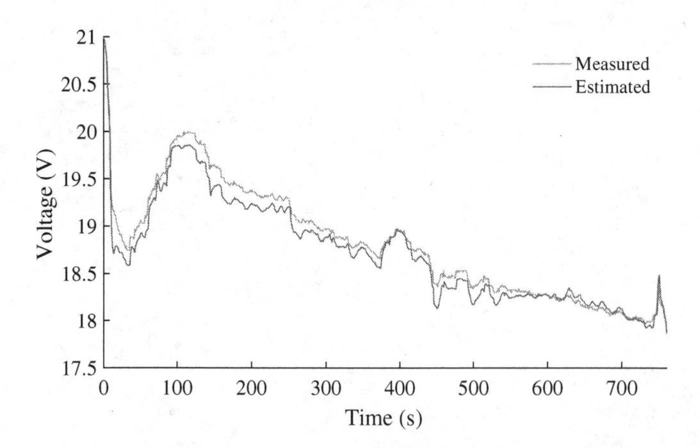

Fig. 5.9 Measured and estimated battery voltage for Flight 2.

transform is implemented with kappa of −5 and no scaling. The Q and R covariance matrices are assumed to be diagonal.

Given that the battery state can be estimated accurately, and given the defined distributions for process noise and future input trajectories, predictions can be made. Remaining flying time predictions are made once per minute, based on the state estimate computed at that time, and using Monte Carlo sampling with 100 samples. That is, for each prediction, 100 different realizations of the future state trajectory are computed, and for each future time instant, the algorithm checks whether EOD has been met. A time horizon of 5000 s is used. Because each flight lasts less than 20 minutes, this guarantees that every realization will end eventually at EOD. In total, 20 experimental flights were flown [20], and to demonstrate the implementation of the framework, data from two flights have been used.

For each battery, EOD is predicted, and the minimum one determines by when the aircraft must land. Normally, EOD is defined by 15 V. Of course, the aircraft must land *before* that happens, accounting for any possible load increases that could immediately bring the aircraft voltage down to that level. So, the pilot must land sometime before this voltage is reached.

Figure 5.10 shows the predicted remaining flying time for the first flight. As EOD is approached, the remaining flying time prediction decreases steadily. The minimum and maximum predictions provide an indication of the uncertainty in the predictions. When the flight landed around 775 s, the prediction was that still around 200 s of flight remained, on average. The pilot landed with sufficient time to spare as per indicated in the requirements.

Since EOD was not actually reached, the EOD threshold is moved up to 18 V, so that prediction accuracy can be validated. In this case, a ground truth can then be computed for remaining flying time with which to compare predictions. Figure 5.11 shows the same results with this value. Clearly, the predictions are

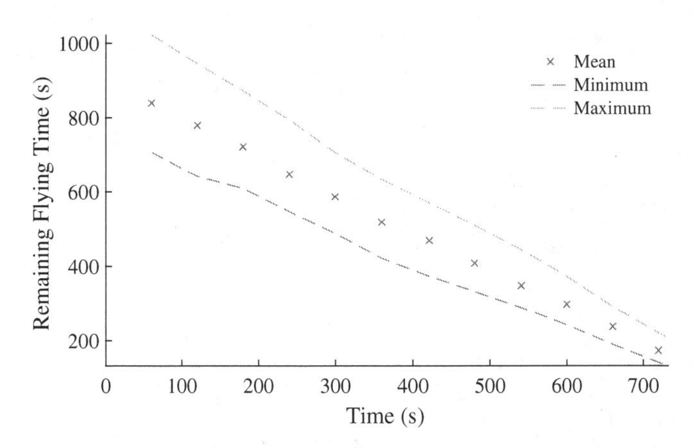

Fig. 5.10 Predicted remaining flying time for Flight 1 with EOD 15V.

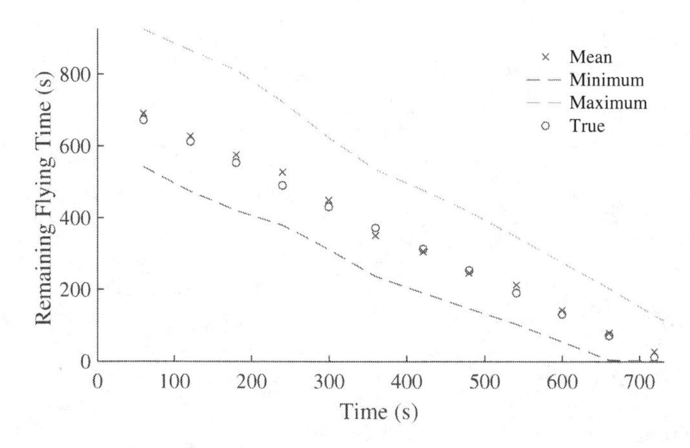

Fig. 5.11 Predicted remaining flying time for Flight 1 with EOD 18V.

very accurate in this case, because the flight follows the average anticipated behavior well in terms of the required battery power. The average relative accuracy, based on the mean prediction, is 98 percent.

Figure 5.12 shows the same results for a different flight. Here, the predictions are not as accurate, because the actual flight is more in the tail of the distribution of possible power demands. In addition, the battery packs used in the second flight are not the same as those used in the previous flight. It is still considered within the considered range of uncertainty. This emphasizes the need to correctly capture uncertainty within the framework. If the prediction is that there are two minutes remaining flying time on average, but the future loading will be higher than average, then a pilot would want to know the most conservative prediction, to

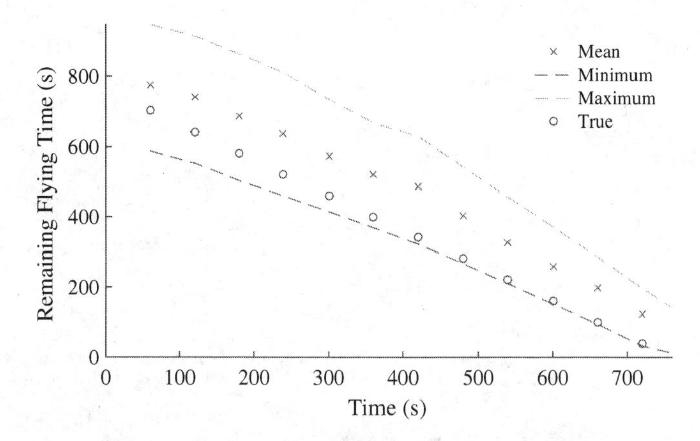

Fig. 5.12 Predicted remaining flying time for Flight 2.

ensure that the aircraft can be landed with a very low probability of reaching EOD before the aircraft lands. The average relative accuracy, based on the mean prediction, for the second flight is 87 percent.

5.5 CONCLUSIONS

This chapter delineates a technique for prognostics as exemplified on remaining charge estimation on an electric UAV. The approach chosen here is model-based, where the form of the model has been linked to the internal processes of the battery and validated using experimental data. The model was used in a Kalman filter framework to make predictions of end of discharge. Particular emphasis was given to the overall architecture, uncertainty management, and requirement definition. For the application shown, prediction results were used by the operator to make decisions about when to return to the landing strip.

Generally, a similar process for estimating remaining life can be followed for other components where a damage progression model is built for the components of interest. Obviously, the model would be different if, say, a bearing, an electronics component, or a structural element were the object of interest.

While a human operator was the recipient of the information in the example shown, the information could in principle also be integrated into a higher level decision-making algorithm that can provide actionable information within an autonomous vehicle context. In that case, and depending on the prognostic horizon, the reaction could be to autonomously change controller settings (maybe flying slower or at a lower altitude), reconfigure system resources to ensure primary mission goals (perhaps switching off secondary components), invoke a replanning or rescheduling routine (perhaps omitting waypoints), or provide the information for later processing in a maintenance setting.

REFERENCES

[1] Goebel, K., and Saha, B., "Prognostics Applied to Electric Propulsion UAV," *In*: K. Valavanis, and G. Vachtsevanos (eds) *Handbook of Unmanned Aerial Vehicles. Springer, Dordrecht*, Springer, 2015, pp. 1053–1070.

[2] Daigle, M., and Goebel, K., "Multiple Damage Progression Paths in Model-Based Prognostics," *Proceedings of the 2011 IEEE Aerospace Conference*, Big Sky, MO, March 2011.

[3] Orchard, M., and Vachtsevanos, G., "A Particle Filtering Approach for On-Line Fault Diagnosis and Failure Prognosis," *Transactions of the Institute of Measurement and Control*, No. 3–4, June 2009, pp. 221–246.

[4] Oliva, J., Weihrauch, C., and Bertram, T., "A Model-Based Approach for Predicting the Remaining Driving Range in Electric Vehicles," *Annual Conference of the Prognostics and Health Management Society 2013*, New Orleans, LA, Oct. 2013, pp. 438–448.

[5] Daigle, M., Sankararaman, S., and Kulkarni, C., "Stochastic Prediction of Remaining Driving Time and Distance for a Planetary Rover," *2015 IEEE Aerospace Conference*, Big Sky, MO, March 2015.

[6] Sankararaman, S., "Significance, Interpretation, and Quantification of Uncertainty in Prognostics and Remaining Useful Life Prediction," *Mechanical Systems and Signal Processing*, Vol. 52, 2015, pp. 228–247.

[7] Bole, B., Daigle, M., and Gorospe, G., "Online Prediction of Battery Discharge and Estimation of Parasitic Loads for an Electric Aircraft," *Second European Conference of the Prognostics and Health Management Society*, Nantes, France, July 2014, pp. 23–32.

[8] Daigle, M., and Sankararaman, S., "Advanced Methods for Determining Prediction Uncertainty in Model-Based Prognostics with Application to Planetary Rovers," *Annual Conference of the Prognostics and Health Management Society*, New Orleans, LA, Oct. 2013, pp. 262–274.

[9] Daigle, M., and Goebel, K., "Model-Based Prognostics with Concurrent Damage Progression Processes," *IEEE Transactions on Systems, Man, and Cybernetics: Systems*, Vol. 43, No. 4, May 2013, pp. 535–546.

[10] Daigle, M., Sankararaman, S., and Roychoudhury, I., "System-Level Prognostics for the National Airspace," *Annual Conference of the Prognostics and Health Management Society*, Denver, CO, October 2016, pp. 397–405.

[11] Sankararaman, S., Daigle, M., and Goebel, K., "Uncertainty Quantification in Remaining Useful Life Prediction using First-Order Reliability Methods," *IEEE Transactions on Reliability*, Vol. 63, No. 2, June 2014, pp. 603–619.

[12] Julier, S. J., and Uhlmann, J. K., "A New Extension of the Kalman Filter to Nonlinear Systems," *Proceedings of the 11th International Symposium on Aerospace/Defense Sensing, Simulation and Controls*, Orlando, FL, 1997, pp. 182–193.

[13] Julier, S. J., and Uhlmann, J. K., "Unscented Filtering and Nonlinear Estimation," *Proceedings of the IEEE*, Vol. 92, No. 3, 2004, pp. 401–422.

[14] Arulampalam, M., Maskell, S., Gordon, N., and Clapp, T., "A Tutorial on Particle Filters for Online Nonlinear/Non-Gaussian Bayesian Tracking," *IEEE Transactions on Signal Processing*, Vol. 50, No. 2, 2002, pp. 174–188.

[15] Saxena, A., Celaya, J., Balaban, E., Goebel, K., Saha, B., Saha, S., and Schwabacher, M., "Metrics for Evaluating Performance of Prognostic Techniques," *International Conference on Prognostics and Health Management*, Denver CO, Oct. 2008.

[16] Saxena, A., Roychoudhury, I., Celaya, J., Saha, B., Saha, S., and Goebel, K., "Requirements Flowdown for Prognostics Health Management," *Proceedings of AIAA Infotech Aerospace 2012 Conference*, Garden Grove, CA, June 2012.

[17] Daigle, M., and Kulkarni, C., "Electrochemistry-Based Battery Modeling for Prognostics," *Annual Conference of the Prognostics and Health Management Society*, New Orleans, LA, Oct. 2013, pp. 249–261.

[18] Rahn, C. D., and Wang, C.-Y., *Battery Systems Engineering*, Wiley, 2013.

[19] Karthikeyan, D. K., Sikha, G., and White, R. E., "Thermodynamic Model Development for Lithium Intercalation Electrodes," *Journal of Power Sources*, Vol. 185, No. 2, 2008, pp. 1398–1407.

[20] Hogge, E., Kulkarni, C., Vazquez, S., Smalling, K., Storm, T., Hill, B., and Quach, C., "Flight Tests of a Remaining Flying Time Prediction System for Small Electric Aircraft in the Presence of Faults," *Annual Conference of the Prognostics and Health Management Society*, St. Petersburg, FL, PHM Society, 2017, p. 12.

CHAPTER 6

Model Reference Adaptive Control for General Aviation Aircraft: Development, Simulation, and Flight Test

Melvin Rafi*, James E. Steck†, Venkatasubramani S. R. Pappu‡ and John M. Watkins§

Wichita State University, Wichita, KS, 67260

Bryan S. Steele¶

Textron Aviation, Wichita, KS, 67206

NOMENCLATURE

C_D, C_L, C_M	Coefficient of drag, lift, and pitching moment
C_y, C_l, C_n	Coefficient of side force, rolling moment, and yawing moment
F_{A_X}, F_{A_Y}, F_{A_Z}	Aerodynamic forces in forward, lateral, and vertical direction
F_{T_X}, F_{T_Y}, F_{T_Z}	Thrust forces in the forward, lateral and vertical directions
I_{xx}, I_{yy}, I_{zz}, I_{xz}	Aircraft moments and product of inertia
K_p, K_d	Proportional and derivative controller gains
K_1, K_2	Kalman and MSO gains
L, M, N	Aircraft rolling moment, pitching moment, and yawing moment
P, Q, R	Aircraft roll rate, pitch rate, and yaw rate
PLA	Aircraft power lever angle
T	Aircraft thrust
U, V, W	Forward, lateral, and vertical aircraft velocities
$\dot{U}$, $\dot{V}$, $\dot{W}$	Forward, lateral, and vertical aircraft accelerations
$\bar{c}$	Mean geometric chord
g	Acceleration due to gravity
m	Aircraft mass
$\hat{p}$, $\hat{q}$, $\hat{r}$	Nondimensionalized roll, pitch, and yaw rates

*Graduate Research Assistant, Aerospace Engineering, and AIAA Student Member.
†Professor, Aerospace Engineering, and AIAA Associate Fellow.
‡Graduate Research Assistant, Aerospace Engineering, and AIAA Student Member.
§Professor and Chair, Electrical Engineering and Computer Science.
¶Senior Engineer, Aerosciences, and AIAA Professional Member.

$\overline{q}$	Dynamic pressure
S	Wing reference area
θ, γ, α	Aircraft pitch angle, flight path angle, and angle of attack
β, ψ, ϕ	Aircraft sideslip angle, heading angle, and bank angle
δ_e, δ_a, δ_r	Elevator, aileron, and rudder deflection
ϕ_T	Thrust line angle relative to fuselage axis
η	Neural network learning rate
ω	Aircraft angular velocity
ω_n	Natural frequency
ζ	Damping ratio

6.1 INTRODUCTION

The area of adaptive control systems has seen tremendous progress and advancements in recent times. These advancements have allowed control systems in aircraft to be "smarter," to better adapt to the operating environment, and ultimately, to give the pilot a lower workload in the event of failure or damage. One subset of adaptive control is the model reference adaptive control (MRAC) architecture.

Utilizing the MRAC architecture in a control system enhances its ability to respond to unexpected changes in the plant dynamics and to uncertainties and unmodeled dynamics. The research documented in this chapter details MRAC development at Wichita State University, applied specifically to general aviation (GA) aircraft. The research culminates in a flight-tested MRAC system developed for the Textron Aviation CJ-144 fly-by-wire test bed, incorporating a dynamic inverse controller and adaptive neural network, together with a model follower controller. Subsequently, extensions are added to enhance the ability of the MRAC system in handling situations involving atmospheric disturbances and sensor-induced noise. Results from simulation, illustrating the MRAC system's performance in both favorable and adverse conditions, are presented for various scenarios, and the validation processes used to test the MRAC system prior to flight testing are described. Results from flight testing various versions of the controller on the Textron Aviation CJ-144 fly-by-wire test bed are then presented, and these demonstrate the effectiveness of the MRAC system at handling real-world flight conditions and simulated control failures.

One advantage of an MRAC system is its ability to adapt to uncertainties in or changes to the plant dynamics or operating environment. Because it can adapt well to failure conditions, the reliance on mechanical redundancies within a traditional Fly-By-Wire (FBW) system is greatly reduced. These redundancies are often required for certification purposes and have to be built into the control system. Partially because of this, implementing FBW systems in GA aircraft has long been a cost-prohibitive exercise. By being able to adapt very well to failure scenarios and reduce this reliance, MRAC has the potential to reduce these costs. From an aircraft design or development perspective, this could then make it easier for manufacturers to implement and certify FBW systems on board GA aircraft, and could help make FBW systems more commonly available on GA aircraft.

The MRAC architecture differs from a conventional control system in its operation. In a conventional open-loop control setup, an input to the controls results in a direct change in the deflection of the control surfaces. The pilot directly commands an elevator deflection, an aileron deflection, a rudder deflection, and a throttle setting. Taking the longitudinal axis, for example, if the pilot were to only provide a pitch input and leave the throttle position constant, that input would not only lead to a change in the aircraft's pitch angle and rate, but would also lead to an accompanying change in airspeed. Conversely, assuming the throttle position is moved and the pitch input is left constant, the resulting change in airspeed would also lead to a change in pitch angle and rate (due to the change in lift due to thrust).

In the MRAC architecture described here, the controls are instead decoupled, and the pilot may separately command a flight path angle, a bank angle, an amount of side force, and an airspeed. Considering the longitudinal axis once again, a single input in the pitch axis alone would then result in not only a change in elevator deflection, but also an accompanying change in thrust setting, as the controller would attempt to simultaneously maintain both the commanded flight path angle and commanded airspeed. Conversely, if only the power lever is moved and the pitch input is left constant, the controller accompanies the change in thrust setting with a change in elevator deflection in an attempt to maintain level flight. In a simplified sense, the pilot may "point" the aircraft in the desired direction, and the MRAC system will take the airplane there.

Much work has been devoted to the development of MRAC for military and civilian aircraft. MRAC was initially applied to aviation through the work of Kim and Calise [1], with promising results. Since then, the National Aeronautics and Space Administration (NASA) has successfully flight-tested Calise's algorithms on an F-15 demonstrator [2, 3], and has also utilized MRAC in developing controllers for the Generic Transport Model (GTM). Nguyen et al. have developed various modifications to the MRAC architecture, including an optimal control modification for fast adaptation [4] and a predictor model-based least-squares MRAC modification [5, 6]. Stepanyan and Krishnakumar [7] have also developed a Modified Reference Model MRAC (M-MRAC), which provides tracking error feedback to the reference model, reducing high-frequency oscillations typical for large adaptation rates.

6.2 MRAC DEVELOPMENT AT WSU

Wichita State University (WSU) has been actively conducting research into adaptive control techniques [8] since the early 2000s. This research initially began as a decoupled adaptive control system comprising a linear compensator and an artificial neural network (ANN) [9], and was simulated with a generic aircraft configuration.

Over the course of the last decade, the MRAC system developed at WSU has progressed from this simple decoupled setup for a generic Three-Degrees-of-Freedom (3DOF) aircraft configuration [10, 11, 12], to an advanced Six-Degrees-of-Freedom (6DOF) MRAC system with an adaptive bias corrector (ABC) and a dynamic inverse controller [13, 14], specific to GA applications. The adaptive bias corrector was developed as a replacement for the existing ANN and represents a simplified (and possibly certifiable) adaptation method including only the bias term of the neural network. Improvements in recent times have also included higher accuracy, nonlinear aerodynamic models, and the development of a new predictive inverse controller. An Optimal Control Modification (OCM) [15, 16] for the MRAC system was also developed, and the MRAC system has further been adapted to handle control reversal situations [17]. A Single Network Adaptive Critic (SNAC) [18, 19, 20] has additionally been developed for fast adaptation in the MRAC system.

Simulated and actual flight testing have also been key evaluators of the MRAC system's performance over the years, and since 2001, partnerships with Textron Aviation/Beechcraft have allowed for evaluation of the controller's real-world performance on the Textron Aviation CJ-144 Small Aircraft Transportation System (SATS) FBW test bed [21]. The CJ-144 is a modified derivative of the Beechcraft F-33C Bonanza. In these efforts, the MRAC system has demonstrated promising results with good command tracking, even in the event of simulated partial control failures.

The ability of the MRAC system to adapt to uncertainties, such as failures, has led WSU to further refine the MRAC architecture to optimize its responses to such conditions. Most recently, one thrust of research at WSU has been devoted to studying the performance of the MRAC system in the presence of atmospheric disturbances [22–24] and in response to sensor-induced noise [25, 26]. Atmospheric disturbances (such as turbulence, wind gusts, microbursts, and wake vortex conditions) and sensor noise present unique challenges in control system design, as they test the robustness of a controller to uncertainties.

This chapter aims to concisely document the overall advancements made through the development of the GA-based MRAC architecture developed at WSU. The chapter details efforts made in the development of the 6DOF MRAC architecture, extensions made to the MRAC architecture to optimize its responses to scenarios with atmospheric disturbances and sensor noise, and efforts to validate the MRAC system through flight simulation and actual flight testing.

6.3 AIRCRAFT MODEL

The Textron Aviation CJ-144 is a piston engine aircraft, based on the Beechcraft F33C Bonanza, with a conventional wing and horizontal and vertical tails, but

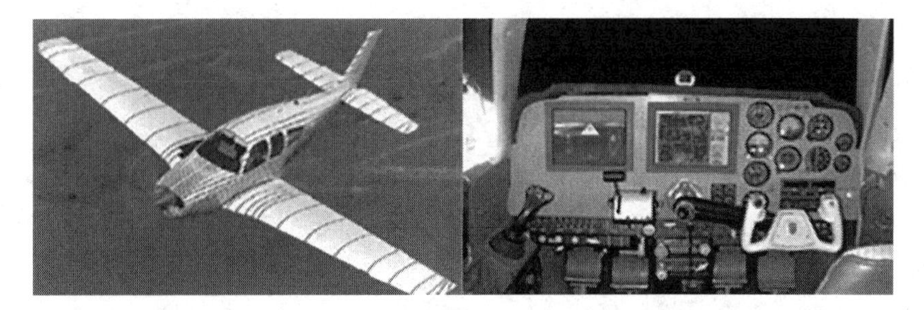

Fig. 6.1 Exterior and interior of the Textron Aviation CJ-144 fly-by-wire test bed. (photo courtesy of Textron Aviation)

modified to incorporate a FBW control system. The left-seat controls are linked to this FBW system, while the right-seat controls are linked to a conventional mechanical control system and are used to override the FBW system if needed. The FBW system is configured with an "EZ-fly" logic that allows command of decoupled flight controls. Consequently, an input to the joystick commands flight path angle and bank angle, an input to the power lever commands airspeed, and an input to the rudder pedals commands side force. An illustration of the aircraft and its cockpit is presented in Fig. 6.1.

6.3.1 GENERAL CHARACTERISTICS

The general properties of the Beechcraft F33C Bonanza [27], upon which the Textron Aviation CJ-144 is based, are presented in Table 1.

TABLE 1 GENERAL CHARACTERISTICS OF THE BEECHCRAFT F33C BONANZA/TEXTRON AVIATION CJ-144

Property	Value
Length	26 ft 8 in.
Height	8 ft 3 in.
Wingspan	32 ft 10 in.
Max. takeoff weight	3400 lb
Power loading	11.9 lb/hp
Max. range (average)	450 nm
Max. cruise speed (typical)	167 knots
Clean stall speed (typical)	63 knots

6.3.2 BASIC AIRCRAFT MODELING

The Textron Aviation CJ-144 described previously was modeled within the MATLAB/Simulink® environment. The base aircraft model is composed of a 6DOF implementation, with the control effectors being elevator (δe), aileron (δa), rudder (δr), and thrust (T). The model uses the nonlinear equations of motion in the body-fixed axis [28], as shown in Eqs. (6.1) to (6.6).

$$m(\dot{U} - VR + WQ) = -mg\sin\theta + F_{A_x} + F_{T_x} \tag{6.1}$$

$$m(\dot{V} - UR - WP) = mg\sin\phi\cos\theta + F_{A_y} + F_{T_y} \tag{6.2}$$

$$m(\dot{W} - UQ + VP) = mg\cos\phi\cos\theta + F_{A_z} + F_{T_z} \tag{6.3}$$

$$I_{xx}\dot{P} - I_{xz}\dot{R} - I_{xz}PQ + (I_{zz} - I_{yy})RQ = L_A + L_T \tag{6.4}$$

$$I_{yy}\dot{Q} + (I_{xx} - I_{zz})PR + I_{xz}(P^2 - R^2) = M_A + M_T \tag{6.5}$$

$$I_{zz}\dot{R} - I_{xz}\dot{P} + (I_{yy} - I_{xx})PQ + I_{xz}QR = N_A + N_T \tag{6.6}$$

The kinematic equations are shown in Eqs. (6.7) to (6.9):

$$P = \dot{\phi} - \dot{\psi}\sin\theta \tag{6.7}$$

$$Q = \dot{\theta}\cos\phi - \dot{\psi}\cos\theta\sin\phi \tag{6.8}$$

$$R = \dot{\psi}\cos\theta\cos\phi - \dot{\theta}\sin\phi \tag{6.9}$$

Textron Aviation supplies proprietary data for a functional linear lift, drag, and moment model below stall. These models generate values for the lift, drag, and moment coefficients based on the specified aircraft state.

6.3.3 STALL MODEL EXTENSION

When aircraft operate during the approach/landing phases of flight, or when they are subject to atmospheric disturbances, it is not uncommon for them to be brought near to stall. It follows that a simple linear aerodynamic model would not suffice in providing an accurate simulated response when evaluating the MRAC system's performance in such conditions. To provide better results when operating close to the edge of the flight envelope, a nonlinear aerodynamic model was devised and incorporated into the existing linear model. The nonlinear model took into account the effects of stall on lift, drag, and moment.

To compensate for the effects of nonlinear aerodynamics, nonlinear lift, drag, and moment components were integrated into the existing linear models. This was done by adding second, third, and fourth order terms to the existing governing equations, denoted by bold type in Eqs. (6.10) to (6.12):

$$C_L = C_{L_0} + C_{L_\alpha}\alpha + \left(\boldsymbol{C_{L_{\alpha 2}}}\right)\boldsymbol{\alpha}^2 + \left(\boldsymbol{C_{L_{\alpha 3}}}\right)\boldsymbol{\alpha}^3 + \left(\boldsymbol{C_{L_{\alpha 4}}}\right)\boldsymbol{\alpha}^4 + C_{L_{d_e}} d_e + C_{L_{d_{e-\text{trimtab}}}} d_{e-\text{trimtab}} + C_{L_{\dot{\alpha}}}\hat{\dot{\alpha}} + C_{L_q}\hat{q} + C_{L_{\text{ground}}} \tag{6.10}$$

$$C_M = C_{M_0} + C_{M_\alpha}\alpha + (\boldsymbol{C_{M_{\alpha 2}}})\boldsymbol{\alpha}^2 + (\boldsymbol{C_{M_{\alpha 3}}})\boldsymbol{\alpha}^3 + (\boldsymbol{C_{M_{\alpha 4}}})\boldsymbol{\alpha}^4 + C_{M_{d_e}} d_e + C_{M_{d_{e-\text{trimtab}}}} d_{e-\text{trimtab}} + C_{M_{\dot{\alpha}}}\hat{\dot{\alpha}} + C_{M_q}\hat{q} + C_{M_{\text{ground}}} \tag{6.11}$$

$$C_D = C_{D_0} + C_{D_{0-\text{gear}}}(gear) + C_{D_{0-\text{flap}}}(flap) + C_{D_{0-\text{TCP}}}(TCP) + C_{D_k}(C_L - C_{L_i})^2 \tag{6.12}$$

The numerical coefficients are evaluated on-the-fly as the simulation runs. The nonlinear equations of motion were implemented and evaluated via the Aerospace Blockset within the MATLAB/Simulink® environment. More details on the development and validation of the nonlinear model are given in Rafi et al. [23].

6.3.4 WING/TAIL MODEL EXTENSION

To expand the aircraft model further, a wing strip theory model was developed to more accurately simulate responses in asymmetric wind conditions. This effort was carried out to facilitate the study of the MRAC system's response to atmospheric disturbances, in particular the wake vortex scenario. While modeling the wing as a whole was sufficient for earlier work, a wake vortex imposes multiple localized effects at various points on the aircraft. Such effects include localized stalling on wings and stabilizers. The resulting three-dimensional (3D) wing strip theory geometry model allowed determination of quantities such as lifting forces and moments imposed on the aircraft due to the wake vortex. A 3D geometry of the aircraft was created, and from this, the coordinates of various points along the wing, horizontal tail, and vertical tail were extracted. The main wing was divided into 10 strips, from wingtip to wingtip, with the point-of-action being the lateral and vertical midpoint of each strip at 25 percent of the chord length. The same was done for the horizontal tail, which was divided into four strips, and the vertical tail, which was divided into two strips. Figure 6.2 shows the points-of-action superimposed on the geometry of the aircraft.

Determination of the coordinates of these points took into account various design characteristics of the aircraft, including dihedral, sweep, incidence, airfoil thickness, taper, and relative locations of the lifting surfaces with respect to the aircraft's Center of Gravity (CG). As such, it is an accurate representation of the aircraft's true geometry.

6.4 BASELINE MRAC ARCHITECTURE

The fundamental elements that make up the WSU 6DOF MRAC architecture are the inverse controller, the adaptive element (artificial neural network), and the model follower and linear controller. Figure 6.3 illustrates a simplified schematic of the 6DOF MRAC architecture.

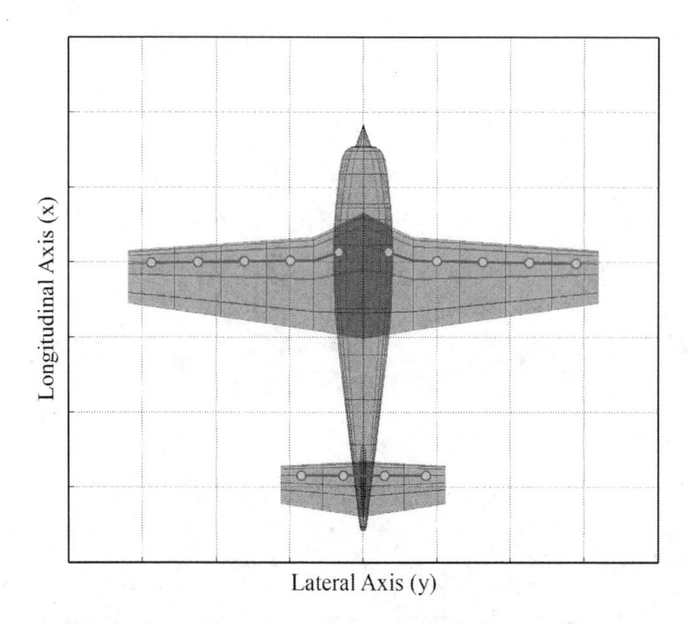

Fig. 6.2 Evaluation points along wing and horizontal stabilizer used for strip theory calculations.

As described previously, the control loops in the MRAC system are decoupled, and the pilot commands flight path angle, bank angle, side force, and airspeed. In brief, the linear PID controllers translate the pilot inputs into accelerations that are fed into the dynamic inverse controller, which then determines the required elevator, aileron, and rudder deflection angles, as well as the required thrust

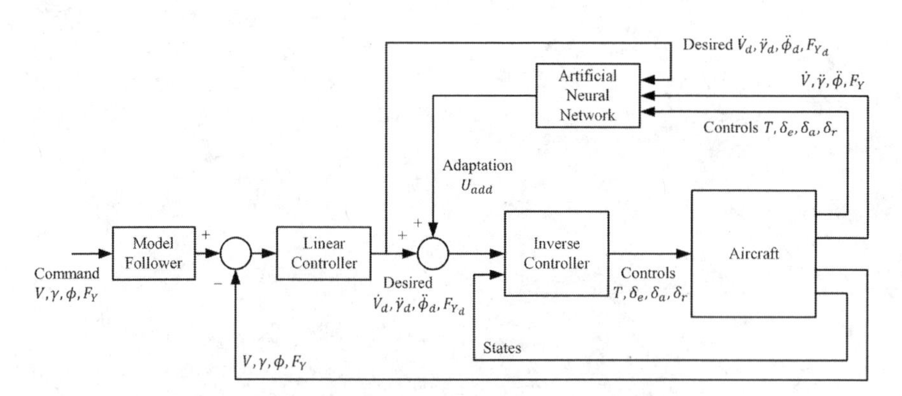

Fig. 6.3 Schematic of the 6DOF flight control architecture.

setting. The response of the aircraft is then fed into the pitch, bank, side force, and velocity ANNs. The ANNs serve to compensate for any error in the inverse controller. These differences may stem from a variety of sources, such as modeling errors or uncertainties in the inverse controller or partial system failures. In case of control saturation, ANN training is halted so as to prevent windup. The sections below describe the functions of the dynamic inverse controller, ANN, and model follower/linear controller in more detail.

6.4.1 INVERSE CONTROLLER

The inverse controller contains a "dynamic inverse" of the nonlinear aircraft model, as given in Eqs. (6.1) to (6.6). It calculates the necessary control surface deflections to achieve the pilot's commanded inputs, based on commanded (or desired) accelerations from the linear controller and the adaptation signal from the ANN. These terms are given the subscript "*d*." As the inversion is computed for the force and moment model at the nominal flight condition, it follows that the inverse becomes less precise as the flight condition deviates from nominal. The role of the ANN is to compensate for this discrepancy. It is important to note that this does not mean that the inverse results from a linearized dynamic model, but from a nonlinear model. While the lift model in Eq. (6.17) is linear, the drag polar in Eq. (6.16) is not, nor are the dynamics in Eqs. (6.13) to (6.15) and Eqs. (6.26) to (6.28). The derivation of the inverse controller is described as follows for both the longitudinal and lateral-directional axes, and Fig. 6.4 provides a schematic of the inverse controller's architecture.

6.4.1.1 LONGITUDINAL AXES

The longitudinal segment of the inverse controller determines the necessary thrust and elevator deflection and consists of a predictor element and a corrector element. Equations (6.1), (6.3), and (6.5) are converted from the body-fixed

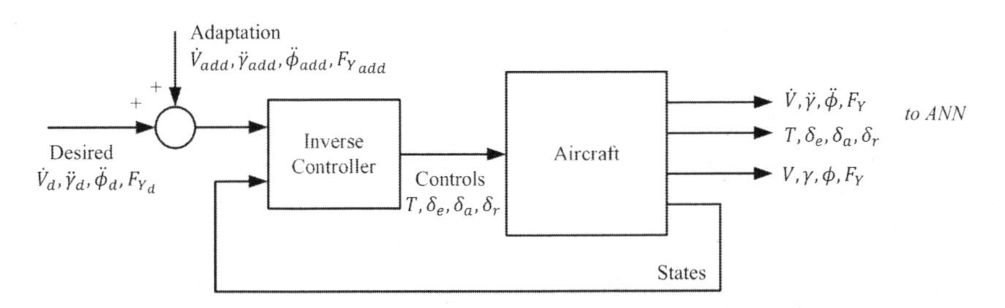

Fig. 6.4 General architecture of inverse controller.

axes to the wind axes, resulting in Eqs. (6.13) to (6.15).

$$m\dot{V}_{\text{true}} = -mg\sin\gamma + \bar{q}SC_D + T\cos(\phi_T + \alpha) \tag{6.13}$$

$$m(V_{\text{true}}\dot{\gamma}) = mg\cos\gamma\cos\phi + \bar{q}SC_L + T\sin(\phi_T + \alpha) \tag{6.14}$$

$$I_{yy}\ddot{\gamma_d} = \bar{q}SC_M + TdT \tag{6.15}$$

Where:

$$C_D = C_{D_0} + C_{D_K}\left(C_{L_0} + C_{L_\alpha}\alpha - C_{L_{\text{trim}}}\right)^2 \tag{6.16}$$

$$C_L = C_{L_0} + C_{L_\alpha}\alpha + C_{L_{\dot{\alpha}}}\dot{\alpha} + C_{L_{\hat{q}}}\hat{q} + C_{L_{\delta e}}\delta_e \tag{6.17}$$

$$C_M = C_{M_0} + C_{M_\alpha}\alpha + C_{M_{\dot{\alpha}}}\dot{\alpha} + C_{M_{\hat{q}}}\hat{q} + C_{M_{\delta e}}\delta_e \tag{6.18}$$

The inverse is calculated in two steps. First, the predictor element solves the equations of motion using the states from the previous time step to predict the next value of angle of attack. Eq. (6.13) is first solved for the predicted thrust, resulting in Eq. (6.19). Next, Eq. (6.15) is solved for the predicted elevator deflection, resulting in Eq. (6.20). Finally, Eq. (6.14) is solved for the predicted angle of attack, which is calculated using the predicted thrust and predicted elevator deflection, resulting in Eq. (6.21). The assumption is made that the second derivative of angle of attack is small, such that $\ddot{\gamma_d} \approx \ddot{\theta_d}$.

$$T_p = \frac{1}{\cos(\phi_T + \alpha)}\left(m\dot{V}_d + mg\sin\gamma + \bar{q}SC_D\right) \tag{6.19}$$

$$\delta_{ep} = \frac{I_{yy}}{\bar{q}S\bar{c}C_{M_{\delta e}}}\ddot{\gamma_d} - \frac{\left(C_{M_0} + C_{M_\alpha}\alpha + C_{M_{\dot{\alpha}}}\dot{\alpha} + C_{M_{\hat{q}}}\hat{q}\right)}{C_{M_{\delta e}}} + \frac{d_T}{\bar{q}S\bar{c}C_{M_{\delta e}}}T_p \tag{6.20}$$

$$\begin{aligned}\alpha_p = {} & \frac{1}{\bar{q}SC_{L_\alpha}}\left(mV_{true}\dot{\gamma} + mg\cos\gamma\cos\phi\right) \\ & - \frac{1}{C_{L_\alpha}}\left(C_{L_0} + C_{L_{\dot{\alpha}}}\dot{\alpha} + C_{L_{\hat{q}}}\hat{q} + C_{L_{\delta e}}\delta_{ep}\right) - \frac{\sin(\phi_T + \alpha)}{\bar{q}SC_{L_\alpha}}T_p\end{aligned} \tag{6.21}$$

The first derivative of flight path angle in Eq. (6.21), $\dot{\gamma}$, is found by adding the time derivative of the aircraft flight path angle from the previous time step and the steady state pitch rate during a coordinated turn [28]. This is shown in Eq. (6.22). During wings-level flight, the second term is zero.

$$\dot{\gamma} = \dot{\gamma}_{\text{from aircraft}} + \frac{g\sin^2\phi}{V_{\text{true}}\cos\phi} \tag{6.22}$$

The predicted angle of attack, α_p, is then used by the corrector element to calculate the desired thrust and elevator deflection, as shown in Eqs. (6.23) and (6.24).

$$T = \frac{1}{\cos(\phi_T + \alpha_p)}\left(m\dot{V}_d + mg\ \sin\gamma + \bar{q}SC_D\right) \tag{6.23}$$

$$\delta_e = \frac{I_{yy}}{\bar{q}S\bar{c}C_{M_{\delta e}}}\ddot{\gamma}_d - \frac{\left(C_{M_0} + C_{M_\alpha}\alpha_p + C_{M_{\dot{\alpha}}}\dot{\alpha} + C_{M_{\hat{q}}}\hat{q}\right)}{C_{M_{\delta e}}} + \frac{d_T}{\bar{q}S\bar{c}C_{M_{\delta e}}}T \tag{6.24}$$

6.4.1.2 LATERAL-DIRECTIONAL AXES

The lateral-directional inverse controller solves Eq. (6.25) for the necessary rudder and aileron deflections, when given the commanded side force F_{Y_d}, roll acceleration P_d, or yaw acceleration R_d. Eq. (6.25) is derived by rearranging Eqs. (6.2), (6.4), and (6.6).

$$\begin{bmatrix} C_{n_\beta} & C_{n_{\delta a}} & C_{n_{\delta r}} \\ C_{l_\beta} & C_{l_{\delta a}} & C_{l_{\delta r}} \\ C_{Y_\beta} & C_{Y_{\delta a}} & C_{Y_{\delta r}} \end{bmatrix}\begin{bmatrix} \beta \\ \delta_a \\ \delta_r \end{bmatrix} = \begin{bmatrix} \dfrac{N_d - N_T}{\bar{q}Sb} - C_{n_0} + C_{n_r}\hat{r} + C_{n_p}\hat{p} \\ \dfrac{L_d - L_T}{\bar{q}Sb} - \left(C_{l_0} + C_{l_r}\hat{r} + C_{l_p}\hat{p}\right) \\ \dfrac{F_Y - F_{Y_T}}{\bar{q}S} - \left(C_{Y_0} + C_{Y_r}\hat{r} + C_{Y_p}\hat{p}\right) \end{bmatrix} \tag{6.25}$$

where

$$N_d = I_{zz}\dot{R}_d - I_{xz}\dot{P}_d + \left(I_{yy} - I_{xx}\right)PQ + I_{xz}QR \tag{6.26}$$

$$L_d = I_{xx}\dot{P}_d - I_{xz}\dot{R}_d + I_{xz}PQ + \left(I_{zz} - I_{yy}\right)QR \tag{6.27}$$

$$F_Y = m\left(\dot{V} + UR - WP\right) - mg\sin\phi\cos\theta \tag{6.28}$$

The side force, F_Y, is fed back from the aircraft. During development of the 6DOF MRAC architecture, an alternative option was the use of the pilot-commanded value of side force, $F_{Y_{\text{command}}}$. This option was investigated and displayed similar performance. However, to maintain consistency with the other control loops, the first derivative of side force was chosen as the inverse controller input. The stability derivatives, thrust effects (N_T, L_T, and F_{Y_T}), wing reference area (S), and wingspan (b) are known constants for the aircraft at the nominal flight conditions. The nondimensional yaw and roll rates ($\hat{r}$ and $\hat{p}$) are fed back from the aircraft. The moments of inertia and mass in Eqs. (6.26) to (6.28) are known constants, and the velocities and accelerations are fed back from the aircraft.

The following derivation expresses N_d, L_d, and F_Y in terms of the commanded accelerations $\ddot{\phi}_d$ and $\dot{F}_{Y_d}$ resulting from the pilot commands. Differentiating the kinematic equations [Eqs. (6.7) to (6.9)] with respect to time yields

Eqs. (6.29) to (6.31):

$$\dot{P}_d = \ddot{\phi}_d - \ddot{\psi}_d \sin\theta - \dot{\psi}\dot{\theta}\cos\theta \tag{6.29}$$

$$\dot{Q}_d = \ddot{\theta}_d \cos\phi - \dot{\theta}\dot{\phi}\sin\phi - \ddot{\psi}_d \cos\theta\sin\phi + \dot{\psi}\dot{\theta}\sin\theta\sin\phi - \dot{\psi}\dot{\phi}\cos\theta\cos\phi \tag{6.30}$$

$$\dot{R}_d = \ddot{\psi}_d \cos\theta\cos\phi - \dot{\psi}\dot{\theta}\sin\theta\cos\phi - \dot{\psi}\dot{\phi}\cos\theta\sin\phi - \ddot{\theta}_d \sin\phi - \dot{\theta}\dot{\phi}\cos\phi \tag{6.31}$$

In Eqs. (6.29) to (6.31), each of the angles and angular rates are fed back from the aircraft to the inverse controller. The second derivative of bank angle, $\ddot{\phi}_d$, is calculated from the pilot's commanded bank angle. The remaining unknown terms, which are the second derivative of pitch angle, $\ddot{\theta}_d$, and the second derivative of heading angle, $\ddot{\psi}_d$, are then subsequently calculated [14]. As this is a fairly involved process, the details of the final derivation are not presented here. The reader is asked to refer to Lemon et al. [14] for more details. To summarize, the inverse controller calculates the necessary aircraft control actuations from the commanded accelerations ($\dot{V}_d$, $\ddot{\gamma}_d$, $\ddot{\phi}_d$, $\dot{F}_{Y_d}$), the linear controller plus adaptation signal, and the aircraft states that are fed back from the previous time step. The reader is asked to refer to Lemon et al. [14] for the full derivation.

6.4.2 ARTIFICIAL NEURAL NETWORK

The ANN compensates for modeling error and uncertainty in the plant inverse. As mentioned previously, the model inverse is computed at the nominal flight condition. When the aircraft deviates from this nominal flight condition, or when the plant dynamics are drastically changed due to damage or component failures, the ANN adaptation signal is updated accordingly to adapt to the current flight conditions. Each of the four control loops has a single neuron adaptive bias corrector. The adaptive bias corrector learns online to reduce the error between the desired and actual values of its respective parameter. The weight update rule is given in Eq. (6.32), where η is a learning rate.

$$\dot{W}(t) = \eta e(t) \tag{6.32}$$

Where the error is given by:

$$e(t) = \dot{x}_d - \dot{x} \tag{6.33}$$

In Eq. (6.33), x represents the state for each of the four control loops. Figure 6.5 illustrates the architecture of the adaptive bias corrector.

For practical purposes, the learning rates (η) within each loop were selected through an observational process. This involved increasing their values independently until the aircraft's response for that particular loop became over-sensitive. The final learning rates were chosen as half of each of these respective values. Table 6.2 lists the learning rates used for each adaptive bias corrector.

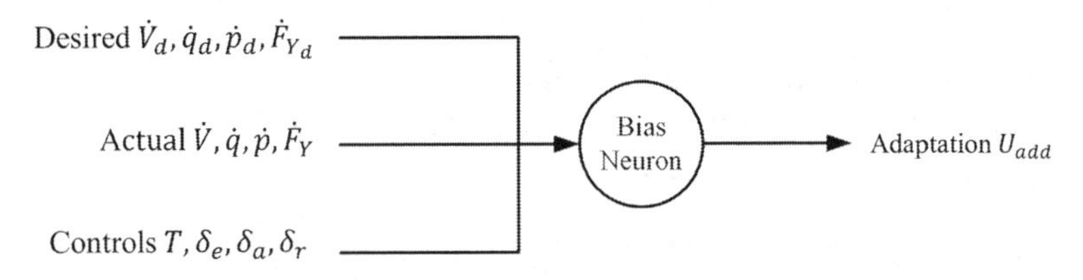

Fig. 6.5 Adaptive bias corrector architecture.

6.4.3 MODEL FOLLOWER AND LINEAR CONTROLLER

The model follower and linear controller were designed to meet specific system response characteristics, such as rise time, overshoot, damping ratio, etc. Figure 6.6 illustrates the controller for the airspeed and side force loops. In the case of no modeling error or when there is sufficient adaptation, the closed loop airspeed and side force systems reduce to first-order systems with a single proportional gain, as given in Eq. (6.34).

$$P = \frac{1}{\tau_{des}} \tag{6.34}$$

Figure 6.7 illustrates the controller for the flight path angle and bank angle loops. Here, the loops reduce to second-order systems when no modeling error exists or sufficient adaptation is present. The flight path angle and bank angle loops have a model follower [Eq. (6.35)] and PD controller [Eq. (6.36)].

TABLE 2 ADAPTIVE BIAS CORRECTOR LEARNING RATES

Control Loop	Learning Rate (η)
Airspeed	0.02
Flight path angle	0.20
Bank angle	0.01
Side force	0.30

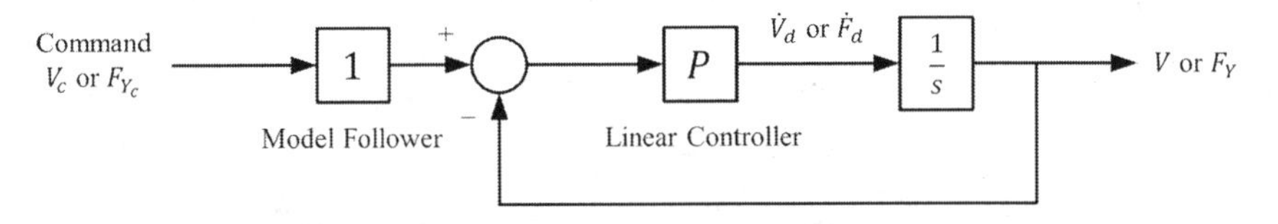

Fig. 6.6 Airspeed and side force loop linear controller.

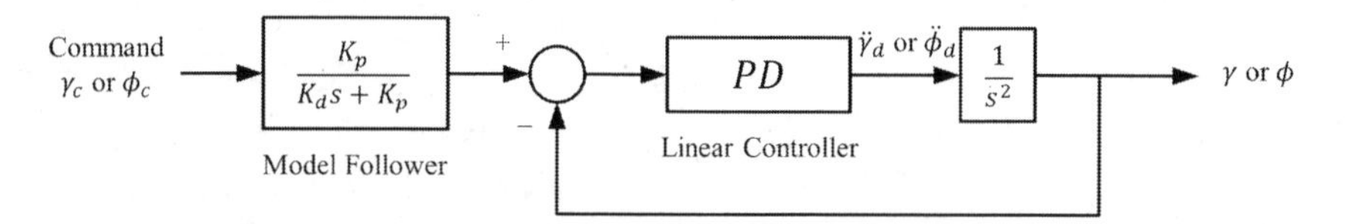

Fig. 6.7 Flight path angle and bank angle loop model follower and linear controller.

Second-order approximations [29] [Eqs. (6.37) and (6.38)] are used to determine K_p and K_d based on desired system response characteristics.

$$\textit{Model Follower} = \frac{K_p}{K_d s + K_p} \tag{6.35}$$

$$PD = K_d s + K_p \tag{6.36}$$

$$K_p = \left(\frac{1 - 0.4167\zeta_{\text{des}} + 2.917\zeta_{\text{des}}^2}{\textit{rise time}_{\text{des}}} \right)^2 = \omega_{n_{\text{des}}}^2 \tag{6.37}$$

$$K_d = 2\zeta \left(\frac{1 - 0.4167\zeta_{\text{des}} + 2.917\zeta_{\text{des}}^2}{\textit{rise time}_{\text{des}}} \right) = 2\zeta_{\text{des}}\omega_{n_{\text{des}}} \tag{6.38}$$

The achievable rise time response and damping of the aircraft's dynamic modes are physically limited by the control surface deflection limits and actuator response times. The elevator actuator on board the Textron Aviation CJ-144 has a maximum slew rate of 4.0 deg/s and was deliberately chosen to be slow for safety reasons. The slow slew rate gives the safety pilot time to override the actuator commands with the mechanical control system, if necessary.

6.5 EXTENSIONS TO THE MRAC: ATMOSPHERIC DISTURBANCE ENVELOPE PROTECTION

Two types of atmospheric disturbances that are dangerous to aircraft are the microburst and wake vortex. History has provided ample evidence of the dangers that these atmospheric disturbances pose to an aircraft and its occupants. Because an MRAC system is suited to adapting to uncertainties in the aircraft or environment, enhancing the MRAC architecture to detect and compensate for sudden and unexpected changes stemming from atmospheric disturbances would aid in improving chances of recovery post-encounter.

6.5.1 MICROBURST CONDITION

A microburst may be qualitatively described as a column of air that flows downward and then outward as the air approaches the ground [30]. A typical microburst encounter entails an aircraft entering, traveling through, and crossing the microburst through its center. The encounter is characterized by a gradual increase in headwind as the aircraft enters the microburst, followed by a "down force" on the aircraft caused by the downward-moving column of air, and finally, followed by a sudden shift to a tailwind as the aircraft passes through the center of the microburst.

Consequently, as an aircraft enters the microburst, a pilot would notice a gradual increase in indicated airspeed. An instinctive reaction to this would be to reduce the amount of thrust applied, so as to bring the aircraft back down to the desired airspeed. The danger arises when the aircraft begins to cross the center of the microburst, where there is a sudden change from headwind to tailwind. Because of the prior reduction in thrust, the aircraft now faces a condition where there may be insufficient ground speed and altitude to safely recover from the tailwind. A stall condition arises, altitude loss may follow, and this may pose a significant danger at lower altitudes, such as during the approach and landing phases.

6.5.1.1 MICROBURST MODEL

To facilitate accurate responses in simulation, a microburst model was designed within the MATLAB/Simulink® environment, based on characteristics of historically recorded microbursts [31–33]. Most of the detail in the development of this model has been covered in prior literature [34], and a brief overview is presented here. The following assumptions and requirements were specified for the model: mathematical simplicity with the fewest possible number of parameters, spatially stationary downdraft and change in headwind, and continuity of at least two derivatives to prevent discontinuities in acceleration.

The parameters used to characterize the simulated microburst were the development length of the microburst (L), maximum forward velocity (Amp_u), and maximum vertical velocity (Amp_w), and the resulting component velocities of the microburst were calculated according to Eqs. (6.39) and (6.40):

$$u(x) = Amp_u \left\{ -10 exp\left(\frac{-10x}{L}\right) \frac{exp\left(\frac{-10x}{L}\right) - 1}{\left[1 + exp\left(\frac{-10x}{L}\right)\right]^3} \right\} \tag{6.39}$$

$$w(x) = Amp_w \left\{ \frac{4 exp\left(\frac{-10x}{L}\right)}{\left[1 + exp\left(\frac{-10x}{L}\right)\right]^2} \right\} \tag{6.40}$$

The calculated velocities due to the microburst were then subtracted from the aircraft's inertial velocities to include the effects of the disturbance on the

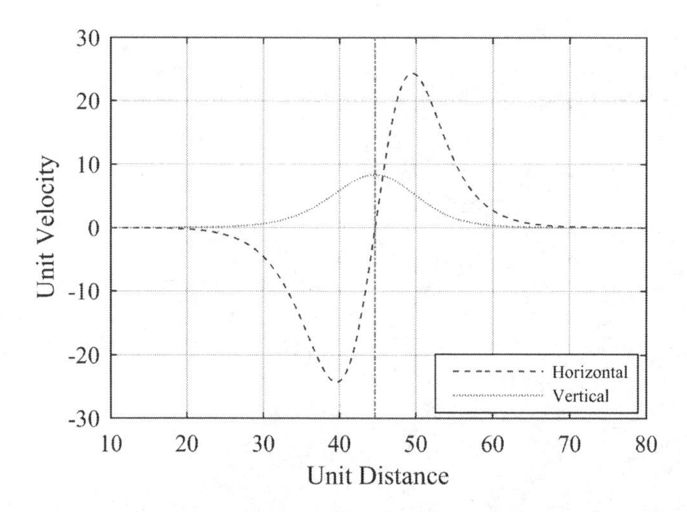

Fig. 6.8 Velocity profile of microburst model.

aerodynamics of the aircraft. A profile of a typical microburst generated based on Eqs. (6.39) and (6.40) is illustrated in Fig. 6.8.

6.5.1.2 ENVELOPE PROTECTION SCHEME

Based on the mechanics of a microburst described previously, any response and recovery scheme has to rely upon recognizing the microburst in its development phase and then reacting accordingly before the shift in wind direction occurs. Rokhsaz et al. [34] determined that a highly successful method of response and recovery to a microburst encounter entailed limiting the aircraft's ground deceleration in relation to its airspeed. A simple formula was derived there to define the maximum allowable inertial deceleration, as shown in Eq. (6.41).

$$a_{\text{limit}} = -0.0250\left(V_{\text{airspeed}} - 100\right) \tag{6.41}$$

This serves as a functional deceleration limit and prevents the aircraft from decelerating too much at low airspeeds typical of approach, [27] while placing no practical limit on the aircraft's maneuverability at higher airspeeds. The logic behind this protection scheme inherently centers around preventing too much of a loss in the aircraft's kinetic energy. As the aircraft enters the microburst, if the pilot keeps the airspeed constant, the ground speed, and hence the kinetic energy, decreases. When the aircraft crosses the center of the microburst, this kinetic energy has to be recovered rapidly so that the airspeed may be quickly brought up and out of the stall regime. The problem arises because thrust is physically limited and the aircraft can only be accelerated so much before thrust saturation is reached. This is the inherent problem that the protection scheme addresses, namely the amount of kinetic energy lost as the aircraft enters the microburst is reduced, as compared to the case where there is no protection scheme in place. Figure 6.9 shows the envelope protection scheme, implemented within the velocity loop of the existing 3DOF MRAC architecture.

6.5.2 WAKE VORTEX CONDITION

As with microbursts, wake vortex encounters have been historically known to pose great dangers to aircraft. These effects are compounded at takeoff and landing, where the leading aircraft may have landing gear and flaps extended, increasing the strength of the generated vortex. Aircraft flying behind vortex-generating aircraft are susceptible to abrupt changes in flight conditions as a direct result of these wake vortices. Often, these changes are sudden and unexpected and occur before the flight crew is able to counteract the disturbance. Roll conditions may be severe enough that the aircraft is placed in an inverted orientation for a period of time.

Qualitatively, a typical wake vortex, as generated by the wing of an aircraft, may be described as two rotating columns of air, separated by a distance approximately similar to the wingspan of the vortex-generating aircraft. These columns rotate in inward but opposite directions and follow a path that is carved out by the vortex-generating aircraft's flight path.

6.5.2.1 WAKE VORTEX MODEL

To obtain a plausible response from the controller, a wake vortex model was created within the MATLAB/Simulink® environment. The reader may refer to prior work [24] for detailed literature on the model's development. The wake vortex model comprises two components: the vortex velocities and vortex moments.

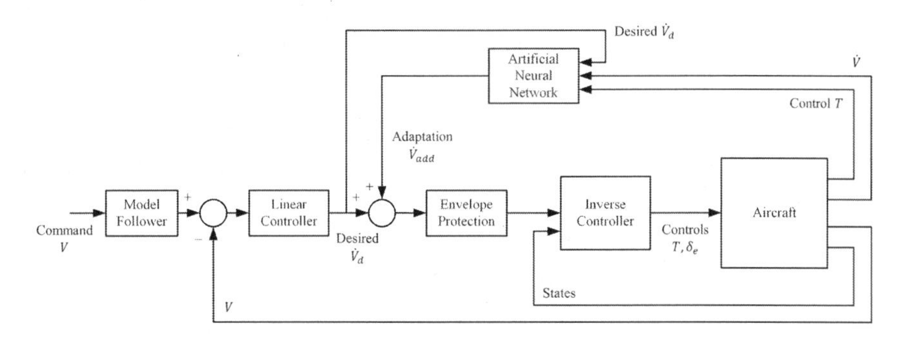

Fig. 6.9 Envelope protection scheme integrated with 3DOF MRAC architecture.

The vortex velocities at each evaluation point i along the wing, horizontal stabilizer, and vertical stabilizer were calculated according to Eq. (6.42):

$$V_{wv}(r_i) = \frac{\Gamma}{2\pi r_i}\left[1 - exp\left(-\frac{r_i^2}{r_c^2}\right)\right] \tag{6.42}$$

where r_i represents the separation radius between vortex trail and evaluation points on vortex-following aircraft and r_c is the core radius of wake vortex. The circulation strength, Γ, was calculated as a function of the weight (W_g), velocity (V_g), and equivalent wingspan (b_{eq_g}) of the vortex-generating aircraft:

$$\Gamma = \frac{W_g}{\rho_g V_g b_{eq_g}} \tag{6.43}$$

The vortex rolling, pitching, and yawing moments due to the wake vortex were calculated at every element on the aircraft, and the total moment obtained by summing up the individual component moments:

$$L_{wv_i} = y_i F_{wv_i} \tag{6.44}$$

The force imposed on each element of the wing as a result of the wake vortex, F_{wv_i}, was determined as a function of the local lift coefficient, the area of the particular wing element, i, and the dynamic pressure. As an example, the rolling moment generated by the wing *solely due to the wake vortex* was calculated using Eq. (6.45):

$$L_{wv} = \sum_{i=1}^{n} L_{wv_i} = \sum_{i=1}^{n} y_i C_{L_{\alpha_i}} \alpha_{wv_i} A_i \bar{q} \tag{6.45}$$

In Eq. (6.45), α_{wv_i}, which is the angle of attack imposed on each element of the wing *solely due to the wake vortex*, was calculated as the inverse tangent of the vertical and longitudinal body-fixed velocities. The lift curve slope at each element, $C_{L_{\alpha_i}}$, was calculated using a nonlinear lift model, incorporating a stall profile, as developed in the microburst model. A nonlinear lift model was critical for obtaining plausible responses from the aircraft, as the wake vortex placed the aircraft in flight conditions with acute angles of attack, past aerodynamic stall.

The process detailed earlier is used for the horizontal and vertical stabilizers and for the pitching and yawing moment coefficients. The completed set of wake vortex moments was then built up as the sum of the delta moment coefficients generated by the wing, horizontal stabilizer, and vertical stabilizer for the rolling, pitching, and yawing moments respectively. This is described in Eqs. (6.46) to (6.48):

$$C_{l_{wv}} = C_{l_{wv_{\text{wing}}}} + C_{l_{wv_{\text{horizontal stabilizer}}}} + C_{l_{wv_{\text{vertical stabilizer}}}} \tag{6.46}$$

$$C_{m_{wv}} = C_{m_{wv_{\text{wing}}}} + C_{m_{wv_{\text{horizontal stabilizer}}}} \tag{6.47}$$

$$C_{n_{\mathrm{wv}}} = C_{n_{wv_{\text{vertical stabilizer}}}} \tag{6.48}$$

The above values were then summed into the aircraft's total moment coefficient buildups (C_l, C_m, C_n). Figure 6.10 illustrates the longitudinal, lateral, and vertical velocities that the vortex-following aircraft would experience if it were to approach, intercept, and emerge from the vortex at a 10-degree angle with respect to the wake vortex trail.

6.5.2.2 ENVELOPE PROTECTION SCHEME

As described earlier, a wake vortex imposes upon the aircraft large magnitudes of velocities. These velocities translate into rolling, pitching, and yawing moments. Typically, the rolling moment imposed is of greatest magnitude and can often place the vortex-following aircraft into configurations with bank angles greater than ± 90 degrees. In such a scenario, the force of weight normally experienced along the vertical body axis becomes applied along the lateral body axis, and the inverse controller is thus made to coordinate the aircraft "sideways" (zero commanded side force), a condition for which it was not designed and for which the aircraft cannot be trimmed. To account for this, a protection scheme was developed to augment the ANN's side force adaptation $\left(F_{y_{\text{add}}}\right)$ and the inverse controller's rudder deflection output (δr_{cmd}), as in Eq. (6.49):

$$if\ (\emptyset_{\text{commanded}} - \emptyset_{\text{actual}}) \geq \pm 40\ degrees \begin{cases} F_{y_{\text{add}}} = 0 \\ \delta r_{\text{cmd}} = 0 \end{cases} \tag{6.49}$$

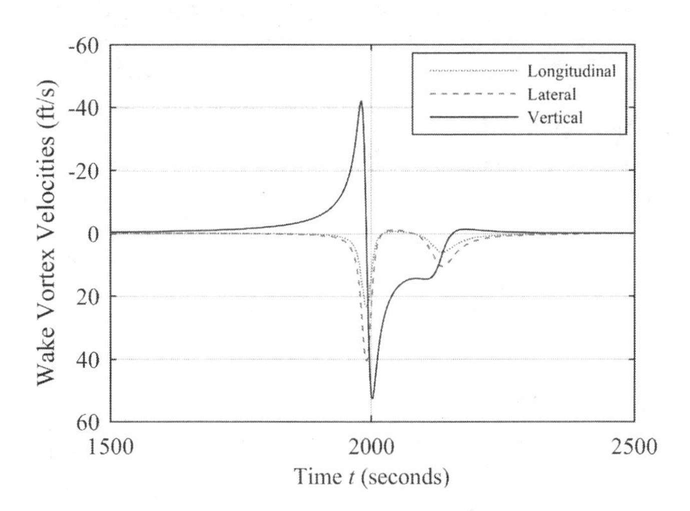

Fig. 6.10 Velocity profile of wake vortex model.

The idea behind this stems from the notion that, in a sudden and uncommanded roll, the actual bank angle will differ significantly from the commanded bank angle. A pilot would typically be commanding zero bank angle when flying straight and level, and a wake vortex may suddenly roll the aircraft excessively. In these scenarios, the actual and commanded bank angles will have a large delta, signifying that the roll is undesired and uncommanded. This is used as the criteria for triggering the augmented directives.

In tests, this roll departure augmentation system was shown to be very effective at contributing toward safe recovery of the aircraft during a wake vortex encounter.

6.6 EXTENSIONS TO THE MRAC: ENHANCING RESILIENCE TO SENSOR-INDUCED NOISE

Other forms of disturbances to a control system may stem from sensor noise and turbulence effects, both of which have been prevalent issues in control system design. Whereas a control system may respond well to "clean" input signals and commands in simulation, this may often not be the case when a controller is deployed in the field. While the MRAC system has continually demonstrated promising results with good command tracking in flight tests on the CJ-144, uncertainties in hardware have led to sensor noise being introduced into the airspeed loop. The net result of this was a throttle surging phenomenon occurring during certain phases of flight.

Through further research and testing, the source of noise was found to come from an airspeed sensor. The ANN adapts on acceleration, $\dot{V}$, which is the derivative of the measured noisy velocity signal. Differentiating a noisy velocity signal led to a noisy acceleration signal. By extension, this resulted in a noisy acceleration signal being fed into the ANN, compromising the ANN's adaptation output.

Two methods were devised to alleviate this issue and enhance the MRAC system's resilience to sensor noise. The first utilized a Modified State Observer (MSO) and the second utilized a Kalman Filter on the sensed airspeed. These efforts are discussed in the sections below.

6.6.1 MODIFIED STATE OBSERVER

The MSO utilized a state estimation and neural network-based nonlinear approximation technique [25] that integrated directly with the nonlinear aircraft dynamics. Rajagopal et al. [35] proposed the use of a general observer structure that separates the nominal closed-loop dynamics from estimation error dynamics, and a similar version of an MSO-based adaptive controller has been used to control the longitudinal dynamics of an aircraft [36]. Figure 6.11 illustrates the architecture of the MSO modification, implemented within the velocity loop of the existing 3DOF MRAC architecture.

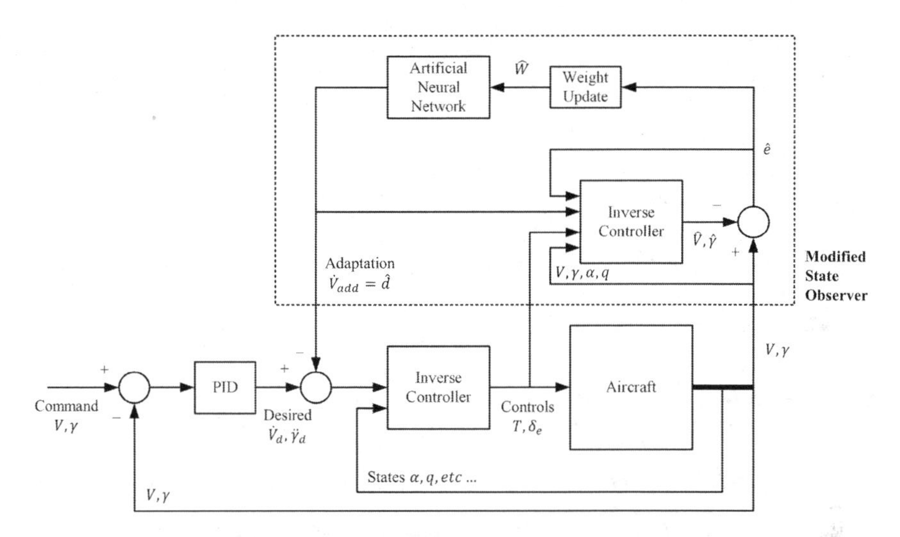

Fig. 6.11 Modified state observer integrated with 3DOF MRAC architecture.

Let the states V and γ be represented by the vector x_T such that $x_T = [V_c, \gamma]^T$, and let the states α and q be represented by the vector x_m such that $x_m = [\alpha, q]^T$. Also, let the input be represented by $u = [T, \delta_e]^T$. The nonlinear aircraft dynamics may then be conceptually represented by Eq. (6.50):

$$[\dot{x}_T, \dot{x}_m] = f(x_T, x_m, u) + d(x_T, x_m, u) \tag{6.50}$$

The term $d(x_T, x_m, u)$ represents the unknown dynamics that need to be estimated. The MSO-based adaptive controller uses an observer structure of the form given in Eq. (6.51) to estimate this unknown uncertainty:

$$\dot{\hat{x}}_T = f(x_T, x_m, u) + \hat{d}(x_T, x_m, u) + K_2(x_T - \hat{x}_T) \tag{6.51}$$

In Eq. (6.51), the term $\hat{d}(x_T, x_m, u) = \hat{W}^T \Phi(x_T, u)$ and represents the approximated uncertainty, where $\hat{W}$ are neural network weights and $\Phi(x_T, u)$ are the basis functions of the neural network. The basis functions are defined as the Kronecker product of the states affecting the commands (which are γ, $\dot{\gamma}$, and V) and are given by Eq. (6.52), where the division by 169 normalizes the respective terms to the nominal cruise speed of 100 knots:

$$\Phi = \begin{bmatrix} 1 & \dot{\gamma} & \frac{V}{169} & \dot{\gamma}\frac{V}{169} & \gamma & \gamma\dot{\gamma} & \gamma\frac{V}{169} & \gamma\dot{\gamma}\frac{V}{169} \end{bmatrix}^T \tag{6.52}$$

The variable K_2 is a design parameter. The estimation error $\hat{e} = (x_T - \hat{x}_T)$ between the measured and estimated states is used for adapting the weights through the use of the Lyapunov theory-based weight update rule given in Eq. (6.53):

$$\dot{\hat{W}} = \Gamma\ \Phi(x_T, u)\hat{e}^T P - \ \sigma\hat{W} \tag{6.53}$$

where Γ is the adaptation rate, σ is the robustness term to ensure boundedness of neural network weights, and P is the solution to the Lyapunov equation [6.26]. It is to be noted that the MSO adapts based on estimation error, not modeling or tracking error, thus making it robust to sensor noise and disturbances. The design of the PD controller gains, parameters K_2 and Γ are described in Pappu et al. [25]. The performance of the MSO in flight test are presented in Section 6.8.

6.6.2 KALMAN FILTER

The Kalman Filter modification was designed to filter out unwanted noise from the aircraft's airspeed sensors, without adversely altering the airspeed readout or rendering it inaccurate, and to output a smooth estimated value of the acceleration. Figure 6.12 illustrates the architecture of the Kalman Filter modification [26], implemented within the velocity loop of the existing 3DOF MRAC architecture:

The output from the Kalman Filter was determined according to Eq. (6.54):

$$\dot{\hat{V}} = \dot{V}_{\text{cmd}}(t-1) + K_1\big(\text{V}(t) - \hat{V}(t-1)\big) \tag{6.54}$$

This estimated acceleration value was then used as the input to the ANN, which adapts on $\dot{\hat{V}}$. This constrasts with the use of the unfiltered and noisy velocity values directly from the sensor, as was the case in the original design without the filter. In choosing a Kalman gain, frequency-response analysis was used to select a value that only removed as much unwanted noise as was necessary. A final value of 0.35 was chosen for K_1.

It is worthwhile to emphasize that the output for the Kalman Filter was chosen as the derivative of the velocity, because the ANN adapts on the acceleration, $\dot{V}$.

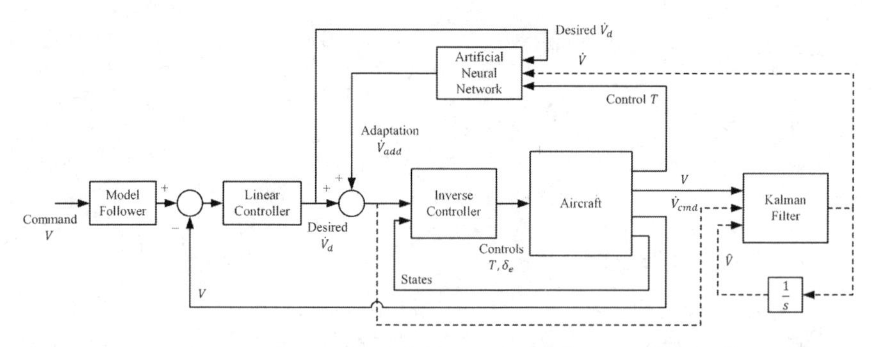

Fig. 6.12 Kalman Filter integrated with 3DOF MRAC architecture.

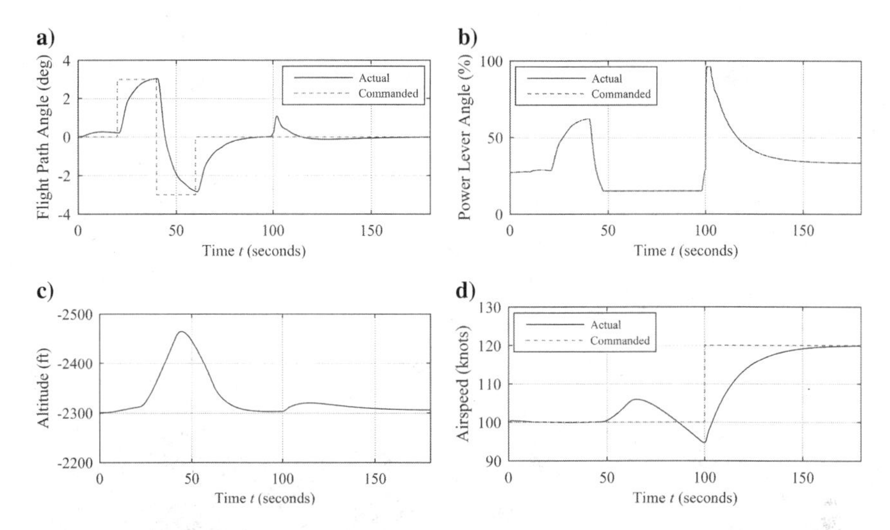

Fig. 6.13 Aircraft response in simulation to pitch doublet and speed step commands (3DOF MRAC architecture).

Differentiating a noisy measured velocity would give a noisy acceleration signal. This was deemed to be the root cause of the surging of the Power Lever Angle (PLA). By using the Kalman Filter to calculate a smooth estimate of the acceleration, and then feeding that smooth estimate into the ANN, one obtained a value of $\dot{V}_{\text{add}}$ that ultimately resulted in less surging. Subsequently, by integrating the estimate of the acceleration, one also obtained a smooth estimate of the velocity, $\hat{V}$.

6.7 ANALYSIS THROUGH SIMULATION

Prior to flight testing aboard the Textron Aviation CJ-144, the MRAC systems are typically run through a series of tests within MATLAB/Simulink®. The maneuvers run in simulation are designed to mimic the maneuvers during flight test and simulate how the controller would respond in the actual flight test. These maneuvers typically include a pitch doublet and airspeed step command. The results presented here are a selection of the maneuvers the controller is usually tasked with, and demonstrate responses to commands in the longitudinal axis.

6.7.1 RESPONSE TO STANDARD COMMANDS (NO FAILURES)

Figure 6.13 shows the MRAC system's response to a pitch doublet and airspeed step command. A flight path angle command of $+3/-3$ degrees was given, followed by a $+20$ knot commanded airspeed increase. This represents a standard test given to the MRAC system in evaluating its performance. The results

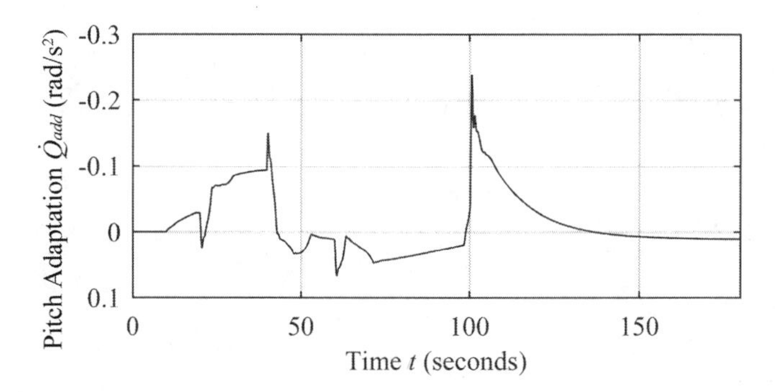

Fig. 6.14 Neural network pitch adaptation (3DOF MRAC architecture).

shown here were generated from the 3DOF version of the MRAC architecture and were conducted with an initial velocity of 100 knots (169 ft/s) at 2,300 ft.

As demonstrated earlier, the MRAC system responded to the pitch doublet and airspeed commands very well. Both flight path angle and airspeed targets were achieved without saturation on the PLA. Through this period, the ANN adapted to the changes in the errors between the commanded and actual states and continually output adaptation values that were fed back to the inverse controller. These adaptations are shown in Figs. 6.14 and 6.15.

6.7.2 RESPONSE TO ATMOSPHERIC DISTURBANCES

In testing the MRAC system's response to microbursts and wake vortex conditions, simulation runs were performed within MATLAB/Simulink®. Due to the dangerous nature of such weather phenomena, testing was limited to simulation.

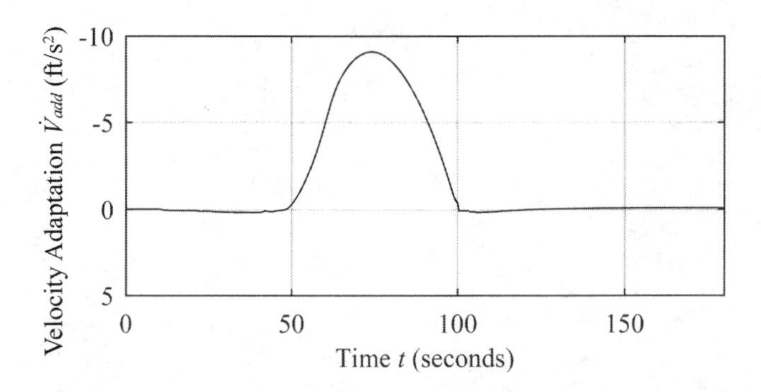

Fig. 6.15 Neural network velocity adaptation (3DOF MRAC architecture).

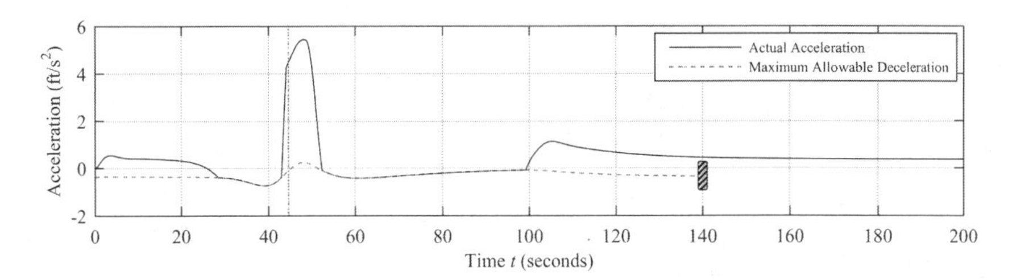

Fig. 6.16 Acceleration profile of aircraft showing envelope protection scheme triggered.

6.7.2.1 MICROBURST CONDITION

Figures 6.16 through 6.20 show the flight profile of the aircraft as it encountered the microburst with and without the envelope protection scheme activated. The simulation runs were conducted with an initial velocity of 68 knots (114 ft/s) at 2300 ft. This was the lowest speed at which the envelope protection scheme was still able to recover the aircraft from the microburst encounter. It is also worth noting that 68 knots (114 ft/s) is close to the Beechcraft Bonanza's clean flaps-up stall speed of 63 knots (106 ft/s) [27].

Figure 6.16 shows the envelope protection scheme being triggered (between 28 s and 42 s, and between 52 s and 99 s). The solid line represents the $\dot{V}_{\mathrm{cmd}}$ acceleration input to the inverse controller, while the dotted line represents the maximum allowable deceleration (lower limit on acceleration), as calculated by Eq. (6.41). This deceleration limit would constantly change, as it is a function of the aircraft's airspeed. An algorithm continuously checked the aircraft's actual acceleration against this limit. If the actual acceleration fell below the deceleration limit, a thrust input was commanded to halt the deceleration. In Fig. 6.16, the envelope protection scheme was triggered twice, once between 28 s and 42 s, and again between 52 s and 99 s. During these two time periods, the inverse controller commanded a thrust input such that the actual acceleration (solid line) would follow that defined by the deceleration limit (dotted line). Figure 6.17 shows that with the scheme inactive, the aircraft progressively lost altitude and

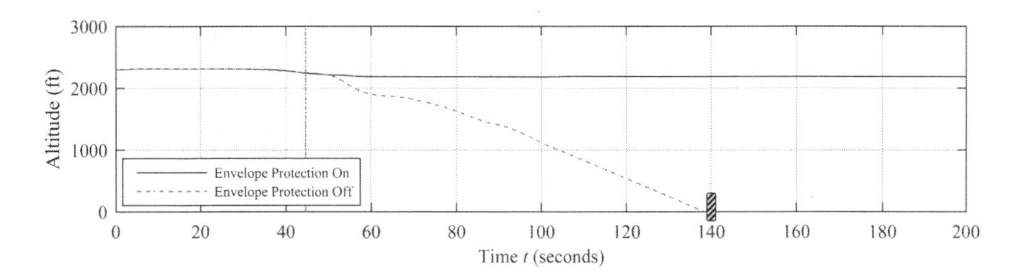

Fig. 6.17 Altitude profile comparison with envelope protection active and inactive.

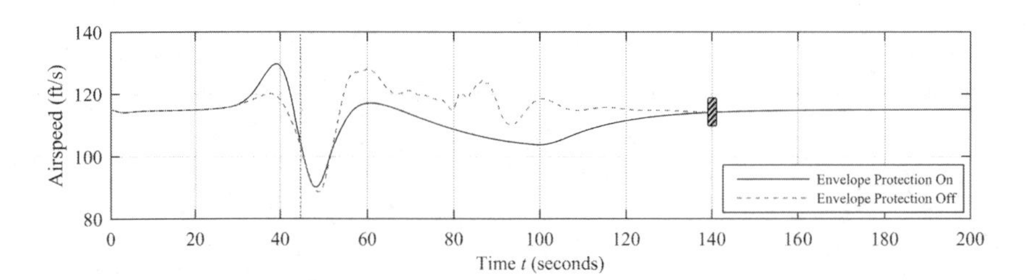

Fig. 6.18 Airspeed profile comparison with envelope protection active and inactive.

was unable to recover once the microburst was encountered. With the scheme active, the thrust setting was kept above a certain minimum throughout the encounter, and enough thrust and airspeed was available as the aircraft crossed the center of the microburst to avert a stall and keep altitude loss to a minimum.

In Fig. 6.18, beginning at approximately 28 s (prior to the microburst centerline, as the aircraft was beginning to encounter the headwind), one can see the increase in airspeed that reached up to 77 knots (130 ft/s). This is a direct effect of the envelope protection scheme attempting to keep the aircraft above the minimum allowable deceleration limit. It is this critical addition of a "buffer" in airspeed that significantly improved the aircraft's chances of recovery past the centerline.

Figure 6.18 also shows that, with the scheme active, the aircraft was kept in a stable flight condition such that the inverse controller was able to return the aircraft's airspeed to the commanded value of 68 knots (114 ft/s). With the scheme inactive, the controller attempted to maintain altitude, but was unsuccessful due to insufficient ground speed prior to crossing the microburst. As a result, the inverse controller commanded a high thrust setting in an attempt to accelerate through. Recovery, however, was not achieved, and simulation results showed that the aircraft impacted the ground shortly before 140 s. Figures 6.19 and 6.20 illustrate why this is so.

With the envelope protection scheme active, the aircraft's angle of attack did not extend far past the stall angle of attack. With the thrust kept above a

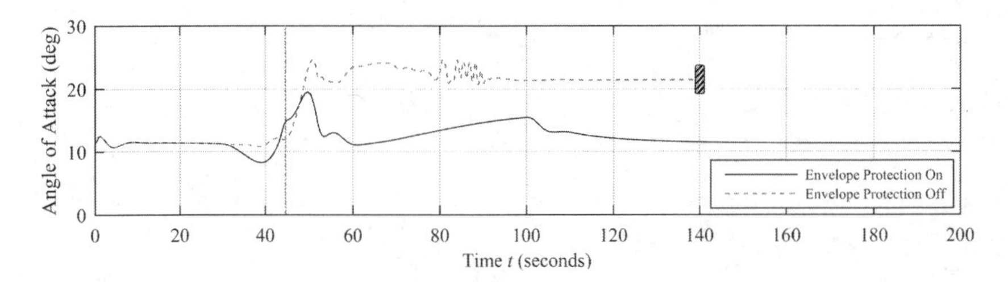

Fig. 6.19 Angle-of-attack profile comparison with envelope protection active and inactive.

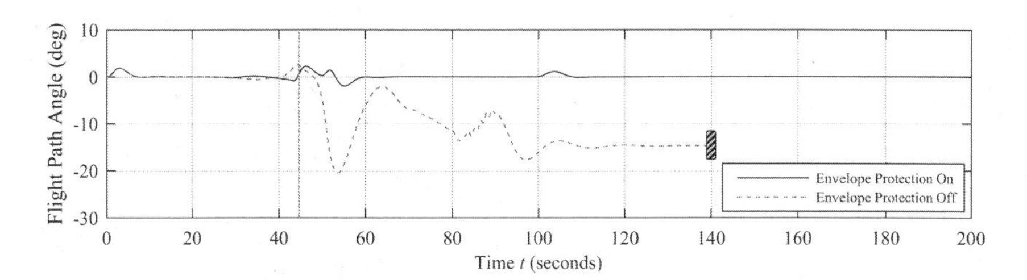

Fig. 6.20 Flight path angle profile comparison with envelope protection active and inactive.

minimum, sufficient ground speed was available, altitude recovery was possible, and the inverse controller did not need to continually increase pitch angle in an effort to stop the descent. The airspeed available allowed swift recovery in altitude, and the inverse controller was able to reduce pitch and return the aircraft to level flight conditions.

With the scheme inactive, the lack of a "buffer" in the available airspeed caused the inverse controller to continually pitch the aircraft up in an effort to regain altitude. Instead of gaining altitude, however, the aircraft lost altitude due to stall. Though airspeed was subsequently regained as a result of this loss in altitude (coupled with a higher thrust setting), the high angle of attack (past stall) impeded recovery. This loss in altitude can be seen in a plot of the flight path angle, as shown in Fig. 6.20.

6.7.2.2 WAKE VORTEX CONDITION

Three scenarios were used to evaluate the effectiveness of the MRAC system in the wake vortex condition. The first scenario involved flying the aircraft into the wake vortex with the MRAC system active but without any pilot input. The second scenario involved repeating this scenario, but with the pilot in the loop. Finally, the third scenario involved flying the aircraft into the wake vortex without the MRAC system active. All three scenarios were set up with an initial altitude of 2300 ft, initial airspeed of 80 knots, wake vortex altitude of 2300 ft, and wake vortex intercept angle of 30 degrees. Tests for the second and third scenarios were conducted with the assistance of certified pilots, with experience ranging from 100 hours to 1200 hours, across private, instrument, and commercial ratings.

Controller-Only Response without Pilot Input (Normal Mode 1)

The first set of results describes the controller's response to the wake vortex encounter without any pilot intervention. The controller was allowed to react on its own and no pilot input was given to the controls.

Figure 6.21 shows a top-down view of the vortex-following aircraft's trajectory in comparison with the location of the wake vortex. The solid line indicates the path

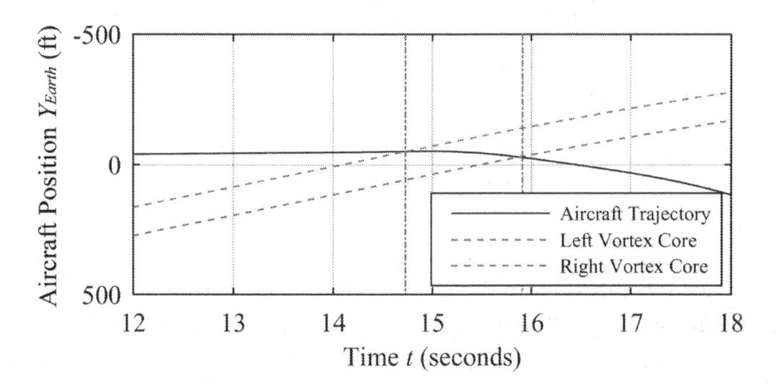

Fig. 6.21 Top-down view of aircraft trajectory and wake vortex trails.

traversed by the vortex-following aircraft, while the upper and lower dotted lines respectively indicate the location of the centers of the left and right cores of the wake vortex trail. In this scenario, the aircraft intercepted the wake vortex at a 30-degree angle, at just under 15 and 16 s for the left and right cores, respectively. The dotted-dashed vertical lines denote the intercept points with each core. Its course was accordingly altered by the wake vortex once it passed the disturbance.

The vortex velocities and rolling moment generated as a result of this wake vortex are shown in Figs. 6.22 and 6.23, where a negative value indicates a left-wing-down roll while a positive value indicates a right-wing-down roll.

The left core of the vortex-generating aircraft, which is a mid-sized regional/ business jet, imposed upon the vortex-following aircraft a vertical velocity of approximately -40 ft/s on the upward rotation and 50 ft/s on the downward rotation. This velocity augmented the trajectory of the vortex-following aircraft, causing it to lose a significant amount of altitude. Note that, due to this altitude loss, the aircraft was further away from the right core, and the rolling moment imposed by the right core is thus less in magnitude than that imposed by the

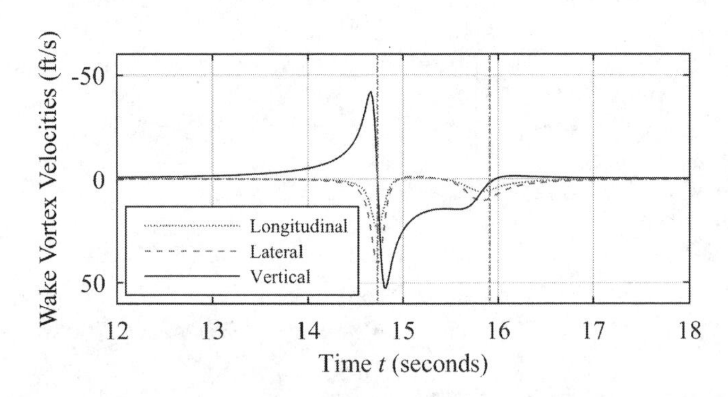

Fig. 6.22 Wake vortex component velocities experienced by following aircraft.

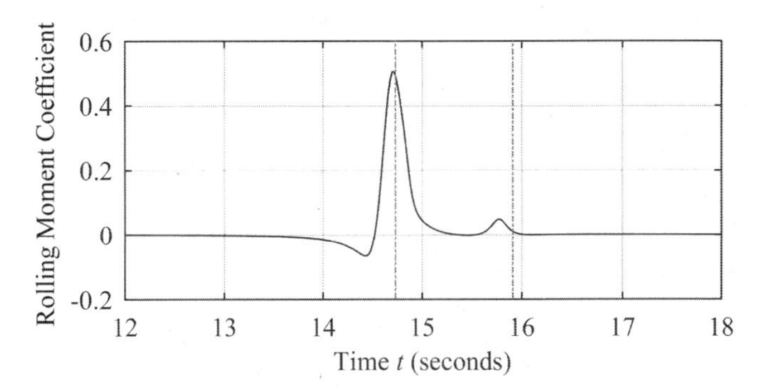

Fig. 6.23 Rolling moment coefficient imposed on following aircraft due to wake vortex.

left core (see Fig. 6.23). The resulting flight path angle deviation and altitude loss are seen in Figs. 6.24 and 6.25.

The vortex-following aircraft cruised initially at 2300 ft, and lost approximately 300 ft of altitude as it encountered the downward rotations of the left and right cores of the wake vortex. At the same time, the vortex-following aircraft also experienced a slight left roll as it initially encountered the upward rotation of the left core. Once it entered the region of the core radius and the vertical velocities of the left core abruptly increased, the roll moment experienced by the aircraft caused it to roll right by approximately 100 degrees, putting it into a wings-vertical condition. This is shown in Fig. 6.26.

The wake vortex also brought about a momentary loss of airspeed, followed by a gradual increase as the loss of altitude progressed, as shown in Fig. 6.27. The initial loss of airspeed could potentially prove to be dangerous, should the airspeed be brought close to or below the stall range, and the subsequent increase in airspeed (due to the kinetic energy gained from the altitude loss) could also place the aircraft in an overspeed condition.

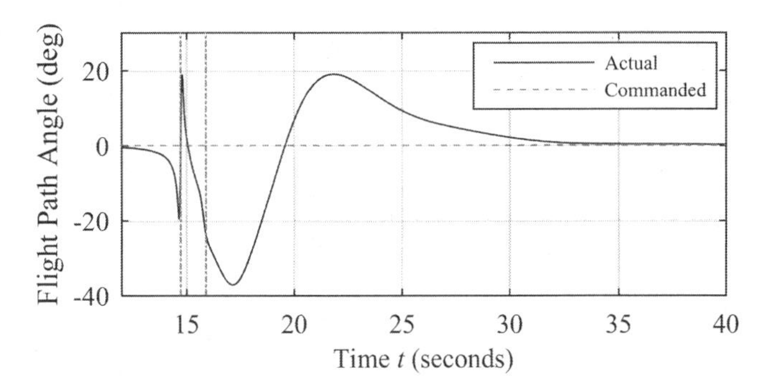

Fig. 6.24 Flight path angle during wake vortex encounter.

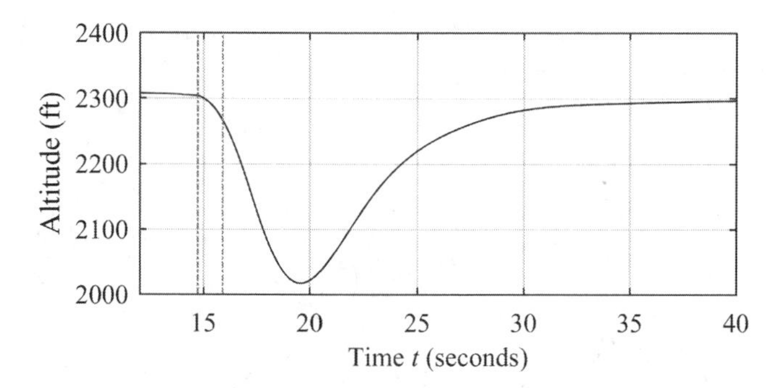

Fig. 6.25 Altitude loss experienced as a result of wake vortex disturbance.

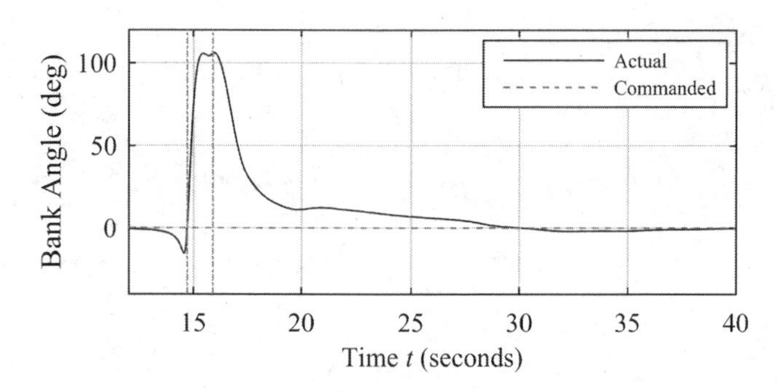

Fig. 6.26 Bank angle during wake vortex encounter.

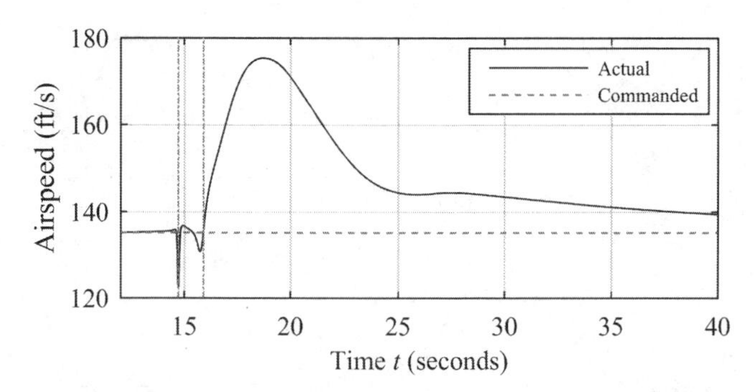

Fig. 6.27 Airspeed change experienced as a result of wake vortex disturbance.

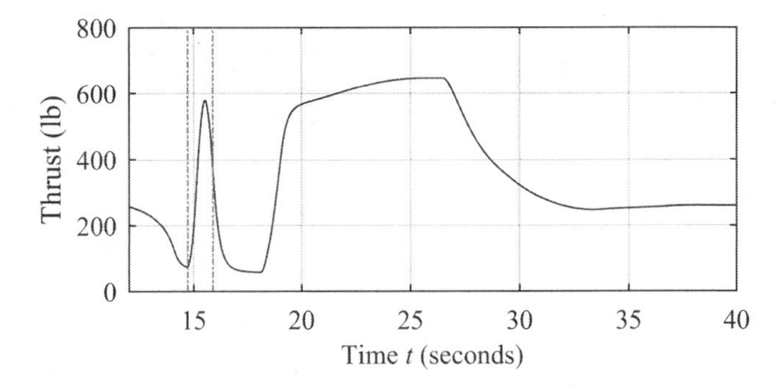

Fig. 6.28 Controller thrust response to wake vortex.

Through this entire encounter, the inverse controller reacted accordingly with inputs to the thrust setting, elevator, aileron, and rudder. These responses are shown in Figs. 6.28 through 6.31. In each case, the control surface deflections acted to oppose the disturbance and return the aircraft to the commanded configuration. The elevator deflected upward to compensate for the sudden loss in altitude, the ailerons deflected to bank the aircraft in the opposite direction of the vortex-induced roll, and the thrust setting was first increased to compensate for the initial loss in airspeed and then rapidly reduced to eliminate the build-up of airspeed during the descent.

It may also be seen in Fig. 6.31 that the rudder was centered with zero deflection. This is the roll departure augmentation scheme being activated between 15 and 27 s. This occurred during the period where bank angle exceeded 40 degrees and remained for an additional 10 s before normal control law directives resumed. During this period, the side force ANN output was augmented and the rudder deflection returned to zero. It is worth noting too that the control surfaces did not reach their saturation limits throughout the entire encounter.

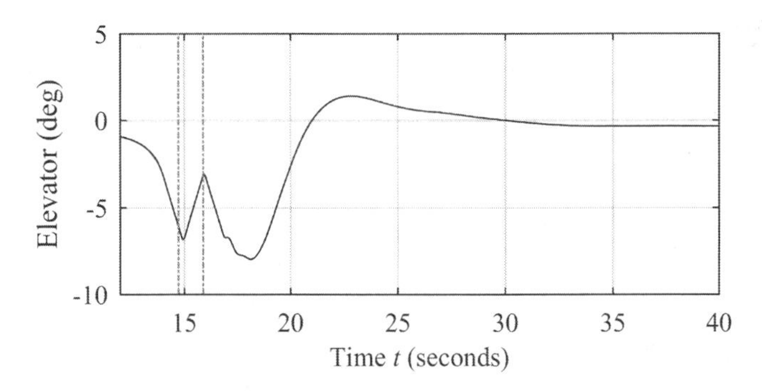

Fig. 6.29 Controller elevator response to wake vortex.

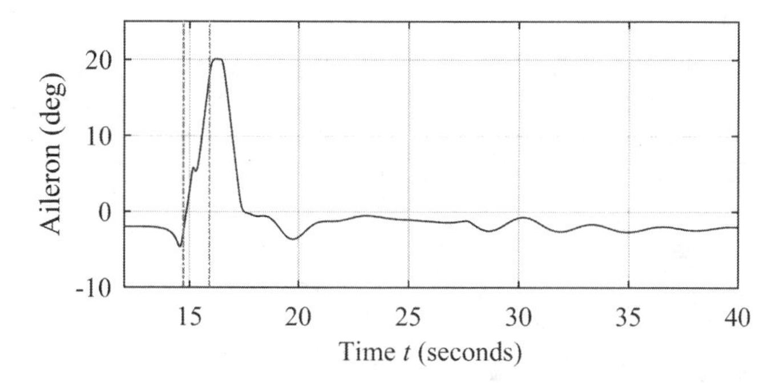

Fig. 6.30 Controller aileron response to wake vortex.

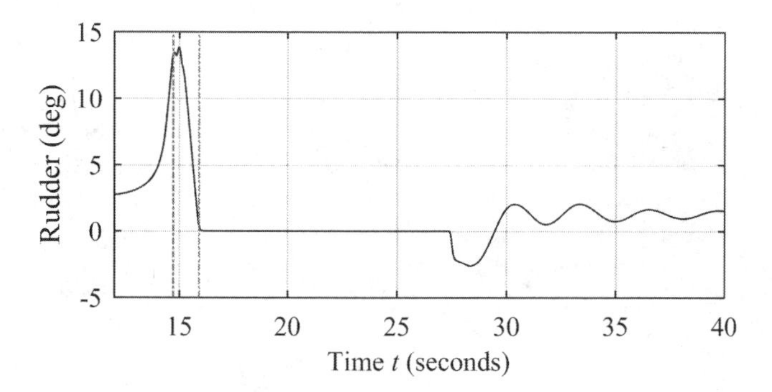

Fig. 6.31 Controller rudder response to wake vortex.

Pilot Response with Controller Input (Normal Mode 2)

The second scenario tested involved having the pilot in the loop, actively responding to and recovering from the wake vortex encounter with the controller's assistance. A selection of five datasets from the simulated flight tests is presented in Fig. 6.32. Three performance indices used to evaluate the controller's performance and make comparisons were the maximum altitude lost, maximum bank angle induced, and maximum airspeed lost. A comparison of these performance indices is shown in Fig. 6.32.

The amount of bank angle induced was observed to be on par when the pilot was in the loop, as with the controller acting independently with no pilot input. In a wake vortex encounter, the aircraft will initially roll slightly in one direction before abruptly and sharply rolling in the opposite direction. In these test cases, the pilots experienced a slight roll to the left for several seconds before experiencing a sharp roll to the right. The initial left bank caused a number of pilots to correctively roll to the right. This presented a danger, as the pilots' right roll input exacerbated the sharp right bank induced by the wake vortex a few seconds later.

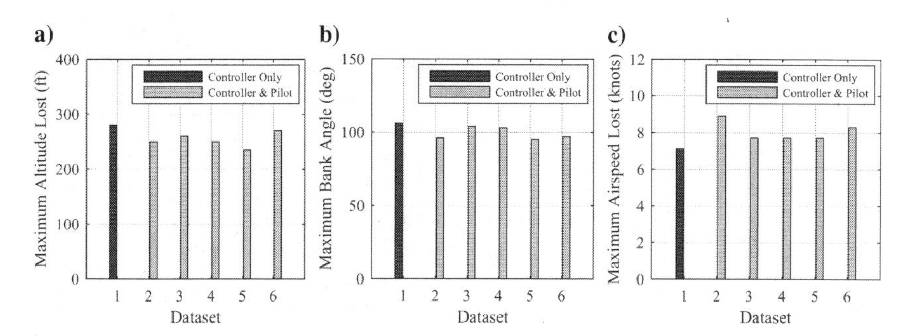

Fig. 6.32 Maximum altitude, bank angle, and airspeed lost (Normal Mode 2).

Pilot-Only Response without Controller Input (Direct Mode)

The third scenario tested involved having the pilot responding to and recovering from the wake vortex encounter with the MRAC system *deactivated*. In Direct Mode, the simulated aircraft was configured to operate with conventional, coupled controls only, as would be found on a typical GA aircraft. To provide a valid comparison, the same five datasets shown in the second scenario are presented in Fig. 6.33. As before, the performance indices used to evaluate performance and make comparisons were the maximum altitude lost, maximum bank angle induced, and maximum airspeed lost. Without the MRAC system active, the differences in performance during and after the wake vortex encounter were significant. A comparison of these performance indices is shown in Fig. 6.33.

For the majority of flight tests, altitude loss was shown to be greater without the MRAC system active. In cases where Direct Mode yielded poorer performance, the amount of altitude lost ranged from as little as 200 ft more to as much as 350 ft more, when compared to Normal Mode 2. While 200 ft of lost altitude may not seem much, this may not be the case if one considers an aircraft on approach to land, where unexpectedly losing several hundred feet of altitude may cause the aircraft to impact the underlying terrain.

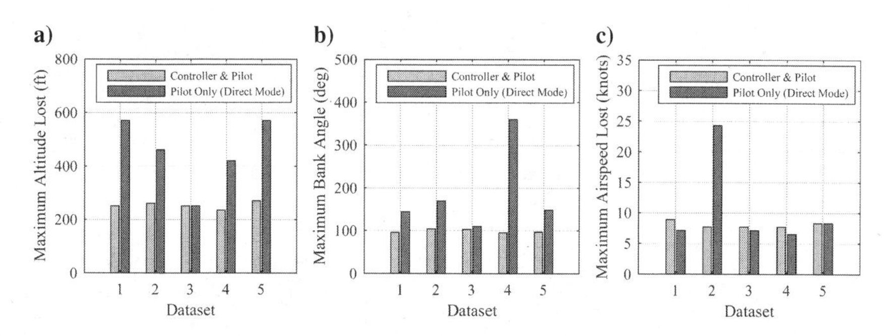

Fig. 6.33 Maximum altitude, bank angle, and airspeed lost (Direct Mode).

The greater amount of altitude lost in Direct Mode may also be, in part, due to the significantly greater amount of bank angle induced as part of the wake vortex encounter. From Fig. 6.33, a much greater bank angle was induced in Direct Mode immediately following the wake vortex encounter when compared to Normal Mode 2. The differences ranged from as little as 10 degrees more to as much as 260 degrees more. Looking at dataset 4, in particular, the amount of bank angle induced in Direct Mode was approximately 360 degrees. In this case, the aircraft was forced into a right roll, temporarily placed in an full-inverted flight configuration, and then returned upright again through a continued right roll. By comparison, in Normal Mode 2, the greatest amount of bank angle induced hovered around 100 degrees. Without swift action to recover to a wings-level configuration, such drastic bank angles could result in an accelerated descent and subsequent collision with the ground.

To further add to the dangerous configurations seen in Direct Mode, the amount of airspeed lost immediately following the wake vortex encounter was also occasionally seen to be greater than in Normal Mode 2. In Normal Mode 2, the aircraft typically lost 7 to 8 knots of airspeed. In Direct Mode, the maximum amount of airspeed lost reached 24 knots. At cruise speeds, this may not pose a problem. However, when an aircraft is on approach and is already flying close to the stall speed range, losing such large amounts of airspeed may be sufficient to send the aircraft into the stall speed range. When close to the ground, this is another situation that is highly undesirable, and could lead to loss-of-control and a potential impact with the ground.

6.8 VALIDATION THROUGH FLIGHT TEST

Flight tests conducted by Textron Aviation cover a series of maneuvers that exercise the various control loops. The results presented in the following sections are once again a selection from the collection of flight tests over recent years and show the responses of the MRAC system to pitch doublet and airspeed step commands, similar to those given in simulation. It is worthwhile to note that some of the results shown belong to tests conducted on different flights and on different days. As such, the reader is asked to keep in mind that slight variations in responses to the same command input may be attributed to differing flight conditions between tests.

6.8.1 RESPONSE TO STANDARD COMMANDS (NO FAILURES)

The flight test results presented illustrate the aircraft's response to commands typical of those given in practice. A flight path angle command of approximately $+1.5/-2$ degrees was given, followed by a $+20$-knot commanded airspeed increase from 110 knots to 130 knots. The resulting changes in altitude and airspeed, as well as the ANN's reactions to these changes, are presented in Fig. 6.34.

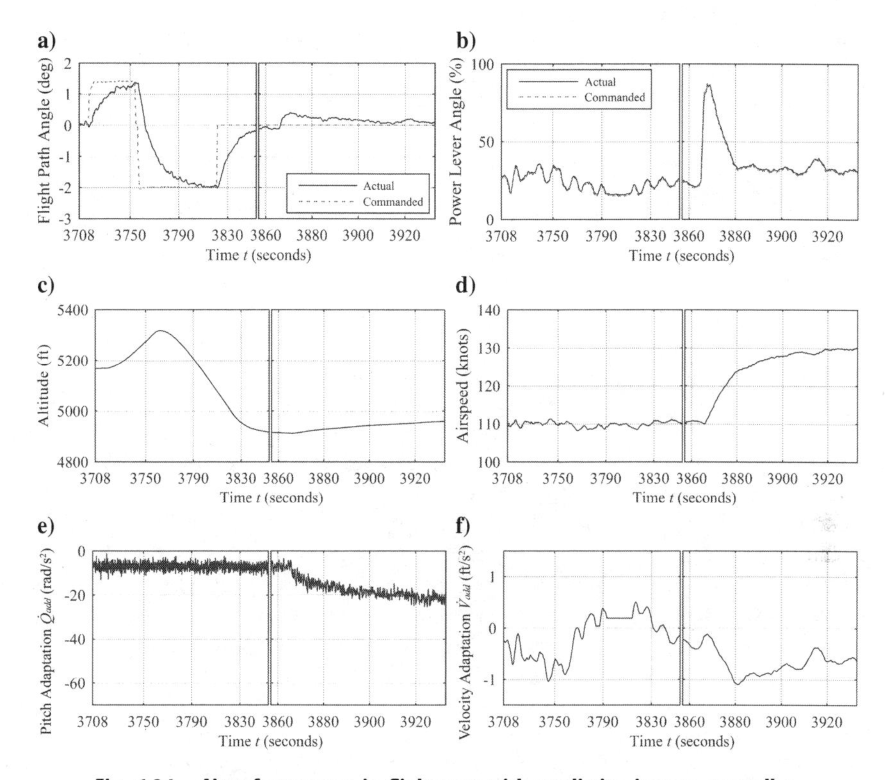

Fig. 6.34 Aircraft response in flight test with predictive inverse controller.

The controller demonstrated its ability to reach the commanded values of flight path angle and airspeed without reaching the physical limits of the aircraft. The PLA did not saturate during the airspeed step command. The ANN was also seen to react to the changes in the aircraft states. For example, when the airspeed step command was given and the airspeed increased, the tendency for the aircraft to climb presented itself through a slight but sudden increase in flight path angle. Immediately, the ANN pitch adaptation increased in magnitude to maintain the commanded flight path angle of 0 degrees. All of this happened "behind the scenes" without the pilot having to exercise additional control of the aircraft beyond what he or she would normally do.

6.8.2 RESPONSE TO AIRCRAFT SYSTEM FAILURES

While the MRAC system makes the aircraft easier to fly and control under normal flight conditions, the benefits and advantages of this adaptive control system stand out even more under off-nominal conditions. The ANN's ability to adapt to changes in the aircraft's dynamics allows for a much more composed response in situations where failures have occurred within various aircraft systems.

Considering the longitudinal flight axis, two significant types of failures involve the loss of ability to control the elevator and the loss of ability to control the engine. In flight test, the MRAC system is tasked to respond to these two failures through a simulated partial elevator failure and a simulated engine failure. These simulated failures mimicking reductions in control effectiveness are software-implemented. For example, in the case of a 50 percent reduction in control effectiveness, the nominal command calculated by the inverse controller is multiplied by a factor of 0.5 prior to being fed to the respective control servo.

6.8.2.1 PARTIAL ELEVATOR FAILURE

In the partial elevator failure scenario, the effectiveness of the elevator was reduced by 50 percent. Once again, a flight path angle command of approximately $+1.5/-2$ degrees was given, followed by a $+20$-knot commanded airspeed increase from 110 knots to 130 knots. The resulting changes in altitude and airspeed, as well as the ANN's reactions to these changes, are presented in Fig. 6.35.

Despite the reduction in elevator effectiveness, the commanded flight path angles took similar amounts of time to achieve, as compared to the no-failure scenario. Because the aircraft was operating in a failure condition, the ANN pitch adaptation was greater in magnitude than in the no-failure scenario, compensating for the reduced ability of the aircraft to pitch up or down. This pitch adaptation acted to restore the ability of the aircraft to perform as if it were operating under nominal conditions with no failures. Similarly, when the airspeed step command was given and the resulting increase in flight path angle appeared, the pitch adaptation increased significantly to bring the aircraft back to level flight without requiring pilot input.

6.8.2.2 PARTIAL ENGINE FAILURE

In the partial engine failure scenario, the effectiveness of the thrust was reduced by 50 percent. To provide an equivalent comparison, the results from the same set of maneuvers are again presented. A flight path angle command of approximately $+1.5/-2$ degrees was given, followed by a $+20$-knot commanded airspeed increase from 110 knots to 130 knots. The resulting changes in altitude and airspeed, as well as the ANN's reactions to these changes, are presented in Fig. 6.36.

As with the partial elevator failure scenario, even in the presence of reduced thrust, the commanded pitch doublet and airspeed step maneuvers were achieved in similar amounts of time as compared to the no-failure scenario. The ANN velocity adaptation was greater in average magnitude than in the no-failure scenario, compensating for the lowered ability of the aircraft to accelerate. This velocity adaptation once again acted to restore the ability of the aircraft to perform as if

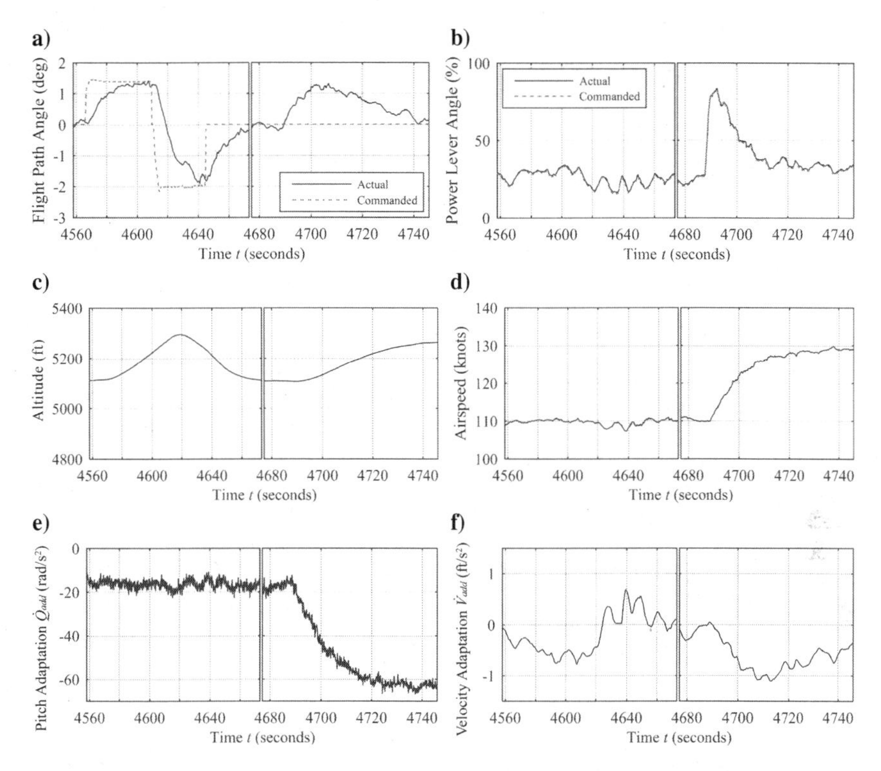

Fig. 6.35 Aircraft response to simulated partial elevator failure with predictive inverse controller.

it were operating under nominal conditions with no failures, and the +20-knot airspeed increase was achieved without strain on the physical limits of the aircraft or the MRAC system.

6.8.3 RESPONSE TO SENSOR-INDUCED NOISE

While the MRAC system had consistently exhibited good performance in simulation and in initial flight tests, external uncertainties in certain phases of later flight tests had led to the occasional presence of a throttle-surging phenomenon. Further investigation and testing attributed this to sensor noise being introduced into the airspeed loop. The ANN adapts on acceleration, $\dot{V}$, which is the derivative of the measured noisy velocity signal. Differentiating a noisy velocity signal led to a noisy acceleration signal. By extension, this resulted in a noisy acceleration signal being fed into the ANN, compromising the ANN's adaptation output.

Prior to implementation of the Kalman Filter and Modified State Observer methods, the net effect of this was a rapid, high-frequency surging of the PLA. Figure 6.37 shows the controller's response to a $+1.5/-2$ degree flight path angle command followed by a $+20$-knot airspeed increase from 110 knots to 130 knots.

The flight path angle and airspeed commands were easily achieved by the controller. However, the PLA exhibited extreme amounts of surging, and at a large magnitude. During the pitch doublet command, the PLA surged from 20 percent to 80 percent within a span of 2 to 3 s when at its worst. During the airspeed step command, a PLA surge from 20 percent to 60 percent was seen, with an oscillatory pattern.

In general, this surging was seen anytime a throttle input was required. For example, if the aircraft was tasked to perform a climb, and the throttle had to be increased to maintain airspeed, surging was observed. If a simple airspeed increase was commanded, resulting in a throttle increase, surging was also observed.

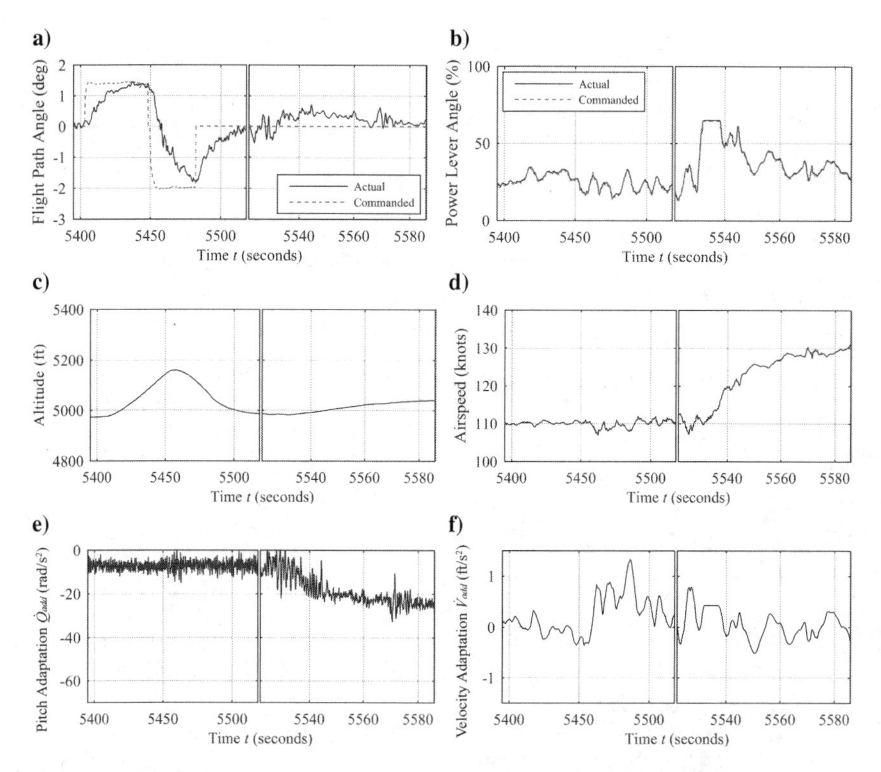

Fig. 6.36 Aircraft response to simulated partial engine failure with predictive inverse controller.

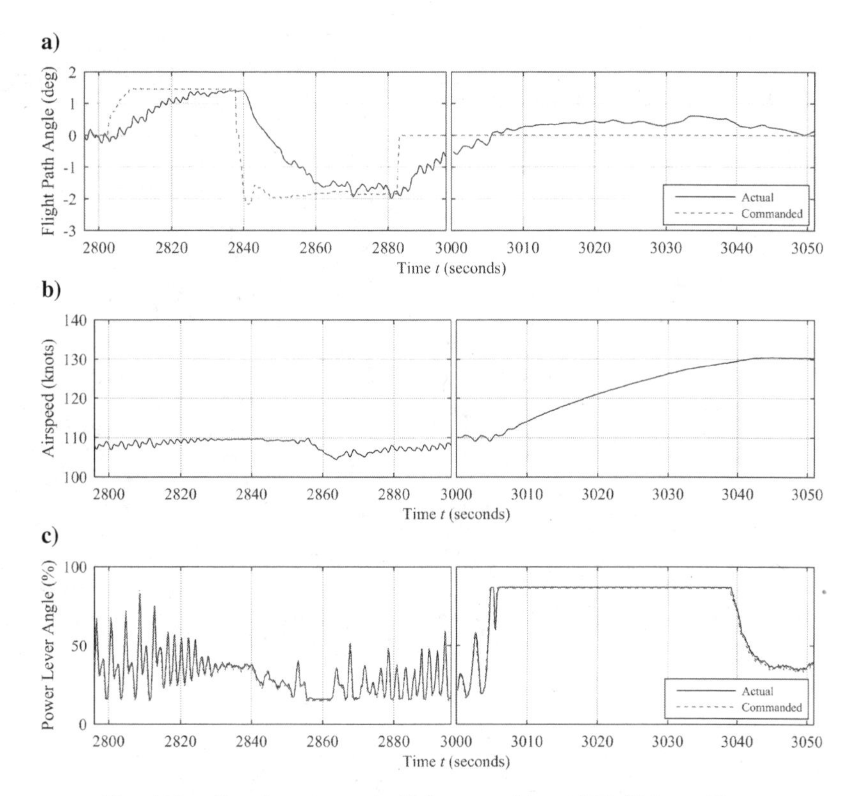

Fig. 6.37 Aircraft response in flight test without MSO/Kalman Filter.

6.8.3.1 MODIFIED STATE OBSERVER

Figure 6.38 illustrates the flight test results with the Modified State Observer implemented, showing responses that were much better than in the unaugmented non-MSO case in Fig. 6.37.

Through the pitch doublet and airspeed step, the controller was able to track the commanded values with far reduced PLA surging, as compared to non-MSO controller depicted in Fig. 6.37. The magnitude and frequency of the PLA surge was significantly reduced, and variations in the PLA during the pitch doublet command ranged approximately between 30 percent and 50 percent, as opposed to 20 percent and 80 percent in the non-MSO case shown in Fig. 6.37.

6.8.3.2 KALMAN FILTER

Figure 6.39 illustrates the flight test results with the Kalman Filter implemented. As with the MSO, the responses were much better than in the unaugmented no-filter case.

Surging of the PLA was again significantly reduced, and variations in the PLA during the pitch doublet command ranged between 20 percent and 40 percent, as opposed to 20 percent and 80 percent in the no-filter case (Fig. 6.37). The frequency of the oscillations was also greatly reduced. During the airspeed step command, the PLA increased in a single, smooth stroke, and the oscillatory behavior observed before was no longer present. Tracking of the desired flight path angle and airspeed was also excellent, and there were no discernible differences in the rise and settling times in both the no-filter and added-filter cases.

Aside from overcoming hardware uncertainties and sensor-induced noise, the addition of the Kalman Filter and MSO modifications yield an additional important benefit, in that the resilience of the MRAC system to variations in the operating environment/equipment is greatly strengthened and reinforced. While both methods yield good results, the Kalman Filter modification has a slight advantage, being the more straightforward architecture to implement in a practical sense. In

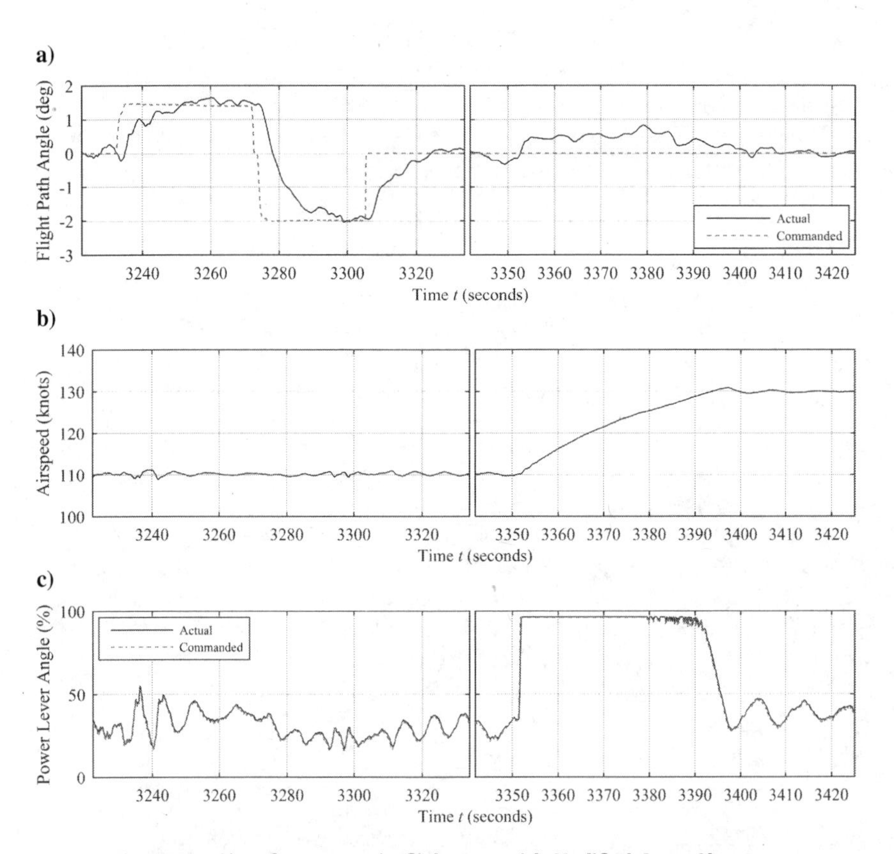

Fig. 6.38 Aircraft response in flight test with Modified State Observer.

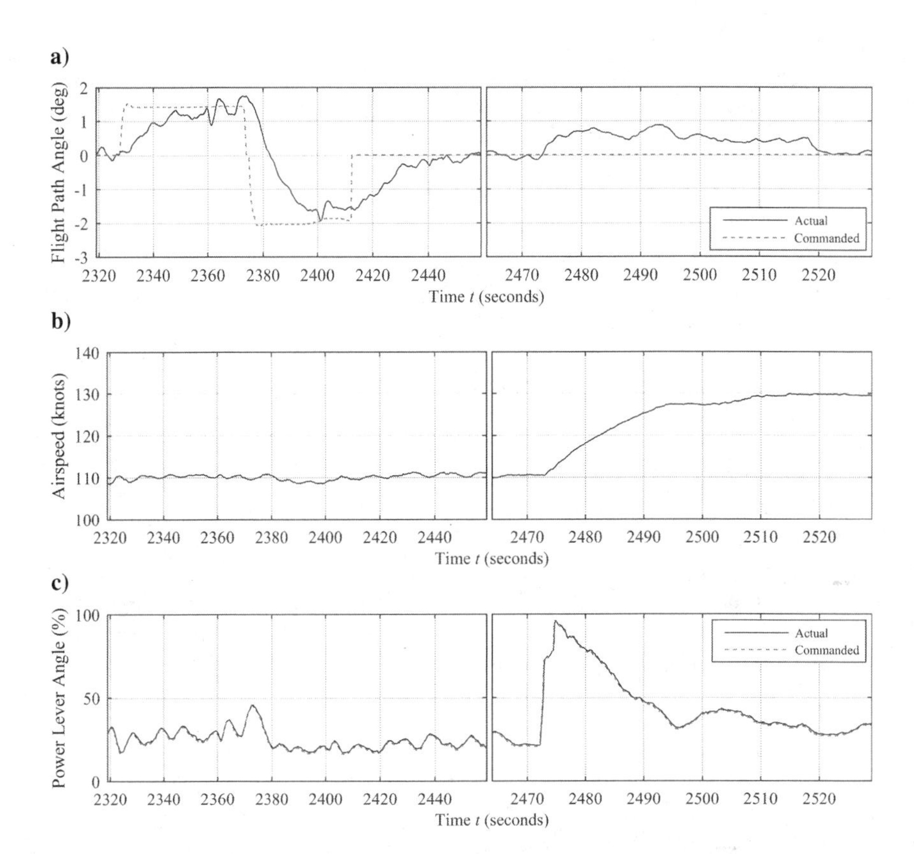

Fig. 6.39 Aircraft response in flight test with Kalman Filter.

all, these additions expand the ability of the MRAC system to handle a large set of dynamic real-world conditions.

6.9 SUMMARY

The application of Model Reference Adaptive Control (MRAC) to General Aviation (GA) aircraft is a novel and recent concept. The research into resilient adaptive control systems at the Wichita State University over the course of the last decade has culminated in a full Six-Degrees-of-Freedom predictive inverse controller that has been proven in simulation, in ground testing, and in real-world flight testing. In this chapter, the authors have discussed the development of the WSU MRAC system, detailed extensions to the architecture aimed at improving resilience to atmospheric disturbances and uncertainties stemming from sensor-induced noise, and validated the controller's positive performance through results from simulation and actual flight test.

Given the performance seen in flight testing, one may conclude that the benefits of MRAC in a GA setup are very apparent. One of the initial motivations for this research was to create control systems for GA aircraft that would ultimately make the genre easier to fly for both novice and experienced pilots and that would provide useful and unobtrusive assistance in situations involving aircraft system failures. Loss-of-control events have been deemed to be a significant contributing factor to aircraft accidents, and the utilization of MRAC to react and respond to variations in plant dynamics and disturbances in the environment is an effective method of mitigating risk factors involved in flight. Under normal flight conditions, the adaptive elements of the ANN were shown to provide assistance in achieving desired pilot commands. In response to simulated failures, the MRAC system demonstrated the ability to adapt "behind the scenes" to reduce pilot workload and restore the aircraft's performance to a level comparable to a no-failure scenario. Further, with the addition of extensions that increase the MRAC system's resilience to atmospheric disturbances/operational uncertainties, the real-world capability of the MRAC GA setup was greatly expanded.

From these perspectives, the MRAC architecture is a good candidate for a control system capable of making GA aircraft easier and safer to fly, and the additional redundancies provided by the MRAC system show potential in aiding an aircraft in its certification process. It is the hope of the research team that the benefits of this adaptive control architecture would eventually be realized through implementation in production GA aircraft.

ACKNOWLEDGEMENTS

This material is based upon work supported by NASA under award number NNXO9AP20A. Any opinions, findings, and conclusions or recommendations expressed in this material are those of the authors and do not necessarily reflect the views of the National Aeronautics and Space Administration.

REFERENCES

[1] Kim, B., and Calise, A., "Nonlinear Flight Control Using Neural Networks," *Journal of Guidance, Control, and Dynamics*, Vol. 20, No. 1, Jan.–Feb. 1997, pp. 26–33.

[2] Bosworth, J., and Williams-Hayes, P., "Flight Test Results from the NF-15B Intelligent Flight Control System (IFCs) Project with Adaptation to a Simulated Stabilator Failure," AIAA-2007-2818, *AIAA Infotech@Aerospace*, Rohnert Park, CA, May 2007.

[3] Burken, J., and Kaneshige, J., "Reconfigurable Control with Neural Network Augmentation for a Modified F-15 Aircraft," AIAA-2007-2823, *AIAA Infotech@Aerospace Conference*, Rohnert Park, CA, May 2007.

[4] Nguyen, N., Krishnakumar, K., and Boskovic, J., "An Optimal Control Modification to Model-Reference Adaptive Control for Fast Adaptation," *AIAA Guidance,*

Navigation, and Control Conference and Exhibit, Honolulu, HI, Aug. 2008, AIAA 2008 - 7283.

[5] Nguyen, N., "Predictor-Model-Based Least-Squares Model-Reference Adaptive Control with Chebyshev Orthogonal Polynomial Approximation," *AIAA Infotech@Aerospace Conference*, Boston, MA, Aug. 2013, AIAA 2013 - 5134.

[6] Nguyen, N., "Least-Squares Model-Reference Adaptive Control with Chebyshev Orthogonal Polynomial Approximation," *Journal of Aerospace Information Systems*, Vol. 10, No. 6, June 2013, pp. 268 - 286.

[7] Stepanyan, V., and Krishnakumar, K., "Modified Reference Model MRAC (M-MRAC): An Application to a Generic Transport Aircraft," *Advances in Intelligent and Autonomous Aerospace Systems*, AIAA, 2012.

[8] Steck, J. E., Rokhsaz, K., Pesonen, U. J., and Duerksen, N., "An Advanced Flight Control System for General Aviation Application," *General Aviation Technology Conference and Exhibit*, Wichita, KS, April, 2004, SAE-2004-01-1807.

[9] Steck, J., Rokhsaz, K., and Shue, S.-P., "Linear and Neural Network Feedback for Flight Control Decoupling," *IEEE Control Systems Magazine*, Vol. 16, No. 4, Aug. 1996, pp. 22 - 30.

[10] Steck, J. E., Rokhsaz, K., Pesonen, U. J., and Duerksen, N., "An Advanced Flight Control System for General Aviation Application," *General Aviation Technology Conference and Exhibit*, Wichita, KS, April 2004, SAE-2004-01-1807.

[11] Pesonen, U. J., Steck, J. E., Rokhsaz, K., Bruner, S., and Duerksen, N., "Adaptive Neural Network Inverse Controller for General Aviation Safety," *AIAA 41st Aerospace Sciences Meeting and Exhibit*, Reno, NV, Jan. 2003, AIAA 2003 - 0573.

[12] Steck, J., Rokhsaz, K., Namuduri, K., and Bruner, S., "Exploring Critical Flight Conditions, Controller Modes, and Parameter Estimation for Adaptive Flight Controls in General Aviation Aircraft," *FAA Final Report*, Wichita State University, Wichita, KS, May 2006, DOT/FAA/AR-01-C-AW-WISU-50.

[13] Hinson, B. T., Steck, J. E., and Rokhsaz, K., "Adaptive Control of an Elastic General Aviation Aircraft," *AIAA Guidance, Navigation, and Control Conference and Exhibit*, Portland, OR, Aug. 2011, AIAA 2011 - 6560.

[14] Lemon, K. A., Steck, J. E., and Hinson, B. T., "Model Reference Adaptive Fight Control Adapted for General Aviation: Controller Gain Simulation and Preliminary Flight Testing On a Bonanza Fly-By-Wire Testbed," *AIAA Guidance, Navigation, and Control Conference and Exhibit*, Toronto, ON, Aug. 2010, AIAA-2010-8278.

[15] Reed, S., Steck, J. E., and Nguyen, N., "Demonstration of the Optimal Control Modification for General Aviation: Design and Simulation," *AIAA Guidance, Navigation, and Control Conference*, Portland, OR, Aug. 2011, AIAA 2011 - 6254.

[16] Reed, S., Steck, J. E., and Nguyen, N., "Demonstration of the Optimal Control Modification for 6-DOF Control of a General Aviation Aircraft," *AIAA Guidance, Navigation, and Control Conference*, National Harbor, MD, Jan. 2014, AIAA 2014 - 1292.

[17] Reed, S., and Steck, J. E., "$U_{add} + U_{mult}$ MRAC Compensation For Control Reversal: Elevator Miss-rigging in a General Aviation Aircraft," *AIAA Guidance, Navigation, and Control Conference*, Minneapolis, MN, Aug. 2012, AIAA 2012 - 4550.

[18] Steck, J. E., Lakshmikanth, G. S., and Watkins, J. M., "Adaptive Critic Optimization of Dynamic Inverse Control," *AIAA Infotech@Aerospace Conference*, Garden Grove, CA, June, 2012, AIAA 2012 - 2408.

[19] Lakshmikanth, G. S., Reed, S., Watkins, J. M., and Steck, J. E., "Single Network Adaptive Critic aided Nonlinear Dynamic Inversion with Optimal Control Modification for Fast Adaptation of MRAC Flight Control," *AIAA Infotech@Aerospace Conference*, Boston, MA, Aug. 2013, AIAA 2013-5208.

[20] Lakshmikanth, G. S., Watkins, J. M., and Steck, J. E., "Adaptive Flight-Control Design Using Neural-Network-Aided Optimal Nonlinear Dynamic Inversion," *Journal of Aerospace Information Systems*, Vol. 11, No. 11 (2014), pp. 785-806.

[21] Steck, J. E., Rokhsaz, K., Pesonen, U. J., Mochrie, S., and Maxfield, M., "Pilot Evaluation of An Adaptive Controller On A General Aviation SATS Testbed Aircraft," *AIAA Guidance, Navigation, and Control Conference and Exhibit*, Providence, RI, Aug. 2004, AIAA-2004-5239.

[22] Steck, J. E., Rokhsaz, K., Pesonen, U. J., Singh, B., and Chandramohan, R., "Effect of Turbulence on an Adaptive Dynamic Inverse Flight Controller," *AIAA Infotech@Aerospace Conference*, Arlington, VA, Sept. 2005, AIAA 2005-7038.

[23] Rafi, M., Steck, J. E., and Rokhsaz, K., "A Microburst Response and Recovery Scheme Using Advanced Flight Envelope Protection," *AIAA Guidance, Navigation, and Control Conference and Exhibit*, Minneapolis, MN, Aug. 2012, AIAA-2012-4444.

[24] Rafi, M., and Steck, J. E., "Response and Recovery of an MRAC Advanced Flight Control System to Wake Vortex Encounters," *AIAA Guidance, Navigation, and Control Conference and Exhibit*, Boston, MA, Aug. 2013, AIAA 2013-5209.

[25] Pappu, V. S. R., Steck, J. E., Rajagopal, K., and Balakrishnan, S. N., "Modified State Observer Based Adaptation of a General Aviation Aircraft - Simulation and Flight Test," *AIAA Guidance, Navigation, and Control Conference and Exhibit*, National Harbor, MD, Jan. 2014, AIAA 2014-1297.

[26] Rafi, M., Steck, J. E., and Watkins, J. M., "Application of a Kalman Filter for Reduction of Sensor/Turbulence-Induced Noise Within a Model Reference Adaptive Controller," *AIAA Guidance, Navigation, and Control Conference and Exhibit*, San Diego, CA, Jan. 2016, AIAA 2016-1625.

[27] Raytheon Aircraft Company, *Beechcraft Bonanza F33A and F33C Acrobatic: Pilot's Operating Handbook and FAA Approved Airplane Flight Manual*, 1996.

[28] Roskam, J., *Airplane Flight Dynamics and Automatic Flight Controls*, DARcorporation, Lawrence, KS, 2001.

[29] Kuo, B. C., *Automatic Control Systems*, 7th ed. Upper Saddle River, NJ, Prentice Hall, 1995, p. 398.

[30] Wolfson, M. M., "Characteristics of Microbursts in the Continental United States," *The Lincoln Laboratory Journal*, Vol. 1, No. 1, 1988.

[31] Proctor, F. H., Bracalente, E. M., Harrah, S. D., Switzer, G. F., and Britt, C. L., "Simulation of the 1994 Charlotte Microburst with Look-Ahead Windshear Radar," *27th Conference on Radar Meteorology, American Meteorological Society*, Vail, CO, Oct. 1995, pp. 530-532.

[32] Federal Aviation Administration, "Flight Simulation Device Qualification Guidance, Windshear - Training and Simulator Requirements," *FAA National Simulator Program*, 2003, FSDQG 03-05, ASF-205.

[33] Tony Lambregts, FAA CSTA Flight Guidance and Control, Private communication, June 2004.

[34] Rokhsaz, K., Steck, J. E., Chandramohan, R., and Singh, B., "Response of an Advanced Flight Control System to Microburst Encounters," SAE AeroTech Congress and Exhibit, Grapevine, TX, Oct. 2005, SAE-2005-01-3420.

[35] Rajagopal, K., Mannava, A., Balakrishnan, S., N., Nguyen, and Krishnakumar, K., "Neuroadaptive Model Following Controller Design for Non-affine Non-square Aircraft System," *AIAA Guidance, Navigation, and Control Conference*, Chicago, IL, Aug. 2009, AIAA 2009-5737.

[36] Balakrishnan, S. N., Rajagopal, Karthikeyan, Steck, J. E., and Dwayne, Kimball, "Robust Adaptive Control of a General Aviation Aircraft," *AIAA Atmospheric Flight Mechanics Conference*, Toronto, ON, Aug. 2010, AIAA-2010-7942.

CHAPTER 7

Handling Inlet Unstart in Hypersonic Vehicles Using Nonlinear Dynamic Inversion Adaptive Control with State Constraints

Douglas Famularo*, Sean G. Whitney† and John Valasek‡

Texas A&M University, College Station, TX 77843-3141

Jonathan A. Muse§ and Michael A. Bolender¶

U.S. Air Force Research Laboratory, Wright-Patterson Air Force Base, OH 45433

NOMENCLATURE

b	span
g	acceleration due to gravity
L	lift force
m	vehicle mass
p	body-axis roll rate
q	body-axis pitch rate
r	body-axis yaw rate
S	planform area
Y	side force
α	angle-of-attack
β	sideslip angle
γ	flight path angle
μ	aerodynamic bank angle
C_{ℓ}	rolling moment coefficient
$C_{\ell,\text{base}}$	baseline aerodynamic rolling moment coefficient
C_{ℓ_p}	roll damping coefficient
C_m	pitching moment coefficient

*Graduate Research Assistant, Vehicle Systems & Control Laboratory, AIAA Student Member, *dif5@tamu.edu*.

†Undergraduate Research Assistant, Vehicle Systems & Control Laboratory, AIAA Student Member, *sgwhitney@tamu.edu*.

‡Professor and Director, Vehicle Systems & Control Laboratory, AIAA Fellow, *valasek@tamu.edu*, *http://vscl.tamu.edu/valasek*.

§Research Aerospace Engineer, AFRL/RQQA, and AIAA Member.

¶Senior Aerospace Engineer, AFRL/RQQA, and AIAA Associate Fellow.

$C_{m,\text{base}}$	baseline aerodynamic pitching moment coefficient
C_{m_q}	pitch damping coefficient
C_n	yawing moment coefficient
$C_{n,\text{base}}$	baseline aerodynamic yawing moment coefficient
C_{n_r}	yaw damping coefficient
F_{T_x}	component of thrust force along body fixed x-axis
F_{T_y}	component of thrust force along body fixed y-axis
F_{T_z}	component of thrust force along body fixed z-axis
I_x	moment of inertia about body fixed x-axis
I_y	moment of inertia about body fixed y-axis
I_z	moment of inertia about body fixed z-axis
I_{xz}	xz product of inertia
M_T	pitching moment due to thrust
V_T	total velocity
$\delta_{f,l}$	deflection of left elevon
$\delta_{f,r}$	deflection of right elevon
$\delta_{t,l}$	deflection of left ruddervator
$\delta_{t,r}$	deflection of right ruddervator
$\overline{\delta}_i$	trim condition for control surface δ_i
$\overline{c}$	mean aerodynamic chord length
$\overline{q}$	dynamic pressure

7.1 INTRODUCTION

The need to restrict dynamical systems within certain regions of the state-space is a common situation in flight control and has led to the integration of state constraint methods with existing control architectures. State constraining augmentation systems have been shown to be capable of preventing pilots from commanding an aircraft to enter the stall region by limiting angle-of-attack [1], limiting the amount of g-force that can be applied to a vehicle [2], or taking other corrective action to ensure the vehicle remains within its flight envelope. In addition, it has been shown that supersonic and hypersonic vehicles are susceptible to a phenomenon known as inlet unstart if the vehicle flies at too steep of an angle-of-attack or sideslip angle [3]. State constraint control techniques that can restrict what values of angle-of-attack and sideslip angle can be achieved could potentially be used to prevent inlet unstart. This hypothesis is the prime motivation of this chapter.

Air-breathing high-speed flight vehicles, whether supersonic or hypersonic, must decelerate the freestream flow down to Mach numbers that are low enough to permit combustion. In the case of conventional gas turbines, the flow in the combustor is subsonic. In the case of supersonic combustion powerplants, such as supersonic combustion ramjets, the flow in the combustor is supersonic. In both cases, this reduction in Mach number can be achieved by a system of either internal or external oblique shocks that form a compression system according to the inlet geometry. The method of obtaining efficient supersonic compression is basically the same for internal as for external compression: that is to say, a staged compression system is created by the use of discrete wedges or conical surfaces, or using isentropic contours, to reduce the Mach number of flow to a suitably low value for the terminal, normal shock as shown in Fig. 7.1 [4]. The inlet is considered to be "started" when the vehicle is operating at a freestream Mach number for which all shocks are formed and positioned to produce the designed value of pressure recovery. A variable-geometry inlet is often used to produce a desired value of pressure recovery at more than one freestream Mach number.

The condition of supersonic flow breakdown is called "inlet unstart" and results in the maximum flow being 20 to 30% less than that of the started inlet, and its equivalent operating point is a supercritical condition [4]. For hypersonic vehicles in particular, inlet unstart is a major safety concern because it can lead to departure from controlled flight that results in loss of the vehicle. Inlet unstart occurs in one of two ways. In a started inlet with a normal shock at the throat of the nozzle, any increase in back pressure (see Fig. 7.2) will drive the shock forward into the contraction section, where it cannot be stabilized because the shock Mach number is greater than the throat value. Pressure recovery is decreased and the shock moves rapidly through the contraction section and is expelled into the freestream, tending to settle at a position consistent with the increased back pressure and producing a reduced flow ratio. This leads to a dangerous loss of thrust. The back pressure that causes the unstart can be caused by airflow through the inlet that is operating below the design Mach number unless variable geometry is used. A decrease in engine flow demand can trigger this situation. The second mechanism for unstart is flow asymmetries that prevent the flow from passing through the throat of the inlet. This is usually caused by the vehicle operating at an excessive angle-of-attack or sideslip angle. The inlet can be restarted only by going through whatever process is built into the design.

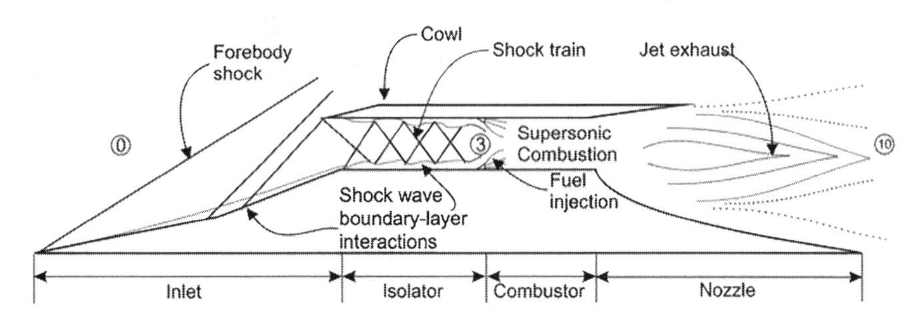

Fig. 7.1 A generic schematic of a scramjet engine working principle where, typically, compression is provided between Station 0 and 3 before fuel injection and combustion [5].

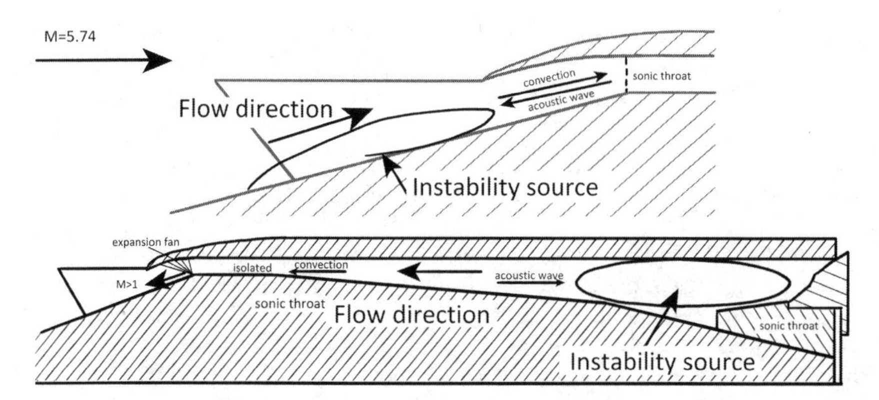

Fig. 7.2 Flow mechanisms that produce inlet unstart [6].

A controller that is effective in handling inlet unstart will have two important features. First, it will function as a preventative measure by not allowing the vehicle to reach values of angle-of-attack or sideslip angle that can trigger an unstart. Second, it will function as a mitigation measure. If an unstart does occur, the controller should be capable of preventing the vehicle from departing controlled flight. The departure from controlled flight can occur because an unstart can lead to a complex sequence of interactions between the propulsion system and airframe aerodynamics that produces a rapid onset of unbalanced forces and moments. This effect is particularly severe in multi-engine vehicles because inlets do not tend to unstart simultaneously, resulting in large yawing and rolling moments in addition to pitching moments. The key idea for the work detailed in this chapter is that state constraint mechanisms can be used as the preventative measure in designing controllers capable of preventing and surviving inlet unstart for air-breathing hypersonic vehicles.

State constraint enforcement mechanisms have been developed in the fields of linear control theory [7], optimal control [8–10], model-predictive control [11], and adaptive control [12–15]. This chapter extends such techniques to an nonlinear dynamic inversion (NDI) adaptive control framework allowing for provably safe control of a nonlinear hypersonic vehicle model. NDI control has been well developed in the past and has been shown to successfully handle dynamical systems that can accurately be modeled using nonlinear equations of motion. Often, in order to deal with model uncertainties, the NDI concept is combined with an adaptive controller to add robustness to the closed-loop system. Therefore, NDI adaptive control allows for the stabilization of nonlinear systems over a wide range of operating conditions without the need for excessive gain scheduling. Several studies have been done to test the robustness properties of these types of controllers when used on air vehicles [16, 17]. In addition, NDI adaptive control has been shown to be effective on highly nonlinear rotorcraft applications [18, 19].

This chapter introduces a full state feedback adaptive control architecture that is capable of enforcing state constraints on nonlinear systems. The adaptive mechanism in the control law allows for the controller to handle parametric uncertainty in the model. The control objective is to stabilize the system tracking error while maintaining system states in a predefined constraint set. If the reference trajectory exceeds these limits, the control and adaptive laws alter their behavior and restrict the system. It is proven through Lyapunov analysis that such alterations restrict the state from exceeding its constraint. Furthermore, if the system is given a reference command that is well within the safe region of the state-space, the controller reduces to an NDI adaptive controller, which can asymptotically track a reference trajectory.

This state constraint mechanism is based on bounding functions, a technique first developed by Muse [14] for linear systems with unknown nonlinear disturbances. The bounding functions used in this technique allow for significant flexibility in designing how the controller transitions between "tracking mode" and "state constraining mode." Such flexibility allows for this technique to potentially lessen chattering in the control signal. Because the bounding functions depend on the state measurements, this technique is capable of enforcing state constraints both when an unacceptably large command is given and in the presence of disturbances.

This chapter is organized as follows. Section 7.2 presents the derivation and definition of the control law and adaptive laws for this state constraining control technique. In Section 7.3, a Lyapunov-based stability analysis is presented, which establishes the boundedness of the system state and tracking error. Section 7.4 derives the necessary conditions for achieving perfect tracking and asymptotic stability, and Section 7.5 derives an analytical bound for the system states. Insight into achieving tighter state constraint enforcement is also provided. Section 7.6 discusses how to construct a bounding function with the desired properties. In Section 7.7, a nonlinear model of a hypersonic vehicle is introduced and the proposed state constraint mechanism is demonstrated in simulation, focusing primarily on inlet unstart avoidance and prevention. Finally, conclusions are given in Section 7.8.

7.2 NONLINEAR DYNAMIC INVERSION ADAPTIVE CONTROL EXTENDED TO INCLUDE STATE CONSTRAINTS

In this section, an NDI adaptive control framework capable of enforcing state constraints is presented. The state constraint mechanism is made up of specifically designed functions known as bounding functions, which activate only when a state approaches a predefined constraint value. The definition of a bounding function follows:

Definition 7.1: *$f_b : \mathbb{R} \mapsto \mathbb{R}$ is a bounding function if it is continuously differentiable and satisfies the following properties:*

- $f_b(a) = 0, \forall\, a \leq \bar{b}$
- $f_b(a) > 0, \forall\, a > \bar{b}$
- $f_b(a) \to \infty, a \to \infty$
- $f_b^{'}(a)$ *is locally Lipschitz and monotonically increasing.*
- $f_b^{'}(a) = 0, \forall\, a \leq \bar{b}$
- $f_b^{'}(a) > 0, \forall\, a > \bar{b}$

where a is a scalar dependent variable, $f_b^{'}(a)$ is the derivative of $f_b(a)$ with respect to a, and $\bar{b}$ is a designer-chosen constraint.

Remark 7.1: A simple choice for a bounding function is

$$f_b(a) = \begin{cases} 0 & a \leq \bar{b} \\ (a - \bar{b})^2 & a > \bar{b} \end{cases}$$

These bounding functions are augmented to previously developed NDI control and adaptive laws capable of tracking a reference model. If all of the bounding functions in all of the closed signals are equal to zero, the proposed framework reduces to the tracking controller. To begin the derivation of this control framework, consider the dynamic system described by the general nonlinear differential equation

$$\dot{x} = f(x) + g(x)\Lambda u \tag{7.1}$$

where $x \in \mathbb{R}^n$ is the state, $u \in \mathbb{R}^m$ is the control, and $f(x) : \mathbb{R}^n \mapsto \mathbb{R}^n$ is an unknown function of the state vector that is assumed to be locally Lipschitz continuous. The function $g(x) : \mathbb{R}^n \mapsto \mathbb{R}^{n \times m}$ is also locally Lipschitz continuous, and is assumed to have full row rank. The matrix $\Lambda \in \mathbb{R}^{m \times m}$ is a constant, unknown, invertible control effectiveness matrix. It is assumed, without loss of generality, that Λ can be decomposed as

$$\Lambda = I_m + \delta\Lambda \tag{7.2}$$

where I_m is the $m \times m$ identity matrix and $\delta\Lambda \in \mathbb{R}^{m \times m}$. It is desired that the system follows a stable, linear reference model with dynamics given by

$$\dot{x}_m = A_m x_m + B_m r \tag{7.3}$$

where $x_m(t) \in \mathbb{R}^n$ is the reference model state, $r(t) \in \mathbb{R}^n$ is a bounded reference input signal, $A_m \in \mathbb{R}^{n \times n}$ is Hurwitz, and $B_m \in \mathbb{R}^{n \times n}$. The tracking error is defined as

$$e = x_m - x \tag{7.4}$$

To analyze the stability of the closed-loop system, it will be necessary to understand the dynamics of the tracking error. Taking a derivative of Eq. (7.4) with respect to time and substituting in Eq. (7.1), the error dynamics can be written as

$$\dot{e} = \dot{x}_m - \dot{x} = \dot{x}_m - f(x) - g(x)\Lambda u \tag{7.5}$$

Next, $g(x)\hat{\Lambda}u$ is added and subtracted from the right-hand side of Eq. (7.5). Defining $\tilde{\Lambda} = \hat{\Lambda} - \Lambda$, this results in

$$\dot{e} = \dot{x}_m - f(x) - g(x)\hat{\Lambda}u + g(x)\tilde{\Lambda}u \tag{7.6}$$

where $\hat{\Lambda} \in \mathbb{R}^{m \times m}$ is an estimate of the control effectiveness matrix that will be updated through an adaptive law. This matrix can also be decomposed as

$$\hat{\Lambda} = I_m + \delta\hat{\Lambda} \tag{7.7}$$

where $\delta\hat{\Lambda} \in \mathbb{R}^{m \times m}$. It will be shown that through the use of the projection operator in the adaptive law for $\delta\hat{\Lambda}$, the total estimate $\hat{\Lambda}$ can be guaranteed to be invertible at all times.

Because $g(x)$ is assumed to have full rank, there will always exist an inverse for this matrix if $g(x)$ is square or a right pseudo-inverse if it is nonsquare. For the case where $g(x)$ is nonsquare, if no bounds are assumed on the control effectors, there is an infinite number of combinations of control variables that will achieve the desired accelerations at any instant in time. Therefore, a constrained optimization problem is set up to minimize the control effort subject to the constraint $g(x)\hat{\Lambda}u = \ell$ where ℓ is given by,

$$\ell = \dot{x}_m - \hat{f}(x) + Ke + \nu - \ell_r \tag{7.8}$$

The function $\hat{f}(x) : \mathbb{R}^n \mapsto \mathbb{R}^n$ is an assumed model of the plant dynamics, $K \in \mathbb{R}^{n \times n}$ is a positive definite diagonal matrix of feedback gains on the tracking error, $\nu \in \mathbb{R}^n$ is an adaptive parameter signal, and $\ell_r \in \mathbb{R}^n$ is a hedge signal responsible for enforcing state constraints. The cost function to be minimized is

$$J = u^T Q u \tag{7.9}$$

where $Q = Q^T > 0$ and $Q \in \mathbb{R}^{m \times m}$. In order for this control law to be continuous, Q and $g(x)\hat{\Lambda}Q^{-1}\hat{\Lambda}^T g^T(x)$ must be invertible. The full solution to this optimization problem can be found in work by Rollins [20], which leads to the control law:

$$u = Q^{-1}\hat{\Lambda}^T g^T(x)(g(x)\hat{\Lambda}Q^{-1}\hat{\Lambda}^T g^T(x))^{-1}\ell \tag{7.10}$$

In the alternate case, where $g(x)$ is square, the control law formulation is simpler. As previously mentioned, it will be shown that for certain projection operator parameters, $\hat{\Lambda}$ is always invertible. Therefore, the control law can be expressed as

$$u = \hat{\Lambda}^{-1} g^{-1}(x)\ell \tag{7.11}$$

Plugging the optimal control law, Eq. (7.10), or when $g(x)$ is square the simplified expression, Eq. (7.11), into the error dynamics allows for Eq. (7.6) to be rewritten as

$$\dot{e} = -f(x) + \hat{f}(x) - Ke - \nu + \ell_r + g(x)\tilde{\Lambda}u \tag{7.12}$$

By recognizing that $\tilde{\Lambda} = \delta\hat{\Lambda} - \delta\Lambda$ and defining $\delta\tilde{\Lambda} = \delta\hat{\Lambda} - \delta\Lambda$, Eq. (7.12) can be rewritten as

$$\dot{e} = -f(x) + \hat{f}(x) - Ke - \nu + \ell_r + g(x)\delta\tilde{\Lambda}u \tag{7.13}$$

Assume that the difference $\hat{f}(x) - f(x)$ can be parameterized as

$$\hat{f}(x) - f(x) = W^T\beta(x) \tag{7.14}$$

where $W \in \mathbb{R}^{p\times n}$ is a set of unknown bounded weights and $\beta(x) \in \mathbb{R}^{p\times 1}$ is a set of known basis functions that are dependent on the state vector. The adaptive parameter signal is defined as $\nu = \widehat{W}^T\beta(x)$ where $\widehat{W} \in \mathbb{R}^{p\times n}$ is an estimate of the weights. Defining $\widetilde{W} = \widehat{W} - W$, Eq. (7.13) becomes

$$\dot{e} = -Ke + \ell_r - \widetilde{W}^T\beta(x) + g(x)\delta\tilde{\Lambda}u \tag{7.15}$$

Since $x = x_m - e$, the plant dynamics can be rewritten as

$$\dot{x} = \dot{x}_m + Ke - \ell_r + \widetilde{W}^T\beta(x) - g(x)\delta\tilde{\Lambda}u \tag{7.16}$$

With the system dynamics defined, the state constraint mechanism, which relies on the use of bounding functions, can be introduced.

Given a bounding function that satisfies Definition 7.1, the adaptive laws and the reference hedge signal are defined as

$$\dot{\widehat{W}} = \Gamma_W \mathrm{Proj}_M(\widehat{W}, \beta(x)(e^T - x^T F'(x))) \tag{7.17}$$

$$\delta\dot{\hat{\Lambda}} = \Gamma_\Lambda \mathrm{Proj}_M(\delta\hat{\Lambda}, g(x)^T(-eu^T + F'(x)xu^T)) \tag{7.18}$$

$$\ell_r = F'_*(x)[\dot{x}_m + Kx_m] \tag{7.19}$$

where $F'(x) \in \mathbb{R}^{n\times n}$ and $F'_*(x) \in \mathbb{R}^{n\times n}$ are diagonal matrices of the form

$$F'(x) = \begin{bmatrix} f'_{b,1}(x_1^2) & 0 & \dots & 0 \\ 0 & f'_{b,2}(x_2^2) & \dots & 0 \\ \vdots & \vdots & \ddots & \vdots \\ 0 & 0 & \dots & f'_{b,n}(x_n^2) \end{bmatrix} \tag{7.20}$$

$$F'_*(x) = \begin{bmatrix} \frac{f'_{b,1}(x_1^2)}{1+f'_{b,1}(x_1^2)} & 0 & \dots & 0 \\ 0 & \frac{f'_{b,2}(x_2^2)}{1+f'_{b,2}(x_2^2)} & \dots & 0 \\ \vdots & \vdots & \ddots & \vdots \\ 0 & 0 & \dots & \frac{f'_{b,n}(x_n^2)}{1+f'_{b,n}(x_n^2)} \end{bmatrix} \tag{7.21}$$

The function $f'_{b,i}(x_i^2)$ is the derivative of the bounding function for the ith state with respect to x_i^2, that is

$$f'_{b,i}(x_i^2) = \frac{df_{b,i}}{dx_i^2} \tag{7.22}$$

Defined as it is in Eq. (7.19), the reference hedge signal ℓ_r will attenuate the input command when state constraints are close to being exceeded by taking advantage of the properties of the bounding functions, which are described in detail shortly. The matrix projection operator, given in Eq. (7.17) and (7.18) by Proj_M, is defined as

$$\text{Proj}_M(\Theta, Y) = [\text{Proj}(\theta_1, y_1), \dots \text{Proj}(\theta_b, y_b)] \tag{7.23}$$

where $\Theta \in \mathbb{R}^{a\times b} = [\theta_1, \theta_2, \dots \theta_b]$, $Y \in \mathbb{R}^{a\times b} = [y_1, y_2, \dots y_b]$, and Proj represents the vector projection operator. This is defined as,

$$\text{Proj}(\theta_i, y_i) = \begin{cases} y_i - \frac{\nabla h(\theta_i)(\nabla h(\theta_i))^T}{||\nabla h(\theta_i)||^2} y_i h(\theta_i), & \text{if } h(\theta_i) > 0, \quad y_i^T \nabla h(\theta_i) > 0 \\ y_i, & \text{otherwise} \end{cases} \tag{7.24}$$

where $\theta_i \in \mathbb{R}^a$, $y_i \in \mathbb{R}^a$, $h(\theta_i)$:$\mathbb{R}^a \rightarrow \mathbb{R}$ is a convex function, and $\nabla h(\theta_i) = [\partial h(\theta_i)/\partial\theta_{i,1} \ \dots \ \partial h(\theta_i)/\partial\theta_{i,a}]$ [21, 22]. The projection operator is constructed so that a known maximum bound can be set for each of the vectors, θ_i. For the set of general nonlinear systems that is the focus of this chapter, this corresponds to known bounds on the norm of the columns of the unknown parameter matrices, $\widehat{W}$ and $\delta\hat{\Lambda}$. The bounds are selected based on one's knowledge of the uncertainty in the system. If the uncertainty is assumed to be unbounded there may not be enough control effectiveness to guarantee that control objectives are met. Therefore, it is assumed that there is some understanding of the uncertainty in the system, and that knowledge is reflected in the controller by bounding the weights through the projection operator. In addition, proper choice of the bounds on the columns of $\delta\hat{\Lambda}$ allows for restriction on how large the matrix can grow. If the bounds are selected in such a way that none of the eigenvalues of $\delta\hat{\Lambda}$ equal -1 at any instant of time, then the total estimate $\hat{\Lambda}$ is guaranteed to be invertible at all times. The following lemma shows that this is always possible.

Lemma 7.1: *For all $\delta\hat{\Lambda} \in \mathbb{R}^{m\times m}$, if the 2-norm of each column is less than $1/m$ at all times, which can be guaranteed by the projection operator, then $\hat{\Lambda}$ will always be invertible.*

Proof: It can be shown that if λ is an eigenvalue of $\delta\hat{\Lambda}$ then $\lambda + 1$ is an eigenvalue of $\hat{\Lambda}$. Let each element of the matrix $\delta\hat{\Lambda}$ have a magnitude smaller than $1/m$. Suppose that $\hat{\Lambda}$ is not invertible. This implies that $\hat{\Lambda}$ has a zero eigenvalue, which implies that $\delta\hat{\Lambda}$ has an eigenvalue equal to -1 and that there exists at least one eigenvector v that satisfies

$$\delta\hat{\Lambda}v = -v \tag{7.25}$$

Let v_i correspond to the element of the eigenvector v with the largest magnitude. The ith row of Eq. (7.25) can be written as

$$\delta\hat{\Lambda}_{i1}v_1 + \delta\hat{\Lambda}_{i2}v_2 + \cdots + \delta\hat{\Lambda}_{im}v_m = -v_i \tag{7.26}$$

This implies that

$$|v_i| = |\delta\hat{\Lambda}_{i1}v_1 + \delta\hat{\Lambda}_{i2}v_2 + \cdots + \delta\hat{\Lambda}_{im}v_m| \tag{7.27}$$

$$|v_i| \leq |\delta\hat{\Lambda}_{i1}v_1| + |\delta\hat{\Lambda}_{i2}v_2| + \cdots + |\delta\hat{\Lambda}_{im}v_m| \tag{7.28}$$

$$|v_i| \leq |\delta\hat{\Lambda}_{i1}||v_1| + |\delta\hat{\Lambda}_{i2}||v_2| + \cdots + |\delta\hat{\Lambda}_{im}||v_m| \tag{7.29}$$

$$|v_i| < (|\delta\hat{\Lambda}_{i1}| + |\delta\hat{\Lambda}_{i2}| + \cdots + |\delta\hat{\Lambda}_{im}|)|v_i| \tag{7.30}$$

Since each element of $\delta\hat{\Lambda}$ has magnitude less than $1/m$, $(|\delta\hat{\Lambda}_{i1}| + |\delta\hat{\Lambda}_{i2}| + \cdots + |\delta\hat{\Lambda}_{im}|) < 1$. This implies that

$$|v_i| < |v_i| \tag{7.31}$$

which is a contradiction. Therefore, $\hat{\Lambda}$ must be invertible whenever each element of $\delta\hat{\Lambda}$ is less than $1/m$. If the projection operator is utilized to bound the 2-norm of each column of $\delta\hat{\Lambda}$ to be less than $1/m$, this will ensure that each element is less than $1/m$ as well and guarantees that $\hat{\Lambda}$ is invertible at all times.

The conclusion of Lemma 7.1 is the reason that the decomposition given in Eq. (7.7) and the assumption that the true unknown parameter matrix Λ can be decomposed, as in Eq. (7.2), are necessary.

As was mentioned previously, the state constraint mechanism is reliant on bounding functions that are found in certain terms in the control and adaptive laws. As one or more of the states approach their pre-specified limits, the corresponding bounding function terms, found in the matrices $F'(x)$ and $F'_*(x)$, change appropriately and are fed into the controller. These terms lessen the effect of the reference model and work to drive the state away from the constraint and toward the origin or some other designated safe region of the state-space. It should be noted that if a state is not approaching a constraint (i.e., $x_i^2 < b_i$) then the bounding function for that state is zero, $\ell_{r,i} = 0$, and the adaptive and control laws work to track that state's reference model signal. If this were true for all of the states at a given time, then the control law reduces to the form of an NDI model reference adaptive controller given by Rollins and Rollins et al. [20, 23]. In the following section a stability analysis is performed, providing the necessary conditions for e and x to be bounded.

7.3 LYAPUNOV STABILITY ANALYSIS

In this section it is shown that, using the control architecture previously defined, the system states and the tracking error are bounded for a bounded input signal. The size of this bound will depend on, among other things, the steady state value of the reference model. This is a direct result of the state constraint mechanism. If the steady state value of the reference model is outside of a predetermined constraint set, perfect tracking is not desired. Instead, the system should remain at the boundary of its constraint set. Therefore, in this section boundedness is proven regardless of whether the steady state value of the reference model is inside or outside of the constraint set. In Section 7.4, it is proven that *if* the steady state value of the reference model is inside of the constraint set, perfect tracking will occur. Section 7.5 once again examines the system regardless of where the steady state value of the reference model lies, and a conservative, analytical bound for the system state and tracking error is produced.

Theorem 7.1: Consider the nonlinear dynamical system given in Eq. (7.1), the reference model defined in Eq. (7.3), the control law defined in Eq. (7.10) with the reference hedge signal defined in Eq. (7.19), and the adaptive laws defined in Eq. (7.17) and (7.18). Suppose that the reference input signal r is bounded, then the tracking error e and the system states x are bounded.

Proof: To show stability, the following candidate Lyapunov function is chosen:

$$V = e^T e + tr(\widetilde{W}^T \Gamma_W^{-1} \widetilde{W}) + tr(\delta\tilde{\Lambda} \Gamma_\Lambda^{-1} \delta\tilde{\Lambda}^T) + \sum_{i=1}^{n} f_{b,i}(x_i^2) \tag{7.32}$$

where $\Gamma_W \in \mathbb{R}^{p \times p}$ with $\Gamma_W = \Gamma_W^T > 0$, $\Gamma_\Lambda \in \mathbb{R}^{m \times m}$ with $\Gamma_\Lambda = \Gamma_\Lambda^T > 0$, and $f_{b,i}(x_i^2)$ is the bounding function for the ith state. The time derivative of Eq. (7.32) is taken along the system trajectories, resulting in

$$\dot{V} = 2e^T \dot{e} + 2tr(\widetilde{W}^T \Gamma_W^{-1} \dot{\widehat{W}}) + 2tr(\delta\tilde{\Lambda} \Gamma_\Lambda^{-1} \delta\dot{\hat{\Lambda}}^T) + \sum_{i=1}^{n} 2x_i f'_{b,i}(x_i^2) \dot{x}_i \tag{7.33}$$

Using the matrix $F'(x)$, defined earlier, the summation can be rewritten as,

$$\dot{V} = 2e^T \dot{e} + 2tr(\widetilde{W}^T \Gamma_W^{-1} \dot{\widehat{W}}) + 2tr(\delta\tilde{\Lambda} \Gamma_\Lambda^{-1} \delta\dot{\hat{\Lambda}}^T) + 2x^T F'(x) \dot{x} \tag{7.34}$$

Substituting the error dynamics from Eq. (7.15) and the plant dynamics from Eq. (7.16) leads to

$$\begin{aligned}\dot{V} &= 2e^T[-Ke + \ell_r] - 2e^T\widetilde{W}^T\beta(x) + 2e^Tg(x)\delta\tilde{\Lambda}u + 2x^TF'(x)[\dot{x}_m + Ke - \ell_r] \\ &+ 2x^TF'(x)\widetilde{W}^T\beta(x) - 2x^TF'(x)g(x)\delta\tilde{\Lambda}u \\ &+ 2tr(\widetilde{W}^T\Gamma_W^{-1}\dot{\widehat{W}}) + 2tr(\delta\tilde{\Lambda}\Gamma_\Lambda^{-1}\delta\dot{\hat{\Lambda}}^T)\end{aligned} \tag{7.35}$$

Applying the trace identity $a^Tb = tr(ba^T)$, Eq. (7.35) is rewritten

$$\begin{aligned}\dot{V} &= 2e^T[-Ke + \ell_r] + 2tr(\widetilde{W}^T(\Gamma_W^{-1}\dot{\widehat{W}} - \beta(x)e^T + \beta(x)x^TF'(x))) \\ &+ 2x^TF'(x)[\dot{x}_m + Ke - \ell_r] + 2tr(\delta\tilde{\Lambda}(\Gamma_\Lambda^{-1}\delta\dot{\hat{\Lambda}}^T + ue^Tg(x) - ux^TF'(x)g(x)))\end{aligned} \tag{7.36}$$

The adaptive laws, given in Eq. (7.17) and (7.18), guarantee that the second and fourth terms in Eq. (7.36) will be less than or equal to zero. Therefore, the following inequality holds:

$$\dot{V} \le 2e^T[-Ke + \ell_r] + 2x^TF'(x)[\dot{x}_m + Ke - \ell_r] \tag{7.37}$$

By utilizing the definition of the tracking error, $e = x_m - x$, and the resulting relationship, $x = x_m - e$, the inequality (7.37) can be rearranged as

$$\begin{aligned}\dot{V} &\le -2e^TKe - 2x^TF'(x)Kx + 2e^T\left[\ell_r - F'(x)(\dot{x}_m + Kx_m - \ell_r)\right] \\ &+ 2x_m^TF'(x)(\dot{x}_m + Kx_m - \ell_r)\end{aligned} \tag{7.38}$$

Substituting Eq. (7.19) into the inequality (7.38) results in

$$\dot{V} \le -2e^TKe - 2x^TF'(x)Kx + 2x_m^TF'_*(x)(\dot{x}_m + Kx_m) \tag{7.39}$$

Let x_{mss} be the steady state value of the reference model state. The following is equivalent to Eq. (7.39):

$$\begin{aligned}\dot{V} &\le -2e^TKe - 2x^TF'(x)Kx + 2x_{mss}^TF'_*(x)Kx_{mss} + 2x_m^TF'_*(x)\dot{x}_m \\ &+ 2x_m^TF'_*(x)Kx_m - 2x_{mss}^TF'_*(x)Kx_{mss}\end{aligned} \tag{7.40}$$

Recall that since $F'(x)$ is a positive semidefinite diagonal matrix and K is a positive definite diagonal matrix, their product will also be a positive semidefinite matrix. Therefore, if

$$\begin{aligned}x^TF'(x)Kx &\ge x_{mss}^TF'_*(x)Kx_{mss} \\ &+ ||x_m^TF'_*(x)\dot{x}_m + x_m^TF'_*(x)Kx_m - x_{mss}^TF'_*(x)Kx_{mss}||_2\end{aligned} \tag{7.41}$$

then $\dot{V} \leq 0$. Since x_m is bounded for a bounded input signal and $F'_*(x)$ is bounded by definition, this implies that x is bounded and therefore e is also bounded, as intended.

7.4 CONDITIONS REQUIRED FOR PERFECT TRACKING AND ASYMPTOTIC STABILITY

In this section, it is shown that if the steady state value of the reference model does not violate any state constraint, perfect tracking will occur as $t \longrightarrow \infty$. Note that a similar analysis could be performed to show that the same conclusion can be drawn on a state by state basis. In other words, perfect tracking is ensured for any individual state whose steady state reference signal value does not violate its individual constraint. The following theorem provides the conditions on the steady state value of the reference model and the parameters of the bounding function such that perfect tracking will occur.

Theorem 7.2: Suppose all of the conditions in Theorem 7.1 are met. Let $r_{ss} \in \mathbb{R}^m$ and $x_{mss} = -A_m^{-1} B_m r_{ss}$, where r_{ss} is the steady state value of the reference input signal and x_{mss} is the steady state value of the reference model. If $\exists \alpha^*, \beta^* \in \mathbb{R}^+$ such that

$$||r(t) - r_{ss}||_2 \leq \alpha^* \exp{(-\beta^* t)}, \quad t \in [0, \infty) \tag{7.42}$$

and if $x \in \mathbb{R}^n$ such that $x_i^2 < (\lambda_{\max}(K)/\lambda_{\min}(K)) x_{i,mss}^2$ implies that $f'_{b,i}(x_i^2) = 0$, then $e \longrightarrow 0$ as $t \longrightarrow \infty$.

Proof: Since $\exists \alpha^*, \beta^* \in \mathbb{R}$ such that $||r(t) - r_{ss}||_2 \leq \alpha^* \exp(-\beta^* t)$ at all times and the reference dynamics in Eq. (7.3) are stable, it is straightforward to show that $\exists \alpha_0, \beta_0 \in \mathbb{R}^+$ such that

$$||x_m^T F'_*(x) \dot{x}_m + x_m^T F'_*(x) K x_m - x_{mss}^T F'_*(x) K x_{mss}||_2 \leq \alpha_0 \exp(-\beta_0 t), \quad t \in [0, \infty) \tag{7.43}$$

The expression for the Lyapunov function derivative given by the inequality (7.40) can now be rewritten as

$$\dot{V} \leq -2e^T K e - 2x^T F'(x) K x + 2x_{mss}^T F'_*(x) K x_{mss} + 2\alpha_0 \exp(-\beta_0 t) \tag{7.44}$$

Evaluating the second and third terms of the inequality (7.44), the following relationship can be established

$$-2x^T F'(x) K x + 2x_{mss}^T F'_*(x) K x_{mss} \leq -2\lambda_{\min}(K) x^T F'(x) x + 2\lambda_{\max}(K) x_{mss}^T F'_*(x) x_{mss} \tag{7.45}$$

Because $F'_{ij}(x) \geq 0$, it is also true that $F'_{ij}(x) \geq F'_{*,ij}(x)$ where $(\cdot)_{ij}$ is the (i, j)th element of the given matrix. Therefore, the following inequality holds:

$$\begin{aligned} &-2\lambda_{\min}(K)x^T F'(x)x + 2\lambda_{\max}(K)x_{mss}^T F'_*(x)x_{mss} \\ &\quad \leq -2\lambda_{\min}(K)x^T F'(x)x + 2\lambda_{\max}(K)x_{mss}^T F'(x)x_{mss} \end{aligned} \tag{7.46}$$

The right-hand side of the inequality (7.46) can now be expressed as

$$\begin{aligned} &-2\lambda_{\min}(K)x^T F'(x)x + 2\lambda_{\max}(K)x_{mss}^T F'(x)x_{mss} \\ &\quad = -2\lambda_{\min}(K)\sum_{i=1}^{n} f'_{b,i}(x_i^2)x_i^2 + 2\lambda_{\max}(K)\sum_{i=1}^{n} f'_{b,i}(x_i^2)x_{i,mss}^2 \end{aligned} \tag{7.47}$$

Since $x_i^2 < (\lambda_{\max}(K)/\lambda_{\min}(K))x_{i,mss}^2$ implies that $f'_{b,i}(x_i^2) = 0$, the second and third terms in the inequality (7.44) always combine to be less than or equal to 0. Therefore, it can be reduced to

$$\dot{V} \leq -2e^T Ke + 2\alpha_0 \exp(-\beta_0 t) \tag{7.48}$$

It follows that,

$$V = V_0 + \int_0^t \dot{V}\, d\tau \tag{7.49}$$

where $V_0 = V(t = 0)$. Utilizing the inequality (7.48) leaves,

$$V \leq V_0 + \int_0^t (-2e^T Ke + 2\alpha_0 \exp(-\beta_0 t))\, d\tau \tag{7.50}$$

which, upon integration, results in

$$V \leq V_0 - \int_0^t 2e^T Ke\, d\tau + 2\frac{\alpha_0}{\beta_0}(1 - \exp(-\beta_0 t)) \tag{7.51}$$

Because $V \geq 0$, the following inequality holds:

$$0 \leq V \leq V_0 - \int_0^t 2e^T Ke\, d\tau + 2\frac{\alpha_0}{\beta_0}(1 - \exp(-\beta_0 t)) \tag{7.52}$$

$$0 \leq \int_0^t 2e^T Ke\, d\tau \leq V_0 + 2\frac{\alpha_0}{\beta_0}(1 - \exp(-\beta_0 t)) \tag{7.53}$$

Since $\int_0^t 2e^T K e\, d\tau$ is nondecreasing and finite, $\lim_{t\to\infty} \int_0^t 2e^T K e\, d\tau$ exists and is finite. Because x and e are bounded, $\dot{e}$ is bounded for all $t \geq 0$. Therefore, e is uniformly continuous in t on $[0, \infty)$. This continuity implies $e^T K e$ is also continuous in t on $[0, \infty)$. Hence, Barbalat's Lemma implies that $e^T K e \longrightarrow 0$ as $t \longrightarrow \infty$, and, since K is positive definite, $e \longrightarrow 0$ as $t \longrightarrow \infty$, as intended.

7.5 ESTABLISHING AN ANALYTICAL BOUND ON THE SYSTEM STATES

In this section, a theorem is developed that places an upper bound on the system states regardless of the steady state value of the reference model. From this theorem, it will be shown that altering the bounding functions $f_{b,i}(x_i^2)$ can tighten the bounds on the states x_i. This insight can be used to improve constraint enforcement performance.

Theorem 7.3: Suppose all of the conditions in Theorem 7.1 are met, the reference input signal r is bounded, and the reference dynamics in Eq. (7.3) are stable and bounded. Also, suppose that the projection operator bounds are chosen such that the unknown ideal weights satisfy $||W||_F \leq ||\widehat{W}||_{F,\max}$ and $||\delta\Lambda||_F \leq ||\delta\hat{\Lambda}||_{F,\max}$, where $||\widehat{W}||_{F,\max}$ is the maximum allowable Frobenius norm of $\widehat{W}$ and $||\delta\hat{\Lambda}||_{F,\max}$ is the maximum allowed Frobenius norm of $\delta\hat{\Lambda}$ set by each respective projection operator defined by Eq. (7.23) and (7.24). If, at $t = 0$, $e = 0$ and $x_i^2 \leq b_i$, then each of the bounding functions $f_{b,i}(x_i^2)$ implemented in the control and adaptive laws, Eq. (7.10), (7.17), and (7.18), are bounded $\forall t \geq 0$ by

$$f_{b,i}(x_i^2) \leq \frac{4}{\lambda_{\min}(\Gamma_W)}||\widehat{W}||_{F,\max}^2 + \frac{4}{\lambda_{\min}(\Gamma_\Lambda)}||\delta\hat{\Lambda}||_{F,\max}^2 + \phi^T F'(\phi)\phi + \sum_{i=1}^{n} \max_{x_i^2 \leq \phi_i} \left[f_{b,i}(x_i^2)\right] \tag{7.54}$$

where $\phi \in \mathbb{R}^n$ is defined such that

$$\lambda_{\min}(K)\phi^T F'(\phi)\phi = |x_m^T \dot{x}_m + \lambda_{\max}(K) x_m^T x_m|_{\max} \tag{7.55}$$

and where $|\cdot|_{\max}$ is the maximum value of the argument $\forall t \geq 0$. Since, by definition, $f_{b,i}(x_i^2) \longrightarrow \infty$ as $x_i^2 \longrightarrow \infty$, the bound (7.54) implies an upper bound on the system states.

Proof: Consider the inequality (7.39), which describes a relationship regarding the derivative of the Lyapunov function used in Theorems 7.1 and 7.2:

$$\dot{V} \leq -2e^T K e - 2x^T F'(x) K x + 2x_m^T F'_*(x)(\dot{x}_m + K x_m)$$

Because K is a diagonal positive definite matrix, the following inequality also holds:

$$\dot{V} \leq -2\lambda_{\min}(K)(e^T e + x^T F'(x) x) + 2x_m^T F'_*(x)\dot{x}_m + 2\lambda_{\max}(K) x_m^T F'_*(x) x_m \tag{7.56}$$

Now, conditions are derived that show when $\dot{V}$ is less than or equal to zero. First, when

$$\lambda_{\min}(K)x^T F'(x)x > |x_m^T F_*'(x)\dot{x}_m + \lambda_{\max}(K)x_m^T F_*'(x)x_m|_{\max} \tag{7.57}$$

holds, $\dot{V} \leq 0$. Considering the definition of ϕ in Eq. (7.55), $x_i \geq \phi_i$ for all i ensures that $\dot{V} \leq 0$. Using this definition, it can be seen that if

$$e^T e \geq \phi^T F'(\phi)\phi \tag{7.58}$$

then $\dot{V} \leq 0$. The projection operator ensures that the adaptive weights are bounded by

$$||\widetilde{W}||_F \leq ||\widetilde{W}||_{F,\max} \quad \text{and} \quad ||\widetilde{\delta\Lambda}||_F \leq ||\widetilde{\delta\Lambda}||_{F,\max} \tag{7.59}$$

This implies that $\dot{V} \leq 0$ outside of the compact set Ω, where

$$\begin{aligned}\Omega(e, x, \widetilde{W}, \delta\tilde{\Lambda}) = &\left\{(e, x, \widetilde{W}, \delta\tilde{\Lambda}): ||e||_2 \leq \sqrt{\phi^T F'(\phi)\phi}\right\} \\ &\bigcap \{(e, x, \widetilde{W}, \delta\tilde{\Lambda}): x_i \leq \phi_i, \ \text{for all } i\} \\ &\bigcap \{(e, x, \widetilde{W}, \delta\tilde{\Lambda}): ||\widetilde{W}||_F \leq ||\widetilde{W}||_{F,\max}\} \\ &\bigcap \{(e, x, \widetilde{W}, \delta\tilde{\Lambda}): ||\delta\tilde{\Lambda}||_F \leq ||\delta\tilde{\Lambda}||_{F,\max}\}\end{aligned}$$

Because $\dot{V} \leq 0$ outside of the compact set Ω, V cannot grow outside of this set. Therefore, V has the following upper bound:

$$V \leq \max_{(e,x,\widetilde{W},\delta\tilde{\Lambda}) \in \Omega} V, \quad t \geq 0 \tag{7.60}$$

From the definition of V in Eq. (7.32), it can be stated that

$$V \leq \max_{(e,x,\widetilde{W},\delta\tilde{\Lambda}) \in \Omega} \left[e^T e + tr(\widetilde{W}^T \Gamma_W^{-1} \widetilde{W}) + tr(\delta\tilde{\Lambda}^T \Gamma_\Lambda^{-1} \delta\tilde{\Lambda}) + \sum_{i=1}^{n} f_{b,i}(x_i^2)\right] \tag{7.61}$$

The properties of the trace operator ensure that the following relationships are true:

$$tr(\widetilde{W}^T \Gamma_W^{-1} \widetilde{W}) \leq \frac{1}{\lambda_{\min}(\Gamma_W)} tr(\widetilde{W}^T \widetilde{W}) \tag{7.62}$$

$$tr(\delta\tilde{\Lambda} \Gamma_\Lambda^{-1} \delta\tilde{\Lambda}^T) \leq \frac{1}{\lambda_{\min}(\Gamma_\Lambda)} tr(\delta\tilde{\Lambda}\delta\tilde{\Lambda}^T) \tag{7.63}$$

Therefore, using the inequalities from the definition of the set Ω, it follows that

$$V \leq \frac{||\widetilde{W}||^2_{F,\max}}{\lambda_{\min}(\Gamma_W)} + \frac{||\delta\tilde{\Lambda}||^2_{F,\max}}{\lambda_{\min}(\Gamma_\Lambda)} + \phi^T F'(\phi)\phi + \sum_{i=1}^{n} \max_{x_i^2 \leq \phi_i} \left[f_{b,i}(x_i^2) \right] \tag{7.64}$$

Note that

$$||\widetilde{W}||_F = ||\widehat{W} - W||_F \leq ||\widehat{W}||_F + ||W||_F \leq 2||\widehat{W}||_{F,\max}, \quad \forall t \geq 0 \tag{7.65}$$

$$||\delta\tilde{\Lambda}||_F = ||\delta\hat{\Lambda} - \delta\Lambda||_F \leq ||\delta\hat{\Lambda}_F|| + ||\delta\Lambda||_F \leq 2||\delta\hat{\Lambda}||_{F,\max}, \quad \forall t \geq 0 \tag{7.66}$$

Therefore,

$$\begin{aligned} V \leq & \frac{4}{\lambda_{\min}(\Gamma_W)}||\widehat{W}||^2_{F,\max} + \frac{4}{\lambda_{\min}(\Gamma_\Lambda)}||\delta\hat{\Lambda}||^2_{F,\max} + \phi^T F'(\phi)\phi \\ & + \sum_{i=1}^{n} \max_{x_i^2 \leq \phi_i} \left[f_{b,i}(x_i^2) \right] \end{aligned} \tag{7.67}$$

Since Eq. (7.32) implies that $f_{b,i}(x_i^2) \leq V$, it can be stated that

$$\begin{aligned} f_{b,i}(x_i^2) \leq & \frac{4}{\lambda_{\min}(\Gamma_W)}||\widehat{W}||^2_{F,\max} + \frac{4}{\lambda_{\min}(\Gamma_\Lambda)}||\delta\hat{\Lambda}||^2_{F,\max} + \phi^T F'(\phi)\phi \\ & + \sum_{i=1}^{n} \max_{x_i^2 \leq \phi_i} \left[f_{b,i}(x_i^2) \right] \end{aligned} \tag{7.68}$$

for all i, as intended.

Remark 7.2: *Because a finite bound for the bounding functions has been established, a bound for the system states x_i is also established in the form of the vector ϕ. It can be seen from Eq. (7.55) that the size of this bound depends on the derivative of the respective bounding function, $f_{b,i}(x_i^2)$. The faster the bounding function grows after $x_i^2 \geq b_i$, the tighter the state will be bound to its constraint.*

7.6 CONSTRUCTING BOUNDING FUNCTIONS

Although alternative methods are available, one process for constructing bounding functions that produce the desired behavior is now presented. When the constraint is either a positive maximum or a negative minimum, the proposed bounding function is given by

$$f_{b,i}(x_i) = \begin{cases} 0 & \text{if } x_i \leq b_i \\ f_{t,i}(x_i) & \text{if } b_i < x_i \leq b_i + c_i \\ \exp(\rho_i x_i) - \phi_{0,i} & \text{if } x_i > b_i + c_i \end{cases} \tag{7.69}$$

where $f_{t,i}(x_i)$ is a polynomial that is chosen to transitionally activate the bounding function when the state is between the predetermined regions b_i and $b_i + c_i$. The parameter c_i is chosen based on how quickly one desires the bounding function to transition from 0 to an exponential function. The parameter $\rho_i > 0$ affects the growth rate of the bounding function, and therefore can be used to tighten the constraint according to Remark 7.2. As $\rho_i > 0$ increases, the bound on the state x_i tightens.

The transition polynomial, $f_{t,i}(x_i)$, takes the form

$$f_{t,i}(x_i) = \phi_{1,i}x_i^4 + \phi_{2,i}x_i^3 + \phi_{3,i}x_i^2 + \phi_{4,i}x_i + \phi_{5,i} \tag{7.70}$$

The coefficients $\phi_{j,i}$ as well as the term $\phi_{0,i}$ are solved for by looking at the desired boundary conditions for the bounding function. These boundary conditions are

$$f_{b,i}(b_i) = f'_{b,i}(b_i) = f''_{b,i}(b_i) = 0 \tag{7.71}$$

$$f_{b,i}(b_i + c_i) = \exp(\rho_i x_i) - \phi_{0,i} \tag{7.72}$$

$$f'_{b,i}(b_i + c_i) = \rho_i \exp(\rho_i x_i) \tag{7.73}$$

$$f''_{b,i}(b_i + c_i) = \rho_i^2 \exp(\rho_i x_i) \tag{7.74}$$

where the prime symbol in Eqs. (7.71)–(7.74) represents differentiation with respect to x_i. By taking two derivatives of the general form of the transitional function $f_{t,i}(x_i)$,

$$f'_{t,i}(x_i) = 4\phi_{1,i}x_i^3 + 3\phi_{2,i}x_i^2 + 2\phi_{3,i}x_i + \phi_{4,i} \tag{7.75}$$

$$f''_{t,i}(x_i) = 12\phi_{1,i}x_i^2 + 6\phi_{2,i}x_i + 2\phi_{3,i} \tag{7.76}$$

and applying the six given boundary conditions, Eqs. (7.71)–(7.74), one is left with six linear equations which can be used to solve for the six $\phi_{j,i}$'s as long as $c_i \neq 0$.

Remark 7.3: More complicated constraint sets are easily applicable with this bounding function technique. This can mean including both a maximum and minimum bound or a more sophisticated structure where the desired bounds are determined based on the current operating condition. In this scenario, a bounding function with the same behavior but different coefficients would be needed to handle the other constraint boundaries. One way to deal with this is to calculate the bounding function in real time based on which constraint, maximum or minimum for example, is closer to being exceeded. Because the process described in Section 7.6 is nothing more than a set of six linear equations, calculating the bounding functions in real time is computationally inexpensive.

7.7 NUMERICAL EXAMPLE: PREVENTION AND RECOVERY FROM AN INLET UNSTART

7.7.1 GENERIC HYPERSONIC VEHICLE MODEL INTRODUCTION

The generic hypersonic vehicle (GHV) is a Simulink-based model of a hypersonic aircraft, developed at the Air Force Research Laboratory for the purpose of testing control algorithms. The equations of motion were derived using a Lagrangian approach by Bilimoria and Schmidt [24]. The derivation utilizes an Earth-centered inertial frame with a rotating, spherical Earth. For simplification, the vehicle is assumed to be inelastic and contain no rotors, terms involving fluid flow were dropped, and wind-related terms were neglected as well. The aerodynamic and thrust forces and moments acting on the vehicle are modeled using look-up tables that were generated using shock-expansion methods with a viscous correction.

The control complement is made up of two elevons ($\delta_{f,r}$ and $\delta_{f,l}$), two rud-dervators ($\delta_{t,r}$ and $\delta_{t,l}$), and throttle (δ_{th}). The vehicle's velocity is controlled using a proportional-integral-derivative (PID)controller, which is briefly discussed here. The input to the controller is the desired velocity, and the output is the air-fuel equivalence ratio. This equivalence ratio is then translated into a throttle command, δ_{th}. For the remainder of this chapter, the PID controller is considered separate from the adaptive controller of interest. The throttle control will be manifested implicitly in the thrust force and moment terms of the equations of motion. Therefore, the control law development will be only for the control surface deflections:

$$u = \begin{bmatrix} \delta_{f,r} & \delta_{f,l} & \delta_{t,r} & \delta_{t,l} \end{bmatrix}^T \tag{7.77}$$

The control architecture in this simulation model is made up of two adaptive dynamic inversion controllers. The purpose of using two controllers is to separate the position-level states from the velocity-level states of the aircraft. For the GHV, the position-level states are the aerodynamic angles: angle-of-attack α, sideslip angle β, and bank angle μ. The aerodynamic bank angle was chosen as a state instead of the bank attitude angle ϕ to ensure that the dynamic inversion is singular only at $\beta = \pm 90^o$. The control variables for this loop of the controller are the body-axis angular rates. The velocity-level states are the body-axis angular rates: p, q, and r. These states are controlled using the control surface deflections.

Overall, the control architecture first translates commanded inputs for angle-of-attack, sideslip angle, and bank angle into commands for the body-axis angular rates p, q, and r. Second, those rates are passed into another adaptive dynamic inversion controller that determines the corresponding control surface deflections to achieve the desired angular rate commands.

A full derivation of the vehicle dynamics is shown by Rollins et al. and Stevens and Lewis [23, 25]. The nonlinear equations of motion for the velocity-level states are given by

$$\begin{bmatrix} \dot{p} \\ \dot{q} \\ \dot{r} \end{bmatrix} = \begin{bmatrix} I_x & 0 & -I_{xz} \\ 0 & I_y & 0 \\ -I_{xz} & 0 & I_z \end{bmatrix}^{-1} \times \left(- \begin{bmatrix} -I_{xz}pq + (I_z - I_y)qr \\ (I_x - I_z)pr + I_{xz}(p^2 - r^2) \\ I_{xz}qr + (I_y - I_x)pq \end{bmatrix} + M_T + \bar{q}SG + \bar{q}SH \begin{bmatrix} \delta_{f,r} \\ \delta_{f,l} \\ \delta_{t,r} \\ \delta_{t,l} \end{bmatrix} \right) \tag{7.78}$$

where G and H are defined by

$$G = \begin{bmatrix} b\big(C_{\ell,\text{base}} + \dfrac{b}{2V_T}(C_{\ell_p}p)\big) \\ \bar{c}\big(C_{m,\text{base}} + \dfrac{\bar{c}}{2V_T}C_{m_q}q + \overline{C}_{m,\delta}\big) \\ b\big(C_{n,\text{base}} + \dfrac{b}{2V_T}(C_{n_r}r)\big) \end{bmatrix} \tag{7.79}$$

$$\overline{C}_{m,\delta} = \frac{\partial C_m}{\partial \delta_{f,r}}(\delta_{f,r} = \bar{\delta}_{f,r}) + \frac{\partial C_m}{\partial \delta_{f,l}}(\delta_{f,l} = \bar{\delta}_{f,l}) + \frac{\partial C_m}{\partial \delta_{t,r}}(\delta_{t,r} = \bar{\delta}_{t,r}) + \frac{\partial C_m}{\partial \delta_{t,l}}(\delta_{t,l} = \bar{\delta}_{t,l}) \tag{7.80}$$

$$H = \begin{bmatrix} b\dfrac{\partial C_\ell}{\partial \delta_{f,r}} & b\dfrac{\partial C_\ell}{\partial \delta_{f,l}} & b\dfrac{\partial C_\ell}{\partial \delta_{t,r}} & b\dfrac{\partial C_\ell}{\partial \delta_{t,l}} \\ \bar{c}\dfrac{\partial C_m}{\partial \delta_{f,r}} & \bar{c}\dfrac{\partial C_m}{\partial \delta_{f,l}} & \bar{c}\dfrac{\partial C_m}{\partial \delta_{t,r}} & \bar{c}\dfrac{\partial C_m}{\partial \delta_{t,l}} \\ b\dfrac{\partial C_n}{\partial \delta_{f,r}} & b\dfrac{\partial C_n}{\partial \delta_{f,l}} & b\dfrac{\partial C_n}{\partial \delta_{t,r}} & b\dfrac{\partial C_n}{\partial \delta_{t,1}} \end{bmatrix} \tag{7.81}$$

Using the notation that C_x represents $\cos(x)$, S_x represents $\sin(x)$, and T_x represents $\tan(x)$ where x is an angle, the position-level state dynamics used are given by

$$\begin{bmatrix} \dot{\beta} \\ \dot{\alpha} \\ \dot{\mu} \end{bmatrix} = \begin{bmatrix} \frac{1}{mV_T}((Y + F_{T_y})C_\beta + mgS_\mu C_\gamma - F_{T_x}C_\alpha S_\beta - F_{T_x}S_\alpha S_\beta \\ \frac{1}{mV_T C_\beta}(-L + mgC_\mu C_\gamma - F_{T_x}S_\alpha + F_{T_x}C_\alpha) \\ (*) \end{bmatrix} + \begin{bmatrix} S_\alpha & 0 & -C_\alpha \\ -T_\beta C_\alpha & 1 & -T_\beta S_\alpha \\ sec(\beta)C_\alpha & 0 & sec(\beta)S_\alpha \end{bmatrix} \begin{bmatrix} p \\ q \\ r \end{bmatrix} \tag{7.82}$$

where

$$(*) = \frac{1}{mV_T}(L(T_\beta + T_\gamma S_\mu) + (Y + F_{T_y})T_\gamma C_\mu C_\beta - mgC_\gamma C_\mu T_\beta + (F_{T_x}S_\alpha - F_{T_z}C_\alpha)(T_\beta + T_\gamma S_\mu) - (F_{T_x}C_\alpha + F_{T_z}S_\alpha)T_\gamma C_\mu S_\beta) \tag{7.83}$$

Note that both sets of dynamics, Eqs. (7.78) and (7.82), are in the form of Eq. (7.1) as required. In the next section, the inlet unstart model implemented in the simulation is discussed.

7.7.2 MODELING INLET UNSTART

A simplified inlet unstart model was included in the GHV simulation in order to test the performance of the state constraint mechanism in the event of an inlet unstart. The model is hysteresis based and depends solely on angle-of-attack, sideslip angle, and freestream Mach number. Starting in a nominal flight condition, the inlet will unstart if the vehicle leaves the region of the $\alpha - \beta$ space marked by the outer ellipsoid in Fig. 7.3. In order to regain normal engine function,the vehicle must reenter the region marked by the inner ellipsoid. Once restarted, the vehicle is again free to traverse the inter-ellipsoid region with full thrust and nominal performance.

In this model, the boundary of the outer ellipsoid was determined by wind tunnel testing of the 10.5-inch HIFiRE 6 model at the Department of Aeronautics, U.S. Air Force Academy [3]. From their experiments, the recorded values at which an unstart will occur are 14.5° and −5.7° for angle-of-attack and ±5.4° for sideslip angle. These values are consistent across each of the cases presented in Section 7.7.4. This is of particular note, as the parameters for this ellipsoid, and even perhaps its very shape, are highly dependent upon specific vehicle geometries. As a result, the susceptibility of any given vehicle to inlet unstart will vary significantly depending on the specifics of the multistate trajectories the vehicle is subjected to.

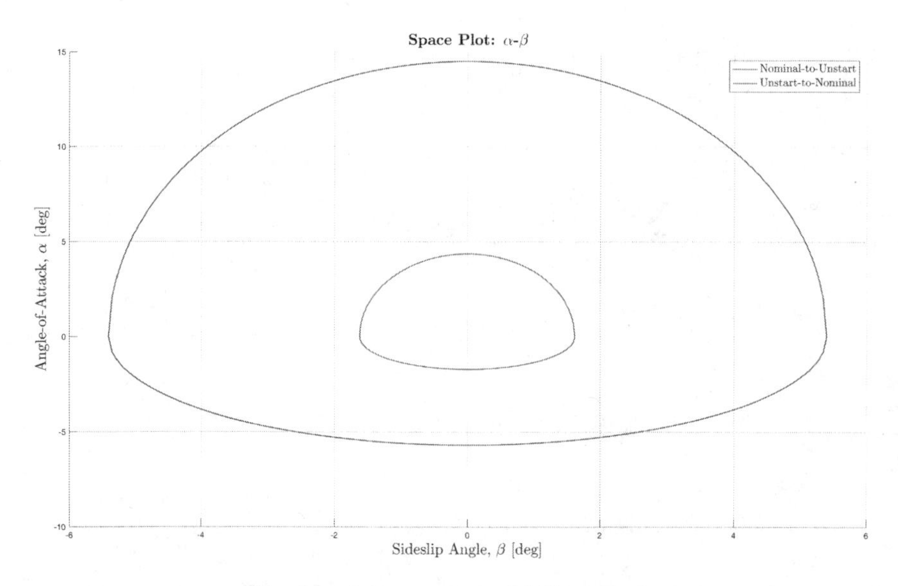

Fig. 7.3 Inlet unstart model illustration.

In contrast, due to the comparative lack of aerodynamic research on the conditions necessary for inlet restart, the size of the inner ellipsoid is chosen arbitrarily but conservatively. Consequentially, recovering from an unstart becomes a more difficult task because the state constraint mechanism is required to drive the system further from its reference model to restart the engine. Because the inlet can reasonably be expected to perform as intended when the vehicle is straight and level, the inner ellipsoid is chosen to be 70% smaller than the outer ellipsoid. This ellipsoid is limited to 4.35° and -1.71° in angle-of-attack and $\pm 1.62^\circ$ in sideslip angle. It should also be noted that these values correspond to Mach 6 flight. The model is set up such that the size of the ellipsoids vary linearly with Mach number. The simulation cases presented, however, are all associated with an initial flight condition of Mach 6 and the ellipsoid sizes already described. Once again, more aerodynamic research on how freestream Mach number affects these unstart parameters is required for a higher fidelity model.

The shape of these ellipsoids is an attempt to capture the unstart characteristics in the more likely circumstance where the desired dynamics are not simply a nonzero command in one specific state, but time-varying trajectories in multiple states. That is, although the limits experimentally determined from the HIFiRE experiment were produced by an uncoupled, independent study of the effects of angle-of-attack and sideslip angle, this model allows for the triggering of an inlet unstart as a function of both angles, the values of which may not exceed their respective, individual experimental limits. Although the specific implementation used here is concerned directly with two particular states, similar concepts could easily be extended for higher dimensional models.

Finally, the aerodynamic and propulsive consequences of the inlet unstart model should be noted. First, when an unstart is triggered, the vehicle experiences a complete and instantaneous loss of thrust and an increase in the magnitude of the aerodynamic drag coefficient. Conversely, when the inlet is restarted, the vehicle experiences a complete and instantaneous restoration of thrust according to the equivalence ratio (throttle) commanded by the controller. Because the apparent lack of thrust prompts an increase in the commanded equivalence ratio, this command is typically saturated at maximum throttle by the time the vehicle is posed for restart. Because of insufficient information on the geometry-specific flow characteristics and resulting changes in aerodynamic force and moment coefficients and stability derivatives, significant post-unstart changes in the plant dynamics are not modeled. These changes are difficult to predict and, as a result, are not addressed in the simulation studies that follow. Although a more thorough testing of the state constraint mechanism's performance in the event of an inlet unstart would include these effects, quantifiably substantiated by experiment or computational fluid dynamics, the cases presented here still demonstrate the benefits of this mechanism as applied to this problem, substantiating the need for further testing.

7.7.3 SIMULATION STUDY DESCRIPTION

A simulation study was designed to demonstrate how enforcing state constraints can potentially be used for both prevention of and recovery from an inlet unstart. In this study, it is assumed that the only way to cause or recover from an inlet unstart is to leave or enter the appropriate regions of the $\alpha - \beta$ space as described in Section 7.7.2. A complete solution to the inlet unstart problem would, in addition to the vehicle states, focus on the inlet itself and would likely include some type of throttle constraints or actuation in the inlet. The goal here is simply to prevent the vehicle from leaving the outer ellipsoid from Fig. 7.3 and, in the extreme conditions where the prevention technique fails, to maintain control effectiveness and quickly drive the vehicle into the inner ellipsoid of Fig. 7.3 so that the engine restarts. The state constraint technique described previously can be used in both situations.

Given the inlet unstart model described in Section 7.7.2, the use of state constraints for the prevention of inlet unstart is very intuitive. If the control law and bounding functions are designed such that the angle-of-attack and sideslip angle of the GHV can never exit the outer ellipsoid of Fig. 7.3, then theoretically, inlet unstart will not occur. This prevention approach is based on two assumptions. The first assumption is that there is a known bound on any external disturbances that can affect the system, a common assumption in controls literature. This known bound on disturbances is reflected in the bounds used by the projection operator, discussed in detail in Section 7.2. The second assumption is perfect knowledge of the inlet unstart model (i.e., perfect knowledge of the sizes and shapes of the ellipsoids of Fig. 7.3). This requires extensive wind tunnel testing and modeling of each specific vehicle and inlet. Other aerodynamic causes of inlet unstart add another layer of uncertainty. In this study, it is assumed that the ellipsoids given earlier are correct and known perfectly, and, as mentioned, only exiting the outer ellipsoid (i.e., flying at large aerodynamic angles) can cause inlet unstart.

The use of state constraints for recovery from inlet unstart is developed based on an analysis of the control technique presented in this chapter, when the system is initialized at a point in state-space that violates a constraint. It is straightforward to show both mathematically and computationally that in these situations the system is very quickly driven within the constraint bounds before any attempt to track the reference model is carried out. If, at the moment an inlet unstart occurs, the bounding function used in the control is switched to reflect the inner ellipsoid of Fig. 7.3 instead of the outer ellipsoid, it is analogous to initializing the system in violation of a constraint. If the vehicle maintains control effectiveness, the behavior described previously (i.e., quickly driving the system inside of the constraint set) will occur and the engine will restart. This concept is demonstrated in Fig. 7.4, which represents a fictitious one-dimensional system with state x. This figure is not based on actual dynamic simulation but can be looked at as a diagram demonstrating how this technique is *supposed* to affect

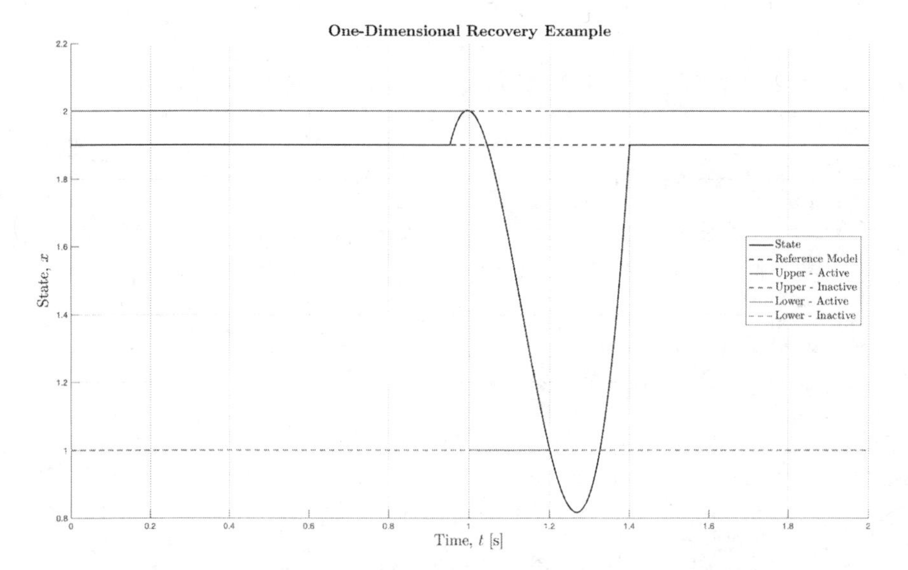

Fig. 7.4 One-dimensional example of the inlet unstart recovery technique. The bounding function corresponds to whichever constraint value is represented by a solid line at the current time.

the system. For a one-dimensional system, the outer and inner ellipsoids of Fig. 7.3 are analogous to the scalars, shown by the 'Upper' and 'Lower' lines in Fig. 7.4, and arbitrarily set to 2 and 1, respectively. In Fig. 7.4, when the ellipsoid values are represented by a solid line, it indicates that the bounding function is being calculated based on that constraint value. Therefore, until $t = 1$ s, the upper ellipsoid is used to calculate the bounding function as a preventative measure, and the state tracks its reference model. At $t = 0.95$ s, a disturbance is introduced, driving the state above its upper limit, which corresponds to an unstart at $t = 1$ s. The bounding function then switches to correspond to the lower ellipsoid value, the state is quickly driven below this new constraint, the inlet restarts, and the state can once again track its reference model.

One cause for concern with this recovery technique is whether the stability analysis will still hold with the instantaneous change in the bounding function and, therefore, infinite derivative value. In general, switching the bounding function should be done at times when the constraints are not in danger of being exceeded and the bounding function is inactive (i.e., $f_b(x_i^2) = 0$). This will ensure that the bounding function and its derivative are continuous and differentiable at all times. In the presence of an unstart, this approach is implausible, as corrective action must be taken extremely quickly. However, any real-life digital flight controller (even with an extremely fast sampling time as would be expected on a hypersonic vehicle) would calculate the bounding function derivative based on its analytical polynomial or exponential form given in Section 7.6 and therefore never produce an unbounded derivative; no actual differentiation would be done online. For this reason, it can be assumed that switching the bounding function in this way is valid and does not affect the results of the stability analysis in Sections 7.3–7.5.

Four cases are presented to demonstrate the concepts described previously. In the first case, the vehicle is commanded to remain within the constraint boundaries for the entire simulation. In addition to showcasing the baseline performance of the vehicle,this case demonstrates an important feature of any successful state constraint controller: that the control law should reduce to a tracking controller if no constraints are encountered. The second case demonstrates the use of the state constraint mechanism as a preventative measure for inlet unstart. The vehicle is commanded to track the same trajectory as in case 1, but a reduction of the entire constraint set compared with the inlet unstart parameters limits the constrained states and momentarily precludes tracking. With the hard constraints again corresponding to the outer ellipsoid in Fig. 7.3, the third case demonstrates the ability of the controller to prevent inlet unstart in the presence of a disturbance within expected bounds (500 lbf). In the fourth and final case, a large disturbance, outside of the expected bounds (2500 lbf), is introduced while the vehicle is close to the boundary of the outer ellipsoid, causing an unstart. The bounding function is then immediately switched to correspond to the inner ellipsoid, and the vehicle quickly recovers. Note that in all four cases, the exact same command signals are sent to the vehicle.

7.7.4 RESULTS

7.7.4.1 BASELINE TRAJECTORY AND CONTROLLER SETTINGS

The baseline trajectory and various controller settings are carefully chosen so as to unambiguously illustrate the concepts outlined in Section 7.7.3 in a consistent manner. That is, in subsequent cases, the inlet unstart model, state constraint mechanism, and adaptive laws (including the projection operator parameters), as well as the trajectory, are unchanged unless otherwise specified. For reference, case 1 contains the results corresponding strictly to this baseline situation, where no unwarranted external disturbances or state constraining mechanisms interact with the vehicle.

Consider the GHV described in Section 7.7.1 and recall the control architecture proposed therein. Specifically, note that both Eqs. (7.78) and (7.82) are of the form of Eq. (7.1). As a result, they can each be written in the form of Eq. (7.16) and can be subjected to the analyses of Sections 7.2–7.5. The reference model dynamics are defined as:

$$\begin{bmatrix} \dot{\alpha}_m \\ \dot{\beta}_m \\ \dot{\mu}_m \end{bmatrix} = \begin{bmatrix} -5 & 0 & 0 \\ 0 & -5 & 0 \\ 0 & 0 & -5 \end{bmatrix} \begin{bmatrix} \alpha_m \\ \beta_m \\ \mu_m \end{bmatrix} + \begin{bmatrix} 5 & 0 & 0 \\ 0 & 5 & 0 \\ 0 & 0 & 5 \end{bmatrix} r_{\alpha,\beta,\mu} \tag{7.84}$$

$$\begin{bmatrix} \dot{p}_m \\ \dot{q}_m \\ \dot{r}_m \end{bmatrix} = \begin{bmatrix} -13 & 0 & 0 \\ 0 & -13 & 0 \\ 0 & 0 & -13 \end{bmatrix} \begin{bmatrix} p_m \\ q_m \\ r_m \end{bmatrix} + \begin{bmatrix} 13 & 0 & 0 \\ 0 & 13 & 0 \\ 0 & 0 & 13 \end{bmatrix} r_{p,q,r} \tag{7.85}$$

where $r_{\alpha,\beta,\mu} \in \mathbb{R}^3$ and $r_{p,q,r} \in \mathbb{R}^3$ denote the commanded input values for their respective subsystems. Based on tuning, the error gains are set as

$$K_{p,q,r} = \begin{bmatrix} 34 & 0 & 0 \\ 0 & 26 & 0 \\ 0 & 0 & 14 \end{bmatrix} \quad K_{\alpha,\beta,\mu} = \begin{bmatrix} 1 & 0 & 0 \\ 0 & 2 & 0 \\ 0 & 0 & 5 \end{bmatrix} \tag{7.86}$$

Consider the inlet unstart model described in Section 7.7.2, the concepts and applications developed in Section 7.7.3, and the form of the bounding function in Section 7.6. In addition to the inlet unstart model parameters,the hard constraints for the bounding function $(b_i + c_i)$ use the same parameters as discussed in Section 7.7.2. The sideslip angle constraint tolerance (c_β) is selected as 60% of the soft constraint (b_β), whereas c_α is selected as 30% of b_α. Given the 70% reduction in the size of the inner ellipsoid constraint set, the state constraint mechanism is fully defined and given by ordered pairs of intended negative minimums and positive maximums for both the nominal and unstarted (denoted by a superscript asterisk) cases:

$$\begin{bmatrix} b_\alpha \\ c_\alpha \\ b_\beta \\ c_\beta \end{bmatrix} = \begin{bmatrix} (-4.385^\circ,\ 11.154^\circ) \\ (-1.315^\circ,\ 3.346^\circ) \\ (-3.375^\circ,\ 3.375^\circ) \\ (-2.025^\circ,\ 2.025^\circ) \end{bmatrix} \quad \begin{bmatrix} b^*_\alpha \\ c^*_\alpha \\ b^*_\beta \\ c^*_\beta \end{bmatrix} = \begin{bmatrix} (-1.315^\circ,\ 3.346^\circ) \\ (-0.395^\circ,\ 1.004^\circ) \\ (-1.013^\circ,\ 1.013^\circ) \\ (-0.607^\circ,\ 0.607^\circ) \end{bmatrix} \tag{7.87}$$

It should be noted that the constraints are not symmetric about zero in angle-of-attack as a generalization of the HIFiRE 6's lack of symmetry in the body-fixed xy-plane.

For the velocity-level control loop, two sets of gains are required for the two adaptive values: the weight estimate, $\widehat{W}_{p,q,r}$, which operates on the known basis vector given by $\beta(x)_{p,q,r} = [1,\ p,\ q,\ r,\ \alpha,\ \beta,\ M]^\top$, where M is freestream Mach number, and the control effectiveness matrix estimate $\hat{\Lambda}_{p,q,r}$. These gains are denoted by $\Gamma_{\widehat{W}_{p,q,r}} = 2.5I_7$ and $\Gamma_{\hat{\Lambda}_{p,q,r}} = 0.1I_4$. To reflect the systemic uncertainty discussed in Section 7.2, the initial value of $\hat{\Lambda}_{p,q,r}$ is chosen as $\hat{\Lambda}_{p,q,r}(t=0) = \text{diag}([1.023, 0.930, 1.052, 1.065])$.

For the position-level control loop, it is assumed that $\Lambda_{\alpha,\beta,\mu} = I_3$ and is known. As a result, the adaptive law corresponds only to the weight estimate, $\widehat{W}_{\alpha,\beta,\mu}$. This weight estimate operates on a different known basis vector given by $\beta(x)_{\alpha,\beta,\mu} = [1,\ \beta,\ \alpha,\ \mu,\ M]^\top$. Thus, this control loop has only one adaptive gain, denoted by $\Gamma_{\widehat{W}_{\alpha,\beta,\mu}} = 0.02I_5$.

The projection operator bounds for $\widehat{W}$ are set as $||\widehat{W}_{p,q,r}||_{\max} = 1.0$ with a tolerance of 0.1, while $||\widehat{W}_{\alpha,\beta,\mu}||_{\max} = 0.004$ with a tolerance of 0.001. In addition, in accordance with Lemma 7.1, the projection operator bound for $||\hat{\Lambda}_{p,q,r}||_{\max}$ is set at 1.0 with a tolerance of 0.24999 to ensure that the elements of $\delta\Lambda_{p,q,r}$ remain less than $1/m$, where $m = 4$ is the number of control surfaces. Finally, the convex function required by Eq. (7.24) is the following:

$$h(\theta_i) = \frac{\theta_i^\top \theta_i - \theta_{i,\max}^2}{\epsilon_{\theta_i} \theta_{i,\max}^2} \tag{7.88}$$

To demonstrate the application of the state constraint theory to *multiple* important states, angle-of-attack is commanded as a singlet with a constant value of -4.3°, while sideslip angle is similarly commanded to a constant value of 3.3° for all cases. With the parameter sets in Eq. (7.87) and absent any sort of external disturbance, this trajectory is not sufficiently large in either state to activate the bounding function.

7.7.4.2 CASE 1: BASELINE PERFORMANCE

The first case, representing baseline performance, is initialized by the cost-based trim of a linear model at the flight condition of zero degrees in both angle-of-attack and sideslip angle at Mach 6 and an altitude of 80,000 ft. In Fig. 7.5, initial perturbations in the aerodynamic angles and the comparatively large magnitude angular rate responses are an artifact of the linear model inaccuracies in this initialization scheme. At the beginning and end of the simulation, during the brief time both angle-of-attack and sideslip angle are changing, dynamic coupling causes small perturbations in bank angle that are corrected in steady state. The slight reduction in the total velocity exhibited in the final seconds of the simulation is the dynamical consequence (via drag) of increasing angle-of-attack, an error that the PID controller mentioned in Section 7.7.1 attempts to correct by saturating the commanded equivalence ratio (see Fig. 7.5).

In the upper-left pane of Fig. 7.6, the time histories of the entire control compliment are shown, with control surfaces deflections labeled on the right and the commanded equivalence ratio labeled on the left. Regarding the adaptive parameters, the upper-right and lower-left panes show smooth variation in time, indicating their effect in accounting for uncertainty as the system states evolve. The lower-right pane shows the adaptive weights corresponding to the control effectiveness matrix estimate ($\hat{\Lambda}_{p,q,r}$) from Eq. (7.11). The evolution of this weight, in accordance with the projection operators described in Section 7.7.4, is also demonstrated.

The trajectory described earlier and shown in Fig. 7.5 nearly causes an unstart, a fact that is made clear by the $\alpha - \beta$ space plot in Fig. 7.7 and will be leveraged in cases 3 and 4. Recall that exceeding the outer ellipsoid triggers an unstart, while returning to the inner ellipsoid restarts the inlet and restores thrust. By placing the trajectory near the boundary of the nominal performance envelope, a useful situation for studying the abilities of the controller in subsequent cases is created. Figure 7.8 shows the time histories of the forces and moments that act on the vehicle and is provided as a reference for future cases. Finally, the remarks given previously regarding initial perturbations, bank angle coupling, and final equivalence ratio are equally applicable in the subsequent cases.

7.7.4.3 CASE 2: PREVENTION OF UNSTART WHEN COMMANDING AN EXCESSIVELY LARGE TRAJECTORY

In case 2, the state constraint mechanism is applied in its originally designed manner by commanding a trajectory that intentionally violates a constraint set. The trajectory is the same as that of of case 1, but the constraint values are reduced by 30% from those listed in Eq. (7.87). As a result, this new, "safer" set

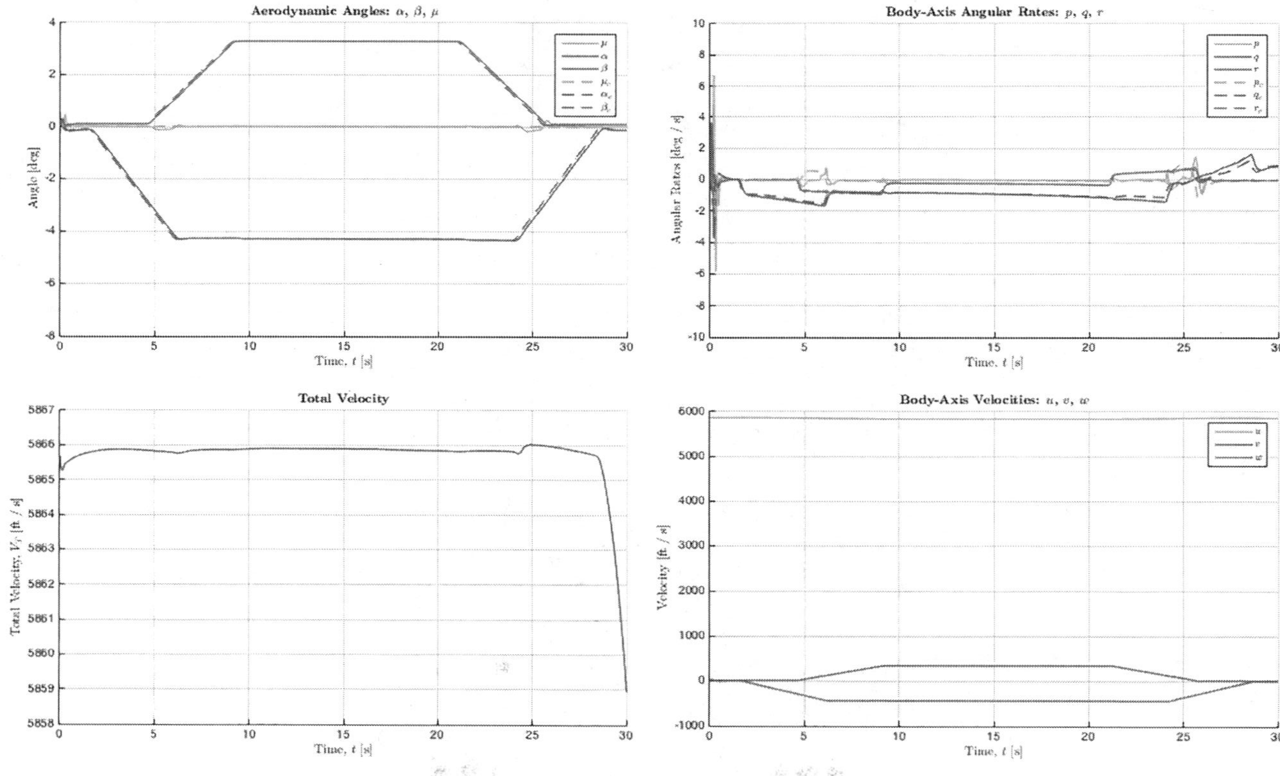

Fig. 7.5 State histories (body-axis velocities, angular rates; aerodynamic angles; total velocity) for case 1.

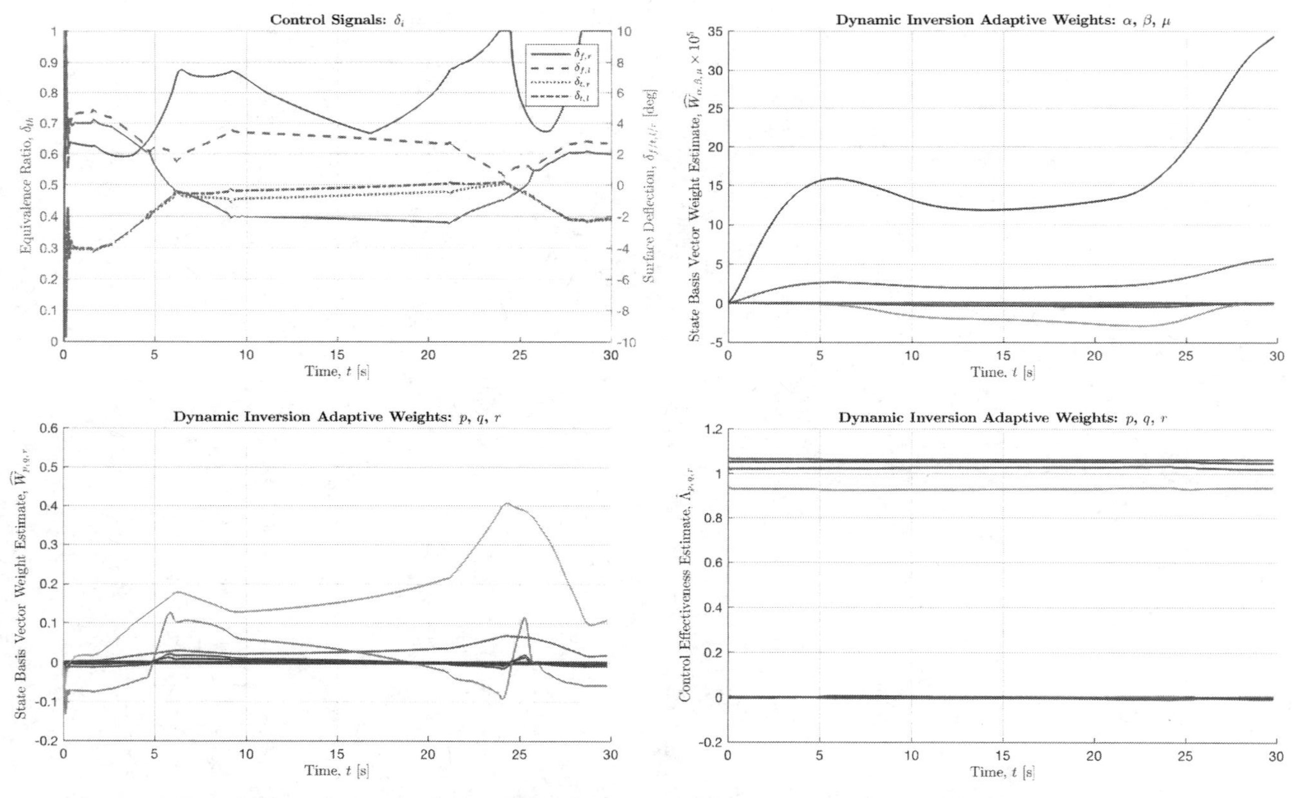

Fig. 7.6 Control surface, equivalence ratio, and dynamic inversion adaptive weight time histories for case 1.

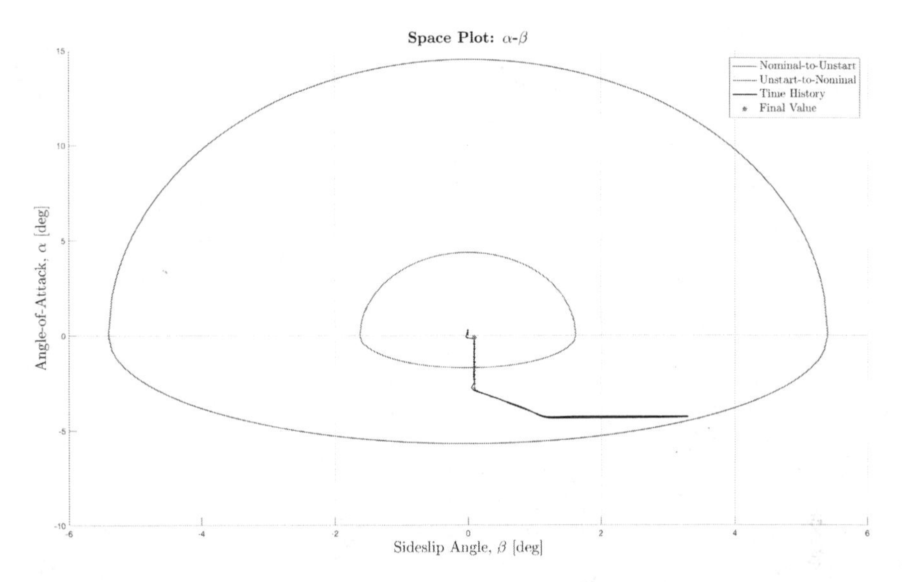

Fig. 7.7 $\alpha - \beta$ space plot, with inlet unstart model parameters and α and β histories for case 1.

of constraints is given by

$$\begin{bmatrix} b_\alpha \\ c_\alpha \\ b_\beta \\ c_\beta \end{bmatrix} = \begin{bmatrix} (-3.070^\circ,\ 7.808^\circ) \\ (-0.921^\circ,\ 2.342^\circ) \\ (-2.363^\circ,\ 2.363^\circ) \\ (-1.418^\circ,\ 1.418^\circ) \end{bmatrix} \tag{7.89}$$

This difference is made most clear by Fig. 7.9, where the new hard constraints are represented in the $\alpha - \beta$ space plot by a dashed rectangle. As intended, the system states never leave this constraint set, despite the reference model commanding them to do so.

The sensitivity parameter (ρ_i), as seen in the bounding function form in Eq. (7.69), still affects the potency of the transitional polynomial active between b_i and $b_i + c_i$ through the algebraic system of boundary conditions found in Eqs. (7.71–7.74). Qualitatively, higher sensitivities correspond to more quickly bounding a state more tightly to its constraint, despite a reference command signal (shown by the dashed lines in the upper panes of Fig. 7.10) that exceeds b_i. Excessively high values for ρ_i can produce an oscillatory response about b_i, whereas excessively low values for ρ_i often result in an ineffectual use of the state constraint mechanism. As a result of these undesirable behaviors, the nominal sensitivities are tuned to be $\rho_\alpha = 1.5$ and $\rho_\beta = 5.3$ throughout the remainder of the simulation cases.

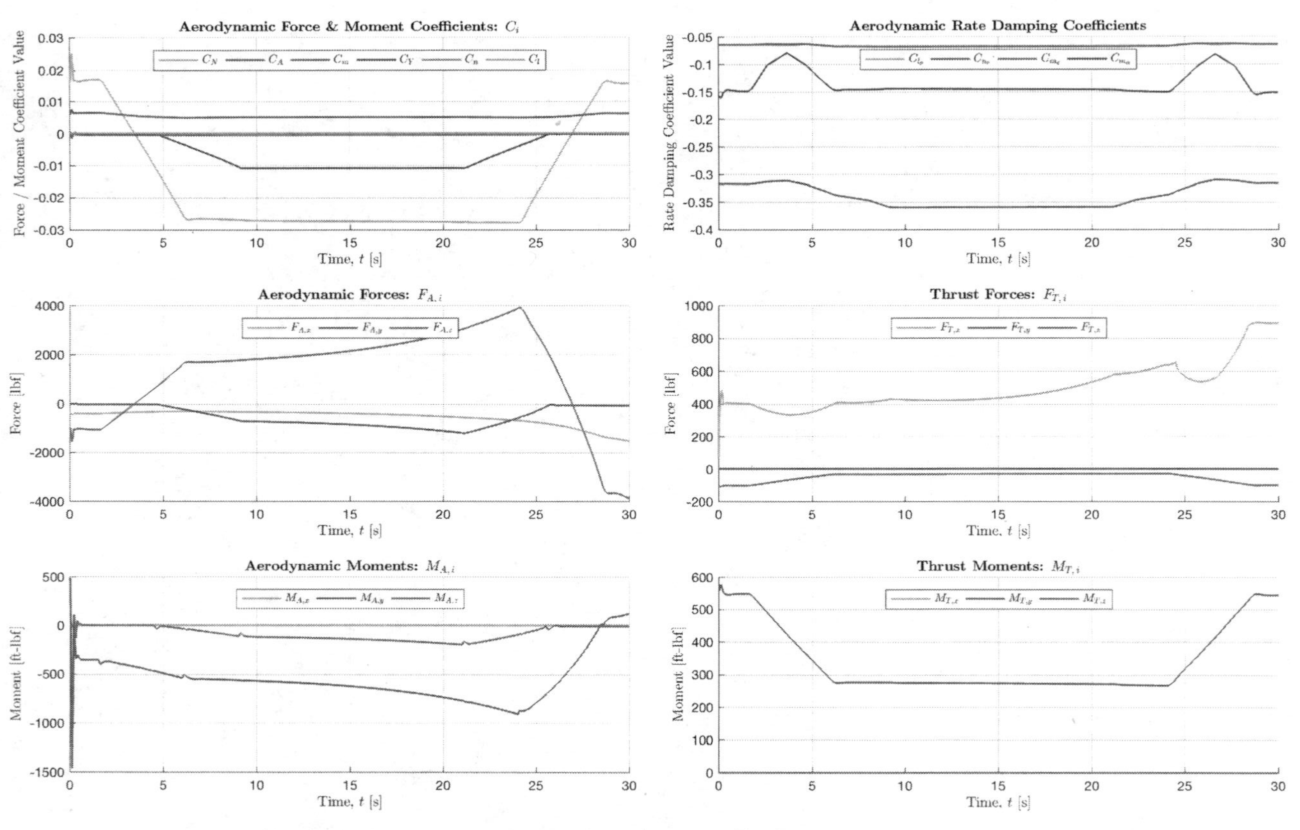

Fig. 7.8 Aerodynamic coefficient, aerodynamic and thrust force, and aerodynamic and thrust moment time histories for case 1.

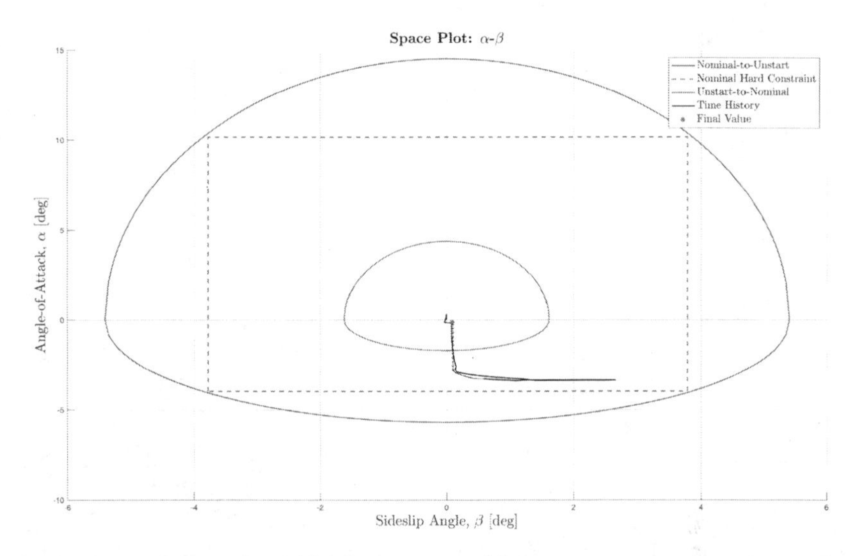

Fig. 7.9 $\alpha - \beta$ space plot, with inlet unstart model parameters, state constraints, and α and β histories for case 2.

It should be noted that the comparatively larger weights at $t \approx 5$ s and $t \approx 25$ s in the upper-right pane of Fig. 7.11 are emblematic of the bounding function's increasing effect on the adaptive law. As the constrained states continue to exceed the soft constraint, the bounding function hedge term (ℓ_r) overpowers the baseline adaptive controller, which accounts for the unexpected performance accordingly by changing certain elements of the adaptive weights quickly in an effort to regain tracking. However, the state constraint mechanism ultimately succeeds in driving the constrained states to constant values satisfying the new constraint set, precluding tracking of the original command. Note that as the reference trajectory returns to within constraint boundaries, tracking can once again be achieved as the bounding function values (and therefore, hedge term) go to zero. Finally, applications of the state constraint mechanism in this manner are the most preventative and safe, but may overconstrain vehicle maneuverability by unnecessarily limiting the angle-of-attack and sideslip angle. The subsequent case will demonstrate that some desirable prevention qualities are still maintained, even with zero safety margin. Figure 7.12 shows the time histories of the forces and moments to be consistent for the situation of a large trajectory.

7.7.4.4 CASE 3: PREVENTION OF UNSTART BY AN EXTERNAL DISTURBANCE

For case 3, again consider the exact trajectory and state constraint mechanism employed in case 1. Similarly, the state constraint mechanism applied here has the same parameters as those listed in Section 7.7.4.1. The only modification

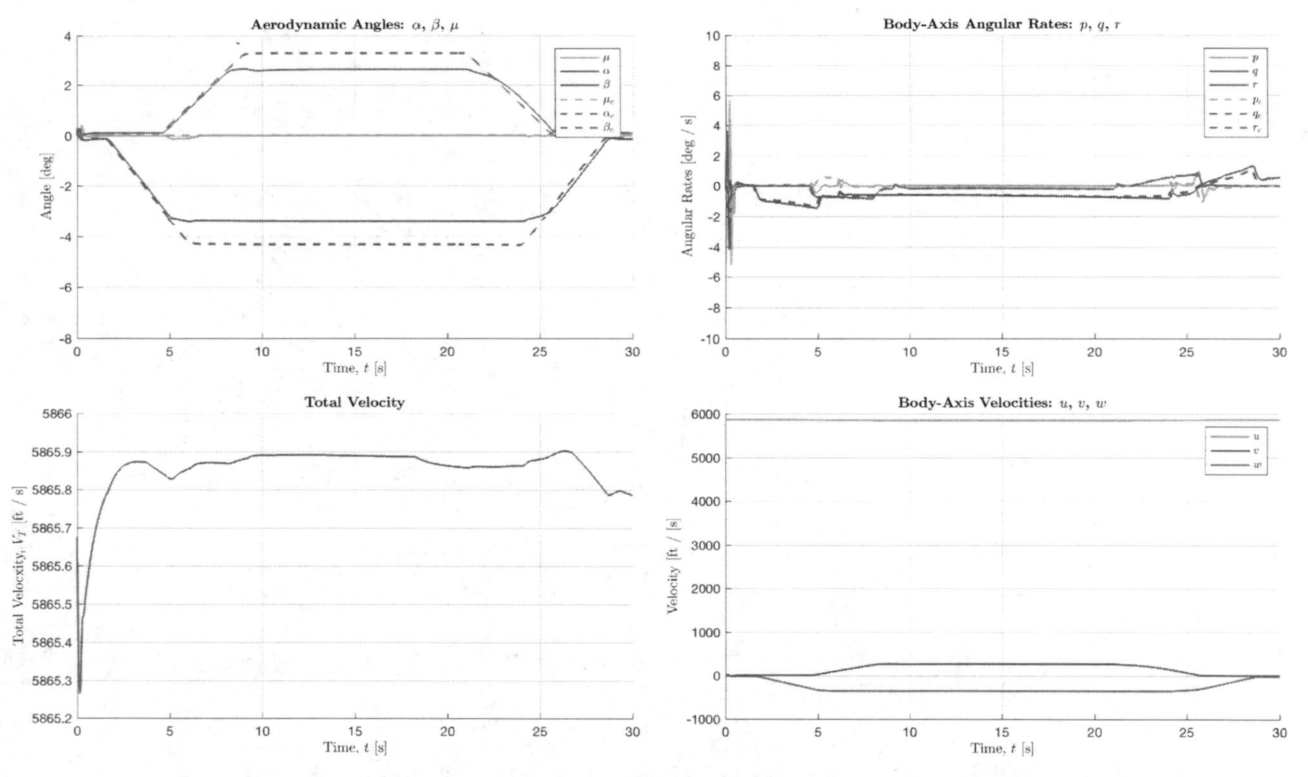

Fig. 7.10 State histories (body-axis velocities, angular rates; aerodynamic angles; total velocity) for case 2.

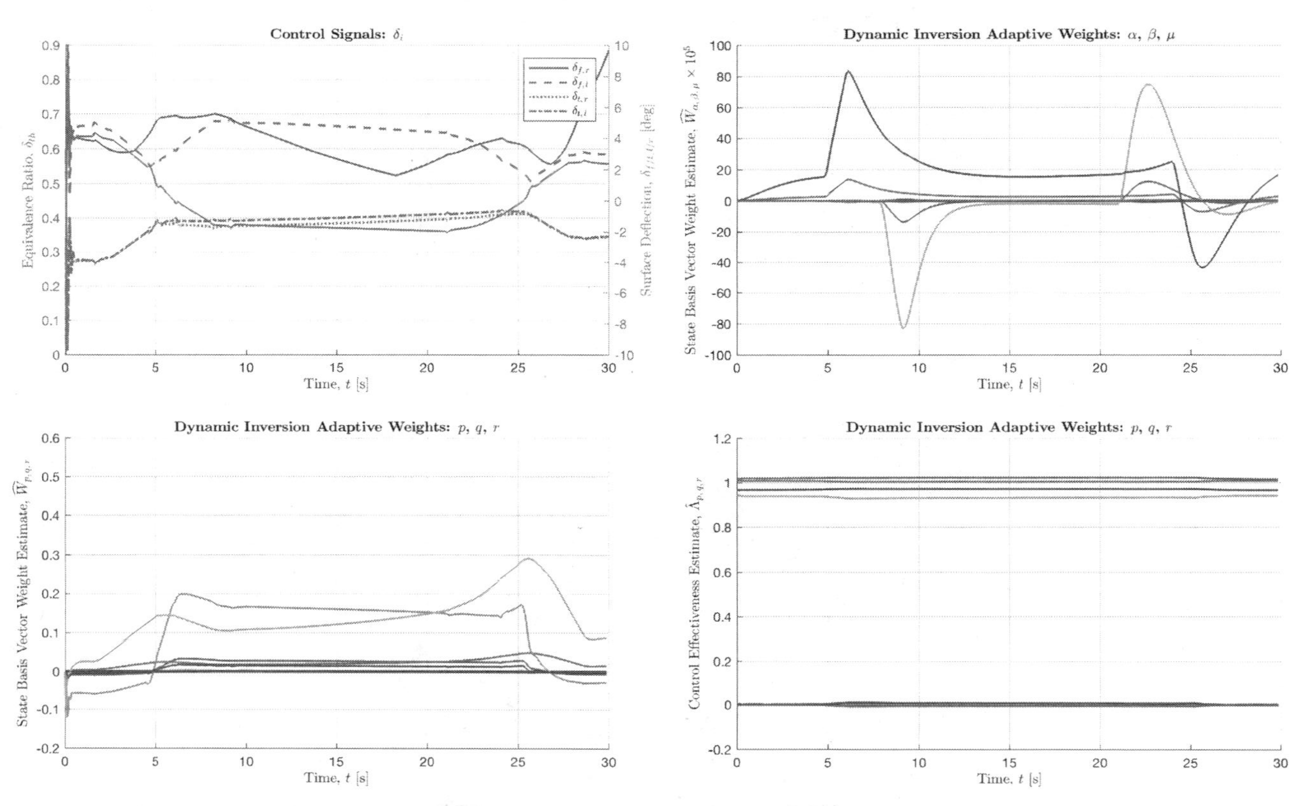

Fig. 7.11 Control surface, equivalence ratio, and dynamic inversion adaptive weight time histories for case 2.

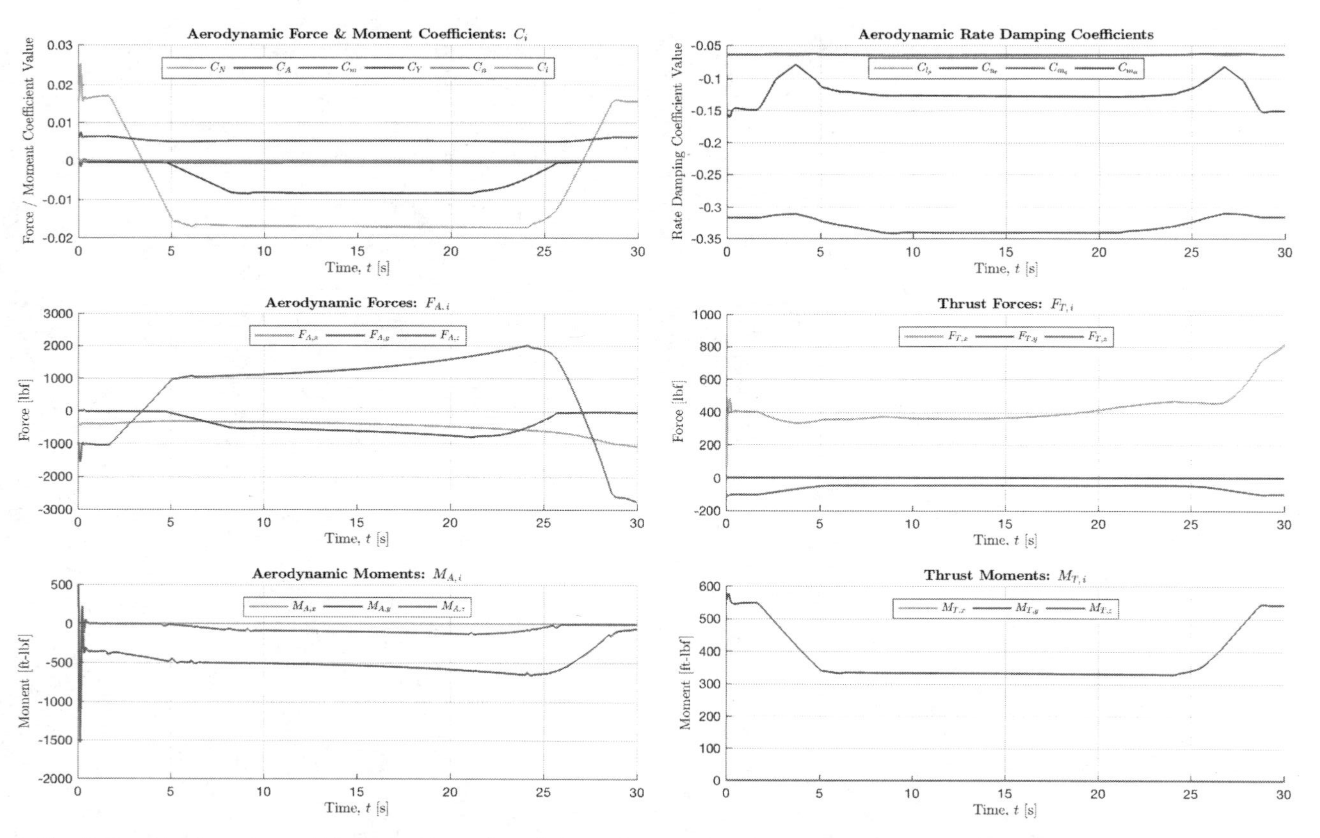

Fig. 7.12 Aerodynamic coefficient, aerodynamic and thrust force, and aerodynamic and thrust moment time histories for case 2.

made here to the baseline example is the introduction of a 500 lbf disturbance applied in the body-axis positive ("down") z-direction that occurs at 10.8 s and lasts 0.9 s, as demonstrated in the middle-left pane of Fig. 7.13. Furthermore, by comparison with Fig. 7.7, Fig. 7.14 demonstrates that the value of the disturbance is selected with the intention to bring the inlet even closer to unstarting by making angle-of-attack more negative. Finally, it is important to note that this exogenous input is appropriately implemented only in the plant dynamics, that is, without the knowledge of either dynamic inversion controller.

As expected, the upper-left pane of Fig. 7.15 exhibits a slight deviation from the baseline trajectory in all three angle time histories at approximately 11 s, again the result of dynamic coupling. These deviations are also reflected in the angular roll rates.

In the upper-left pane of Fig. 7.16, there is a significant decrease in equivalence ratio coinciding with the duration of the disturbance. Because the force is applied in a direction that tends to make angle-of-attack more negative, it also tends to increase the total velocity of the vehicle. In an attempt to return the vehicle to the lower total velocity for which it was initially trimmed, the PID controller reduces the equivalence ratio command.

Concerning the adaptive weights, a similar remark as was made in case 2 regarding the large increase in some weights during deviation from the baseline trajectory applies equally here. Indeed, the adaptive controller recognizes the disturbance and attempts to account for it by varying certain elements of the adaptive weights. This similarly fast variation, shared between the two cases, serves here to validate the preventative action of the state constraint technique.

7.7.4.5 CASE 4: RECOVERY FROM UNSTART BY AN UNEXPECTEDLY LARGE EXTERNAL DISTURBANCE

In the fourth and final case, the ability of the state constraint technique to help the vehicle recover from an inlet unstart is demonstrated. The simulation for case 4 is identical to that of case 3, with the only exception that the magnitude of the applied disturbance is increased to 2500 lbf, shown by the middle-left pane of Fig. 7.17.

Examination of Fig. 7.18 reveals that, at some point in the course of the 30 s simulation, the constrained states lie outside the region of $\alpha - \beta$ space outlined by the outer ellipsoid. As a result, the composite metric based on angle-of-attack and sideslip angle triggers an unstart and the vehicle loses thrust, in accordance with Section 7.7.2 and as shown in the middle- and lower-right panes of Fig. 7.17. During the unstart event, some undesirable, yet understandable, oscillatory behavior is exhibited in the roll rate and control surface time histories. However, their magnitudes are still within an acceptable range, and so this performance is tolerable, especially as the vehicle is operating in such a severely off-nominal condition.

When the inlet unstart occurs, there is an instantaneous change in the value of the bounding function hedge term (ℓ_r) as the parameters now used to calculate the

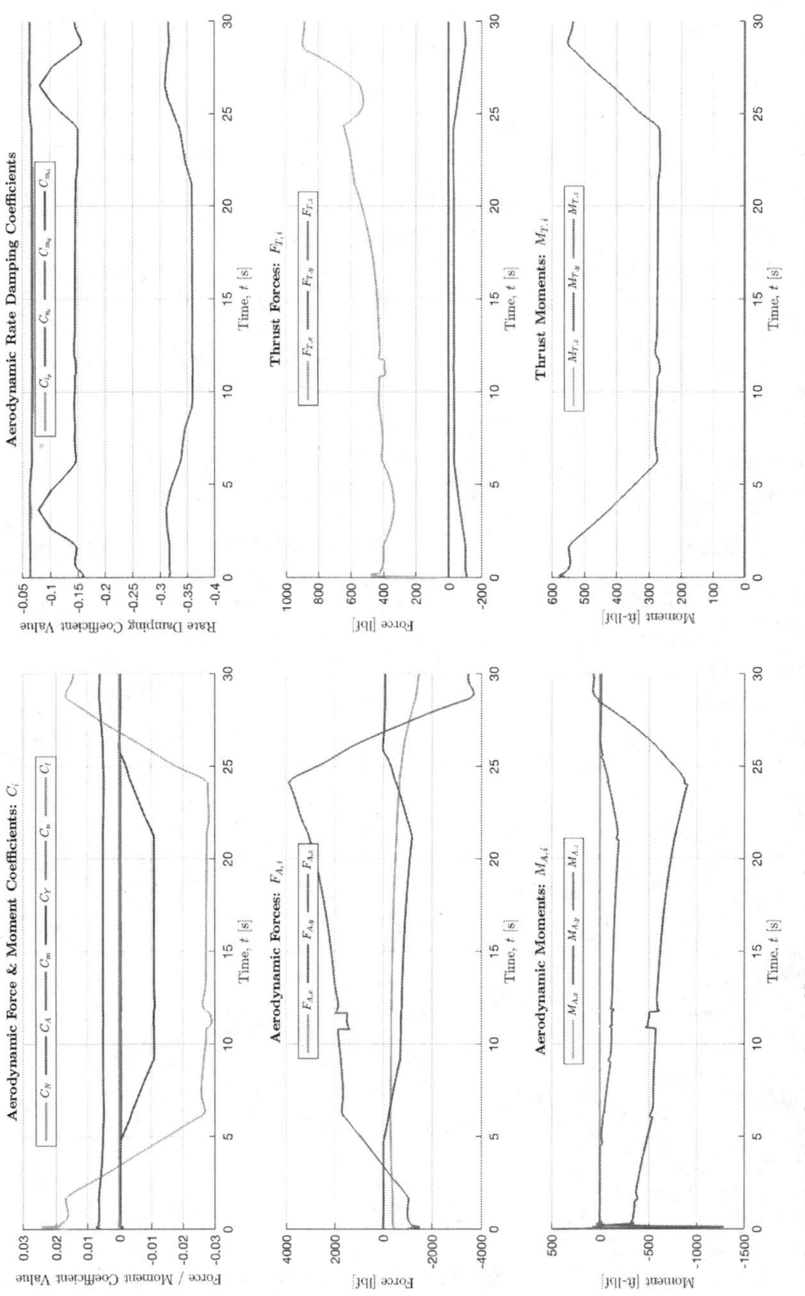

Fig. 7.13 Aerodynamic coefficient, aerodynamic and thrust force, and aerodynamic and thrust moment time histories for case 3.

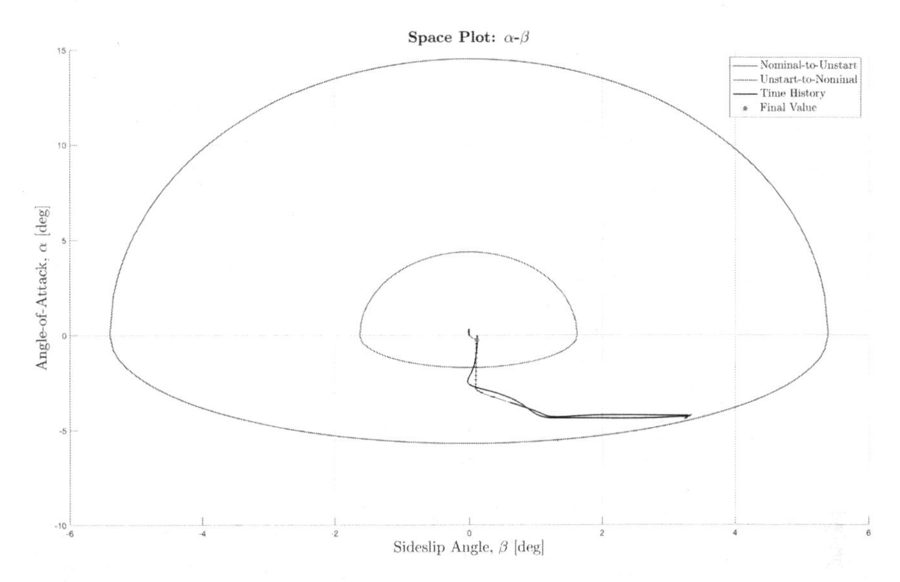

Fig. 7.14 $\alpha - \beta$ space plot, with inlet unstart model parameters and α and β histories for case 3.

bounding functions are described by the *inner* ellipsoid of Fig. 7.18. Furthermore, because during the inlet unstart the vehicle is in complete violation of this new constraint set, the bounding functions are not only partially, transitionally active (as in case 2), but far exceed their hard constraints and are instantaneously fully active, i.e., they take the form

$$f_{b,i}(x_i^2) = \exp(\rho_i^* x_i^2) - \phi_{0,i}^* \tag{7.90}$$

where, once again, the superscript asterisk is used to differentiate the bounding function associated with the unstarted constraints from the nominal bounding function parameters. The sensitivities for this new set of bounding functions are given by $\rho_\alpha^* = 2.0$ and $\rho_\beta^* = 2.35$, while the other parameters associated with this bounding function, $(b_\alpha^*, c_\alpha^*, b_\beta^*, c_\beta^*)$ are given in Eq. (7.87).

Similar to cases 2 and 3, the value of the weights for the α, β, μ adaptation parameters, seen in the upper-right pane of Fig. 7.19, change rapidly at a time consistent with the unexpected change in plant dynamics. In fact, due to the significant dynamical changes associated with inlet unstart, some of the elements in the matrix $\widehat{W}_{\alpha,\beta,\mu}$ are saturated by the projection operator at $||\widehat{W}_{\alpha,\beta,\mu}||_{\max}$. This is seen in the upper-right pane of Fig. 7.19.

Cross-referencing Figs. 7.19 and 7.18 shows the significant effect of the new bounding function parameters on the vehicle. The constrained states are rapidly driven to satisfy these new constraints, which is seen clearly in the upper-left

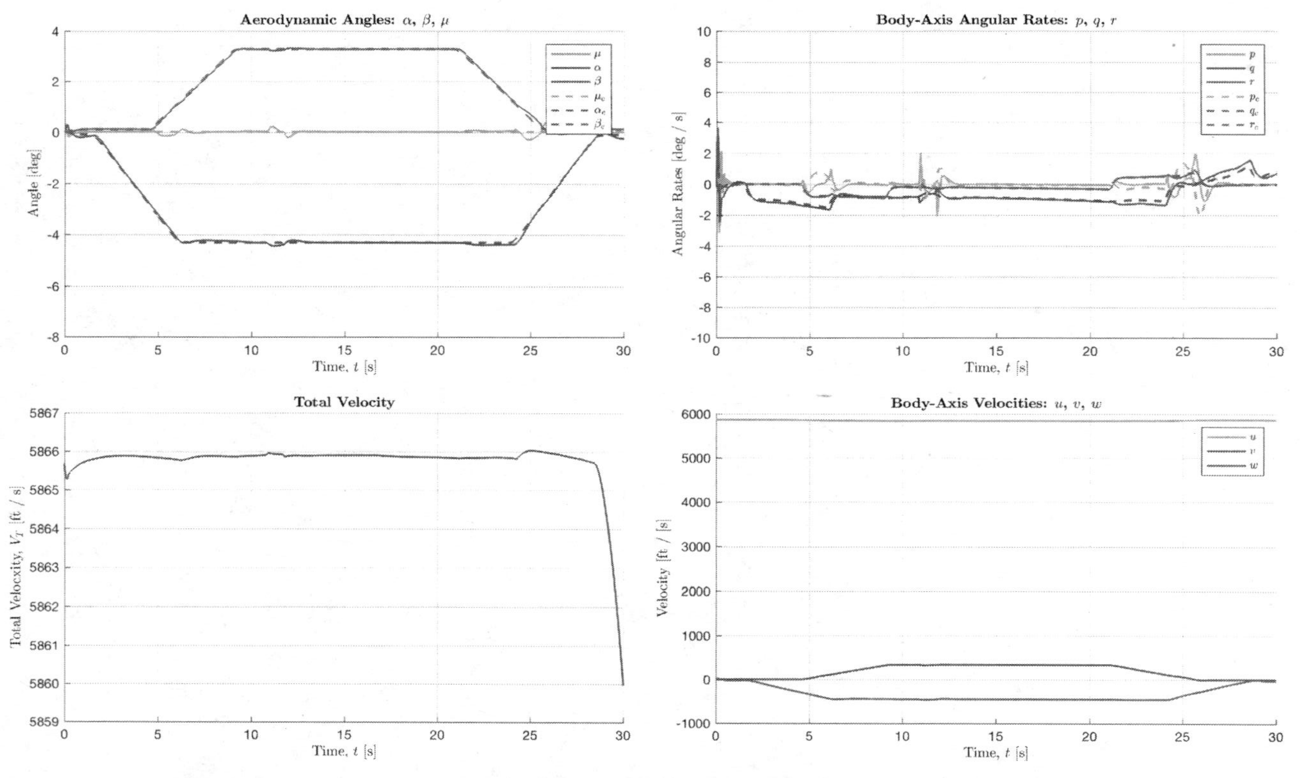

Fig. 7.15 State histories (body-axis velocities, angular rates; aerodynamic angles; total velocity) for case 3.

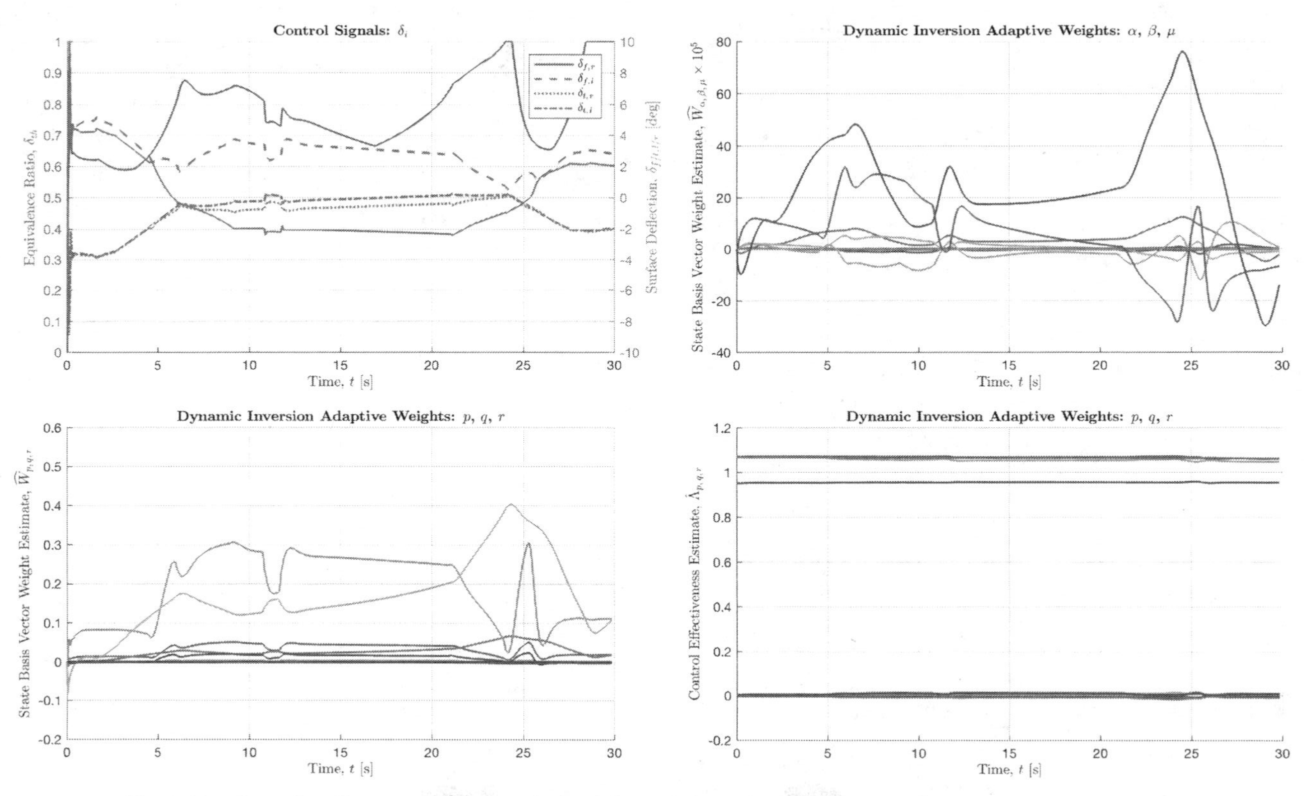

Fig. 7.16 Control surface, equivalence ratio, and dynamic inversion adaptive weight time histories for case 3.

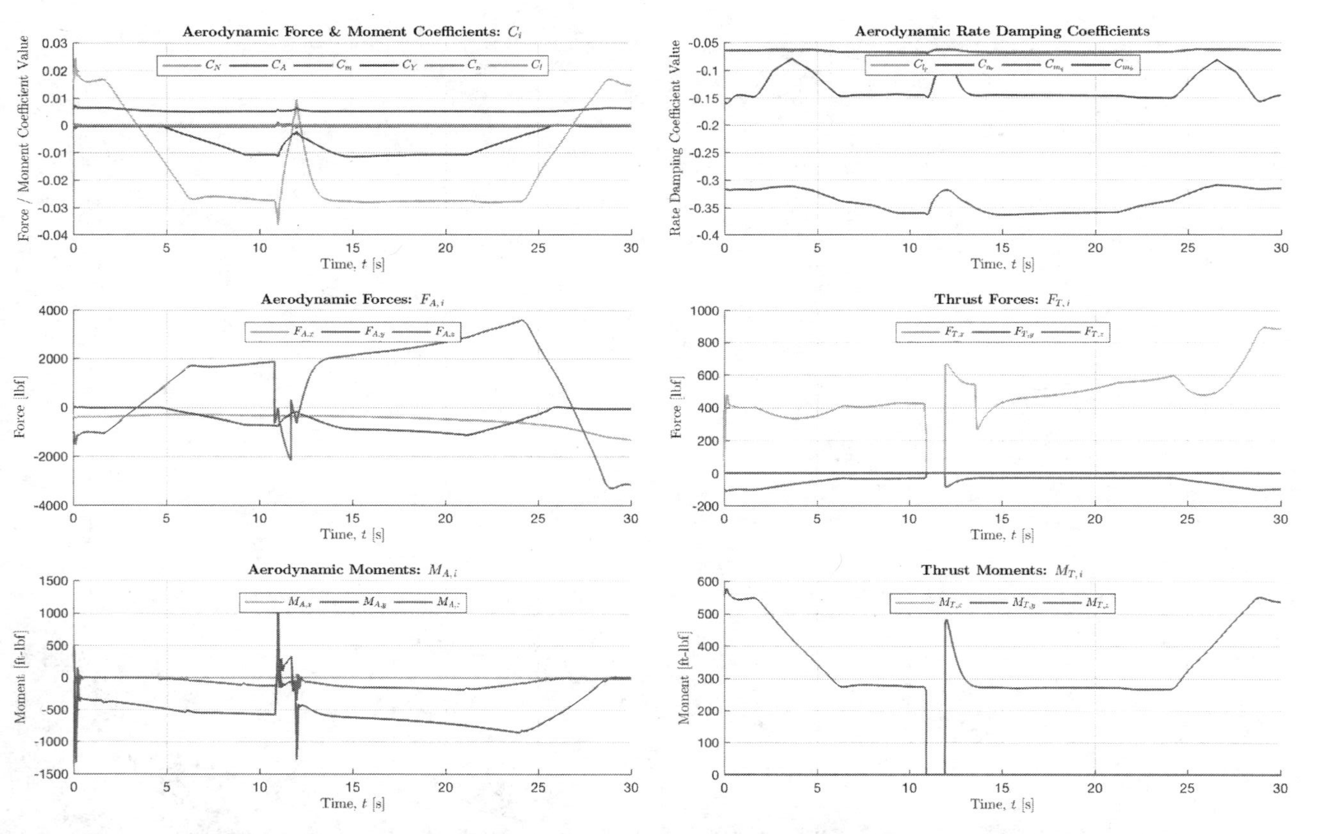

Fig. 7.17 Aerodynamic coefficient, aerodynamic and thrust force, and aerodynamic and thrust moment time histories for case 4.

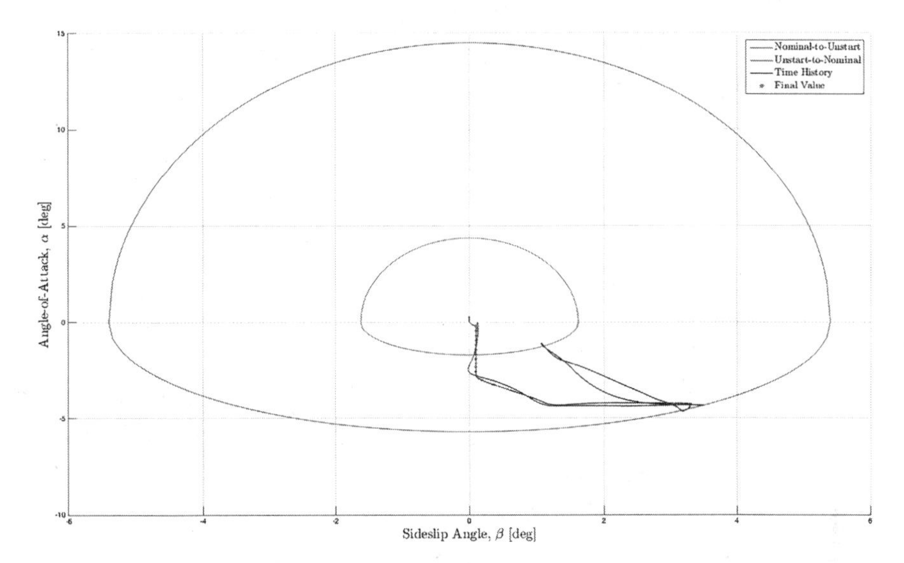

Fig. 7.18 $\alpha - \beta$ **space plot, with inlet unstart model parameters and** α **and** β **histories for case 4.**

pane of Fig. 7.19. The controller is able to drive the vehicle back into a restarted state in 0.83 s, leading to restoration of thrust and the bounding function parameters to their nominal values. Despite the instantaneous restoration of thrust, the elimination of the disturbance, and a significantly different control objective all occurring almost simultaneously, the vehicle response is generally smooth after restart in all states and controls. Most important, Fig. 7.19 reveals that both angle-of-attack and sideslip angle successfully track the baseline trajectory until the end of the simulation. Indeed, as set out in Section 7.7.3, this case validates the unconventional application of a state constraint mechanism as a recovery technique.

7.8 CONCLUSION

This chapter derived a method for enforcing state constraints on nonlinear plants in a dynamic inversion adaptive control framework. The concept of bounding functions was explained, and it was proven that these functions can be used as a constraint enforcement mechanism. It was shown that either an individual constraint can be set for each state or a composite metric involving multiple states can be implemented. As a state begins to approach its constraint value, the bounding functions smoothly adjust the control and adaptive laws to lessen the effect of the reference model and to drive the system to a safer region of the state-space. Insight

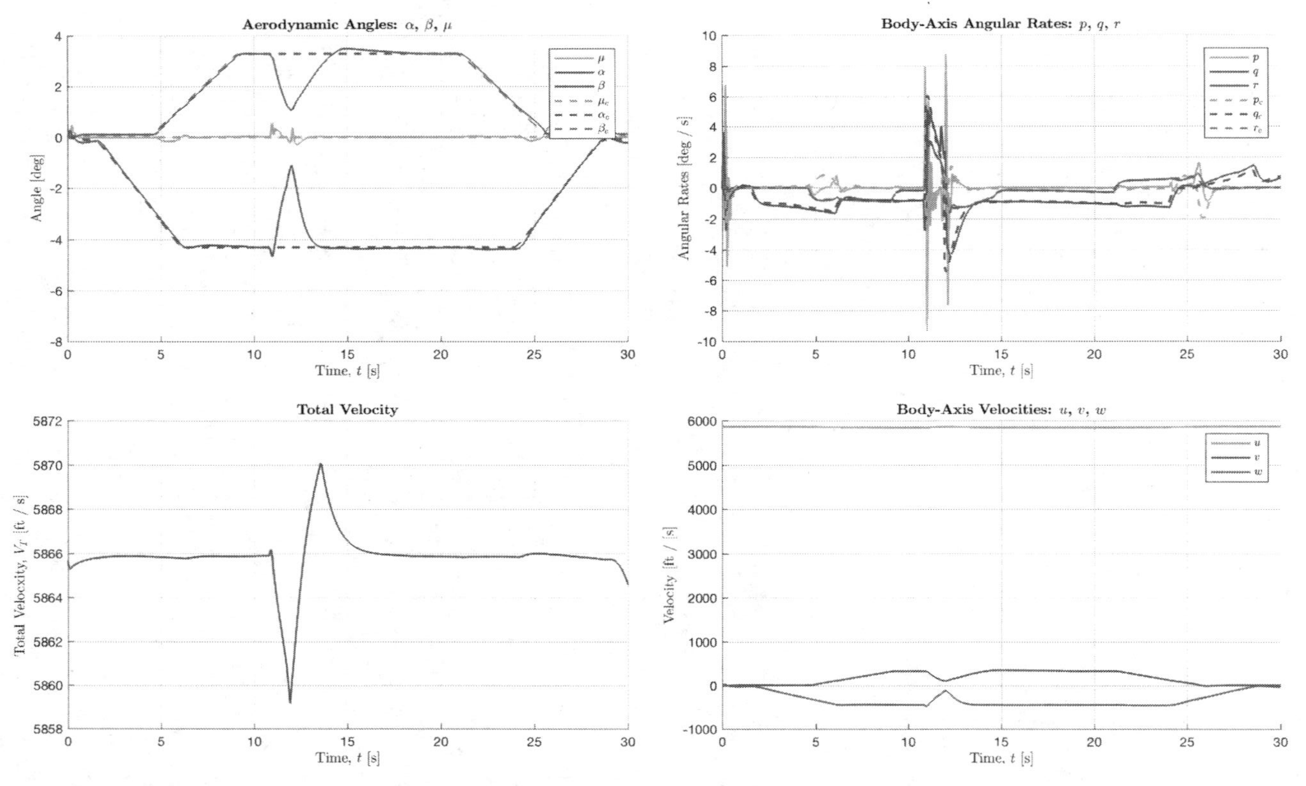

Fig. 7.19 State histories (body-axis velocities, angular rates; aerodynamic angles; total velocity) for case 4.

into how the choice of bounding function can tighten the constraint value was provided.

It was also demonstrated that if none of the states approach a constraint, the control and adaptive laws reduce to the form of a previously developed dynamic inversion adaptive controller capable of driving the system to track a command signal. Furthermore, this chapter proved that, as long as the input command signal is bounded, the closed-loop system will be stable. It was also proven that if the steady state value of the reference model does not call for the violation of any state constraints, perfect tracking will be achieved.

The potential for this state constraining adaptive control technique to be used both for the prevention of and recovery from an inlet unstart was demonstrated on a nonlinear simulation of a hypersonic vehicle. The concept of inlet unstart was described in detail. To summarize, it is a phenomenon seen only in supersonic jet inlets where airflow through the inlet is altered in such a way that the shock wave system is disrupted and then is expelled from the inlet, leading to a significant and almost instantaneous change in the vehicle's performance. The challenges associated with modeling inlet unstart were explained; therefore, a simplified model was implemented in the simulation. Given the limitations of the model, it was shown that the presented controller can potentially prevent inlet unstart from occurring while maintaining a smooth control signal. Furthermore, a way to use the same state constraint mechanism as a recovery technique if an unstart were to occur was presented. The potential of this technique was demonstrated by presenting a test case in which the GHV recovered from an inlet unstart in 0.83 seconds. In summary, this control technique shows promise as a way to handle inlet unstart and warrants further testing as better aerodynamic data become available.

REFERENCES

[1] Falkena, W., Borst, C., and Mulder, J. A., "Investigation of Practical Flight Envelope Protection Systems for Small Aircraft," *AIAA Guidance, Navigation, and Control Conference*, Toronto, Canada, Aug. 2010. doi:10.2514/1.53000.

[2] Chaung, C.-H., and Morimoto, H., "Periodic Optimal Cruise for a Hypersonic Vehicle with Constraints," *Journal of Spacecraft and Rockets*, Vol. 34, No. 2, 1997. doi:10.2514/2.3205.

[3] Stephen, E. J., Hoenisch, S. R., Riggs, C. J., Waddel, M. L., McLaughlin, T., and Bolender, M. A., "Hifire 6 Unstart Conditions at Off-Design Mach Numbers," *AIAA SciTech*, Kissimmee, FL, Jan. 2015. doi:10.2514/6.2015-0109.

[4] Seddon, J., and Goldsmith, E., *Intake Aerodynamics*, 2nd ed. AIAA Education Series, AIAA, New York, 1985. doi:10.2514/4.473616.

[5] Idris, A. C., Saad, M. R., Zare-Behtash, M. R., and Kontis, M. R., "Luminescent Measurement Systems for the Investigation of a Scramjet Inlet-Isolator," *Sensors*, Vol. 14, No. 4, 2014. pp. 6606–6632, doi:10.3390/s140406606.

[6] Cang, J., Li, N., Xu, K., Bao, W., and Yu, D., "Recent Research Progress on Unstart Mechanism, Detection and Control of Hypersonic Inlet," *Progress in Aerospace Sciences*, Vol. 89, 2017. pp. 1–22, doi:89.10.1016, j.paerosci.2016.12.001.

[7] Gilbert, E. G., and Tan, K. T., "Linear Systems with State and Control Constraints: The Theory and Application of Maximal Output Admissible Sets," *IEEE Transactions on Automatic Control*, Vol. 36, No. 9, 1991. doi:10.1109/9.83532.

[8] Hartl, R. F., Sethi, S. P., and Vickson, R. G., "A Survey of the Maximum Principles for Optimal Control Problems with State Constraints," *SIAM Review*, Vol. 37, No. 2, 1995. doi:10.1137/1037043.

[9] Elbert, P., Ebbesen, S., and Guzella, L., "Implementation of Dynamic Programming for n-Dimensional Optimal Control Problems with Final State Constraints," *IEEE Transactions on Control Systems Technology*, Vol. 21, No. 3, 2013. doi:10.1109/TCST.2012.2190935.

[10] Kiefer, T., Graichen, K., and Kugi, A., "Trajectory Tracking of a 3DOF Laboratory Helicopter Under Input and State Constraints," *IEEE Transactions on Control Systems Technology*, Vol. 18, No. 4, 2010. doi:10.1109/TCST.2009.2028877.

[11] Vaddi, S. S., and Sengupta, P., "Controller Design for Hypersonic Vehicles Accomodating Nonlinear State and Control Constraints," *AIAA Guidance, Navigation, and Control Conference*, Chicago, IL, Aug. 2009. doi:10.2514/6.2009-6286.

[12] Sanner, R. M., and Slotine, J.-J. E., "Gaussian Networks for Direct Adaptive Control," *IEEE Transactions on Neural Networks*, Vol. 3, No. 6, Nov. 1992. doi:10.1109/72.165588.

[13] Lavretsky, E., and Gadient, R., "Robust Adaptive Design for Aerial Vehicles with State-Limiting Constraints," *Journal of Guidance, Control, and Dynamics*, Vol. 33, No. 6, 2010. doi:10.2514/1.50101.

[14] Muse, J. A., "A Method For Enforcing State Constraints in Adaptive Control," *AIAA Guidance, Navigation, and Control Conference*, Portland, Aug. 2011. doi:10.2514/6.2011-6205.

[15] Liu, Y.-J., and Tong, S., "Barrier Lyapunov Functions for Nussbaum Gain Adaptive Control of Full State Constrained Nonlinear Systems," *Automatica*, Vol. 76, Feb. 2007. doi:10.1016/j.automatica.2016.10.011.

[16] Brinker, J. S., and Wise, K. A., "Stability and Flying Qualities Robustness of a Dynamic Inversion Aircraft Control Law," *Journal of Guidance, Control, and Dynamics*, Vol. 19, No. 6, 1996. doi:10.2514/3.21782.

[17] Reiner, J., Balas, G. J., and Garrard, W. L., "Robust Dynamic Inversion for Control of Highly Maneuverable Aircraft," *Journal of Guidance, Control, and Dynamics*, Vol. 18, No. 1, 1995. doi:10.2514/3.56651.

[18] Rysdyk, R. T., and Calise, A. J., "Adaptive Model Inversion Flight Control for Tilt-Rotor Aircraft," *Journal of Guidance, Control, and Dynamics*, Vol. 22, No. 3, 1999. doi:10.2514/2.4411.

[19] Das, A., Subbarao, K., and Lewis, F., "Dynamic Inversion with Zero-Dynamics Stabilisation for Quadrotor Control," *IET Control Theory and Applications*, Vol. 3, No. 3, 2009. doi:10.1049/iet-cta:20080002.

[20] Rollins, E., "Nonlinear Adaptive Dynamic Inversion Control for Hypersonic Vehicles," Ph.D. Thesis, Aerospace Engineering Department, Texas A & M University, College Station, TX, 2013.

[21] Lavretsky, E., Gibson, T. E., and Annaswamy, A. M., "Projection Operator in Adaptive Systems," *arXiv E-prints*, Dec. 2011.

[22] Pomet, J.-B., and Praly, L., "Adaptive Nonlinear Regulation: Estimation from the Lyapunov Equation," *IEEE Transactions on Automatic Control*, Vol. 37, No. 6, 1992. doi:10.1109/9.256328.

[23] Rollins, E., Valasek, J., Muse, J. A., and Bolender, M. A., "Nonlinear Adaptive Dynamic Inversion Applied to a Generic Hypersonic Vehicle," In *AIAA Guidance, Navigation, and Control Conference*, Boston, Aug. 2013. doi:10.2514/6.2013-5234.

[24] Billamoria, K. D., and Schmidt, D. K., "Integrated Development of the Equations of Motion for Hypersonic Flight Vehicles," *Journal of Guidance, Control, and Dynamics*, Vol. 18, No. 1, 1995. doi:10.2514/3.56659.

[25] Stevens, B. L., and Lewis, F. L., *Aircraft Control and Simulation*, Wiley, Hoboken, NJ, 1992.

CHAPTER 8

Semisupervised Learning of Lift Optimization of Multi-Element Three-Segment Variable Camber Airfoil

Upender K. Kaul* **and Nhan T. Nguyen**†
NASA Ames Research Center, USA

This chapter describes a new intelligent platform for learning optimal designs of morphing wings based on variable camber continuous trailing edge flaps (VCCTEF), in conjunction with a leading edge flap called the variable camber Krueger (VCK). The new platform consists of a computational fluid dynamics (CFD) methodology coupled with a semisupervised learning methodology. The CFD component of the intelligent platform comprises a full Navier–Stokes solution capability (NASA OVERFLOW solver with Spalart–Allmaras turbulence model) [1] that computes flow over a trielement inboard NASA generic transport model (GTM) wing section. Various VCCTEF/VCK settings and configurations were considered while exploring optimal design for high-lift flight during takeoff and landing. To determine a globally optimal design of such a system, an extremely large set of CFD simulations is needed, which in practice is not feasible to achieve. To alleviate this problem, a semisupervised learning (SSL) methodology was employed [2], based on manifold regularization techniques [3]. A reasonable space of CFD solutions was populated [4] and then the SSL methodology was used to fit this manifold in its entirety, including the gaps in the manifold where there were no CFD solutions available.

The SSL methodology, in conjunction with an elastodynamic solver (FiDDLE) [5] was demonstrated in an earlier study involving structural health monitoring [6]. These CFD-SSL methodologies define the new intelligent platform that forms the basis for our search for optimal design of wings. Although the present platform can be used in various other design and operational problems in engineering, this chapter focuses on the high-lift study of the VCK-VCCTEF system.

A top few design configuration candidates were identified by solving the CFD problem in a small subset of the design space. The SSL component was trained on

*Computational Aerosciences Branch, NASA Advanced Supercomputing (NAS) Division; Associate Fellow, AIAA.
†Intelligent Systems Division; Associate Fellow, AIAA.

the design space and it was then used in a predictive mode to populate a selected set of test points outside of that design space. The new design test space, thus populated, was evaluated by using the CFD component that determined the error between the SSL predictions and the true (CFD) solutions, which was found to be small. This demonstrates the effectiveness of the proposed CFD-SSL methodologies for isolating the best design of the VCK-VCCTEF system, specifically, and it holds promise for quantitatively identifying best designs of flight systems in general.

8.1 INTRODUCTION

The Advanced Air Transportation Technologies (AATT) Project is conducting multidisciplinary foundational research to investigate advanced concepts and technologies for future aircraft systems under the Advanced Air Vehicle Program (AAVP) of the NASA Aeronautics Research Mission Directorate. A NASA study entitled "Elastically Shaped Future Air Vehicle Concept" was conducted in 2010 [7, 8] to examine new concepts that can enable active control of wing aeroelasticity to achieve drag reduction. This study showed that highly flexible wing aerodynamic surfaces can be elastically shaped in flight by active control of wing twist and vertical deflection to optimize the local angle-of-attack of wing sections. Thus, aerodynamic efficiency can be improved through drag reduction during cruise and enhanced lift performance during takeoff and landing.

The study shows that active aeroelastic wing shaping control can have a potential drag reduction benefit. Conventional flap and slat devices inherently generate drag as they increase lift. The study shows that, in cruise, conventional flap and slat systems are not aerodynamically efficient for use in active aeroelastic wing shaping control for drag reduction. A new flap concept, the VCCTEF system, was conceived by NASA to address this need [7]. Initial study results indicate that the VCCTEF system may offer a potential payoff in drag reduction in

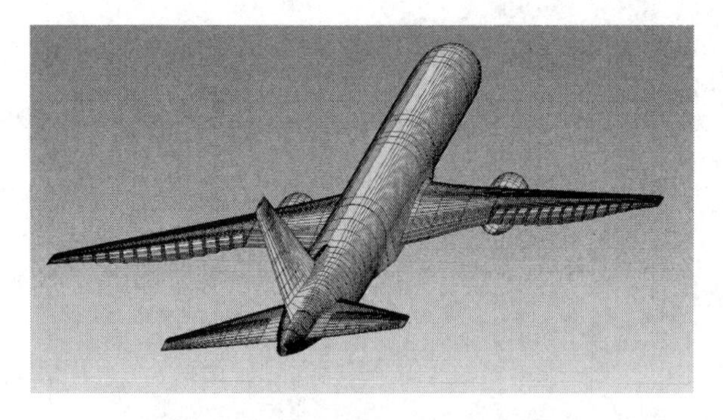

Fig. 8.1 VCCTEF deployed on the GTM.

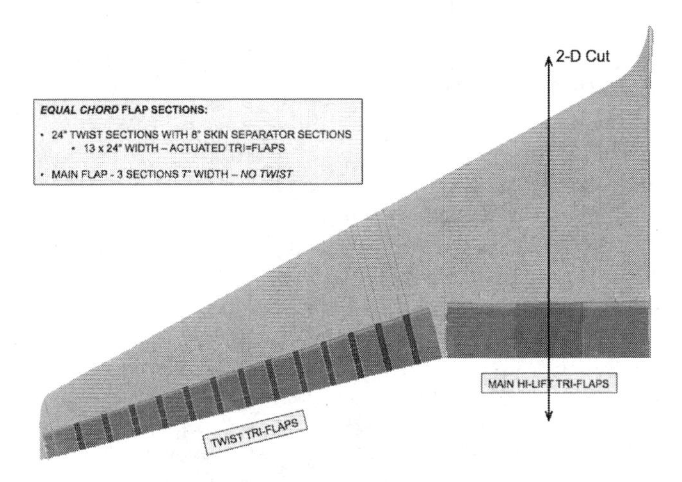

Fig. 8.2 NASA/Boeing VCCTEF configuration.

cruise that could provide significant fuel savings. Figure 8.1 illustrates the VCCTEF deployed on the NASA GTM.

NASA and Boeing are currently conducting further studies of the VCCTEF under the research element Performance Adaptive Aeroelastic Wing (PAAW) within the AATT Project [9, 10]. This study, built upon the development of the VCCTEF system (shown in Fig. 8.2) for the GTM [11], employs lightweight shaped memory alloy (SMA) technology for actuation and three separate chordwise flap segments shaped to provide a variable camber to the flap. Introduction of this camber has potential for drag reduction as compared to a conventional straight, plain flap. The flap is also made up of many 2-foot spanwise sections, which enable different flap settings at each flap spanwise position. This enables wing twist shape control as a function of span to establish the best lift-to-drag ratio (L/D) at any aircraft gross weight or mission segment. Current wing twist on commercial transports is permanently set for one cruise condition, which is usually a 50% fuel loading, or midpoint on the gross weight schedule. The VCCTEF offers different wing twist settings for each gross weight condition and also different settings for climb, cruise, and descent, which is a major factor in obtaining the best L/D for all gross weight conditions and phases of flight. The second feature of VCCTEF is a continuous trailing edge. The individual 2-foot spanwise flap sections are connected with a flexible covering so that no breaks can occur in the flap platform, thus reducing excessive vorticity generation. This can reduce drag and airframe noise. Variable camber, when combined with the continuous trailing edge, results in a further reduction in drag.

The continuous trailing edge flap design, combined with variable camber flap, can result in lower drag. In summary, it can also offer a potential noise reduction benefit due to distinct optimal settings for climb, cruise, and descent [12]. In a previous paper [13], a computational study was conducted to explore the two-dimensional viscous effects in cruise of a number of VCCTEF configurations on lift and drag of the GTM wing section at the wing planform break. The NASA flow solver OVERFLOW was used to conduct this study. The results identified the most aerodynamically efficient VCCTEF configuration among the initial candidates. The study also showed that a three-segment variable camber flap is aerodynamically more efficient than a single-element plain flap. A recent high-lift wind tunnel test conducted in July 2014 at the University of Washington Aeronautical Laboratory [14, 15] confirms this observation.

The present study builds on a recent RANS study [4] that explored the high-lift design space for the trielement airfoil typical of a GTM wing section. The tri-element airfoil comprises VCK, the main airfoil, and the VCCTEF. The design space consists of 224 configurations drawn from various combinations of VCK and VCCTEF settings, as will be described. Limited experimental data [15, 16, 17] are available corresponding to four configurations (VCK65, VCK60, VCK55, and VCK50 – *vck*1), out of the 224 considered here using CFD. In the following paragraphs, details of the CFD results are presented, followed by the details of the SSL methodology and results.

The database of solutions generated by CFD simulations is a subset of a very large manifold of possible solutions, which would be impractical to generate using CFD. Therefore, a semisupervised machine learning methodology [2, 3], which is based on three regularization terms (i.e., least squares, a Laplacian, and a radial basis function) is used to approximate this manifold of solutions. This methodology was demonstrated in the context of structural fault detection [6]. This SSL methodology relies on its Laplacian component to smoothly approximate the manifold over gaps created by a finite number of CFD solutions sparsely populating the solution manifold and is guaranteed to converge with the appropriate choice of kernel parameters [3]. The regularization concept, in general, owes its origin to Tikhonov [18], and it is widely used in learning techniques, both unsupervised and fully supervised. This chapter focuses on the use of the SSL algorithm [3], called the Laplacian regularized least squares algorithm.

8.2 METHODOLOGY

The methodology proposed in the present work is multidisciplinary in nature. The first component is based on modeling and simulation of candidate configurations of the VCK-VCCTEF system under consideration, using CFD. The second component is based on learning the sparse database created in the first component and then providing as large a database as desired through an SSL process, within a small error bound. These two components provide a host of design test points for the experimentalist to validate. The approach adopted here establishes a viable design methodology that converges onto optimal designs of the

VCK-VCCTEF system in a much shorter time than would be possible otherwise. In fact, converging onto an optimal design would not be practically feasible without the SSL component. We shall call the first component RANS simulations and the second component SSL both of which are described in the following sections.

8.2.1 RANS SIMULATIONS

The high-lift flow field was simulated for $M = 0.25$ and $Re = 3.3196 \times 10^7$, where M represents the Mach number and Re the Reynolds number, based on the chord length. RANS simulations, based on the OVERFLOW flow solver with the SA turbulence model, were carried out to generate steady state solutions. Numerous combinations corresponding to 18 *vck* configurations for VCK setting (deflection angle) of 55°, 19 *vck* configurations each for VCK settings of 60° and 65° with respect to the main wing, and 4 VCCTEF settings with the Fowler slot for the inboard wing are considered in this study. These *vck* configurations along with the 4 VCCTEF settings are presented in Table 8.1. Three VCK settings are considered, corresponding to deflection angles of 55°, 60°, and 65°. For each VCK setting, VCK55, VCK60, and VCK65, the *vck* configurations in terms of x and y displacement offset with respect to one experimental configuration (VCK65 + *vck*1) are shown in Fig. 8.3, Fig. 8.4, and Fig. 8.5, respectively. Nineteen *vck* configurations are designated *vck*1–*vck*19. Detailed computational results discussing the lift characteristics are shown in the following discussions for all the 224 configurations, to explore better design than the four studied experimentally.

Figure 8.6 shows the four VCCTEF settings, corresponding to four different flap deflection angles, as shown in Table 8.2. The definitions of various configurations are listed in Table 8.1. The four VCCTEF settings are denoted by 10/10/10, 15/10/5, 20/5/5, and 30/0/0. These settings were selected based on a total deflection angle of 30°, a value that is typically used in landing and takeoff configurations. The VCCTEF settings include a circular arc, a parabolic arc, and a straight deflected flap.

8.2.2 SSL

The purpose of the learning methodology is to fill the gap in the space of a relatively small set of solutions that can be reasonably obtained by CFD, owing to the large amount of computational resources needed. Using the solution space corresponding to 224 configurations populated by CFD, we fit a multidimensional "surface" over this space and thus enable solutions at all intermediate configurations.

We randomly select a subset of these CFD solutions and strip off the $C_{l\max}$ values, where C_l represents the lift coefficient, calling this subset unlabeled (u) data, in the parlance of SSL [3]. We refer to the rest as labeled (l) data. Similarly,

TABLE 8.1 DEFINITION OF *VCK*-VCCTEF CONFIGURATIONS

vck Configuration	10/10/10	15/10/5	20/5/5	30/0/0
*vck*1	*vck*1 + 10/10/10	*vck*1 + 15/10/5	*vck*1 + 20/5/5	*vck*1 + 30/0/0
*vck*2	*vck*2 + 10/10/10	*vck*2 + 15/10/5	*vck*2 + 20/5/5	*vck*2 + 30/0/0
*vck*3	*vck*3 + 10/10/10	*vck*3 + 15/10/5	*vck*3 + 20/5/5	*vck*3 + 30/0/0
*vck*4	*vck*4 + 10/10/10	*vck*4 + 15/10/5	*vck*4 + 20/5/5	*vck*4 + 30/0/0
*vck*5	*vck*5 + 10/10/10	*vck*5 + 15/10/5	*vck*5 + 20/5/5	*vck*5 + 30/0/0
—	—	—	—	—
—	—	—	—	—
—	—	—	—	—
*vck*19	*vck*19 + 10/10/10	*vck*19 + 15/10/5	*vck*19 + 20/5/5	*vck*19 + 30/0/0

vck configurations are represented in Fig. 8.4, Fig. 8.5, and Fig. 8.6

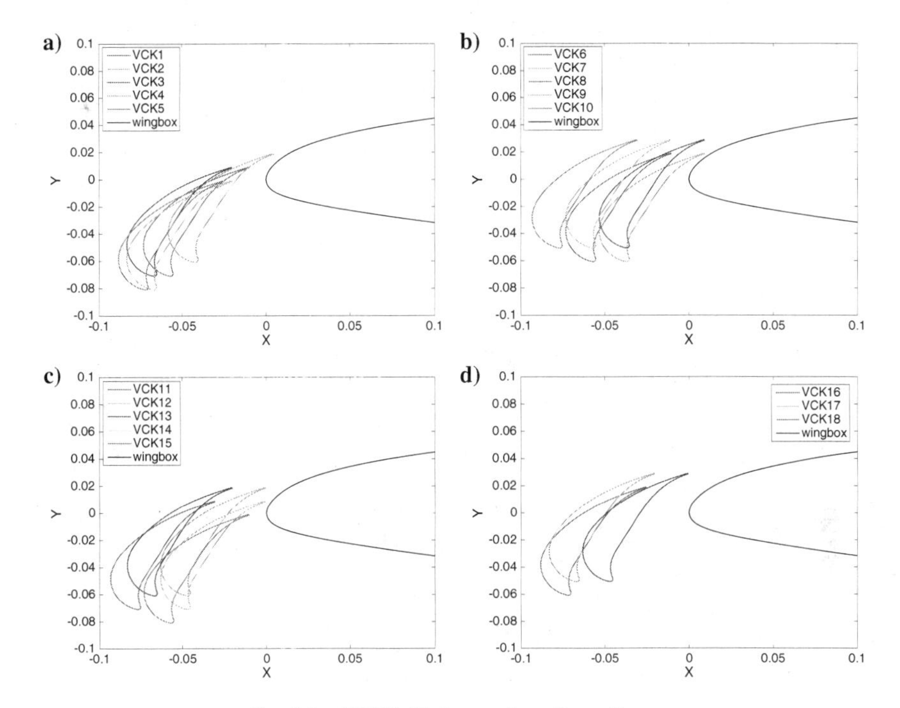

Fig. 8.3 VCK55: Various *vck* configurations.

we select the unlabeled (u) and labeled data (l) for $C_{d\max}$, where C_d represents the drag coefficient. Following [15], we write an expression for our objective function:

$$f^* = \min_{f \in H_K} \frac{1}{l} \Sigma_{i=1}^{l} (y_i - f(x_i))^2 + \gamma_A ||f||_K^2 + \frac{\gamma_I}{(u+l)^2} \mathbf{f}^T L \mathbf{f} \quad (8.1)$$

where for a Mercer kernel K: $X \times X \to R$, $\exists$ a reproducing kernel Hilbert space (RKHS), H_K of functions $X \to R$, with a norm $|| \quad ||_K$, and where the first term $(y_i - f(x_i))^2$ is a squared loss function representing a least squares regularizer, the second term represents a radial basis function based regularizer, and the third term is a Laplacian based regularizer. The optimization problems are solved for different training sets that define the cost function (squared loss function shown in Eq. 8.1) and different choices of the regularization parameters γ_A and γ_I.

The Representer Theorem can be used to show that the minimization problem has a solution in H_K, and it is given by an expansion of kernel functions over both the labeled (l) and the unlabeled (u) data:

$$f^*(x) = \Sigma_{i=1}^{l+u} \alpha_i^* K(x, x_i) \quad (8.2)$$

The minimization process of the objective function constitutes a Laplacian regularized least squares algorithm [3]. The labeled data are $\{(x_i, y_i)\}_{i=1}^{l}$, and the unlabeled data are $\{(x_j)\}_{j=l+1}^{j=l+u}$. By minimizing the objective function given by Eq. (8.1) we arrive at our solution over all the points in the domain of our interest with known error bounds, which is an advantage over other methods such as artificial neural networks (ANN). In addition, the convergence to this global minimum over the considered training set is guaranteed [15]. A convex differentiable objective function of the $(l + u)$ dimensional variable $\alpha = [\alpha_1, \ldots \alpha_{l+u}]^T$ is obtained by substituting Eq. (8.2) in Eq. (8.1) above, which gives

$$\alpha^* = \left(JK + \gamma_A lI + \frac{\gamma_I l}{(u+l)^2} LK\right)^{-1} Y \tag{8.3}$$

where J is a diagonal matrix and K is the Gram matrix of size $(l + u) \times (l + u)$. In J, the first l diagonal entries are 1, and the remaining entries are 0. Y is a

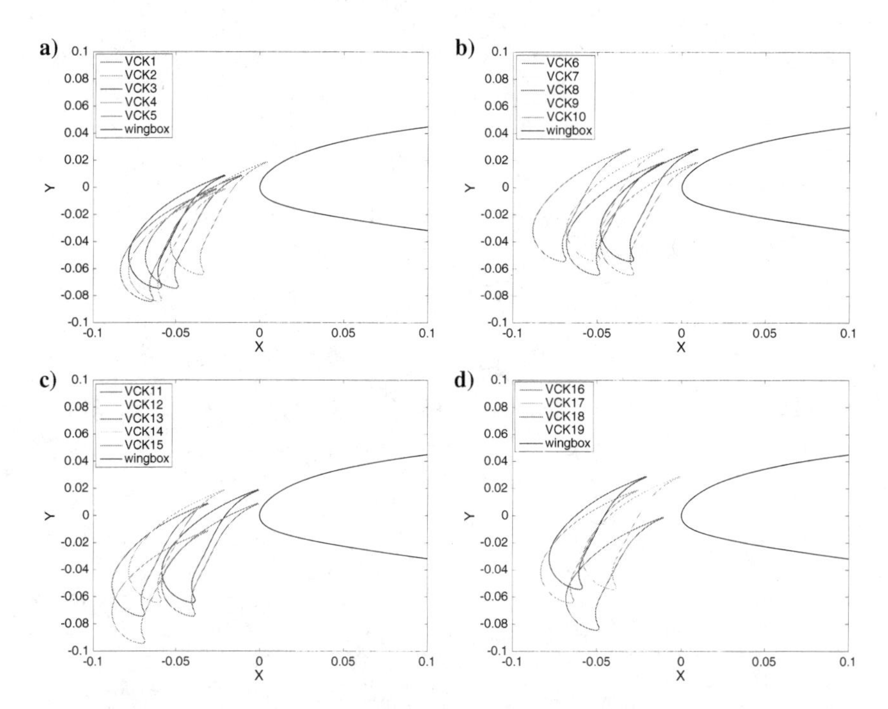

Fig. 8.4 VCK60: Various *vck* configurations.

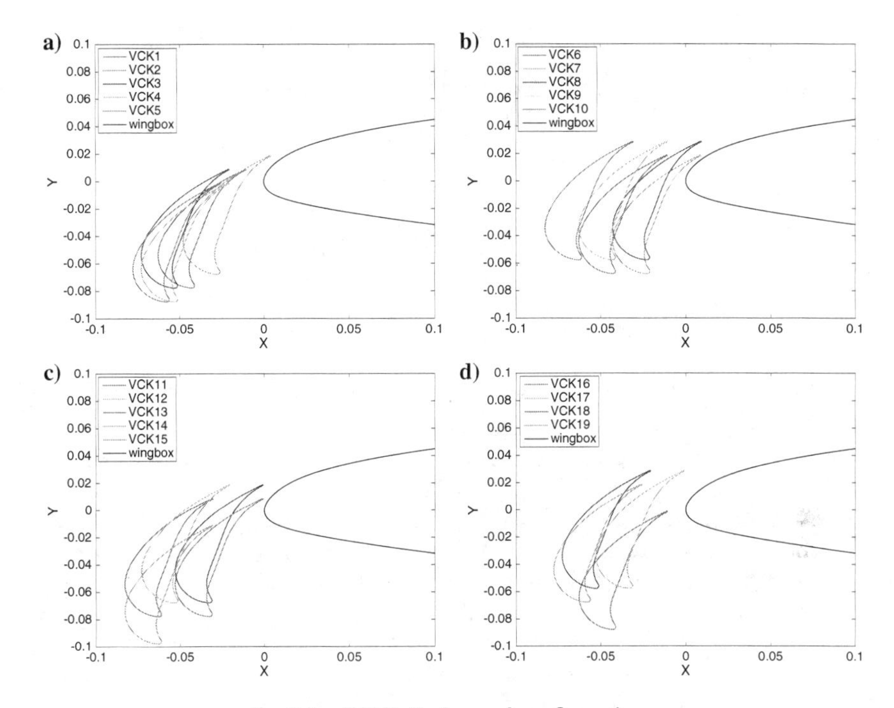

Fig. 8.5 VCK65: Various *vck* configurations.

vector of size $(l + u)$ and L is the graph Laplacian. For further details, see Belkin et al. [3].

Thus, the solution to the optimization problem is obtained via Eq. (8.2) and Eq. (8.3).

TABLE 8.2 DEFINITION OF VCCTEF CONFIGURATIONS

Configuration	Notation	Flap 1, deg	Flap 2, deg	Flap 3, deg
3-segment circular arc camber	10/10/10	10	10	10
3-segment semirigid arc camber	15/10/5	15	10	5
3-segment	20/5/5	20	5	5
1-segment rigid flap	30/0/0	30	—	—
	Flap deflection angles are relative to the upstream segments			

8.3 RESULTS

Results corresponding to CFD simulations are discussed first, followed by the results obtained by the learning methodology.

8.3.1 CFD RESULTS

The results of the CFD study of lift optimization were discussed in an earlier work [4], where a detailed grid sensitivity study was carried out. An optimal grid resolution was determined to be 436 × 106 × 3 for the VCK, 694 × 106 × 3 for the main wing, and 471 × 106 × 3 for the VCCTEF. For further details, see Kaul and Nguyen [4].

A total of 4 × (19 × 2 + 18 × 1) = 224 cases (19 *vck* configurations for each of the two VCK settings, VCK65 and VCK60, and 18 *vck* configurations for the VCK55 setting, all corresponding to four different VCCTEF settings) are considered. A sweep of angle-of-attack ranging from from −5 deg to 20 deg is considered. There are only 18 *vck* configurations in the case of VCK55 because the 19th configuration is unrealistic for this case.

Instead of showing the results for all 224 cases individually, 3-D bar graphs are shown in Fig. 8.7 and Fig. 8.8 for the case of the VCCTEF 10/10/10 setting corresponding to $C_{l\max}$ and $C_{d\max}$, respectively. These 3-D bar graphs give a consolidated view of $C_{l\max}$ and $C_{d\max}$ for all 56 configurations corresponding to the 10/10/10 setting of VCCTEF. Similarly, 3-D bar graphs for the case of VCCTEF 15/10/5 are shown in Fig. 8.9 and Fig. 8.10, for the case of VCCTEF 20/5/5, in Fig. 8.11 and Fig. 8.12, and for the 30/0/0 case, in Fig. 8.13 and Fig. 8.14, respectively. The 3-D bar graphs present the CFD data defining a subset of the manifold of solutions, which the learning methodology will approximate.

For a closer picture of the variation of $C_{l\max}$ and $C_{d\max}$, it will be necessary to show 2-D bar graphs. We will therefore show the 2-D bar graphs for all the VCCTEF settings in the discussion of our results. A 2-D bar graph is first presented showing $C_{l\max}$ for the

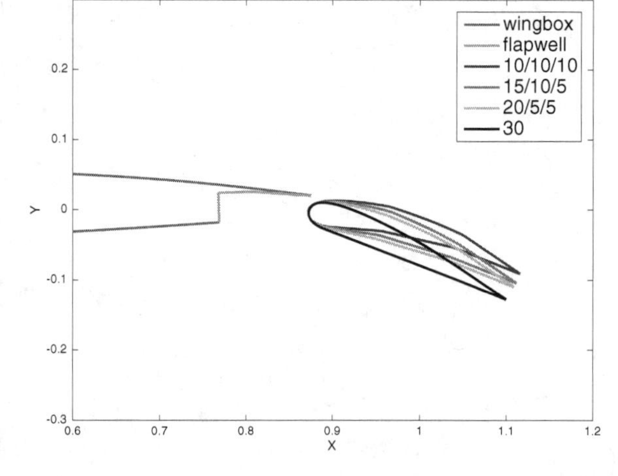

Fig. 8.6 Four VCCTEF settings.

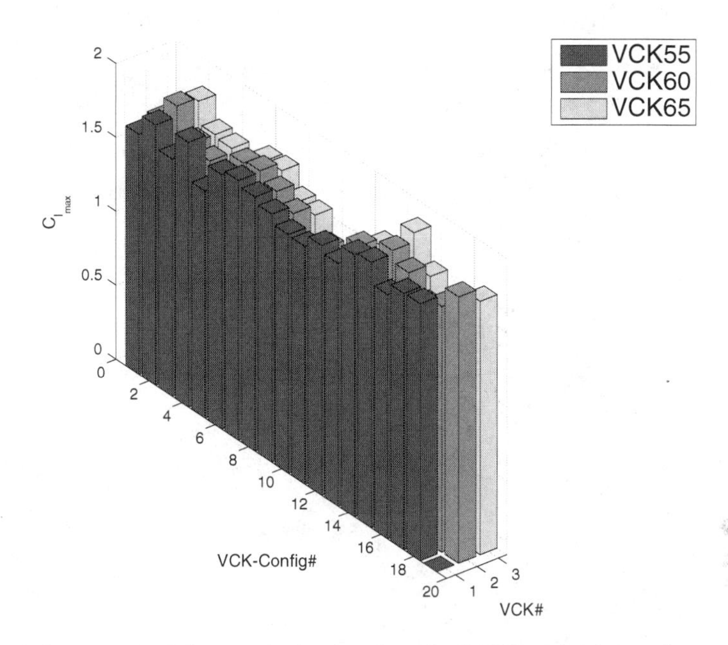

Fig. 8.7 A 3-D bar graph showing $C_{l\max}$ for VCCTEF 10/10/10 setting.

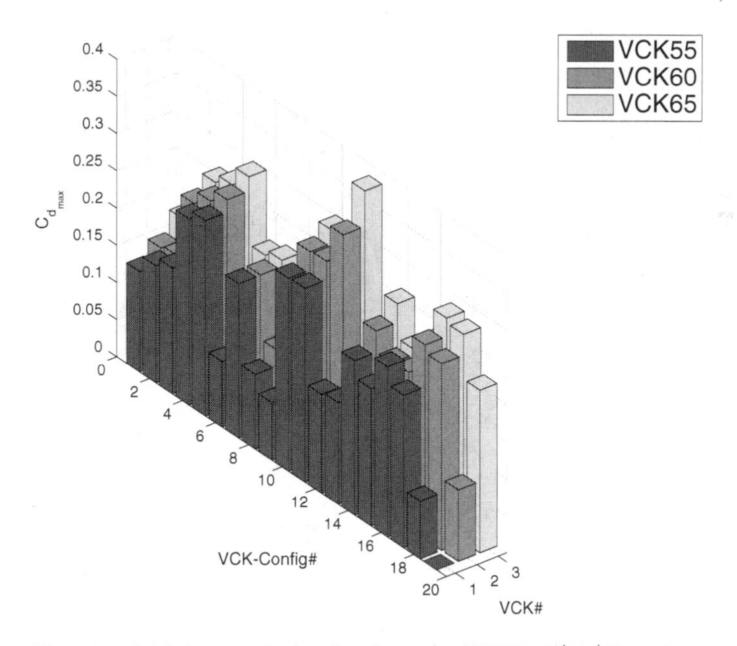

Fig. 8.8 A 3-D bar graph showing $C_{d\max}$ for VCCTEF 10/10/10 setting.

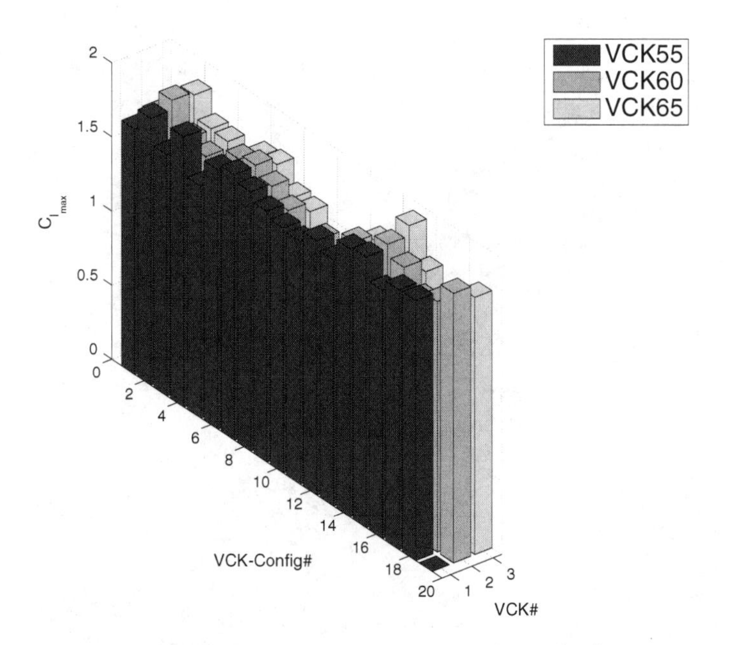

Fig. 8.9 A 3-D bar graph showing $C_{l\max}$ for VCCTEF 15/10/5 setting.

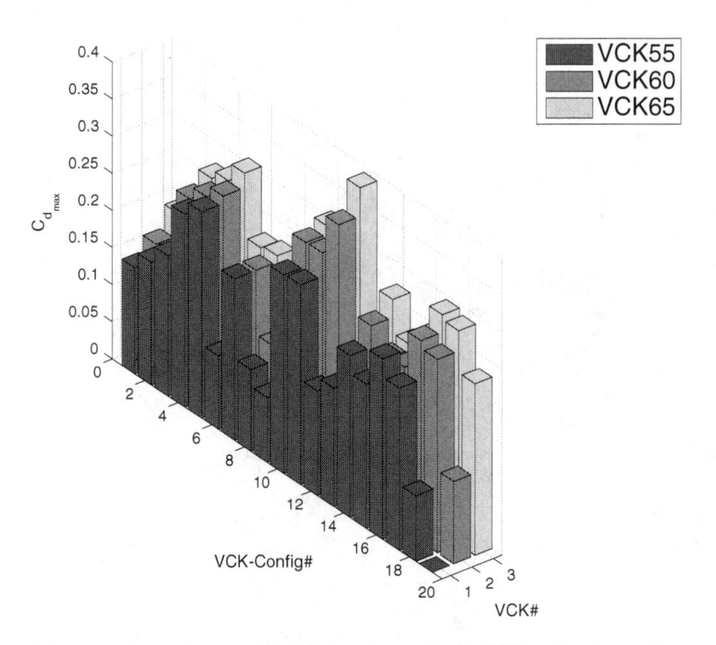

Fig. 8.10 A 3-D bar graph showing $C_{d\max}$ for VCCTEF 15/10/5 setting.

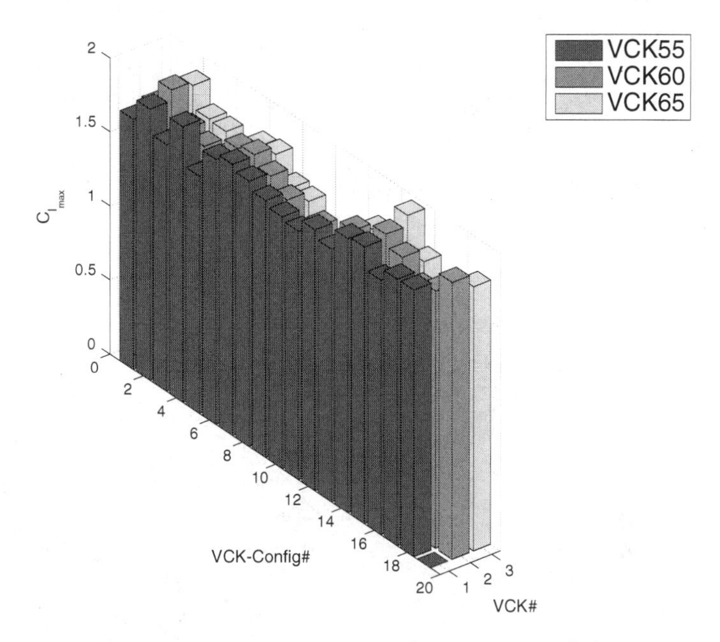

Fig. 8.11 A 3-D bar graph showing $C_{l\max}$ for VCCTEF 20/5/5 setting.

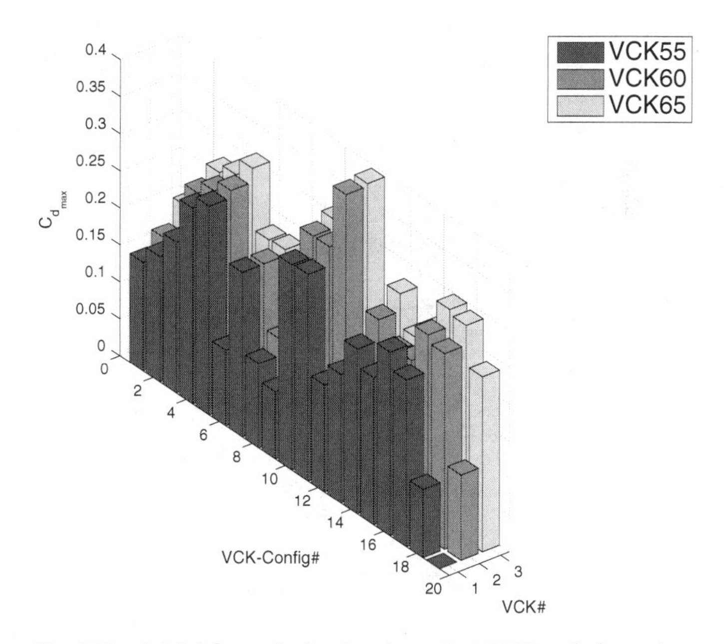

Fig. 8.12 A 3-D bar graph showing $C_{d\max}$ for VCCTEF 20/5/5 setting.

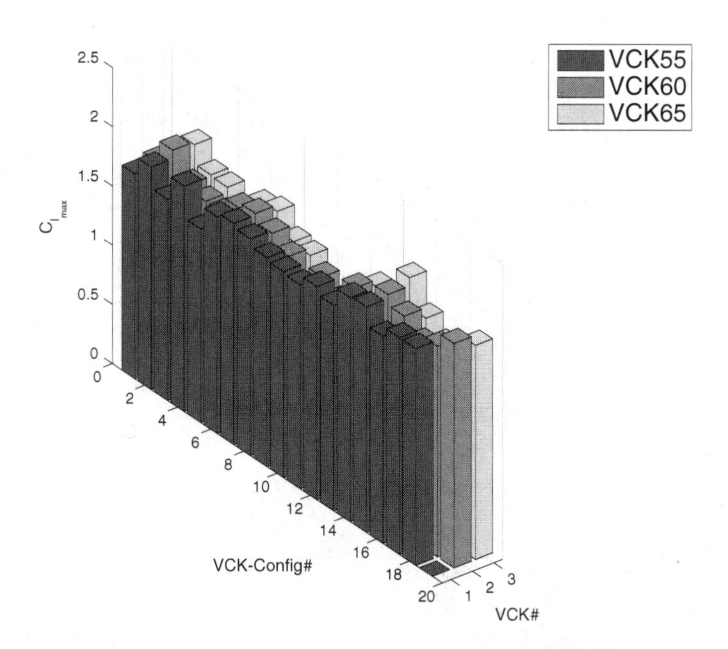

Fig. 8.13 A 3-D bar graph showing $C_{l\max}$ for VCCTEF 30/0/0 setting.

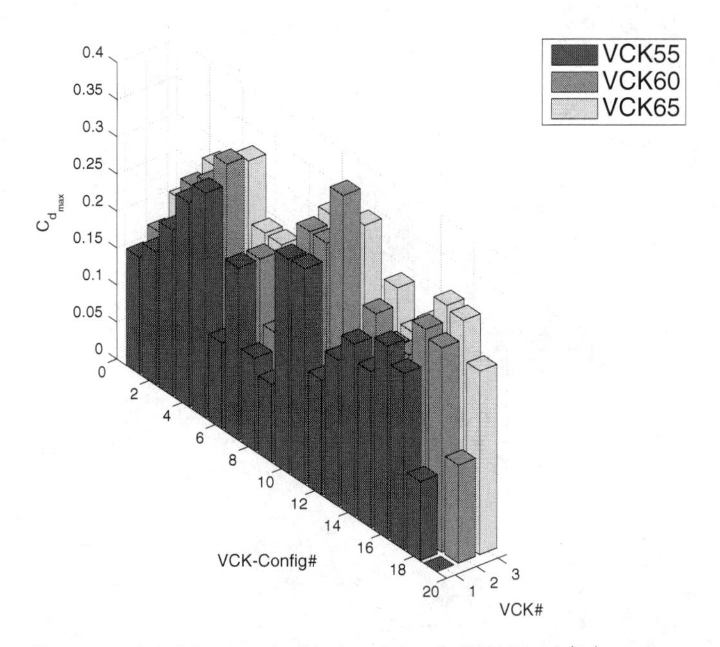

Fig. 8.14 A 3-D bar graph showing $C_{d\max}$ for VCCTEF 30/0/0 setting.

VCCTEF 10/10/10 setting in Fig. 8.15. A corresponding plot for $C_{d\max}$ is shown in Fig. 8.16. Fig. 8.15 gives an overall view of the lift performance of the VCCTEF 10/10/10 setting for all the *vck* configurations corresponding to VCK55, VCK60, and VCK65, and Fig. 8.16 shows corresponding results for $C_{d\max}$. Similarly, Fig. 8.17 and Fig. 8.18 show $C_{l\max}$ and $C_{d\max}$, respectively, for the VCCTEF 15/10/5 setting. Figure 8.19 and Fig. 8.20 show the corresponding results for the VCCTEF 20/5/5 setting, and Figs. 8.21 and 8.22 show the corresponding results for the VCCTEF 30/0/0 setting. Figures 8.15 through 8.22 give an overall view of the $C_{l\max}$ and $C_{d\max}$ results for all the cases considered.

Results for C_l versus α are shown for a subset of these 224 cases. For this purpose, cases giving the four largest values of $C_{l\max}$ are selected from Figs. 8.15 through 8.22. It turns out that for the VCK55 setting, *vck* configurations of 2, 4, 14, and 15 give the largest $C_{l\max}$ for all four VCCTEF settings of 10/10/10, 15/10/5, 20/5/5, and $30 \times 0 \times 0$. For VCK60 and VCK65 settings, *vck* configurations of 2, 7, 15, and 19 give the largest $C_{l\max}$ for all four VCCTEF settings. Therefore, in the discussion of results that follows only these *vck* configurations will be considered.

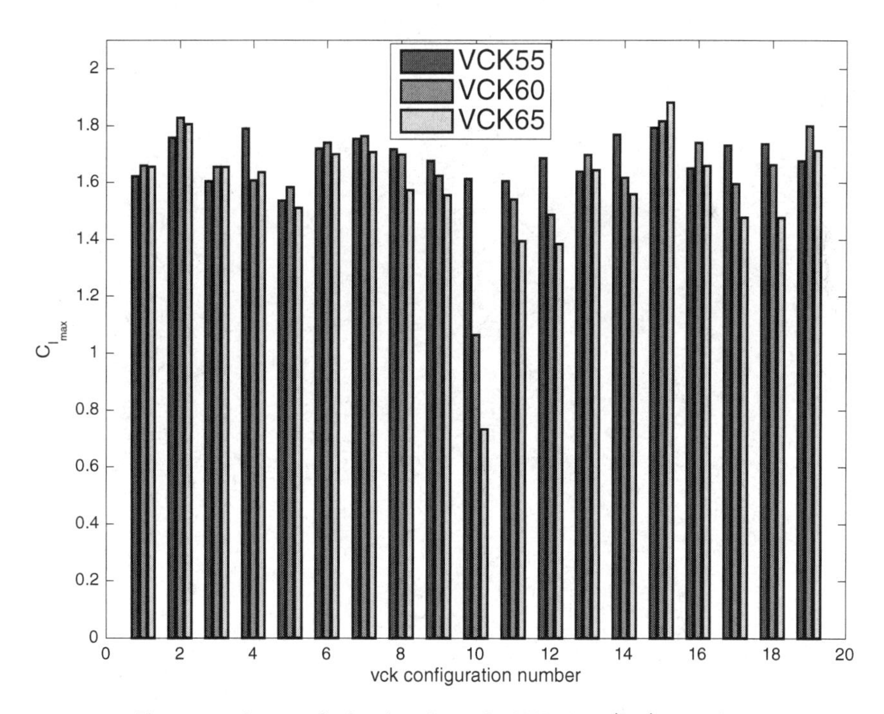

Fig. 8.15 Bar graph showing $C_{l\max}$ for VCCTEF 10/10/10 setting.

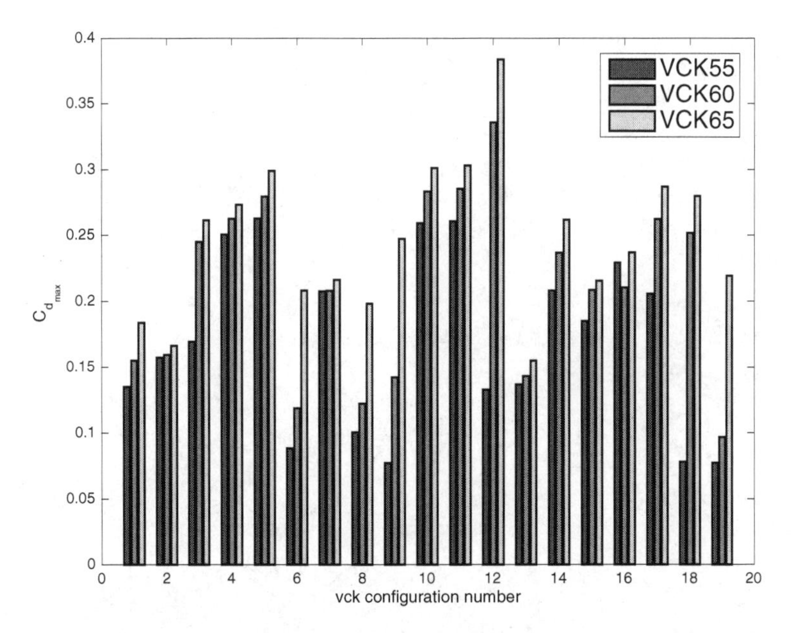

Fig. 8.16 Bar graph showing $C_{d\max}$ for VCCTEF 10/10/10 setting.

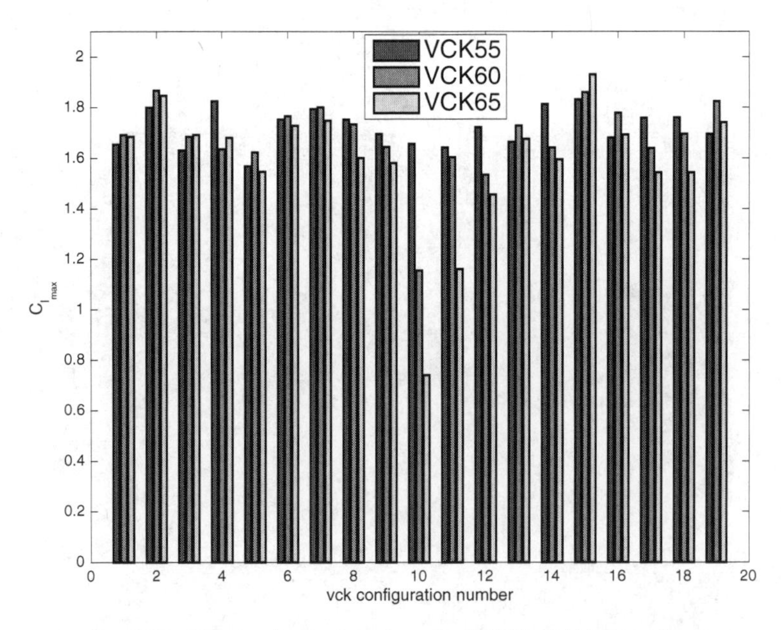

Fig. 8.17 Bar graph showing $C_{l\max}$ for VCCTEF 15/10/5 setting.

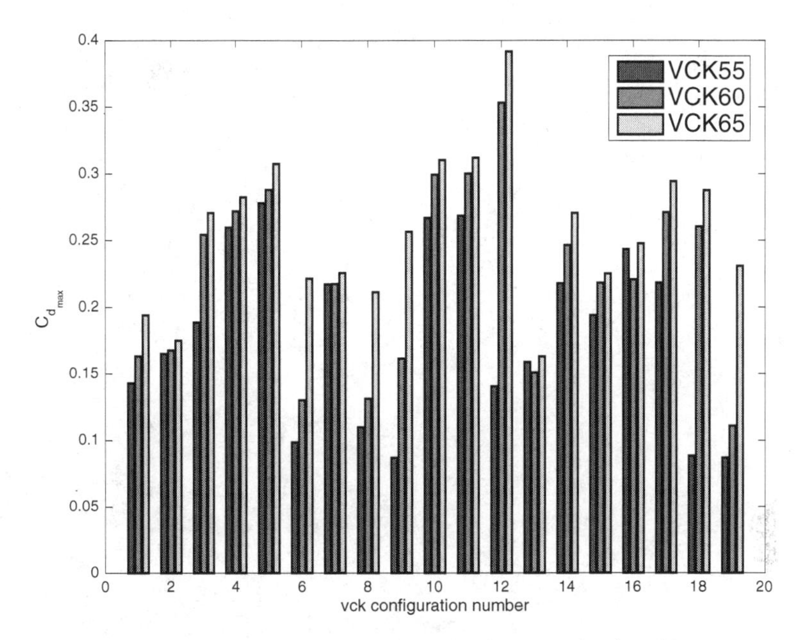

Fig. 8.18 Bar graph showing $C_{d\max}$ for VCCTEF 15/10/5 setting.

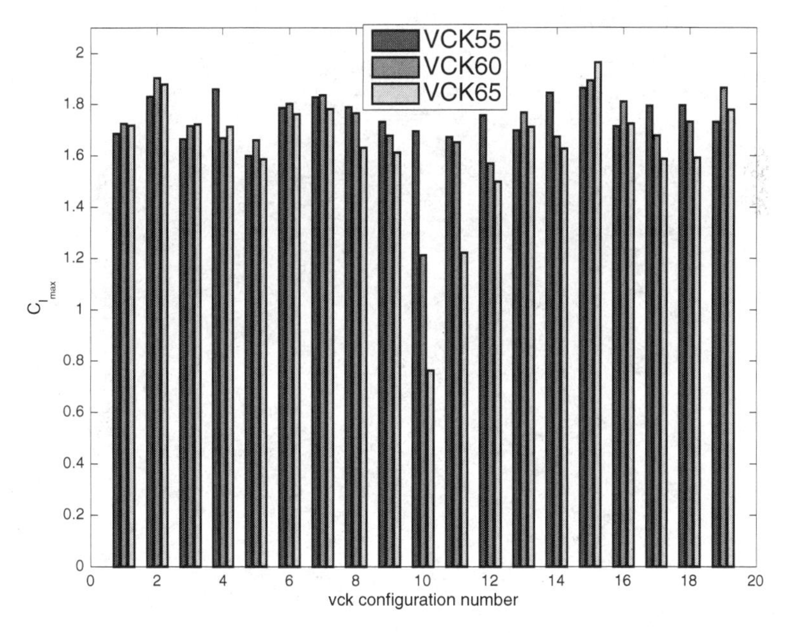

Fig. 8.19 Bar graph showing $C_{l\max}$ for VCCTEF 20/5/5 setting.

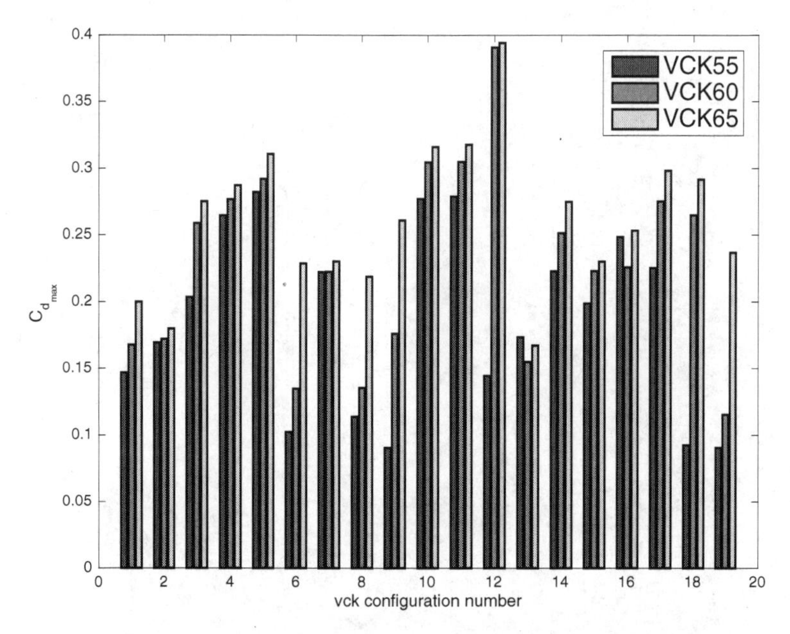

Fig. 8.20 Bar graph showing $C_{d\max}$ for VCCTEF 20/5/5 setting.

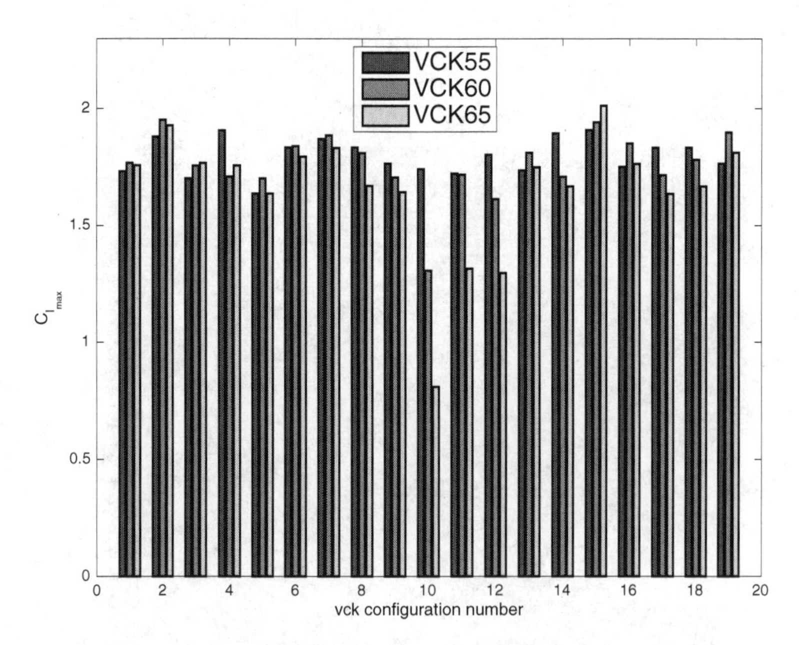

Fig. 8.21 Bar graph showing $C_{l\max}$ for VCCTEF 30/0/0 setting.

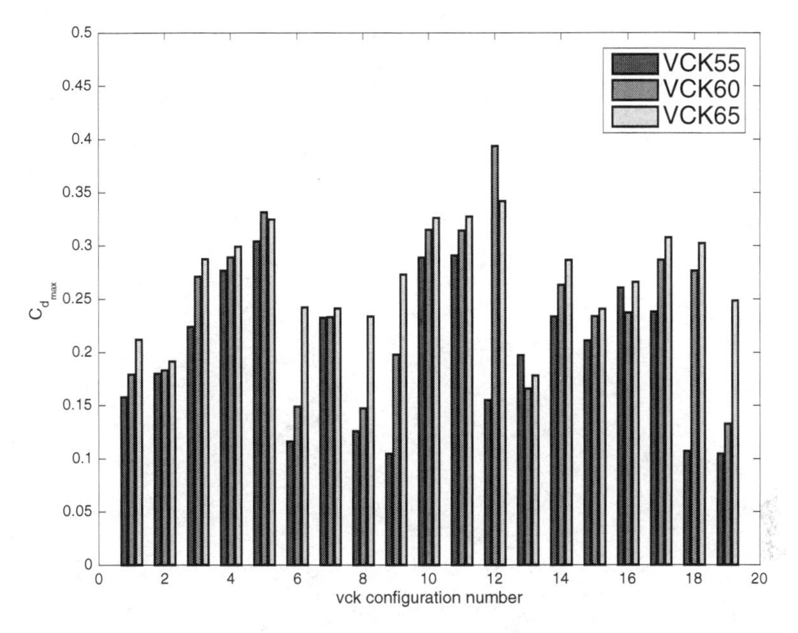

Fig. 8.22 Bar graph showing $C_{d\max}$ for VCCTEF 30/0/0 setting.

The determination of $C_{l\max}$ and $C_{d\max}$ is made by inspecting the lift curve and drag polar results. For example, for the VCK65 setting, $C_{l\max}$ and $C_{d\max}$ are shown in Fig. 8.23a and Fig. 8.23b, respectively. Representative $C_{l\max}$ and $C_{d\max}$ are shown in Fig. 8.24, Fig. 8.25, and Fig. 8.26 for the VCK55, VCK60, and VCK65 settings, respectively, and the case of the VCCTEF 20/5/5 setting.

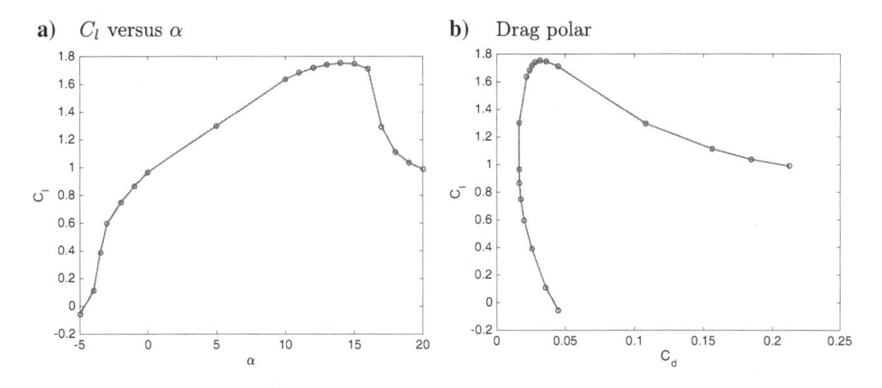

Fig. 8.23 VCK65 results for the *vck*1 configuration and VCCTEF 30/0/0 setting.

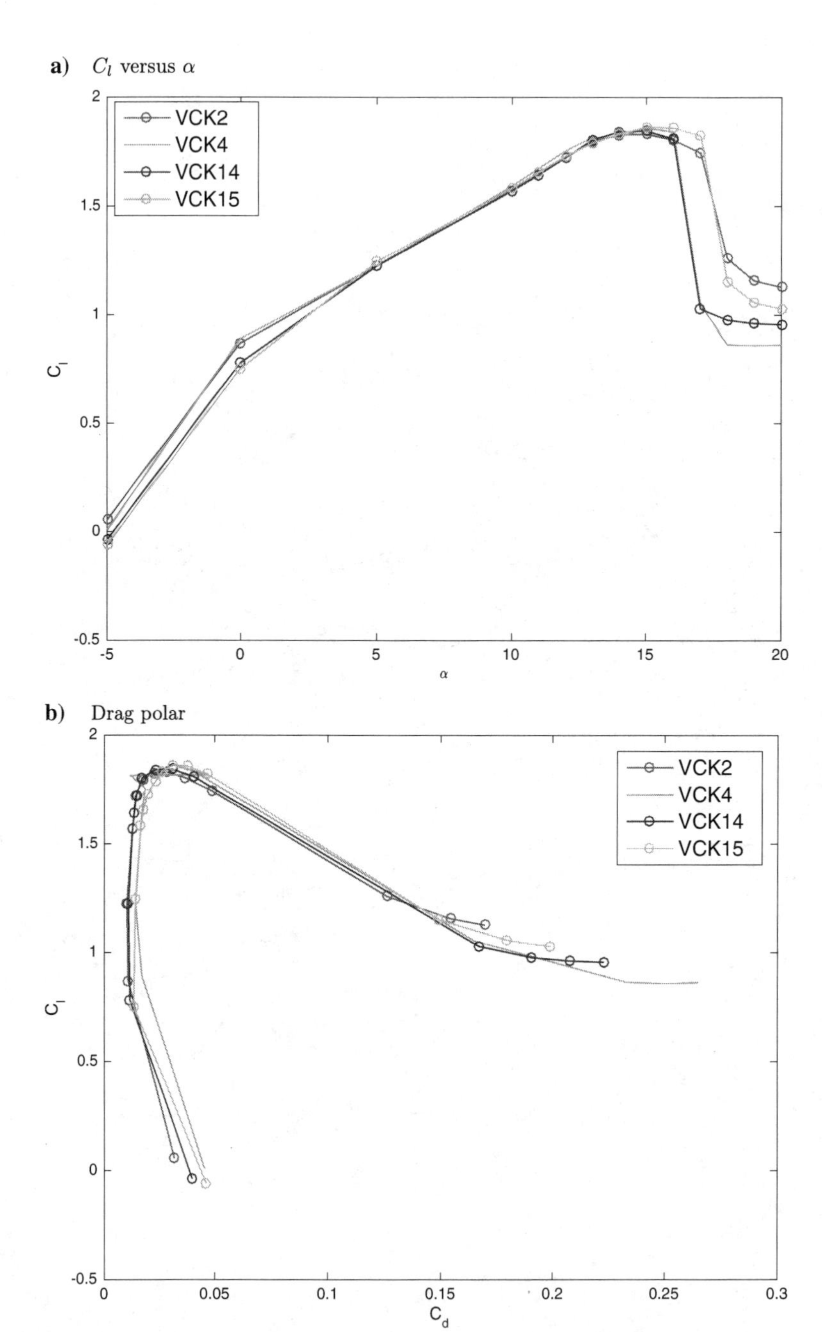

Fig. 8.24 VCK55 results for the VCCTEF 20/5/5 setting.

a) C_l versus α

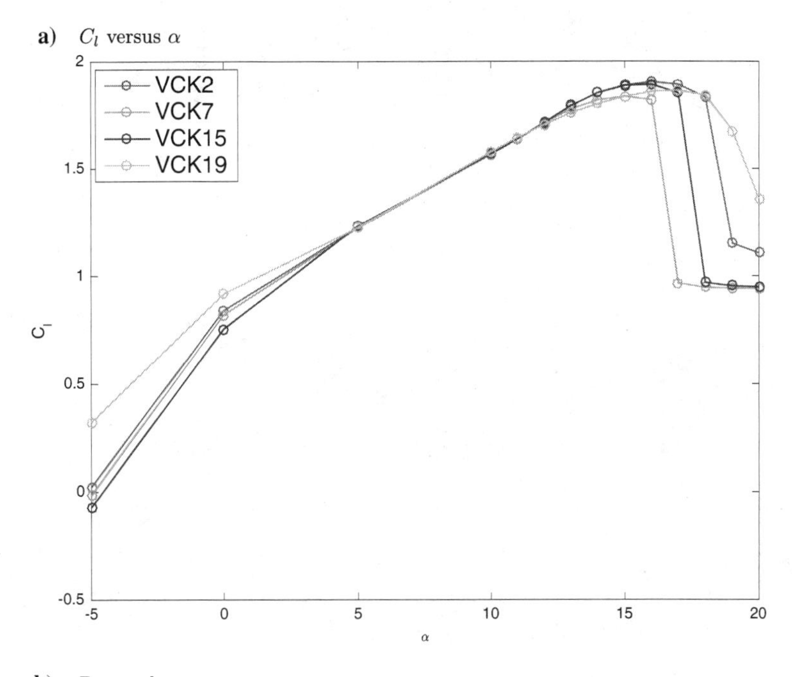

b) Drag polar

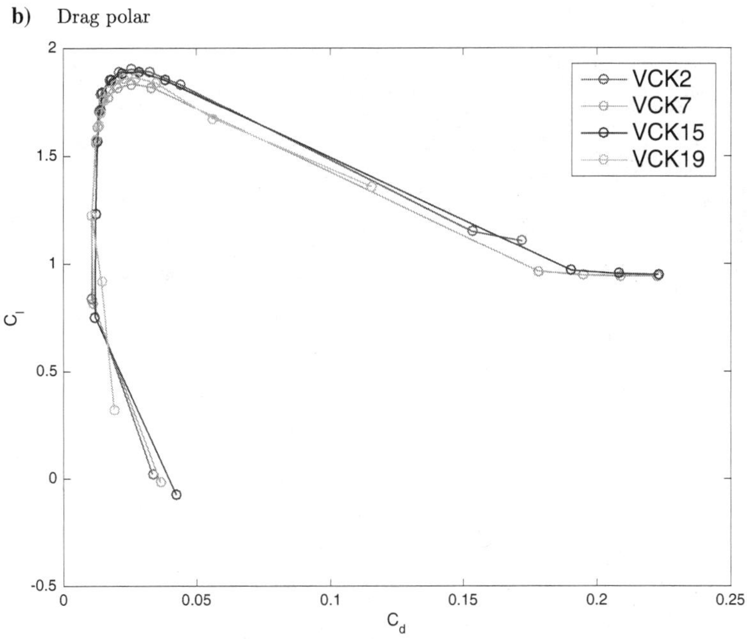

Fig. 8.25 VCK60 results for the VCCTEF 20/5/5 setting.

a) C_l versus α

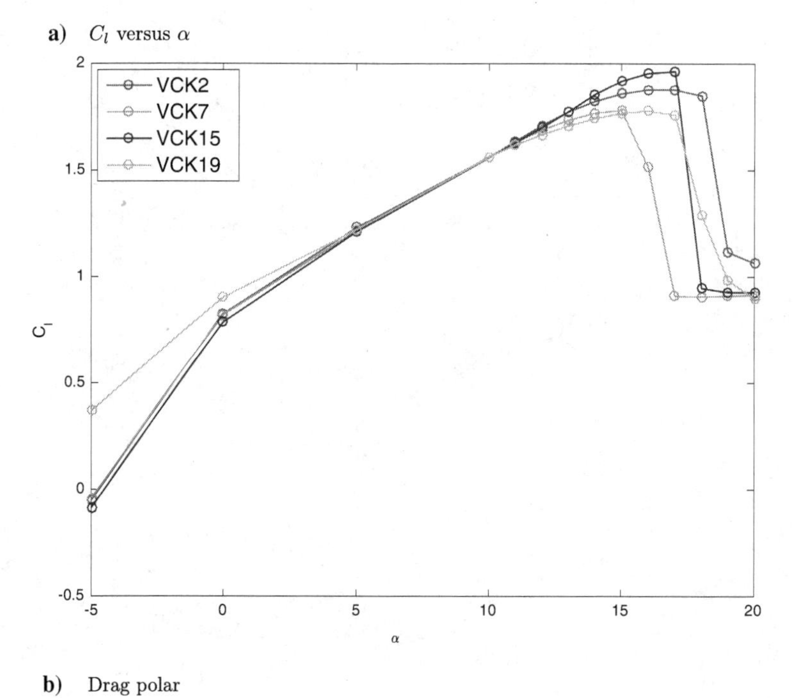

b) Drag polar

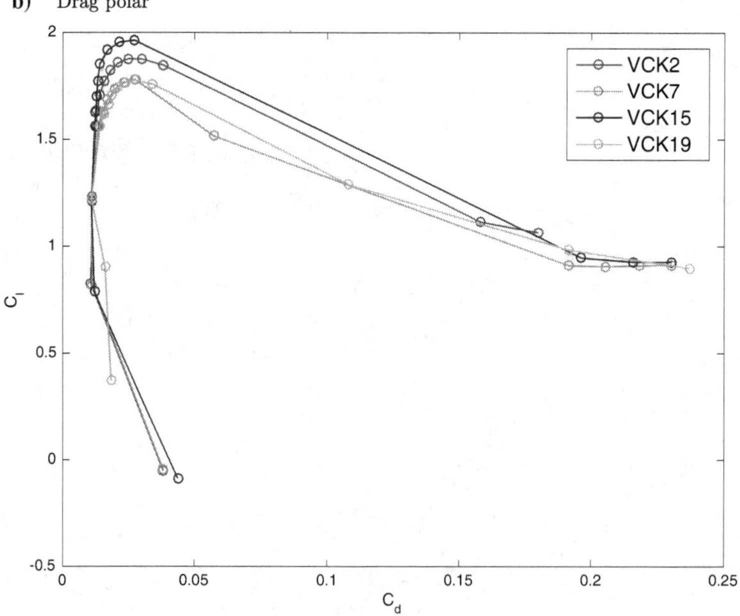

Fig. 8.26 VCK65 results for the VCCTEF 20/5/5 setting.

From Fig. 8.23a and Fig. 8.23b, it is seen that a constant lift curve slope exists only beyond $\alpha = 0$, which shows that at lower angles-of-attack, the lift curve for the tri-element VCK-wing-VCCTEF system does not follow linear theory. This is shown by a nonlinear lift curve in the α range below 0 deg.

Figure 8.24 shows the C_l versus α and drag polar results, respectively, for the VCK55 and VCCTEF 20/5/5 settings corresponding to *vck*2, *vck*4, *vck*14, and *vck*15 configurations. The *vck*15 case outperforms the other three, based on maximum C_l. As mentioned previously, in high-lift flight configuration, we want to minimize the stall speed, which can be accomplished by maximizing C_l. The *vck*15 case also performs the best for the other three VCCTEF settings, 15/10/5, 20/5/5, and 30/0/0 for the VCK55 setting, not shown here.

The situation is different for the VCK60 setting, where the best lift performance $C_{l\max}$ is demonstrated by the *vck*2 case for all the VCCTEF settings. For example, for the VCCTEF 20/5/5 setting, highest $C_{l\max}$ corresponds to the *vck*2 configuration, as shown in Fig. 8.25. This also holds true for the other three VCCTEF settings. For the VCK65 setting, best lift performance is demonstrated again by the *vck*15 configuration for all the VCCTEF settings. For example, this result is shown for the VCCTEF 20/5/5 setting in Fig. 8.26. The overall result for the VCK60 setting is shown in Fig. 8.25a, where *vck*2 and *vck*15 configurations yield the best and the next best high-lift performance. For the VCK65 setting, the overall result is shown in Fig. 8.26a, where *vck*15 and *vck*2 configurations yield the best and the next best high-lift performance, respectively.

It is shown that *vck*15 and *vck*2 configurations are the top two candidates, in terms of overall high-lift performance, out of all the three VCK settings (VCK55, VCK60, and VCK65) for the VCCTEF 20/5/5 setting. Figure 8.27a further shows that the VCCTEF 30/0/0 setting gives the highest lift performance ($C_l - \alpha$), corresponding to *vck*15 and *vck*2 configurations, out of all the four VCCTEF settings.

8.3.2 SSL RESULTS

The CFD solutions for $C_{l\max}$ and $C_{d\max}$ for the 224 VCK-VCCTEF configurations are shown in Figs. 8.7 through 8.27. The SSL is based on these 224 configurations. We now pick a small subset of test configurations outside of these 224 configurations for testing the learning methodology. These test configurations are shown in Fig. 8.28 for the case of VCK55, VCK60, and VCK65 settings, respectively. The learning algorithm is tested on these nine configurations for the three VCK settings corresponding to the VCCTEF 30/0/0 setting. The test metrics quantifying the accuracy of the learning methodology are chosen to be root-mean-square (rms) error, rms error normalized by standard deviation, and rms error normalized by maximum deviation.

First, we report the results corresponding to fully supervised learning (i.e., the complete manifold has no unlabeled points). Figure 8.29 shows the results for $C_{l\max}$

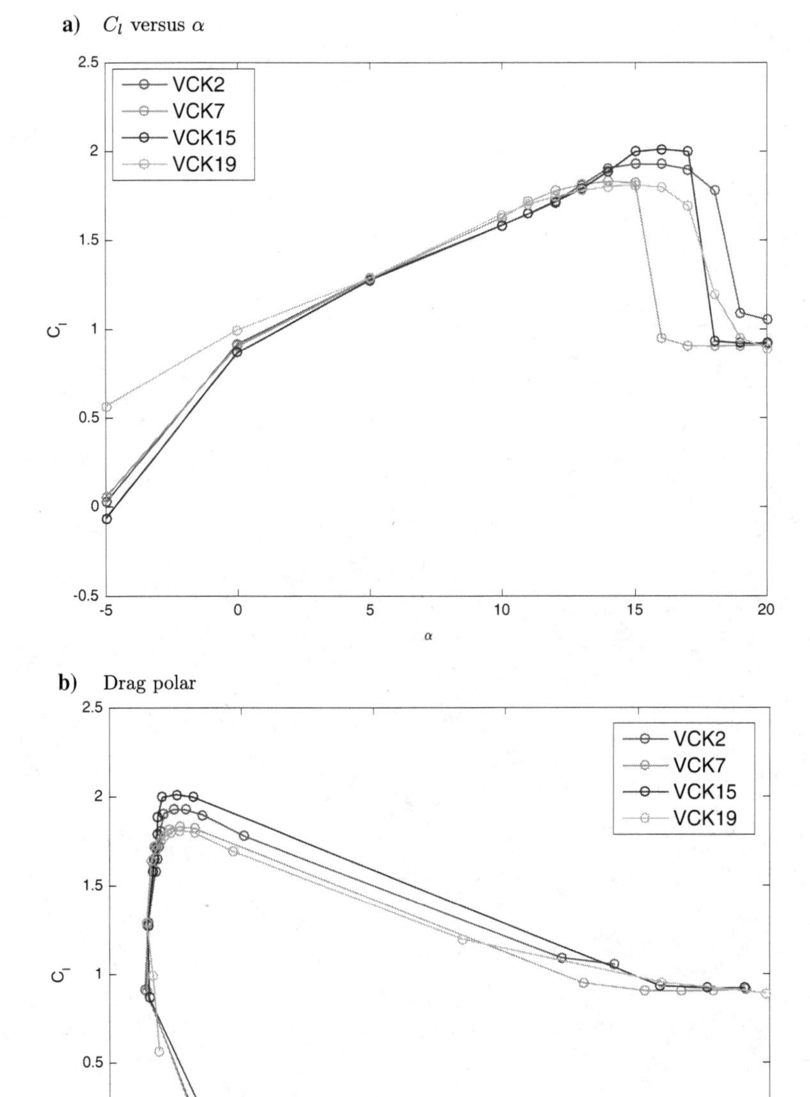

Fig. 8.27 VCK65 results for the VCCTEF 30/0/0 setting.

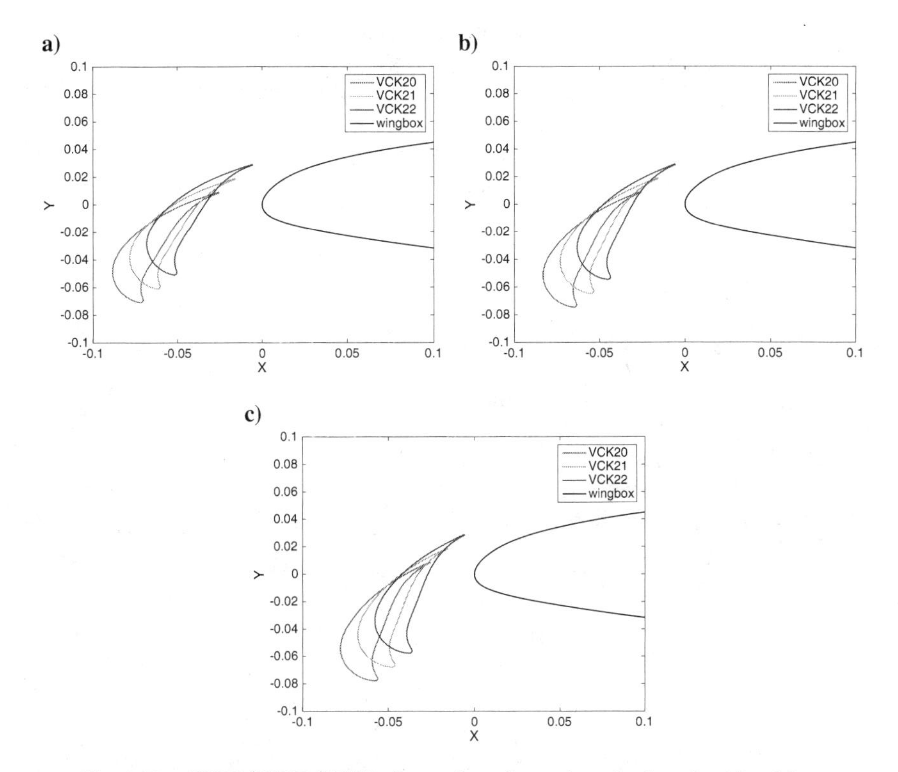

Fig. 8.28 VCK55/VCK60/VCK65: Test *vck* configurations for learning algorithm.

corresponding to VCK55, VCK60, and VCK65 settings, respectively. The predictions approximate the CFD solutions (true solutions) within the rms error bound of about 5%. This is a good approximation, given that the training space is populated with just 224 solutions. For our purpose here, it is sufficient to use this limited training set because our predictions will be used to define a new design space that will eventually be validated by wind tunnel experiments. Knowing the nature of the data from CFD computations, a test set could be chosen selectively to reduce this error further. But, in general, especially for a large database, a deterministic selection of a test set may not be feasible. In light of this, a test set is chosen here deliberately to stress-test the learning algorithm.

Figure 8.30 shows the corresponding $C_{l\max}$ results for SSL, with unlabeled space being a small subset (10%, chosen randomly) of the 224 solutions. Upon comparison of Figs. 8.29 and 8.30, it is shown that the SSL appears to give similar predictions as the fully supervised learning. In fact, the difference in predictions corresponding to the fully supervised learning and SSL falls within the statistical error margin of 5%. But, it should be noted that this error margin is influenced by the choice of random selection of the unlabeled data. Overall, the

SSL yields predictions as accurate as the fully supervised learning, allowing for the statistical error margin of 5%, as stated previously. This is in keeping with the expectation that the Laplacian component of the learning algorithm approximates the manifold over the unlabeled points as well as the fully supervised algorithm.

Similar behavior of the learning model is shown for $C_{d\max}$ through Fig. 8.31 and Fig. 8.32. Figure 8.31 shows results corresponding to the fully supervised learning, and Fig. 8.32 shows results corresponding to the SSL. Again, the rms error between the predictions and the true solution is within 5% for the fully supervised and SSL.

SSL was also used with the unlabeled space as 20% of the 224 solutions. The comparison for $C_{l\max}$ is shown in Table 8.3, where three different measures of error are compared corresponding to fully supervised learning and SSL: rms, rms error normalized by the standard deviation and rms error normalized by the difference in maximum and minimum of $C_{l\max}$. Differences in results

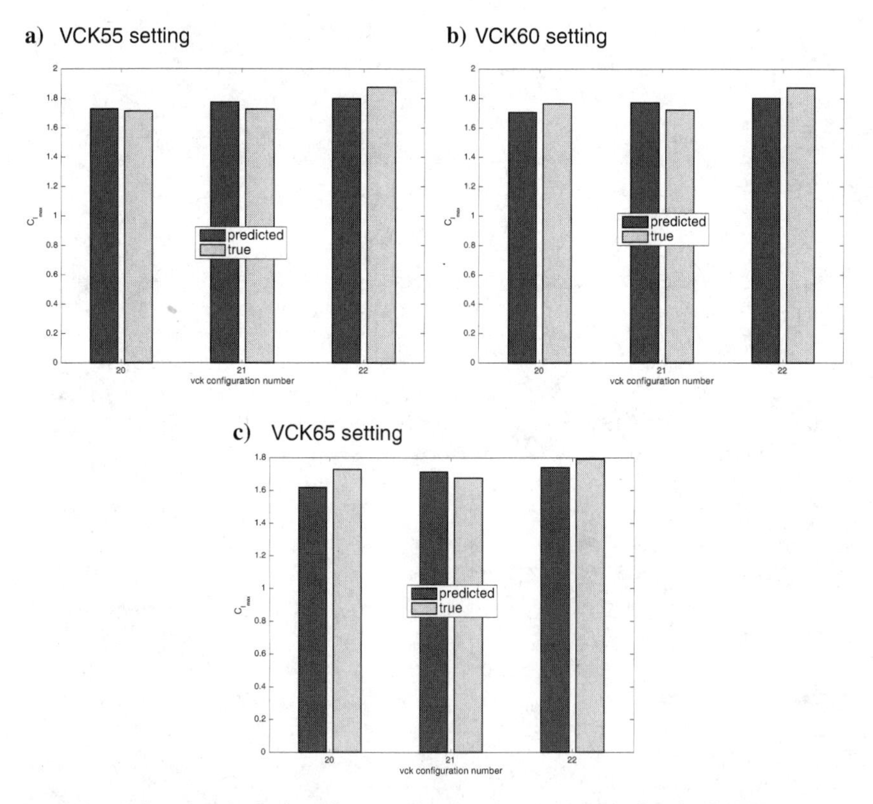

Fig. 8.29 Fully supervised algorithm: predicted versus true $C_{l\max}$ for the VCCTEF 30/0/0 setting.

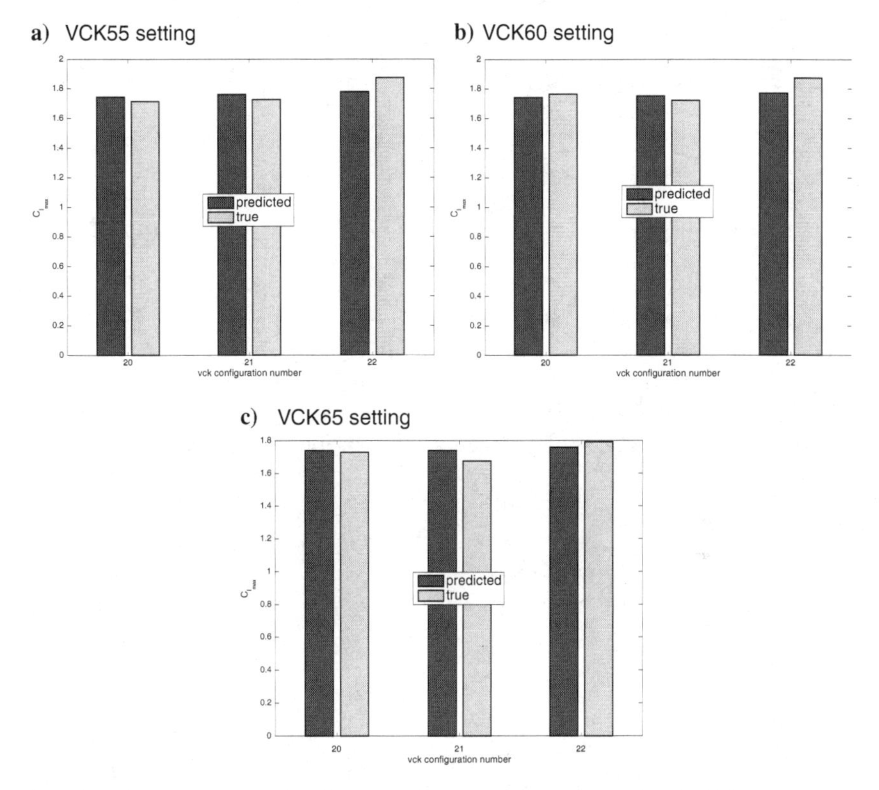

Fig. 8.30 Semisupervised algorithm: predicted versus true $C_{l\max}$ for the VCCTEF 30/0/0 setting.

between fully supervised and semi-supervised learning are within a 5% spread. The difference is not statistically significant. Corresponding results for $C_{d\max}$ are shown in Table 8.4. The predictions with the 20% unlabeled space are slightly degraded as compared to those of the 10% unlabeled space, but they are still close. In general, the errors are statistically similar. It should be mentioned here that the choice of the unlabeled data set introduces a jitter in the results from simulation to simulation, and Laplacian regularizer over unlabeled space may sometimes equal or outperform fully supervised methodology within the statistical error margin, as shown in Table 8.3.

Finally, Table 8.5 shows a comparison of the execution time among the CFD, fully supervised, and semisupervised solution methodologies. Whereas each CFD solution takes about 310 sec on a 480 processor Pleiades supercomputer in the NASA facility, fully supervised learning and SSL methodologies take less than 1 sec on a MacBook Pro laptop.

8.4 SUMMARY

In this chapter, a detailed CFD high-lift study of the VCK-VCCTEF system has been carried out to study the viscous effects in takeoff and landing and explore the best VCK-VCCTEF system designs. For this purpose, a three-segment variable camber airfoil, employed as a performance adaptive aeroelastic wing shaping control effector for a NASA GTM in landing and takeoff configurations, was considered. The objective of the study was to define optimal high-lift VCCTEF settings and VCK settings/configurations. A total of 224 combinations of VCK settings/configurations and VCCTEF settings were considered for the inboard GTM wing, where the VCCTEFs are configured as a Fowler flap that forms a slot between the VCCTEF and the inboard of the main wing. For the VCK settings of deflection angles of 55°, 60°, and 65°, 18, 19, and 19 *vck* configurations, respectively, were considered for each of the four different VCCTEF deflection settings. Different VCK configurations were defined by varying the horizontal and vertical

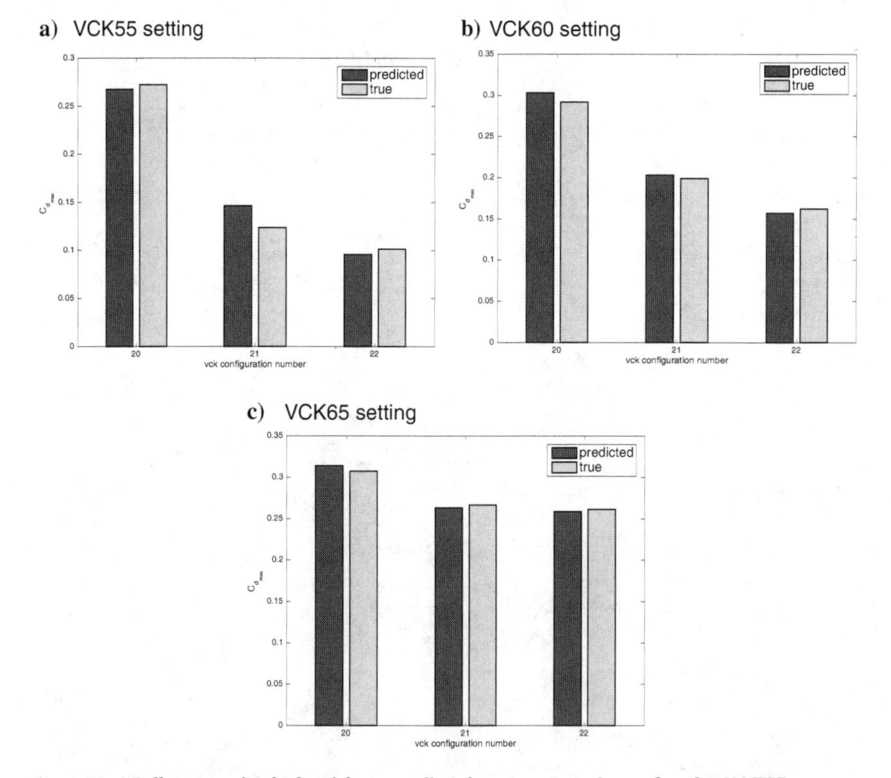

Fig. 8.31 Fully supervised algorithm: predicted versus true $C_{d\max}$ for the VCCTEF 30/0/0 setting.

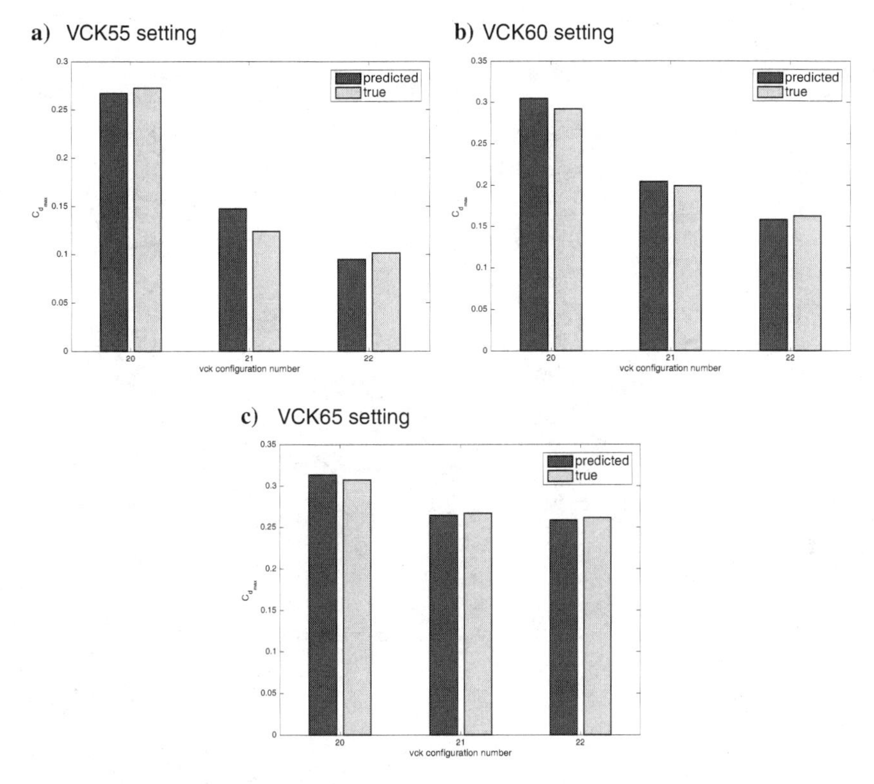

Fig. 8.32 Semisupervised algorithm: predicted versus true $C_{d\max}$ for the VCCTEF 30/0/0 setting.

distance of the *vck* from the main wing. We have identified two topmost *vck* configurations corresponding to each of the three VCK settings, out of the 224 configurations considered. For all the VCK settings, *vck*2 and *vck*15 give the best lift performance, regardless of the VCCTEF setting used. In particular, the VCCTEF 30/0/0 setting gave the highest overall lift performance with the *vck*2 and *vck*15 configurations. Thus, the best configurations for the GTM airfoil have been

TABLE 8.3 DIFFERENT ERROR MEASURES FOR C_L

Algorithm	rms	rms/st	rms/($max(C_{l\max}) - min(C_{l\max})$)
Fully supervised	0.0336	0.8931	0.3161
Semisupervised (U = 10%)	0.0301	0.7984	0.2826
Semisupervised (U = 20%)	0.0256	0.6789	0.2403

TABLE 8.4 DIFFERENT ERROR MEASURES FOR C_D

Algorithm	rms	rms/st	rms/($max(C_{lmax})$ − $min(C_{lmax})$)
Fully supervised	0.0307	0.1236	0.0458
Semisupervised (U = 10%)	0.0324	0.1304	0.0483
Semisupervised (U = 20%)	0.0705	0.2840	0.1051

TABLE 8.5 COMPARISON OF EXECUTION TIME

CFD	Fully Supervised Learning	Semisupervised Learning
310 sec	0.8 sec	0.9 sec

identified from the 224 cases studied. This provides a useful guide for the wind tunnel experiment to verify the best design GTM configurations. Some of these best high-lift configurations offer a counterintuitive design that would not have been considered experimentally a priori.

Because only a small subset of a very large set of possible configurations was considered using CFD, due to a prohibitive computational expense and time limitations, the SSL methodology was adopted as a supplementary tool to circumvent these problems. Using the SSL methodology, a high-lift study of nine additional configurations of the VCK-VCCTEF system was carried out. The predicted $C_{l\max}$ and $C_{d\max}$ results obtained with the SSL methodology were shown to be close to the true (CFD) solutions, within the rms error bound of 5%. This demonstrates the feasibility of the SSL as a design tool for identifying the best performing VCK-VCCTEF configurations very rapidly, in conjunction with CFD simulations and wind tunnel experiments. It should be noted here that the size of the unlabeled data was picked to be 10% by also selecting another number, 20%, which underperformed the number 10%. This number tends to have an inverse relation to the degree of sparsity of the CFD database.

Thus, the SSL methodology for the design problem at hand is well established to rapidly populate an arbitrarily large design space and predict $C_{l\max}$ and $C_{d\max}$ corresponding to the new test points. This reinforces the overall CFD-SSL design methodology adopted here and provides a viable tool for problems in wing design. It is encouraging to note that the SSL methodology used here is independent of a particular design system. In fact, it can be extended to solve a variety of design problems in engineering.

ACKNOWLEDGEMENTS

The authors would like to thank the Advanced Air Transport Technology Project under the Advanced Air Vehicles Program of the NASA Aeronautics Research Mission Directorate (ARMD) for funding support of this work. The authors also would like to acknowledge Boeing Research and Technology and the University of Washington for their collaboration with NASA under NASA contract NNL11AA05B task order NNL12AD09T entitled "Development of Variable Camber Continuous Trailing Edge Flap System for B757 Configured with a More Flexible Wing."

REFERENCES

[1] Spalart, P. R., and Allmaras, S. R., "A One-Equation Turbulence Model for Aerodynamic Flows," *AIAA 30th Aerospace Sciences Meeting and Exhibit*, AIAA 92-0439, Reno, NV, Jan. 1992.

[2] Kaul, U. K., "A Kernel-Based Semi-Supervised Machine Learning Methodology," Internal NASA Report, NASA Ames Research Center, June 20, 2005.

[3] Belkin, M., Niyogi, P., and Sindhwani, V., "Manifold Regularization: A Geometric Framework for Learning from Examples," *Journal of Machine Learning Research*, vol. 7, 2006, 2399–2434; also Internal Report, The University of Chicago, IL, Aug. 2004.

[4] Kaul, U. K., and Nguyen, N. T., "Lift Optimization Study of a Multi-Element Three-Segment Variable Camber Airfoil," *34th Applied Aerodynamics Conference, AVIATION 2016 Forum*, AIAA 2016-3569, June 13–17, 2016, Washington, D.C.

[5] Kaul, U. K., "FiDDLE: A Computer Code for Finite Difference Development of Linear Elasticity in Generalized Curvilinear Coordinates," NASA/TM-2005-213450, Jan. 2005.

[6] Kaul, U. K., and Oza, N. C., "Machine Learning for Detecting and Locating Damage in a Rotating Gear," *SAE Transactions*, Paper 2005-01-3371, 2005; also SAE *J. Aerospace*, 2005, V114-1.

[7] Nguyen, N., "Elastically Shaped Future Air Vehicle Concept," NASA Innovation Fund Award 2010 Report, Oct. 2010, submitted to NASA Innovative Partnerships Program. http://ntrs.nasa.gov/archive/nasa/casi.ntrs.nasa.gov/20110023698.pdf.

[8] Nguyen, N., Trinh, K., Reynolds, K., Kless, J., Aftosmis, M., Urnes, J., and Ippolito, C., "Elastically Shaped Wing Optimization and Aircraft Concept for Improved Cruise Efficiency," *AIAA Aerospace Sciences Meeting*, AIAA-2013-0141, Jan. 2013.

[9] Boeing Report No. 2012X0015, "Development of Variable Camber Continuous Trailing Edge Flap System," Oct. 4, 2012.

[10] Urnes, J., Nguyen, N., Ippolito, C., Totah, J., Trinh, K., and Ting, E., "A Mission Adaptive Variable Camber Flap Control System to Optimize High Lift and Cruise Lift to Drag Ratios of Future N+3 Transport Aircraft," *AIAA Aerospace Sciences Meeting*, AIAA-2013-0214, Jan. 2013.

[11] Jordan, T. L., Langford, W. M., Belcastro, C. M., Foster, J. M., Shah, G. H., Howland, G., and Kidd, R., "Development of a Dynamically Scaled Generic Transport Model

Testbed for Flight Research Experiments," *AUVSI Unmanned Unlimited*, Arlington, VA, 2004.

[12] Nguyen, N. T., Kaul, U. K., Lebofsky, S., Ting, E., Chaparro, D., and Urnes, J., "Development of Variable Camber Continuous Trailing Edge Flap for Performance Adaptive Aeroelastic Wing," SAE Paper 15ATC-0250, 2015.

[13] Kaul, U. K., and Nguyen, N. T., "Drag Optimization Study of Variable Camber Continuous Trailing Edge Flap (VCCTEF) Using OVERFLOW," *AIAA 32nd Applied Aerodynamics Conference*, AIAA 2014-2444, Atlanta, GA, June 2014.

[14] Nguyen, N. T., Precup, N., Livne, L., Urnes, J., Dickey, E., Nelson, C., Chiew, J., Rodriguez, D., Ting, E., and Lebofsky, S., "Wind Tunnel Investigation of a Flexible Wing High-Lift Configuration with a Variable Camber Continuous Trailing Edge Flap Design," *AIAA 33rd Applied Aerodynamics Confererence*, AIAA-2015-2417, June 2015.

[15] Nguyen, N., Precup, N., Urnes, J., Nelson, C., Lebofsky, S., Ting, E., and Livne, E., "Experimental Investigation of a Flexible Wing with a Variable Camber Continuous Trailing Edge Flap Design," *AIAA 32nd Applied Aerodynamics*, AIAA 2014-2441, June 2014.

[16] Nguyen, N., Ting, E., and Lebofsky, S., "Aeroelastic Analysis of Wind Tunnel Test Data of a Flexible Wing with a Variable Camber Continuous Trailing Edge Flap," *56th AIAA/ASME/ASCE/AHS/ASC Structures, Structural Dynamics, and Materials Conference*, AIAA-2015-1405, Jan. 2015.

[17] Precup, N., Mor, M., and Livne, E., "The Design, Construction, and Tests of a Concept Aeroelastic Wind Tunnel Model of a High-Lift Variable Camber Continuous Trailing Edge Flap (HL-VCCTEF) Wing Configuration," *56th AIAA/ASCE/AHS/ASC Structures, Structural Dynamics, and Materials Conference*, AIAA 2015-1406, Kissimmee, FL, Jan. 5–9, 2015.

[18] Tikhonov, A. N., "Regularization of Incorrectly Posed Problems," *Soviet Mathematics Doklady*, 4, 1963, 1624–1627.

CHAPTER 9

Adaptive Architectures for Control of Uncertain Dynamical Systems with Actuator Dynamics

Benjamin C. Gruenwald, Jonathan A. Muse, Daniel Wagner and Tansel Yucelen*

Stability properties of adaptive controllers can be seriously affected by the presence of actuator dynamics. Specifically, if the bandwidth of the actuator dynamics is sufficiently high, then a common practice is to neglect these dynamics in the design of adaptive controllers. However, if this is not the case and/or safety-critical applications of adaptive controllers are considered, then stability verification steps must be taken to show the allowable bandwidth range for actuators such that adaptive controllers lead to stable closed-loop system trajectories. Motivated from this standpoint, the purpose of this chapter is to present safe adaptive architectures for control of uncertain dynamical systems with actuator dynamics. A linear matrix inequalities-based hedging framework, which modifies the ideal reference model trajectories to allow for correct adaptation that is not affected by the presence of actuator dynamics, is presented first. Stability of the closed-loop dynamical system utilizing this framework is discussed using Lyapunov-based analysis tools. Next, convergence properties of the modified reference model trajectories to the ideal reference model trajectories are established, and a generalization of the proposed linear matrix inequalities-based hedging framework to adaptive control is presented. An algorithm is also presented to allow for less conservative computations of the minimum allowable actuator bandwidths, as well as

*Corresponding Author: University of South Florida, Engineering Building C 2209, 4202 East Fowler Avenue, Tampa, Florida, United States of America (Email: yucelen@lacis.team; Phone: +1 813 974 5656; Fax: +1 813 974 3539).

B. C. Gruenwald and T. Yucelen are with the Laboratory for Autonomy, Control, Information, and Systems (LACIS, http://www.lacis.team/) of the Department of Mechanical Engineering at the University of South Florida, Tampa, Florida, United States of America; J. A. Muse is with the Autonomous Control Branch at the Air Force Research Laboratory Aerospace Systems Directorate, Wright Patterson Air Force Base, Ohio, United States of America; and D. Wagner is with the Advanced Algorithms for Control and Communications Laboratory of the Department of Electrical Engineering at the Czech Technical University, Prague, Czech Republic.

All authors contributed equally to the reported findings. This research was supported by the Air Force Research Laboratory Aerospace Systems Directorate under the Universal Technology Corporation Grant 15-S2606-04-C27.

solving for cases in which there are multiple actuators. Finally, the reported findings are illustrated through an example on a linearized hypersonic aircraft model.

9.1 INTRODUCTION

In the absence of actuator dynamics (i.e., actuators with sufficiently high bandwidth), it is well known that adaptive control laws have the capability to suppress the effect of wide classes of system uncertainties to guarantee stability and achieve given command-following specifications. In the presence of actuator dynamics, however, stability verification steps must be taken into account to show the allowable bandwidth range for actuators and the magnitude of tolerable system uncertainties such that adaptive control laws work safely.

To elucidate this point in a simple setting, consider a scalar dynamical system given by

$$\dot{x}(t) = -x(t) + u(t) + wx(t), \quad x(0) = x_0, \tag{9.1}$$

where $x(t) \in \mathbb{R}$ denotes the state available for feedback, $u(t) \in \mathbb{R}$ denotes the control input, and $w \in \mathbb{R}$ is a positive constant. To simplify the control design, assume that w is known and let our objective be to cancel the effect of the term "$wx(t)$" in Eq. (9.1). In the absence of actuator dynamics, this problem can be trivially solved by using $u(t) = -wx(t)$. However, one has the dynamics given by

$$\dot{x}(t) = -x(t) + v(t) + wx(t), \quad x(0) = x_0, \tag{9.2}$$

$$\dot{v}(t) = -m(v(t) - u(t)), \quad v(0) = v_0, \tag{9.3}$$

in the presence of actuator dynamics, where $v(t) \in \mathbb{R}$ denotes the state of the actuator and m denotes the pole of the scalar actuator dynamics. In this case with $u(t) = -wx(t)$, Eq. (9.2) and Eq. (9.3) can be written in a compact form given by

$$\begin{bmatrix} \dot{x}(t) \\ \dot{v}(t) \end{bmatrix} = A \begin{bmatrix} x(t) \\ v(t) \end{bmatrix}, \quad A \triangleq \begin{bmatrix} w-1 & 1 \\ -mw & -m \end{bmatrix}, \quad \begin{bmatrix} x(0) \\ v(0) \end{bmatrix} = \begin{bmatrix} x_0 \\ v_0 \end{bmatrix}, \tag{9.4}$$

such that A must be Hurwitz to achieve closed-loop system stability. In this simple setting, one can easily construct Fig. 9.1 (or resort to any other well-established classical linear control theory tools) to understand the stability interplay with respect to the location of the actuator pole m and the magnitude of the term "$wx(t)$" to be cancelled, where it is clear that m must be large enough for closed-loop system stability provided that w is large.

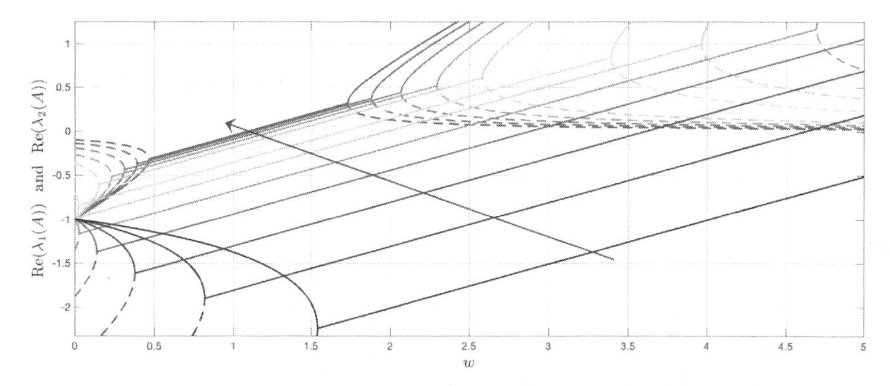

Fig. 9.1 Change in the real parts of the eigenvalues of matrix *A* in Eq. (9.4) with respect to the changes in *w* and *m* (change in the real part of the first eigenvalue, Re(λ_1(*A*)), is depicted by the solid lines, change in the real part of the second eigenvalue, Re(λ_2(*A*)), is depicted by the dashed lines, and arrow indicates the direction that *m* is reduced from 5 to 0.1 gradually).

Now, assume that w is unknown such that one cannot use $u(t) = -wx(t)$. From an adaptive control design point of view, one needs to utilize a control input such as $u(t) = -\hat{w}(t)x(t)$, where $\hat{w}(t) \in \mathbb{R}$ is an estimate of w satisfying an update law of the form $\dot{\hat{w}}(t) = f(x(t)), f : \mathbb{R} \to \mathbb{R}$. For the selection of the function $f(x(t))$, there are several possibilities (see, for example, [1–3]), but they all yield to nonlinear closed-loop dynamical systems. Therefore, determination of a stability interplay like the one in Fig. 9.1 becomes significantly harder even for this simple setting because one cannot readily resort to any well-established classical linear control theory tools. To this end, the following highly practical question arises: *How can one design adaptive controllers for uncertain dynamical systems with actuator dynamics such that the fundamental stability interplay between the allowable system uncertainties and the bandwidth of the actuator dynamics can be rigorously characterized?*

To this end, notable contributions include [4–9] that allow the design of adaptive controllers in the presence of actuator dynamics. In particular, direct approaches to this problem are presented in [4–6] such that the resulting closed-loop dynamical systems that are studied are explicitly affected by the presence of actuator dynamics. Although not explicitly applied to the problem of actuator dynamics, the approach documented in [7] highlights how linear matrix inequalities (LMIs) can be utilized to compute a minimum filter bandwidth for guaranteed closed-loop system stability.

A novel hedging approach is presented in [8] that enables adaptive control laws to be constructed such that their adaptation performance (i.e., their learning performances of the system uncertainties) is not affected by the presence of actuator dynamics. Specifically, this is accomplished by modifying the ideal reference model trajectories with a hedge signal such that standard adaptation dynamics are achieved even in the presence of actuator dynamics. Yet, until recent works by the authors [9], it has not been analyzed that this modification to the ideal reference model trajectories does not yield to unbounded reference model trajectories in the presence of actuator dynamics.

In this chapter, we present safe adaptive architectures for control of uncertain dynamical systems with actuator dynamics. Specifically, after presenting necessary preliminaries on (model reference) adaptive control, we first present an overview of the LMIs-based hedging framework documented in our earlier work [9] and discuss stability of the closed-loop dynamical system using Lyapunov-based analysis tools. Next, we establish convergence properties of the modified reference model trajectories to the ideal reference model trajectories and present a generalization of the proposed framework to the case in which system uncertainties are nonlinear. An algorithm is also presented to allow for less conservative computations of the minimum allowable actuator bandwidths as well as solving for cases in which there are multiple actuators. Finally, we illustrate the reported findings through an example of a linearized hypersonic aircraft model. The safe adaptive architectures presented in this chapter for uncertain dynamical systems with actuator dynamics allow for the computation of the fundamental stability interplay between the allowable system uncertainties and the bandwidth of the actuator dynamics and thus address the earlier question.

A fairly standard notation is used in this chapter. Specifically, $\mathbb{R}$ denotes the set of real numbers, $\mathbb{R}^n$ denotes the set of $n \times 1$ real column vectors, $\mathbb{R}^{n\times m}$ denotes the set of $n \times m$ real matrices, $\mathbb{R}_+$ (resp., $\overline{\mathbb{R}}_+$) denotes the set of positive (resp., non-negative-definite) real numbers, $\mathbb{R}_+^{n\times n}$ (resp., $\overline{\mathbb{R}}_+^{n\times n}$) denotes the set of $n \times n$ positive-definite (resp., non-negative-definite) real matrices, $\mathbb{D}^{n\times n}$ denotes the set of $n \times n$ real matrices with diagonal scalar entries, $(\cdot)^T$ denotes the transpose operator, $(\cdot)^{-1}$ denotes the inverse operator, $\mathrm{tr}(\cdot)$ denotes the trace operator, $\|\cdot\|_2$ denotes the Euclidean norm, $\|\cdot\|_F$ denotes the Frobenius matrix norm, $[A]_{ij}$ denotes the ij-th entry of the real matrix $A \in \mathbb{R}^{n\times m}$, $\lambda_{\min}(A)$ (resp., $\lambda_{\max}(A)$) for the minimum (resp., maximum) eigenvalue of the real matrix $A \in \mathbb{R}^{n\times m}$, and '$\triangleq$' denotes the equality by definition.

9.2 PRELIMINARIES ON ADAPTIVE CONTROL

In this section, we introduce some fundamental results that are needed to develop the main results of this chapter. We begin with the following definition.

Definition 9.1: Let $\Omega = \{\theta \in \mathbb{R}^n : (\theta_i^{\min} \leq \theta_i \leq \theta_i^{\max})_{i=1,2,\ldots,n}\}$ be a convex hypercube in $\mathbb{R}^n$, where $(\theta_i^{\min}, \ \theta_i^{\max})$ represent the minimum and maximum bounds for the ith component of the n-dimensional parameter vector θ. In addition, let $\Omega_\epsilon = \{\theta \in \mathbb{R}^n : (\theta_i^{\min} + \epsilon \leq \theta_i \leq \theta_i^{\max} - \epsilon)_{i=1,2,\ldots,n}\}$ be a second hypercube for a sufficiently small positive constant ϵ, where

$\Omega_\epsilon \subset \Omega$. Then the projection operator $Proj : \mathbb{R}^n \times \mathbb{R}^n \to \mathbb{R}^n$ is defined compenent-wise by

$$\mathrm{Proj}(\theta, y) \triangleq \begin{cases} \left(\dfrac{\theta_i^{\max} - \theta_i}{\epsilon}\right) y_i, & \text{if } \theta_i > \theta_i^{\max} - \epsilon \text{ and } y_i > 0, \\ \left(\dfrac{\theta_i - \theta_i^{\min}}{\epsilon}\right) y_i, & \text{if } \theta_i < \theta_i^{\min} + \epsilon \text{ and } y_i < 0, \\ y_i, & \text{otherwise,} \end{cases} \tag{9.5}$$

where $y \in \mathbb{R}^n$ [3].

It follows from Definition 9.1 that

$$(\theta - \theta^*)^T(\mathrm{Proj}(\theta, y) - y) \leq 0, \qquad \theta^* \in \Omega_\varepsilon \tag{9.6}$$

holds [3, 10]. Throughout the chapter, we use the generalization of this definition to matrices as

$$\mathrm{Proj}_m(\Theta, Y) = (\mathrm{Proj}(\mathrm{col}_1(\Theta), \mathrm{col}_1(Y)), \ldots, \mathrm{Proj}(\mathrm{col}_m(\Theta), \mathrm{col}_m(Y))), \tag{9.7}$$

where $\Theta \in \mathbb{R}^{n\times m}$, $Y \in \mathbb{R}^{n\times m}$, and $\mathrm{col}_i(\cdot)$ denotes the i-th column operator. In this case, for a given Θ^*, it follows from Eq. (9.6) that

$$\mathrm{tr}\left[(\Theta - \Theta^*)^T(\mathrm{Proj}_m(\Theta, Y) - Y)\right]$$
$$= \sum_{i=1}^{m} \left[\mathrm{col}_i(\Theta - \Theta^*)^T(\mathrm{Proj}(\mathrm{col}_i(\Theta), \mathrm{col}_i(Y)) - \mathrm{col}_i(Y))\right] \leq 0. \tag{9.8}$$

Next, we briefly present the standard (model reference) adaptive control problem. In particular, consider a class of uncertain dynamical systems given by

$$\dot{x}(t) = Ax(t) + B[u(t) + \delta(x(t))], \quad x(0) = x_0, \tag{9.9}$$

where $x(t) \in \mathbb{R}^n$ is the state vector available for feedback, $u(t) \in \mathbb{R}^m$ is the control input, $\delta : \mathbb{R}^n \to \mathbb{R}^m$ is an uncertainty, $A \in \mathbb{R}^{n\times n}$ is a known system matrix, $B \in \mathbb{R}^{n\times m}$ is a known input matrix, and the pair (A, B) is controllable.

Consider the reference model capturing a desired, ideal closed-loop dynamical system performance

$$\dot{x}_r(t) = A_r x_r(t) + B_r c(t), \quad x_r(0) = x_{r0}, \tag{9.10}$$

where $x_r(t) \in \mathbb{R}^n$ is the reference state vector, $c(t) \in \mathbb{R}^m$ is a given uniformly continuous bounded command, $A_r \in \mathbb{R}^{n\times n}$ is the Hurwitz reference model matrix, and $B_r \in \mathbb{R}^{n\times m}$ is the command input matrix. The objective of the (model reference) adaptive control problem is to construct an adaptive feedback control law $u(t)$ such that the state vector $x(t)$ asymptotically follows the reference state vector $x_r(t)$.

For the purpose of solving this problem, consider the feedback control law given by

$$u(t) = u_n(t) + u_a(t), \tag{9.11}$$

where $u_n(t)$ and $u_a(t)$ are the nominal feedback control law and the adaptive feedback control law, respectively. Let the nominal feedback control law be given by

$$u_n(t) = -K_1 x(t) + K_2 c(t), \tag{9.12}$$

where $K_1 \in \mathbb{R}^{m\times n}$ and $K_2 \in \mathbb{R}^{m\times m}$ are the nominal feedback and feedforward gains, respectively, such that $A_r \triangleq A - BK_1$ and $B_r \triangleq BK_2$ hold.

Assumption 9.1: The uncertainty in Eq. (9.9) is parameterized as

$$\delta(x) = W^T \sigma(x), \quad x \in \mathbb{R}^n, \tag{9.13}$$

where $W \in \mathbb{R}^{s\times m}$ is an unknown weight matrix and $\sigma : \mathbb{R}^n \to \mathbb{R}^s$ is a known basis function of the form $\sigma(x) = [\sigma_1(x), \sigma_2(x), \ldots, \sigma_s(x)]^T$, which satisfies

$$\sigma(x(t)) = K(x(t))x(t) + b(x(t)), \quad K : \mathbb{R}^n \to \mathbb{R}^{s\times n}, \quad b : \mathbb{R}^n \to \mathbb{R}^s, \tag{9.14}$$

with $b(x(t))$ being a bounded term. In addition, Eq. (9.14) satisfies the inequality given by

$$\|\sigma(x(t)) - \sigma_0\| \leq \alpha \|x(t)\|, \quad x(t) \in \mathbb{R}^n, \tag{9.15}$$

with $\sigma_0 \in \overline{\mathbb{R}}_+$ *and* $\alpha \in \mathbb{R}_+$.

Now, using Eqs. (9.11) and (9.12) in Eq. (9.9) with Assumption 9.1 yields

$$\dot{x}(t) = A_r x(t) + B_r c(t) + B[u_a(t) + W^T \sigma(x(t))], \tag{9.16}$$

Motivated by Eq. (9.16), let the adaptive feedback control law be given by

$$u_a(t) = -\hat{W}^T(t)\sigma(x(t)), \tag{9.17}$$

where $\hat{W}(t) \in \mathbb{R}^{s\times m}$ is the estimate of W that satisfies the (nonlinear) update law

$$\dot{\hat{W}}(t) = \gamma \text{Proj}_m[\hat{W}(t),\ \sigma(x(t))e^T(t)PB], \quad \hat{W}(0) = \hat{W}_0, \tag{9.18}$$

where $\gamma \in \mathbb{R}_+$ is the learning rate gain, $e(t) \triangleq x(t) - x_r(t)$ is the system error state vector, and $P \in \mathbb{R}_+^{n\times n}$ is a solution of the Lyapunov equation

$$0 = A_r^T P + PA_r + R, \tag{9.19}$$

with $R \in \mathbb{R}_+^{n\times n}$. Since A_r is Hurwitz, it follows from the converse Lyapunov theory [11] that there exists a unique P satisfying Eq. (9.19) for a given R. Additionally, the projection bounds, which are necessary for the implementation of Eq. (9.18), are defined such that

$$\left|[\hat{W}(t)]_{ij}\right| \leq \hat{W}_{\max,i+(j-1)s}, \quad i = 1, \ldots, s \text{ and } j = 1, \ldots, m, \tag{9.20}$$

where $\hat{W}_{\max,i+(j-1)s} \in \mathbb{R}_+$ denote (symmetric) element-wise projection bounds. Note that the results documented in this chapter can be readily applied to the case when asymmetric projection bounds are considered.

Using Eq. (9.17) in Eq. (9.16) yields

$$\dot{x}(t) = A_r x(t) + B_r c(t) - B\tilde{W}^T(t)\sigma(x(t)), \tag{9.21}$$

and the system error dynamics follow from Eq. (9.10) and Eq. (9.21) as

$$\dot{e}(t) = A_r e(t) - B\tilde{W}^T(t)\sigma(x(t)), \quad e(0) = e_0, \tag{9.22}$$

where $\tilde{W}(t) \triangleq \hat{W}(t) - W \in \mathbb{R}^{s\times m}$. The weight update law given by Eq. (9.18) can be derived using Lyapunov analysis by considering the Lyapunov function candidate given by (see, for example, [1–3])

$$\mathcal{V}(e, \tilde{W}) = e^T P e + \gamma^{-1} \text{tr}\, \tilde{W}^T \tilde{W}. \tag{9.23}$$

Note that $\mathcal{V}(0, 0) = 0$ and $\mathcal{V}(e, \tilde{W}) > 0$ for all $(e, \tilde{W}) \neq (0, 0)$. Now, differentiating Eq. (9.23) yields

$$\begin{aligned}\dot{\mathcal{V}}(e(t), \tilde{W}(t)) &= -e^T(t)Re(t) - 2e^T(t)PB\tilde{W}^T(t)\sigma(x(t)) \\ &\quad + 2\gamma^{-1}\text{tr}\, \tilde{W}^T(t)\dot{\hat{W}}(t),\end{aligned} \tag{9.24}$$

where using Eq. (9.18) in Eq. (9.24) results in $\dot{\mathcal{V}}(e(t), \tilde{W}(t)) \leq -e^T(t)Re(t) \leq 0$, which guarantees that the system error state vector $e(t)$ and the weight error $\tilde{W}(t)$ are Lyapunov stable, and are therefore bounded for all $t \in \overline{\mathbb{R}}_+$. Since $x(t)$ and $c(t)$ are bounded for all $t \in \overline{\mathbb{R}}_+$, it follows from Eq. (9.22) that $\dot{e}(t)$ is bounded, and hence $\ddot{\mathcal{V}}(e(t), \tilde{W}(t))$ is bounded for all $t \in \overline{\mathbb{R}}_+$. It then follows from Barbalat's lemma [12] that $\lim_{t\to\infty} \dot{\mathcal{V}}(e(t), \tilde{W}(t)) = 0$, which consequently shows that $e(t) \to 0$ as $t \to \infty$.

9.3 AN LMI-BASED HEDGING APPROACH TO ACTUATOR DYNAMICS

Based on the adaptive control framework introduced in Section 9.2, we now provide an overview of the key findings reported in [9] for the actuator dynamics problem, which forms the basis for the results given in Sections 9.4, 9.5, and 9.6. Specifically, consider the uncertain dynamical system given by

$$\dot{x}(t) = Ax(t) + B(v(t) + \delta(x(t))), \quad x(0) = x_0, \tag{9.25}$$

where $v(t) \in \mathbb{R}^m$ is the actuator output given by the dynamics

$$\begin{aligned} \dot{x}_c(t) &= -Mx_c(t) + u(t), \quad x_c(0) = x_{c0}, \\ v(t) &= Mx_c(t), \quad v(0) = v_0, \end{aligned} \tag{9.26}$$

where $x_c(t) \in \mathbb{R}^m$ is the actuator state vector, $M \in \mathbb{R}^{m\times m} \cap \mathbb{D}^{m\times m}$ with diagonal entries $\lambda_{i,i} > 0$, $i = 1, \ldots, m$, represents the actuator bandwidth of each control channel, and $u(t) \in \mathbb{R}^m$ is the control input.

In order to introduce the LMI-based hedging approach in this section, we begin with the case in which the system uncertainties are linear (we refer to Section 9.5 for generalizations). In other words, the basis function is given as $\sigma(x(t)) = x(t)$, such that Assumption 9.1 holds with $b(\cdot) = 0$, $\sigma_0 = 0$, and $\alpha = 1$.

Consider now, by adding and subtracting $Bu(t)$, the following equivalent form of Eq. (9.25) subject to Assumption 9.1 with linear uncertainties,

$$\dot{x}(t) = Ax(t) + B[u(t) + W^T x(t)] + B[v(t) - u(t)]. \tag{9.27}$$

Using Eqs. (9.11) and (9.12) along with the adaptive feedback control law given by Eq. (9.17) with $\sigma(x(t)) = x(t)$, it follows that Eq. (9.27) can be equivalently rewritten as

$$\dot{x}(t) = A_r x(t) + B_r c(t) - B\tilde{W}^T(t)x(t) + B[v(t) - u(t)]. \tag{9.28}$$

Next, using the hedging approach [8], in which the objective is to construct an adaptive feedback control law such that its adaptation performance (i.e., its learning performance of system uncertainty W) is not affected by the presence of actuator dynamics, the reference model is modified to the following

$$\dot{x}_r(t) = A_r x_r(t) + B_r c(t) + B[v(t) - u(t)], \quad x_r(0) = x_{r0}, \tag{9.29}$$

such that the system error dynamics can be given using Eq. (9.28) and Eq. (9.29) as

$$\dot{e}(t) = A_r e(t) - B\tilde{W}^T(t)x(t), \quad e(0) = e_0. \tag{9.30}$$

The following lemma is needed for the results in this section. For this purpose, let $\underline{\lambda} \in \mathbb{R}_+$ be such that $\underline{\lambda} \le \lambda_{i,i}$ for all $i = 1, \ldots, m$ and let $\overline{\omega} \in \mathbb{R}_+$ be such that $\hat{W}_{\max,i+(j-1)n} \le \overline{\omega}$ for all $i = 1, \ldots, n$ and $j = 1, \ldots, m$.

Lemma 9.1: There exists a set $\kappa \triangleq \{\underline{\lambda} : \underline{\lambda} \le \lambda_{i,i},\ i = 1, \ldots, m\} \cup \{\overline{\omega} : \hat{W}_{\max,i+(j-1)n} \le \overline{\omega},\ i = 1, \ldots, n,\ j = 1, \ldots, m\}$ such that if $(\underline{\lambda}, \overline{\omega}) \in \kappa$, then

$$\mathcal{A}(\hat{W}(t), M) = \begin{bmatrix} A + B\hat{W}^T(t) & BM \\ -K_1 - \hat{W}^T(t) & -M \end{bmatrix} \tag{9.31}$$

is quadratically stable.

Proof: The proof is omitted, but it can be shown using similar steps as the proof of Lemma 3.1 in [9].

Theorem 9.1: Consider the uncertain dynamical system given by Eq. (9.25) subject to Assumption 9.1 with $\sigma(x(t)) = x(t)$, the reference model given by Eq. (9.29), the actuator dynamics given by Eq. (9.26), the feedback control law given by Eq. (9.11), with the nominal feedback control law given by Eq. (9.12), and the adaptive feedback control law given by Eq. (9.17) along with the update law given by Eq. (9.18) with $\sigma(x(t)) = x(t)$. If $(\underline{\lambda}, \overline{\omega}) \in \kappa$, then the solution $(e(t), \tilde{W}(t), x_r(t), v(t))$ of the closed-loop dynamical system is bounded for all $t \in \overline{\mathbb{R}}_+$ and $\lim_{t\to\infty} e(t) = 0$.

Proof: The proof can be shown using similar steps as in [9], and hence is omitted.

We would like to highlight that the Lyapunov stability and boundedness of the system error state $e(t)$ and the weight error $\tilde{W}(t)$ follow by considering the Lyapunov function candidate

$$\mathcal{V}(e, \tilde{W}) = e^T Pe + \gamma^{-1}\mathrm{tr}\ \tilde{W}^T \tilde{W}. \tag{9.32}$$

Similar to the last paragraph in Section 9.2, it can be shown that $\dot{\mathcal{V}}(e(t), \tilde{W}(t)) \leq -e^T(t)Re(t) \leq 0$, which guarantees the Lyapunov stability and hence the boundedness of the solution $(e(t), \tilde{W}(t))$. However, unlike the last paragraph in Section 9.2, it can not yet be concluded that $e(t) \to 0$ as $t \to \infty$ due to the modified reference model given by Eq. (9.29), which may yield unbounded trajectories.

The next step then is to show the boundedness of the modified reference model given by Eq. (9.29) by grouping $x_r(t)$ and $x_c(t)$ in compact form as

$$\dot{\xi}(t) = \mathcal{A}(\hat{W}(t), M)\xi(t) + \omega(\cdot), \tag{9.33}$$

with $\xi(t) = [x_r^T(t), x_c^T(t)]^T$ and

$$\omega(\cdot) = \begin{bmatrix} B\hat{W}^T(t)e(t) + BK_1 e(t) \\ -\hat{W}^T(t)e(t) - K_1 e(t) + K_2 c(t) \end{bmatrix}. \tag{9.34}$$

It then follows that since $\omega(\cdot)$ is bounded and $\mathcal{A}(\hat{W}(t), M)$ is quadratically stable by Lemma 9.1, then $x_r(t)$ and $x_c(t)$ (and therefore $v(t)$) are also bounded (see, for example, [12]), and hence, the same conclusion can be made as in the last paragraph in Section 9.2.

For the results given in Theorem 9.1 to hold, it is assumed that Eq. (9.31) is quadratically stable [13, 14]. Lemma 9.1 shows the feasibility of this assumption when $(\underline{\lambda}, \overline{\omega}) \in \kappa$. As shown in the respective proof in [9], this holds when the actuator dynamics are sufficiently fast or the projection bounds on $\hat{W}(t)$ are sufficiently small, revealing a fundamental stability interplay between the allowable system uncertainties and the bandwidths of the actuator dynamics.

We now utilize LMIs to satisfy the quadratic stability of Eq. (9.31) for given projection bounds $\hat{W}_{\max}$ for the elements of $\hat{W}(t)$ and the bandwidths of the actuator dynamics M. For this purpose, let $\overline{W}_{i_1,\ldots,i_l} \in \mathbb{R}^{n\times m}$ be defined as

$$\overline{W}_{i_1,\ldots,i_l} = \begin{bmatrix} (-1)^{i_1}\hat{W}_{\max,1} & (-1)^{i_{1+n}}\hat{W}_{\max,1+n} & \cdots & (-1)^{i_{1+(m-1)n}}\hat{W}_{\max,1+(m-1)n} \\ (-1)^{i_2}\hat{W}_{\max,2} & (-1)^{i_{2+n}}\hat{W}_{\max,2+n} & \cdots & (-1)^{i_{2+(m-1)n}}\hat{W}_{\max,2+(m-1)n} \\ \vdots & \vdots & \ddots & \vdots \\ (-1)^{i_n}\hat{W}_{\max,n} & (-1)^{i_{2n}}\hat{W}_{\max,2n} & \cdots & (-1)^{i_{mn}}\hat{W}_{\max,mn} \end{bmatrix}, \tag{9.35}$$

where $i_l \in \{1, 2\}$, $l \in \{1, \ldots, 2^{mn}\}$, such that $\overline{W}_{i_1,\ldots,i_l}$ represents the corners of the hypercube defining the maximum variation of $\hat{W}(t)$. Following the results in [7, 9, 15], if

$$\mathcal{A}_{i_1,\ldots,i_l} = \begin{bmatrix} A + B\overline{W}^T_{i_1,\ldots,i_l} & BM \\ -K_1 - \overline{W}^T_{i_1,\ldots,i_l} & -M \end{bmatrix} \tag{9.36}$$

satisfies the matrix inequality

$$\mathcal{A}^T_{i_1,\ldots,i_l}\mathcal{P} + \mathcal{P}\mathcal{A}_{i_1,\ldots,i_l} < 0, \quad \mathcal{P} = \mathcal{P}^T > 0, \tag{9.37}$$

for all permutations of $\overline{W}_{i_1,\ldots,i_l}$, then Eq. (9.31) is quadratically stable. Since Eq. (9.31) is quadratically stable for large values of M (see above), we cast Eq. (9.37) as a convex optimization problem as

$$\begin{aligned} &\text{minimize } M, \\ &\text{subject to Eq. (9.37).} \end{aligned}$$

Therefore, we can satisfy Eq. (9.37) by minimizing M for a given projection bound.

It should be noted that several architectures are introduced in [9], including cases with unknown actuator output and unknown control input. Although we consider only one of these cases, the results in the remainder of this chapter can be applied to the other architectures in [9] without loss of generality.

9.4 CONVERGENCE OF REFERENCE MODELS

In Section 9.3, we showed that the uncertain dynamical system given by Eq. (9.25) asymptotically tracks the modified reference model given by Eq. (9.29), which is ensured to be bounded for a verifiable minimum actuator bandwidth. In this section, we now would like to know that the modified reference model, which is used to allow for correct adaptation in the presence of actuator dynamics, has certain convergence properties to the ideal reference model. To analyze this, we begin by rewriting the ideal reference model given by Eq. (9.10) as

$$\dot{x}_{r_i}(t) = A_r x_{r_i}(t) + B_r c(t), \quad x_{r_i}(0) = x_{r0}, \tag{9.38}$$

where $x_{r_i}(t) \in \mathbb{R}^n$ is the ideal reference state vector. Then, the error between the modified reference model given by Eq. (9.29) and the ideal reference model given by Eq. (9.38) follows as $e_r(t) \triangleq x_r(t) - x_{r_i}(t)$.

Now, it follows from the reference model error and the system error that

$$\left\| x(t) - x_{r_i}(t) \right\| \mathcal{L}\infty = \|e(t) + e_r(t)\|_{\mathcal{L}_\infty} \leq \|e(t)\|_{\mathcal{L}_\infty} + \|e_r(t)\|_{\mathcal{L}_\infty}. \tag{9.39}$$

This clearly implies that close tracking of the ideal reference model can be achieved by suppressing the bounds on both error signals in Eq. (9.39).

From Theorem 9.1, note that $\dot{\mathcal{V}}(e(t), \tilde{W}(t)) \leq 0$ for $t \in \overline{\mathbb{R}}_+$ based on the Lyapunov function candidate given by Eq. (9.32). This further yields

$$\mathcal{V}(e(t), \tilde{W}(t)) \leq \mathcal{V}(e(0), \tilde{W}(0)). \tag{9.40}$$

Using the inequalities $\lambda_{\min}(P)\|e(t)\|_2^2 \leq \mathcal{V}(e(t), \tilde{W}(t))$ and $\mathcal{V}(e(0), \tilde{W}(0)) \leq \lambda_{\max}(P)\|e(0)\|_2^2 + \gamma^{-1}\left\|\tilde{W}(0)\right\|_F^2$ in Eq. (9.40) results in

$$\|e(t)\|_2 \leq \sqrt{\frac{1}{\lambda_{\min}(P)}\left(\lambda_{\max}(P)\|e(0)\|_2^2 + \gamma^{-1}\left\|\tilde{W}(0)\right\|_F^2\right)}. \tag{9.41}$$

Letting $x_{r_0} = x_0$ such that $e(0) = 0$ and noting $\|\cdot\|_\infty \leq \|\cdot\|_2$, which is uniform, then Eq. (9.41) yields

$$\|e_\tau(t)\|_{\mathcal{L}_\infty} \leq \frac{\left\|\tilde{W}(0)\right\|_F}{\sqrt{\gamma\lambda_{\min}(P)}}. \tag{9.42}$$

As a direct consequence of Eq. (9.42) holding uniformly in τ, the transient performance bounds on the system error follow as

$$\|e(t)\|_{\mathcal{L}_\infty} \leq \frac{\left\|\tilde{W}(0)\right\|_F}{\sqrt{\gamma\lambda_{\min}(P)}}. \tag{9.43}$$

Since the system error satisfies certain transient performance bounds which can be decreased, for example, by judiciously increasing the learning gain γ, we now introduce the following theorem (with the proof included in Appendix A) to ensure suppression of the reference model error bound. For this purpose, we first note that the actuator bandwidth matrix $M_0 \in \mathbb{R}_+^{m\times m} \cap \mathbb{D}^{m\times m}$ can be related to the actual actuator bandwidth matrix $M \in \mathbb{R}_+^{m\times m} \cap \mathbb{D}^{m\times m}$ as $M \triangleq \phi M_0$ where $\phi \in \mathbb{R}_+$, and $\phi \geq 1$. In addition,

$$\mathcal{P} \triangleq \begin{bmatrix} P & PB \\ B^T P & B^T PB + \rho I \end{bmatrix}, \tag{9.44}$$

where $P \in \mathbb{R}_+^{n\times n}$ is a solution of the Lyapunov equation given by Eq. (9.19) with $R \in \mathbb{R}_+^{n\times n}$, $\rho \in \mathbb{R}_+$, and we note that $\beta \in \mathbb{R}_+$, $\omega^* \in \mathbb{R}_+$, and $\eta \in \mathbb{R}_+$.

Theorem 9.2: Consider the reference model given by Eq. (9.29), the ideal reference model given by Eq. (9.38), the actuator dynamics given by Eq. (9.26), the feedback control law given by Eq. (9.11), with the nominal feedback control law given by Eq. (9.12) and the adaptive feedback control law given by Eq. (9.17) with $\sigma(x(t)) = x(t)$. If $(\underline{\lambda}, \overline{\omega}) \in \kappa$, then $\|e_r(t)\|_{\mathcal{L}_\infty} \rightarrow 0$ as $\lambda_{\min}(M - M_0) \rightarrow \infty$ and is guaranteed to shrink at the rate

$$\|e_r(t)\|_{\mathcal{L}_\infty} \leq \rho\omega^* \sqrt{\frac{\lambda_{\max}(\mathcal{P})}{\lambda_{\min}(\mathcal{P})}} \sqrt{\frac{5\eta + (\phi - 1)\beta}{\eta(\eta + (\phi - 1)\beta)^2}}. \tag{9.45}$$

Corollary 9.1: Consider the uncertain dynamical system given by Eq. (9.25) subject to Assumption 9.1 with $\sigma(x(t)) = x(t)$, the reference model given by Eq. (9.29), the actuator dynamics given by Eq. (9.26), the feedback control law given by Eq. (9.11), with the nominal feedback control law given by Eq. (9.12) and the adaptive feedback control law given by Eq. (9.17) along with the update law given by Eq. (9.18) with $\sigma(x(t)) = x(t)$. If $(\underline{\lambda}, \overline{\omega}) \in \kappa$, then

$$\|x(t) - x_{r_i}(t)\|_{\mathcal{L}_\infty} \leq \|e(t)\|_{\mathcal{L}_\infty} + \|e_r(t)\|_{\mathcal{L}_\infty}. \tag{9.46}$$

Proof: The result follows from Theorems 9.1 and 9.2.

It should be noted that $\|x(t) - x_{r_i}(t)\|_{\mathcal{L}_\infty}$ can be made arbitrarily small through the use of judiciously large learning gain γ and large bandwidth values for the actuator dynamics. Now, we consider the following theorem (with the proof included in Appendix A), which guarantees asymptotic convergence to the ideal reference model for constant reference commands.

Theorem 9.3: Consider the uncertain dynamical system given by Eq. (9.25) subject to Assumption 9.1 with $\sigma(x(t)) = x(t)$, the reference model given by Eq. (9.29), the actuator dynamics given by Eq. (9.26), the feedback control law given by Eq. (9.11), with the nominal feedback control law given by Eq. (9.12) and the adaptive feedback control law given by Eq. (9.17) along with the update law Eq. (9.18) with $\sigma(x(t)) = x(t)$. For a constant reference command, if $(\underline{\lambda}, \overline{\omega}) \in \kappa$, the modified reference model given by Eq. (9.29), will asymptotically converge to the ideal reference model given by Eq. (9.38). In addition, it follows that $x(t) - x_{r_i}(t) \rightarrow 0$ as $t \rightarrow \infty$.

To summarize the findings of this section, we show that the error between uncertain dynamical system and the ideal reference model can be decreased for large adaptation gains and large actuator bandwidths. Additionally, we show for constant reference commands that the uncertain dynamical system asymptotically converges to the ideal reference model capturing a given desired closed-loop adaptive system behavior. In the next section, we further generalize the LMI-based hedging approach for the case in which the system uncertainties are nonlinear (i.e., $\sigma(x(t)) \neq x(t)$).

9.5 GENERALIZATIONS TO NONLINEAR UNCERTAIN DYNAMICAL SYSTEMS

We now consider the nonlinear uncertain dynamical system given by Eq. (9.25) in which the uncertainties $\delta(x(t))$ are indeed nonlinear, unlike what is assumed in Sections 9.3 and 9.4. Thus, we consider Assumption 9.1 for the results in this section. In this case, it follows that the steps taken to get Eq. (9.28) from Eq. (9.25) can also be used with $\delta(x(t)) = W^T\sigma(x(t))$ to get

$$\dot{x}(t) = A_r x(t) + B_r c(t) - B\tilde{W}^T(t)\sigma(x(t)) + B[v(t) - u(t)]. \tag{9.47}$$

Now, using Eq. (9.29) and Eq. (9.47), the system error dynamics follow as

$$\dot{e}(t) = A_r e(t) - B\tilde{W}^T(t)\sigma(x(t)), \quad e(0) = e_0. \tag{9.48}$$

We can now analyze the stability of the closed-loop dynamical system while ensuring the modified reference model dynamics in Eq. (9.29) yield to bounded reference model trajectories in the presence of actuator dynamics. This is accomplished through the following lemma and theorem whose proof can be found in Appendix A. For this purpose, let $\hat{\theta}(t, x(t)) \triangleq \hat{W}^T(t)K(x(t))$, where $K(x(t))$ follows from Eq. (9.14). In addition, let $\underline{\lambda} \in \mathbb{R}_+$, $\underline{\lambda}_0 \in \mathbb{R}_+$, and $\underline{\lambda}_1 \in \mathbb{R}_+$ be such that $\underline{\lambda} \leq \lambda_{i,i}$, $\underline{\lambda}_0 \leq \lambda_{i,i}$, and $\underline{\lambda}_1 \leq \lambda_{i,i}$ for all $i = 1, \ldots, m$, and let $\overline{\omega} \in \mathbb{R}_+$ be such that $\hat{W}_{\max,i+(j-1)s} \leq \overline{\omega}$ for all $i = 1, \ldots, s$ and $j = 1, \ldots, m$.

Lemma 9.2: There exists a set $\kappa_0 \triangleq \{\underline{\lambda}_0 : \underline{\lambda}_0 \leq \lambda_{i,i},\ i = 1, \ldots, m\}$ such that if $\underline{\lambda}_0 \in \kappa_0$, then

$$\mathcal{A}(M) = \begin{bmatrix} A & BM \\ -K_1 & -M \end{bmatrix}, \tag{9.49}$$

is quadratically stable. In addition, there exists a set $\kappa_1 \triangleq \{\underline{\lambda} : \underline{\lambda} \leq \lambda_{i,i},\ i = 1, \ldots, m\} \cup \{\overline{\omega}, \underline{\lambda}_1 : \hat{W}_{\max,i+(j-1)s} \leq \overline{\omega},\ i = 1, \ldots, s,\ j = 1, \ldots, m \text{ and } \underline{\lambda}_1 \leq \lambda_{i,i},\ i = 1, \ldots, m\}$ such that if $(\underline{\lambda}, \overline{\omega}, \underline{\lambda}_1) \in \kappa_1$ and K(x(t)) is bounded, then

$$\mathcal{A}(\hat{\theta}(t, x(t)), M) = \begin{bmatrix} A + B\hat{\theta}(t, x(t)) & BM \\ -K_1 - \hat{\theta}(t, x(t)) & -M \end{bmatrix} \tag{9.50}$$

is quadratically stable.

Proof: The proof is omitted, but it can be shown using similar steps as the proof of Lemma 3.1 in [9].

Theorem 9.4: Consider the uncertain nonlinear dynamical system given by Eq. (9.25) subject to Assumption 9.1, the reference model given by Eq. (9.29), the actuator dynamics given by Eq. (9.26), the feedback control law given by Eq. (9.11), with the nominal feedback control law given by Eq. (9.12) and the adaptive feedback control law given by Eq. (9.17) along with the update law given by Eq. (9.18). If $(\underline{\lambda}, \overline{\omega}, \underline{\lambda}_1) \in \kappa_1$ and $\underline{\lambda}_0 \in \kappa_0$, then the solution $(e(t), \tilde{W}(t), x_r(t), v(t))$ of the closed-loop dynamical system are bounded for all $t \in \overline{\mathbb{R}}_+$ and $\lim_{t\to\infty} e(t) = 0$.

As in Section 9.3, LMIs can be used here to cast a convex optimization problem to compute the minimum allowable actuator bandwidth matrix M for given projection bounds $\hat{W}_{\max}$ and a bound on $K(x(t))$. For this purpose, let $\overline{W}_{i_1,\ldots,i_l} \in \mathbb{R}^{s\times m}$ be defined as

$$\overline{W}_{i_1,\ldots,i_l} = \begin{bmatrix} (-1)^{i_1}\hat{W}_{\max,1} & (-1)^{i_{1+s}}\hat{W}_{\max,1+s} & \cdots & (-1)^{i_{1+(m-1)s}}\hat{W}_{\max,1+(m-1)s} \\ (-1)^{i_2}\hat{W}_{\max,2} & (-1)^{i_{2+s}}\hat{W}_{\max,2+s} & \cdots & (-1)^{i_{2+(m-1)s}}\hat{W}_{\max,2+(m-1)s} \\ \vdots & \vdots & \ddots & \vdots \\ (-1)^{i_s}\hat{W}_{\max,s} & (-1)^{i_{2s}}\hat{W}_{\max,2s} & \cdots & (-1)^{i_{ms}}\hat{W}_{\max,ms} \end{bmatrix}, \tag{9.51}$$

where $i_l \in \{1, 2\}$, $l \in \{1,\ldots,2^{ms}\}$, such that $\overline{W}_{i_1,\ldots,i_l}$ represents the corners of the hypercube defining the maximum variation of $\hat{W}(t)$. Furthermore, since we define $\hat{\theta}(t, x(t)) \triangleq \hat{W}^T(t)K(x(t))$, let $\overline{\theta}_{i_1,\ldots,i_r} \in \mathbb{R}^{m\times n}$ be defined as

$$\overline{\theta}_{i_1,\ldots,i_r} = \begin{bmatrix} (-1)^{i_1}\hat{\theta}_{\max,1} & (-1)^{i_{1+m}}\hat{\theta}_{\max,1+m} & \cdots & (-1)^{i_{1+(n-1)m}}\hat{\theta}_{\max,1+(n-1)m} \\ (-1)^{i_2}\hat{\theta}_{\max,2} & (-1)^{i_{2+m}}\hat{\theta}_{\max,2+m} & \cdots & (-1)^{i_{2+(n-1)m}}\hat{\theta}_{\max,2+(n-1)m} \\ \vdots & \vdots & \ddots & \vdots \\ (-1)^{i_m}\hat{\theta}_{\max,m} & (-1)^{i_{2m}}\hat{\theta}_{\max,2m} & \cdots & (-1)^{i_{nm}}\hat{\theta}_{\max,nm} \end{bmatrix}, \tag{9.52}$$

where $i_r \in \{1, 2\}$, $r \in \{1,\ldots,2^{ms+1}\}$, such that $\overline{\theta}_{i_1,\ldots,i_r}$ represents the corners of the hypercube defining the maximum variation of the product of $\hat{W}^T(t)K(x(t))$. Utilizing the results in [7, 15], if

$$\mathcal{A}_{i_1,\ldots,i_r} = \begin{bmatrix} A + B\overline{\theta}_{1_{i_1,\ldots,i_r}} & BM \\ -K_1 - \overline{\theta}_{1_{i_1,\ldots,i_r}} & -M \end{bmatrix}, \tag{9.53}$$

satisfies the matrix inequality

$$\mathcal{A}^T_{i_1,\ldots,i_r}\mathcal{P} + \mathcal{P}\mathcal{A}_{i_1,\ldots,i_r} < 0, \quad \mathcal{P} = \mathcal{P}^T > 0, \tag{9.54}$$

for all permutations of $\overline{\theta}_{i_1,\ldots,i_r}$, then Eq. (9.50) is quadratically stable. We can then cast Eq. (9.54) as a convex optimization problem and solve it using LMIs.

Example 9.1: To elucidate the proposed framework, consider a second-order system given by

$$\dot{x}_1(t) = x_2(t), \tag{9.55}$$

$$\dot{x}_2(t) = x_1(t) + x_2(t) + v(t) + \delta(x(t)). \tag{9.56}$$

For this example, let $x_1(t)$ represent the angle in radians, $x_2(t)$ represent the angular rate of change in radians per second, and $\delta(x(t))$ represent an uncertainty of the form $\delta(x(t)) = \alpha_1 x_2(t)\sin(x_1(t)) + \alpha_2 x_1(t)\cos(x_2(t))$, where α_i, $i = 1,2$, are unknown parameters. For this example, we set $\alpha_1 = \alpha_2 = 1$ and choose $K_1 = [2,\ 2.4]$ and $K_2 = 1$ for the nominal controller design that yields a reference model with a natural frequency of $\omega_n = 1.0$ rad/s and a damping ratio $\zeta = 0.7$. We use element-wise projection bounds such that $|[\hat{W}(t)]_{1,1}| \leq 1.1$ and $|[\hat{W}(t)]_{2,1}| \leq 1.1$ and set all initial conditions to zero such that Assumption 9.1 is satisfied with $\alpha = 1$. Using this along with the bounds on $\hat{W}(t)$, LMIs are used to compute the minimum allowable actuator bandwidth of $\lambda_{\min} = 3.7$. Furthermore, the tracking reference command c(t) is applied through a first order filter and we use a single actuator bandwidth for the control input such that $M = \lambda$, $\lambda \in \mathbb{R}_+$ and we set $R = I_2$ for the proposed controller design.

Figure 9.2 shows the the proposed controller performance in the presence of actuator dynamics. As it is calculated that the minimum actuator bandwidth allowed for the actuator dynamics is 3.7, it is theoretically expected that the system performances are guaranteed to be bounded for actuator bandwidth values greater than or equal to the calculated minimum, as is evident from Fig. 9.2. In addition, it is shown that the system becomes unstable for $\lambda = 2.2$, which is consistent with the presented theory in that we provide a (conservative) upper bound on the allowable actuator bandwidth such that the closed-loop system remains bounded.

At this point, we have introduced the LMI-based hedging approach to compute the minimum allowable actuator bandwidth limit. It was shown in the previous section that certain convergence properties hold for the modified reference model. In this section, a more general case was shown for a class of uncertain nonlinear dynamical systems. Now, as mentioned in Example 9.1, the proposed approach has a certain level of conservatism in the computed actuator bandwidth limit. In order to reduce this conservatism, an algorithm is presented in the next section that makes use of an affine quadratic stability condition. It is shown that we can obtain less conservative results as well as generalize our results to nonconvex cases.

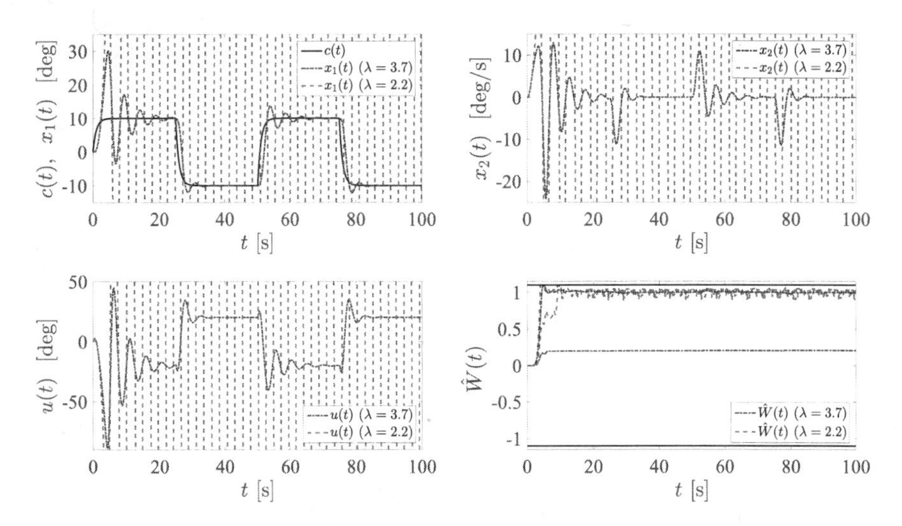

Fig. 9.2 Proposed controller performance with actuator dynamics for Example 9.1 ($\gamma = 10$).

9.6 AN AFFINE QUADRATIC STABILITY CONDITION

Although a general solution for the theoretical stability of hedging in (model reference) adaptive control in the presence of actuator dynamics has been established using LMIs, these solutions are limited to specific cases where convexity is guaranteed. In this section, we establish a new affine quadratic stability condition such that it generalizes our results to generally nonconvex cases. Specifically, we introduce a new change of coordinates and provide a means of convexifying the problem.

We start by introducing an affine transformation for Eq. (9.53). Consider the actuator bandwidth matrix M, which can be rewritten as

$$M(\Delta\lambda) = M_0 + \sum_{j=1}^{m} \Delta\lambda_j M_j, \tag{9.57}$$

where $M_0 = -\lambda_{\text{feas}} I_{m\times m} \in \mathbb{R}^{m\times m} \cap \mathbb{D}^{m\times m}$ is chosen such that M_0 satisfies Eq. (9.54), $M_j \in \mathbb{R}^{m\times m}$ is a matrix such that $M_j(j, j) = 1$ and zero everywhere else, and $\Delta\lambda_j \in \mathbb{R}_+$ belongs to the set $\Delta\lambda = (\Delta\lambda_1, \ldots, \Delta\lambda_m)$, which is defined by the parameter box

$$\Lambda = \left\{\Delta\lambda \in \mathbb{R}^m | \Delta\lambda_j \in \left[\underline{\Delta\lambda_j}, \overline{\Delta\lambda_j}\right]\right\}, \tag{9.58}$$

and is the convex hull, $\Lambda = \text{conv}(\Lambda_0)$, of the corners

$$\Lambda_0 = \left\{\Delta\lambda \in \mathbb{R}^m | \Delta\lambda_j \in \left\{\underline{\Delta\lambda_j}, \overline{\Delta\lambda_j}\right\}\right\}, \tag{9.59}$$

where $\underline{\Delta\lambda_j}$ and $\overline{\Delta\lambda}_j$ are the upper and lower bounds of each actuator bandwidth, respectively. Now, using Eq. (9.57) and letting $B = \text{col}([B_1, B_2, \ldots, B_m])$, we can rewrite Eq. (9.53) as

$$\mathcal{A}(\overline{W}_{i_1,\ldots,i_l}, \Delta\lambda) = \mathcal{A}_{0,i_1,\ldots,i_l} + \sum_{j=1}^{m} \Delta\lambda_j \mathcal{A}_j, \tag{9.60}$$

where

$$\mathcal{A}_{0,i_1,\ldots,i_l} = \begin{bmatrix} A + B\overline{W}^T_{i_1,\ldots,i_l} & -BM_0 \\ -K_1 - \overline{W}^T_{i_1,\ldots,i_l} & M_0 \end{bmatrix} \in \mathbb{R}^{(n+m)\times(n+m)}, \ \mathcal{A}_j \in \mathbb{R}^{(n+m)\times(n+m)}$$

with $\mathcal{A}_j(1:n, n+j) = -B_j M_j$, $\mathcal{A}_j(n+j, n+j) = M_j$, and zero everywhere else. The following theorem establishes an affine quadratic stability condition which generalizes our previous results to generally nonconvex cases. The proof of the theorem can be found in Appendix A.

Theorem 9.5: Consider an affine function described by Eq. (9.60) and the parameter box $\Lambda = \text{conv}(\Lambda_0)$ as defined by Eqs. (9.58) (9.59). Let M_0 to be chosen sufficiently large enough such that $\mathcal{A}_{0,i_1,\ldots,i_l}$ always satisfies

$$\mathcal{A}^T_{0,i_1,\ldots,i_l}\mathcal{P}_0 + \mathcal{P}_0\mathcal{A}_{0,i_1,\ldots,i_l} < 0, \tag{9.61}$$

$$\mathcal{P}_0 = \mathcal{P}_0^T > 0, \tag{9.62}$$

for all $\overline{W}_{i_1,\ldots,i_l}$. If there exists real matrices $\mathcal{P}_0, \mathcal{P}_1, \ldots, \mathcal{P}_m$ where

$$\mathcal{P}(\Delta\lambda) = \mathcal{P}_0 + \sum_{j=1}^{m} \Delta\lambda_j \mathcal{P}_j, \tag{9.63}$$

such that Eq. (9.62) and the linear matrix inequality conditions

$$\mathcal{A}^T(\overline{W}_{i_1,\ldots,i_l}, \Delta\lambda)\mathcal{P}(\Delta\lambda) + \mathcal{P}(\Delta\lambda)\mathcal{A}(\overline{W}_{i_1,\ldots,i_l}, \Delta\lambda) < 0, \quad \forall \Delta\lambda \in \Lambda_0 \tag{9.64}$$

$$\mathcal{P}(\Delta\lambda) > 0, \quad \forall \Delta\lambda \in \Lambda_0 \tag{9.65}$$

$$\mathcal{A}_j^T\mathcal{P}_j + \mathcal{P}_j\mathcal{A}_j \geq 0, \quad \text{for } i = j = 1, \ldots, m, \tag{9.66}$$

are also satisfied, then Eq. (9.60) is affinely quadratically stable $\forall \Delta\lambda \in \Lambda$.

We now utilize LMIs to satisfy the affine quadratic stability conditions given by Eqs. (9.64) through (9.66) for given projection bounds of $\hat{W}_{\max}$ for the elements $\hat{W}(t)$, respectively, and the change in actuator bandwidth limits contained within the parameter box Λ_0. For this purpose, let $\overline{W}_{i_1,\ldots,i_l} \in \mathbb{R}^{n\times m}$ be given by Eq. (9.51). Following the results of Theorem 9.5, if

$$\mathcal{A}(\overline{W}_{i_1,\ldots,i_l}, \Delta\lambda) = \mathcal{A}_{0,i_1,\ldots,i_l} + \sum_{j=1}^{m} \Delta\lambda_j \mathcal{A}_j, \tag{9.67}$$

satisfies the LMIs given by Eqs. (9.64) through (9.66) for all permutations of $\overline{W}_{i_1,\ldots,i_l}$, then Eq. (9.53) is affinely quadratically stable. Since Eq. (9.53) is feasible for large values of M (see Lemma 9.1), we can recast Eqs. (9.64) through (9.66) as the general eigenvalue problem given by

$$\begin{aligned} &\text{maximize } \Lambda_0, \\ &\text{subject to Eqs.(9.64), (9.65), (9.66).} \end{aligned} \tag{9.68}$$

We can therefore satisfy Eqs. (9.64) through (9.66) by maximizing Λ_0.

The following algorithm describes a way to evaluate Eq. (9.68) in a finite number of iterations by expanding the corners Λ_0 of the parameter box Λ. Algorithm 9.1 introduces the new term ϵ_{LMI}, which is a specified step tolerance. The generalized eigenvalue problem given by Eq. (9.68) is solved using YALMIP [16], but other solvers can also be used [17].

Algorithm 9.1 Feasible Region Search Algorithm

Data A, B, K_1, M_0, $\hat{W}_{\max}$, ϵ_{LMI}
  Result $\Delta\lambda_{\max}$
  for $\Delta\lambda_1 = 0 : \epsilon_{\text{LMI}} : \lambda_{\text{feas},1}$ **do**
    for $\Delta\lambda_2 = 0 : \epsilon_{\text{LMI}} : \lambda_{\text{feas},2}$ **do**
      ⋮
      for $\Delta\lambda_m = 0 : \epsilon_{\text{LMI}} : \lambda_{\text{feas},m}$ **do**
        if *Eqs.* (9.64), (9.65), *and* (9.66) *are feasible* **then**
          Continue
        else if *Eqs.* (9.64), (9.65), *and* (9.66) *are infeasible* **then**
          Bisect using Eqs. (9.64), (9.65), and (9.66)
        end
      end
    end
  end

Figure 9.3 describes the process in which a shaded feasible region for three actuators is approximated by Algorithm 9.1 when $\epsilon_{\text{LMI}} = \lambda_{\text{feas}}$. For this case, the search is done in a total of eight steps. Algorithm 9.1 can only approximate a feasible space since there are an infinite amount of combinations of the upper bounds defined in Λ_0, but by decreasing ϵ_{LMI} one can easily find a better estimate for the feasible region at the expense of computation time.

Algorithm 9.1 will produce a less conservative solution than the case where convex optimization problem from Section 9.3 is solved using bisection and a positive definite constraint on the $\mathcal{P}$ matrix. To illustrate this point, consider an example in which $M \in \mathbb{R}_+$, $A = -1$, $K_1 = 0$ and $B = 1$. Figure 9.4 shows the differences when searching for a minimum M when given a range of $\hat{W}_{\max}$. It is clear that Algorithm 9.1 produces less conservative results from the structured $\mathcal{P}$ algorithm.

Example 9.2: [18] To further elucidate the presented algorithm, we consider a case in which the system matrices follow as

$$A = \begin{bmatrix} 0 & 1 \\ -1 & -1.4 \end{bmatrix}, \quad B = \begin{bmatrix} 0 & 1 & 1 \\ 1 & 1 & 1 \end{bmatrix}. \tag{9.69}$$

Note that since the system matrix A is Hurwitz, it is not necessary to include a stabilizing feedback gain K_1, such that for the LMI analysis, we can set $K_1 = 0$. The algorithm tolerance is set to $\epsilon_{\text{LMI}} = 0.2$ and the initial conditions of matrix $M_0 \in \mathbb{R}^{3\times 3}$ are given such that $M_0 = 30I_3$ (sufficiently large).

Figure 9.5 considers two cases in which $|\hat{W}_{\max}| \leq 1$ and $|\hat{W}_{\max}| \leq 2$. The side by side comparison shows that by increasing the bounds on $|\hat{W}_{\max}|$ (due to an increase in the amount of system uncertainty), we decrease the feasible region.

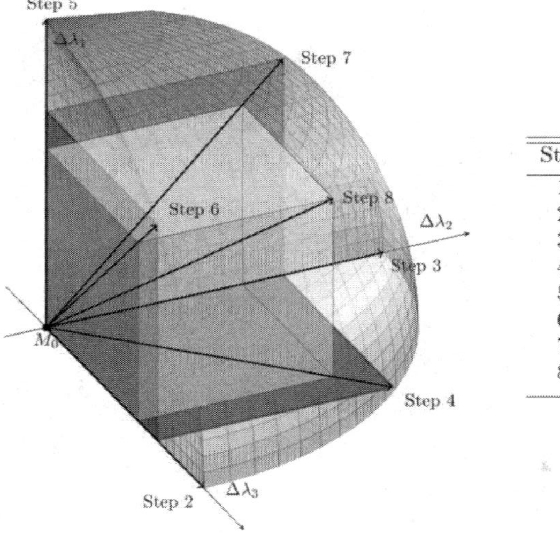

Step	Description
1	If $\Delta\lambda_3 = \Delta\lambda_2 = \Delta\lambda_1 = 0$ is feasible, continue to next step
2	If $\Delta\lambda_3 = \lambda_{\text{feas}}$ and $\Delta\lambda_2 = \Delta\lambda_1 = 0$ is infeasible, then bisect until feasible
3	If $\Delta\lambda_2 = \lambda_{\text{feas}}$ and $\Delta\lambda_3 = \Delta\lambda_1 = 0$ is infeasible, then bisect until feasible
4	If $\Delta\lambda_3 = \Delta\lambda_2 = \lambda_{\text{feas}}$ and $\Delta\lambda_1 = 0$ is infeasible, then bisect until feasible
5	If $\Delta\lambda_1 = \lambda_{\text{feas}}$ and $\Delta\lambda_3 = \Delta\lambda_1 = 0$ is infeasible, then bisect until feasible
6	If $\Delta\lambda_3 = \Delta\lambda_1 = \lambda_{\text{feas}}$ and $\Delta\lambda_2 = 0$ is infeasible, then bisect until feasible
7	If $\Delta\lambda_2 = \Delta\lambda_1 = \lambda_{\text{feas}}$ and $\Delta\lambda_1 = 0$ is infeasible, then bisect until feasible
8	If $\Delta\lambda_3 = \Delta\lambda_2 = \Delta\lambda_1 = \lambda_{\text{feas}}$ is infeasible, then bisect until feasible

Fig. 9.3 Algorithm 9.1 solving three actuators ($\epsilon_{\text{LMI}} = \lambda_{\text{feas}}$).

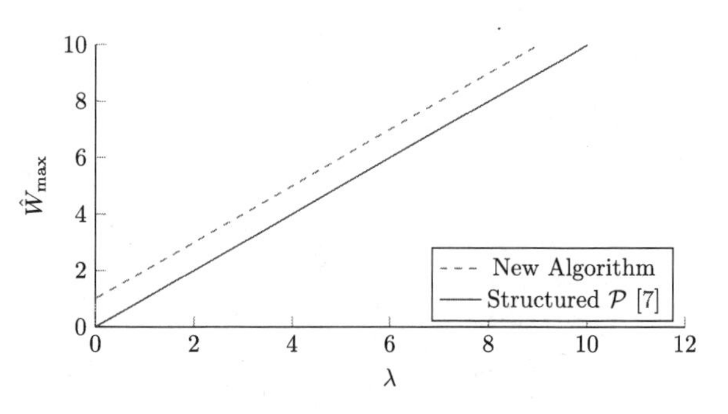

Fig. 9.4 Linear matrix inequality comparison plot (A_r = − 1 and B = 1).

This section introduced a new method for evaluating the robustness of adaptive controls in the presence of multiactuator dynamics. In comparison with other methods, our results extended out to multiactuator systems and solve them efficiently. In the next section, we provide an application of the proposed LMI-based hedging approach, along with the algorithm for multiactuators to a hypersonic aircraft example.

9.7 HYPERSONIC AIRCRAFT EXAMPLE

To further elucidate our proposed approach to the actuator dynamics problem, we provide the following application to a hypersonic vehicle. This section first provides a brief overview of a PI-based nominal control design that is used for command following purposes. It then provides a state-space model of a generic hypersonic vehicle (GHV) and explains how the model is decoupled into longitudinal and lateral-directional dynamics for which separate controllers are designed. The simulation results illustrate nominal performance, a standard adaptive performance, and the proposed adaptive performance with actuator dynamics.

9.7.1 AUGMENTATION OF A PI NOMINAL CONTROLLER

Consider the uncertain dynamical system given by

$$\dot{x}_p(t) = A_p x_p(t) + B_p[u(t) + W^T \sigma(x_p(t))], \quad x_p(0) = x_{p0}, \tag{9.70}$$

where $x_p(t) \in \mathbb{R}^{n_p}$ is the state vector available for feedback, $u(t) \in \mathbb{R}^m$ is the control input, $A_p \in \mathbb{R}^{n_p \times n_p}$ is a known system matrix, $B_p \in \mathbb{R}^{n_p \times m}$ is a known control input matrix, $W \in \mathbb{R}^{s \times m}$ is an unknown weight matrix and $\sigma : \mathbb{R}^{n_p} \to \mathbb{R}^s$ is a known basis function of the form $\sigma(x_p) = [\sigma_1(x_p), \sigma_2(x_p), \ldots, \sigma_s(x_p)]^T$, and the pair (A_p, B_p) is controllable.

To address command following, let $c(t) \in \mathbb{R}^{n_a}$ be a given piecewise continuous command and $x_a(t) \in \mathbb{R}^{n_a}$ be the integrator state satisfying

$$\dot{x}_a(t) = E_p x_p(t) - c(t), \quad x_a(0) = x_{a0}, \tag{9.71}$$

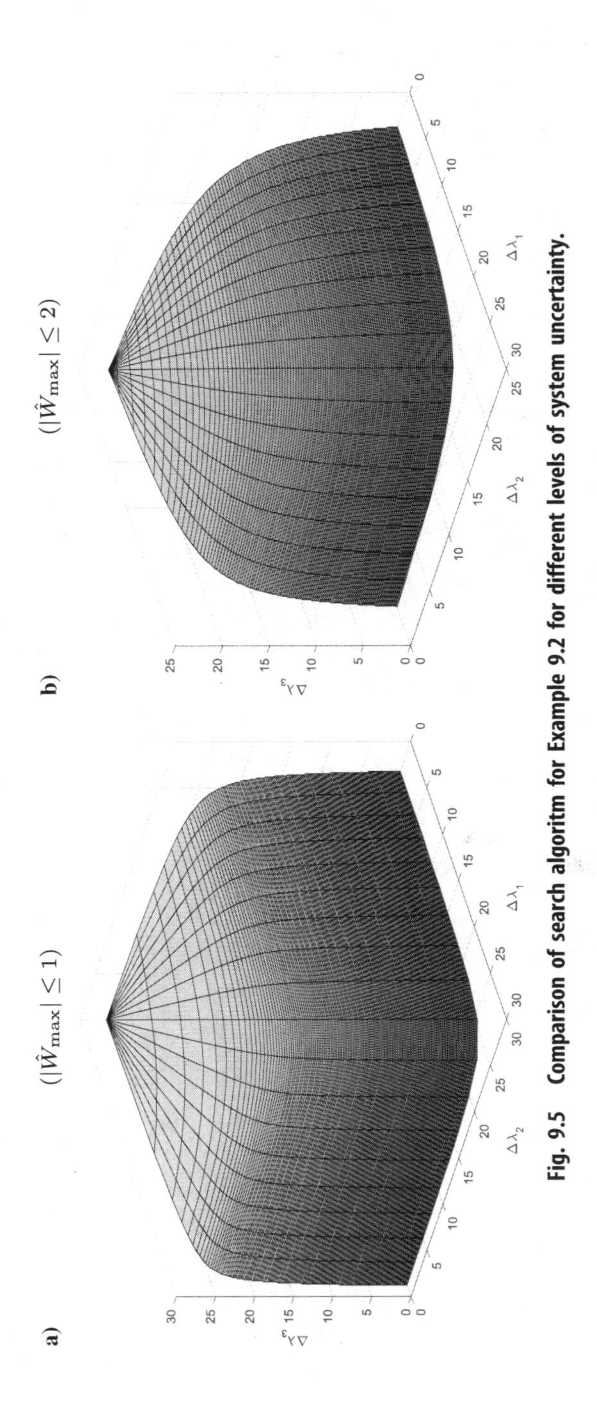

Fig. 9.5 Comparison of search algoritm for Example 9.2 for different levels of system uncertainty.

where $E_p \in \mathbb{R}^{n_a \times n_p}$ allows the selection of a subset of $x_p(t)$ to follow $c(t)$. It then follows that Eq. (9.70) and Eq. (9.71) can be augmented as

$$\dot{x}(t) = Ax(t) + B_r c(t) + B\Big[u(t) + W^T \sigma(x_p(t))\Big], \quad x(0) = x_0, \tag{9.72}$$

where $x(t) \triangleq [x_p^T(t), x_a^T(t)]^T \in \mathbb{R}^n$, $n = n_p + n_a$, is the augmented stated vector, $x_0 = [x_{p0}^T, x_{a0}^T]^T \in \mathbb{R}^n$, $A \triangleq \begin{bmatrix} A_p & 0_{n_p \times n_a} \\ E_p & 0_{n_a \times n_a} \end{bmatrix} \in \mathbb{R}^{n \times n}$, $B \triangleq [B_p^T, 0_{n_a \times m}^T]^T \in \mathbb{R}^{n \times m}$, $B_r \triangleq [0_{n_p \times n_a}^T, \ -I_{n_a \times n_a}]^T \in \mathbb{R}^{n \times n_a}$. Next, consider the feedback control law given by

$$u(t) = u_n(t) + u_a(t), \tag{9.73}$$

and let the nominal feedback control law be given by

$$u_n(t) = -Kx(t), \quad K \in \mathbb{R}^{m \times n}, \tag{9.74}$$

such that $A_r \triangleq A - BK$ is Hurwitz. Using Eq. (9.73) and Eq. (9.74) in Eq. (9.72) yields

$$\dot{x}(t) = A_r x(t) + B_r c(t) + B\Big[u_a(t) + W^T \sigma(x_p(t))\Big]. \tag{9.75}$$

Note that Eq. (9.75) has a similar form as Eq. (9.16) in Section 9.2. It follows that the framework proposed in this chapter can be implemented using this PI-based nominal control, which will be used to design the controllers for the longitudinal and lateral-directional dynamics of the GHV model, can be found in Appendix B.

9.7.2 LONGITUDINAL CONTROL DESIGN

For the decoupled longitudinal dynamics, we consider the state vector defined as $x_{p_{lo}}(t) = [\alpha(t), q(t)]^T$ and the control input defined as $u_{lo}(t) = \delta_e(t)$, with the respective system matrices

$$A_{p_{lo}} = \begin{bmatrix} -2.39 \times 10^{-1} & 1 \\ 4.26 & -1.19 \times 10^{-1} \end{bmatrix}, \tag{9.76}$$

$$B_{p_{lo}} = \begin{bmatrix} -1.33 \times 10^{-4} \\ -1.84 \times 10^{-1} \end{bmatrix}. \tag{9.77}$$

LQR theory [19] is used to design the nominal controller with $E_{p_{lo}} = [\,1, 0\,]$ such that a desired angle-of-attack command is followed. The controller gain matrix K_{lo} is obtained using the highlighted augmented formulation, along with the weighting matrices $Q_{lo} = \text{diag}[\,2{,}000, 25, 400{,}000\,]$ to penalize $x_{lo}(t)$ and $R_{lo} = 12.5$ to penalize $u_{\text{lo}}(t)$, resulting in the following gain matrix

$$K_{lo} = [\,-135.9 \quad -37.7 \quad -178.9\,], \tag{9.78}$$

which has a desirable 60.4° phase margin and a crossover frequency of 6.75 Hz. The solution to $A_{r_{lo}}^T P_{lo} + P_{lo} A_{r_{lo}} + R_{1_{lo}} = 0$, where $A_{r_{lo}} \triangleq A_{lo} - B_{lo} K_{lo}$, is calculated with $R_{1_{lo}} = \text{diag}[\,1{,}000,\, 1{,}000,\, 2{,}000\,]$ that is used for both the standard adaptive control design and the proposed controller.

9.7.3 LATERAL-DIRECTIONAL CONTROL DESIGN

The decoupled lateral-directional dynamics follow similarly. Specifically, we consider the state vector defined as $x_{p_{la}}(t) = [\beta(t), p(t), r(t), \phi(t)]^T$ and the control input vector is defined as $u_{la}(t) = [\delta_a(t), \delta_r(t)]^T$, with the respective system matrices

$$A_{p_{la}} = \begin{bmatrix} -6.97\times 10^{-2} & -1.04\times 10^{-2} & -9.99\times 10^{-1} & -5.35\times 10^{-3} \\ -1.31\times 10^{3} & -2.03 & -7.54\times 10^{-3} & 0 \\ 2.07 & -1.55\times 10^{-3} & -5.31\times 10^{-2} & 0 \\ -2.38\times 10^{-4} & 8.54\times 10^{-1} & -8.84\times 10^{-3} & -3.00\times 10^{-6} \end{bmatrix}, \tag{9.79}$$

$$B_{p_{la}} = \begin{bmatrix} -2.47\times 10^{-5} & -2.18\times 10^{-4} \\ -8.04 & 10.3 \\ 3.17\times 10^{-2} & 2.85\times 10^{-1} \\ 0 & 0 \end{bmatrix}. \tag{9.80}$$

LQR theory [19] is used to design the nominal controller with

$$E_{p_{la}} = \begin{bmatrix} 1 & 0 & 0 & 0 \\ 0 & 0 & 0 & 1 \end{bmatrix} \tag{9.81}$$

such that a desired sideslip angle command and roll angle command are followed. The controller gain matrix K_{la} is obtained using the highlighted augmented formulation along with the weighting matrices $Q_{la} = \text{diag}[10, 0.01, 0.01, 10, 400{,}000, 500]$ to penalize $x_{la}(t)$ and $R_{la} = \text{diag}[\,2.5, 50\,]$ to penalize $u_{la}(t)$, resulting in the following gain matrix

$$K_{la} = \begin{bmatrix} 285.4 & -1.23 & -42.8 & -6.15 & 116.3 & -13.5 \\ 92.3 & -0.1 & -28.2 & 0.6 & 85.6 & 0.9 \end{bmatrix}, \tag{9.82}$$

which has desirable phase margins of 61.4° and 61.9° and respective crossover frequencies of 6.93 Hz and 8.68 Hz. The solution to $A_{r_{la}}^T P_{la} + P_{la} A_{r_{la}} + R_{1_{la}} = 0$, where $A_{r_{la}} \triangleq A_{la} - B_{la} K_{la}$ is calculated with $R_{1_{la}} = \text{diag}[\,10, 10, 2000, 10, 300{,}000, 2000]$ that is used for both the standard adaptive control design and the proposed controller.

9.7.4 NOMINAL SYSTEM AND STANDARD ADAPTIVE CONTROLLER PERFORMANCE

The longitudinal and lateral-directional controllers are augmented and applied to the overall coupled system. We first consider the case when there is no uncertainty in the system to show the nominal performance of the control designs. To destabilize the nominal system, we then introduce uncertainty, which for our simulation we consider $W_{lo} = [\,-100 \quad .01\,]^T$ and $W_{la} = \begin{bmatrix} 100 & 70 & 0.01 & 0.01 & 0.01 & 0.01 \\ 0.01 & 0.01 & 0.01 & 0.01 & 0.01 & 0.01 \end{bmatrix}^T$, which dominantly effect the stability derivatives C_{m_α} and C_{n_β}. Figure 9.6 shows the response of the nominal control performance for both cases with and without system uncertainty.

A standard model reference adaptive control is then augmented with the longitudinal and lateral-directional controllers to stabilize the system. The learning gain matrices used for the longitudinal and lateral-directional adaptive controllers are $\Gamma_{lo} = \text{diag}[\,1000, 1\,]$ and $\Gamma_{la} = \text{diag}[\,1000, 1, 1, 1\,]$, respectively. In addition, the element-wise projection bounds are set such that $\left|[\hat{W}_{lo}(t)]_{1,1}\right| \leq 105$ and $\left|[\hat{W}_{lo}(t)]_{2,1}\right| \leq 0.1$ for the longitudinal adaptive control law and $\left|[\hat{W}_{la}(t)]_{1,1}\right| \leq 105$, $\left|[\hat{W}_{lo}(t)]_{1,2}\right| \leq 75$, and $\left|[\hat{W}_{la}(t)]_{i,j}\right| \leq 0.1$, $i = 2, 3, 4$ and $j = 1, 2$, for the lateral-directional adaptive control law. It can be seen from Fig. 9.7 that the uncertain system follows the reference model trajectory.

9.7.5 LMI-BASED HEDGING APPROACH

Now, using the adaptive control laws designed in the section above, we introduce the effect of actuator dynamics and implement a modified reference model. Since

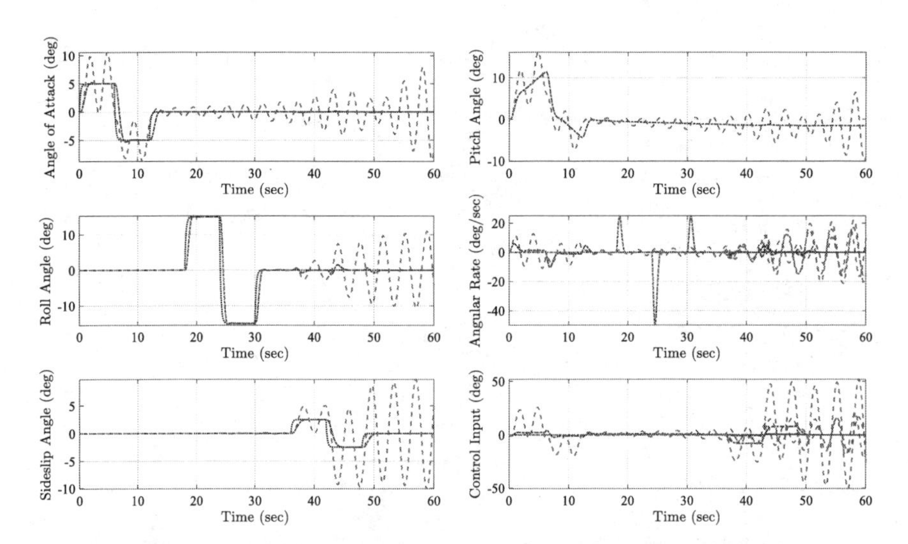

Fig. 9.6 Nominal controller performance with and without uncertainty, where the dash-dot line represents the nominal performance without uncertainty and the dashed line is the nominal performance with uncertainty.

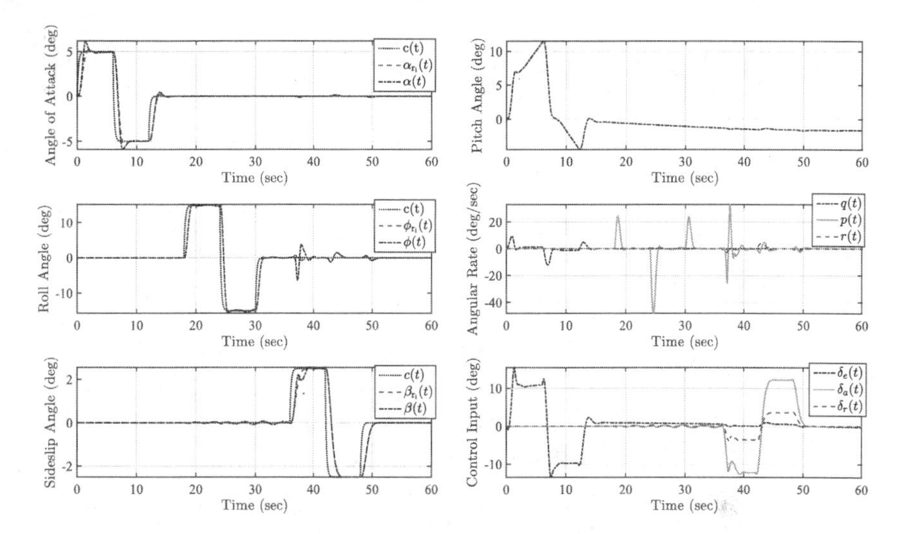

Fig. 9.7 Standard adaptive controller performance with uncertainty.

the longitudinal and lateral-directional controllers are designed separately, we can analyze the effect of the actuator dynamics for each controller.

We begin with the longitudinal controller and use the LMI analysis developed in the chapter to compute the minimum actuator bandwidth for the elevator control surface as $\lambda_e = 7.96$. We set the actuator bandwidth values for the lateral-directional control surfaces sufficiently large such that only the effect of the elevator actuation is analyzed. Figures 9.8 and 9.9 show the proposed controller performance at the LMI calculated minimum actuator bandwidth and a bandwidth value of $\lambda_e = 7.5$. It can be seen from Fig. 9.8 that at the calculated minimum bandwidth value the modified reference model remains bounded. It is not until the bandwidth value reaches $\lambda_e = 7.5$, as shown in Fig. 9.9, that the modified reference model yields an unbounded trajectory.

Finally, to analyze the effect of the lateral-directional actuation, we set the elevator actuator bandwidth sufficiently large and use the LMI analysis to compute the region in which the lateral-directional actuator bandwidths are feasible. This can be seen from Fig. 9.10, which shows the minimum actuator bandwidth values of the lateral-directional control surfaces λ_a and λ_r. Note that Fig. 9.10 provides both the LMI calculated feasible limit as well as the feasible limit provided by the simulation results, which correspond to the command profile, initial conditions, and other parameters for the provided example. We select two points to simulate the proposed controller performance as seen in Figs. 9.11 and 9.12. Since the feasible boundary corresponds to calculated minimum feasible λ_a and λ_r values for the actuator bandwidths, it is expected that the system performances are guaranteed to be bounded for actuator bandwidth values at points greater than

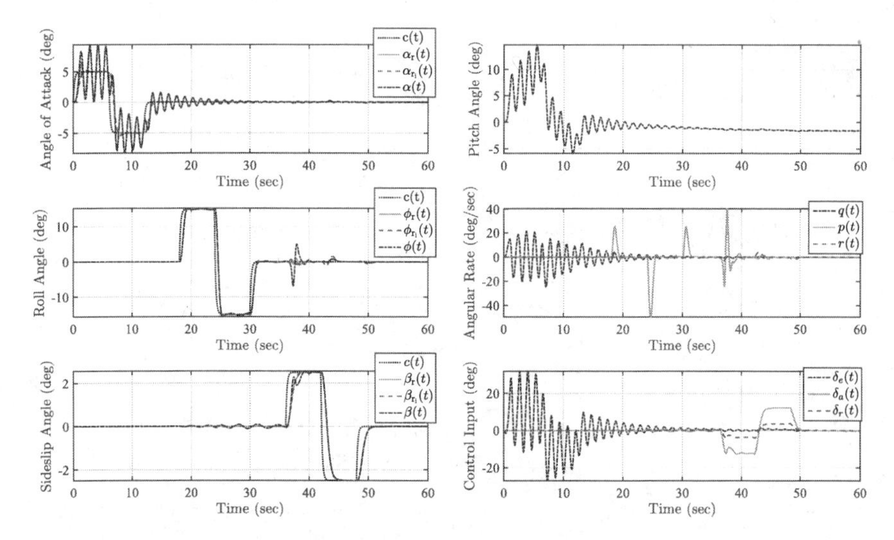

Fig. 9.8 Proposed longitudinal controller performance with actuator dynamics ($\lambda_e = 7.96$).

and equal to the calculated feasible boundary. This can be seen in Fig. 9.11 when the actuator bandwidths are at the minimum point $(\lambda_a, \lambda_r) = (10, 9.4)$, which is located on the feasible boundary. In Fig. 9.12, we let the actuator bandwidths be outside the calculated feasible region to show that the closed-loop system remains

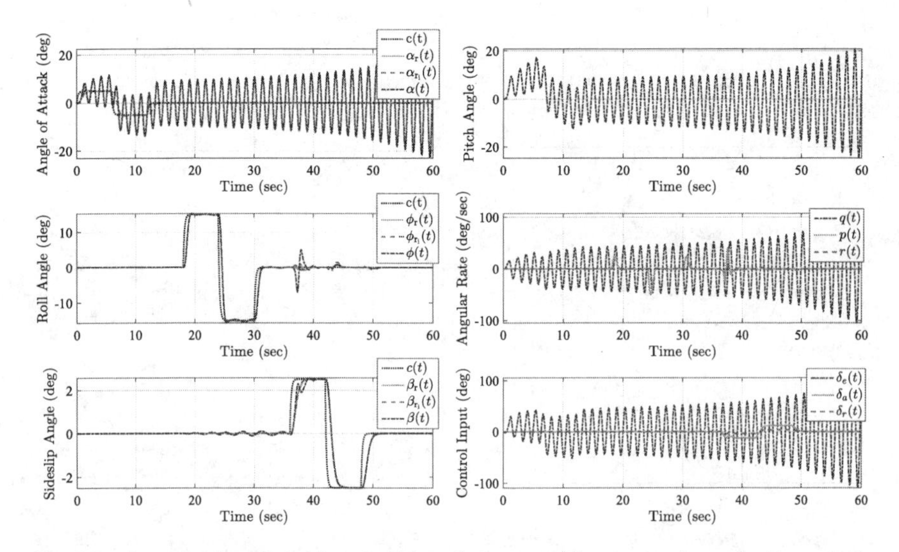

Fig. 9.9 Proposed longitudinal controller performance with actuator dynamics ($\lambda_e = 7.5$).

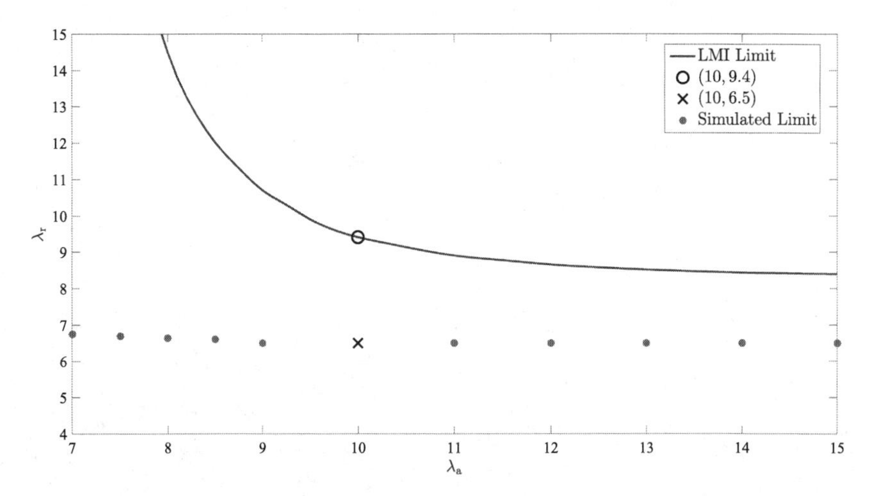

Fig. 9.10 LMI calculated feasible region for lateral-directional actautor bandwidth values.

bounded until the actuator bandwidths reach a value of $(\lambda_a, \lambda_r) = (10, 6.5)$. This is consistent with the presented theory, as we provide a (conservative) upper bound on the allowable actuator dynamics such that the closed-loop system remains bounded.

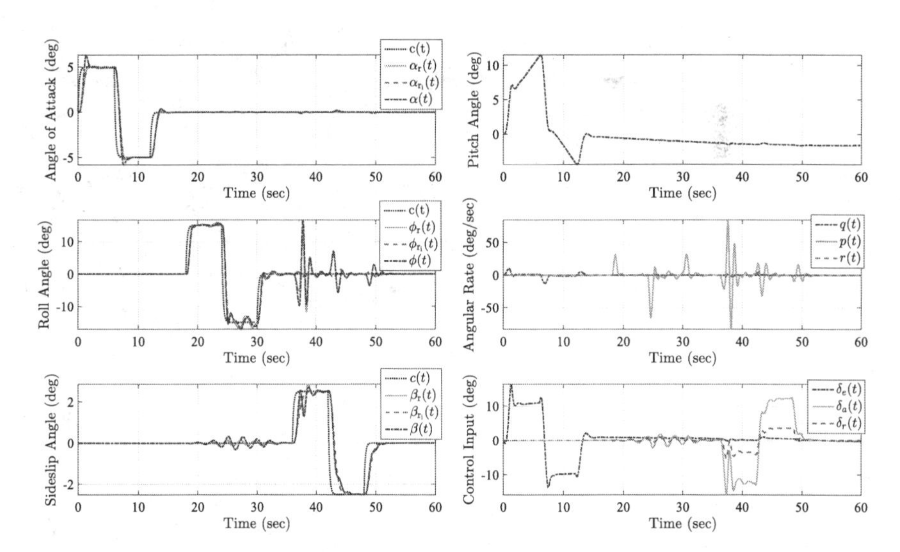

Fig. 9.11 Proposed lateral-directional controller performance with actuator dynamics ($(\lambda_a, \lambda_r) = (10, 9.4)$).

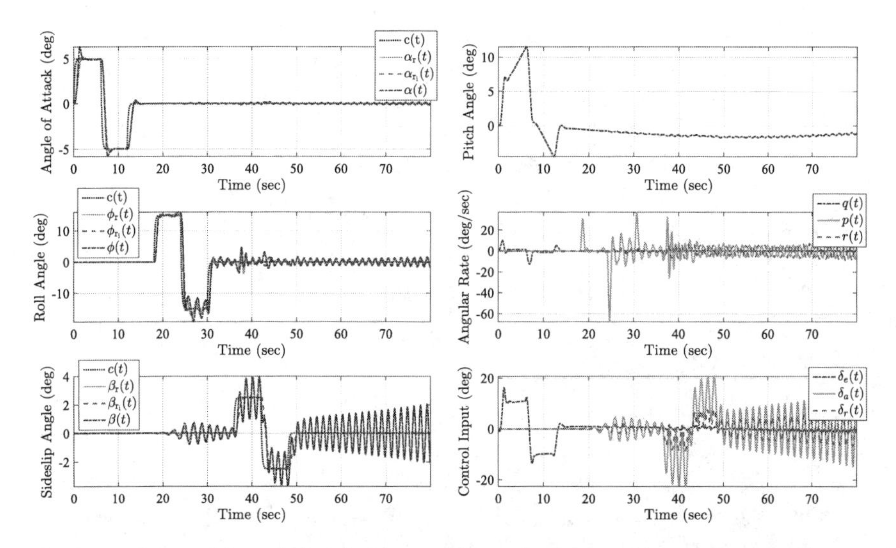

Fig. 9.12 Proposed lateral-directional controller performance with actuator dynamics ($(\lambda_a, \lambda_r) = (10,6.5)$).

9.8 SUMMARY

It is well known that the presence of actuator dynamics can seriously limit the stability and achievable performance of adaptive controllers. This requires additional verification steps to determine a range of actuator bandwidth values such that adaptive controllers perform properly. In this chapter, we presented a framework such that adaptive controllers can be designed for uncertain dynamical systems with actuator dynamics. This framework has the capability to rigorously characterize the fundamental stability interplay between the allowable system uncertainties and the bandwidth of the actuator dynamics. We first introduced an LMI-based hedging framework, which allows for correct adaptation that is not affected by the presence of actuator dynamics. Stability of the closed-loop dynamical system utilizing this framework was discussed using Lyapunov-based analysis tools and an LMI analysis was used to characterize the fundamental stability interplay between the allowable system uncertainties and the bandwidth of the actuator dynamics. Next, we established convergence properties of the reference model trajectories and presented a generalization of the proposed LMI-based hedging framework to adaptive control. An algorithm was also presented to allow for less conservative computations of the minimum allowable actuator bandwidths, as well as solving for cases in which there are multiple actuators. The results of these sections combine to produce a complete analysis of the actuator dynamics problem, which we then illustrate through an example of a linearized hypersonic aircraft model. We showed how the feasible actuator bandwidth values can be computed for both the longitudinal and lateral-directional control surfaces with a small amount of conservatism, especially for the longitudinal controller. Future research will include extensions of the proposed framework to output feedback adaptive control (e.g., by applying the results presented above to [20, 21]), adaptive control in the presence of not only actuator dynamics but also actuator amplitude and rate saturation limits, and applications to large-scale systems.

APPENDIX A: THEOREM PROOFS

Proof of Theorem 9.2: Making use of arguments presented in [22], we begin by considering the reference model error dynamics, which follow from the modified reference model given by Eq. (9.29) and the ideal reference model given by Eq. (9.38) subject to the actuator dynamics given by Eq. (9.26) and the feedback control laws given by Eqs. (9.11), (9.12), and (9.17) as

$$\begin{aligned}\dot{e}_r(t) &= A_r e_r(t) + B[v(t) - u(t)] \\ &= A_r e_r(t) + B[Mx_c(t) + \hat{W}^T(t)x(t) + K_1 x(t) - K_2 c(t)] \\ &= Ae_r(t) + B[Mx_c(t) + \hat{W}^T(t)e(t) + \hat{W}^T(t)e_r(t) + \hat{W}^T(t)x_{r_i}(t) \\ &\quad + K_1 e(t) + K_1 x_{r_i}(t) - K_2 c(t)].\end{aligned} \tag{9A.1}$$

Additionally, the actuator dynamics given by Eq. (9.26) subject to the feedback control laws given by Eqs. (9.11), (9.12), and (9.17) can be written as

$$\begin{aligned}\dot{x}_c(t) &= -Mx_c(t) - \hat{W}_1^T(t)x(t) - K_1 x(t) + K_2 c(t) \\ &= -Mx_c(t) - \hat{W}^T(t)e(t) - \hat{W}^T(t)e_r(t) - \hat{W}^T(t)x_{r_i}(t) \\ &\quad - K_1 e(t) - K_1 e_r(t) - K_1 x_{r_i}(t) + K_2 c(t).\end{aligned} \tag{9A.2}$$

It follows that Eq. (9A.1) and Eq. (9A.2) can be written in compact form as

$$\dot{\bar{e}}(t) = \mathcal{A}(\hat{W}(t), M)\bar{e}(t) + \mathcal{B}\omega_1(\cdot), \tag{9A.3}$$

with $\bar{e} = [e_r^T(t), x_c^T(t)]^T$, $\mathcal{B} = [B^T, \ -I_m]^T$, and

$$\omega_1(\cdot) = \hat{W}^T(t)e(t) + \hat{W}^T(t)x_{r_i}(t) + K_1 e(t) + K_1 x_{r_i}(t) - K_2 c(t). \tag{9A.4}$$

Note that $\omega_1(\cdot)$ in Eq. (9A.3) is bounded as a result of the analysis in Theorem 9.1 and the boundedness of the ideal reference model given by Eq. (9.38). In addition, since $\mathcal{A}(\hat{W}(t), M)$ is quadratically stable for $(\underline{\lambda}, \overline{\omega}) \in \kappa$ by Lemma 9.1, there exists a $P \in \mathbb{R}_+^{(n+m)\times(n+m)}$ satisfying

$$\mathcal{A}^T(\hat{W}(t), M)\mathcal{P} + \mathcal{P}\mathcal{A}(\hat{W}(t), M) + \mathcal{Q} = 0, \tag{9A.5}$$

with $\mathcal{Q} \in \mathbb{R}_+^{(n+m)\times(n+m)}$. Using Eq. (9.44), first note that the positive-definiteness of this selection follows from the positive-definiteness of P and the positive-definiteness of the Schur complement of Eq. (9.44) given by

$$S_1 = B^T PB + \rho I - B^T P(P)^{-1} PB = \rho I > 0. \tag{9A.6}$$

Then, using Eq. (9.44) in Eq. (9A.5) and solving for $\mathcal{Q}$ it follows that

$$\begin{aligned}\mathcal{Q} &= -\left[\mathcal{A}^T(\hat{W}(t), M)\mathcal{P} + \mathcal{P}\mathcal{A}(\hat{W}(t), M)\right] \\ &= \begin{bmatrix} R & -Q_2(t) \\ -Q_2^T(t) & 2\rho M \end{bmatrix}\end{aligned} \tag{9A.7}$$

where $Q_2(t) \triangleq A_r^T PB - \rho(K_1 + \hat{W}(t))$. Since R is a positive-definite matrix, it follows from the Schur complement of Eq. (9A.7) given by

$$S_2 = 2\rho M - (B^T PA_r - \rho(K_1^T + \hat{W}^T(t)))R^{-1}(A_r^T PB - \rho(K_1 + \hat{W}(t))), \tag{9A.8}$$

that Eq. (9A.7) is a positive-definite matrix for large values of M. Furthermore, note that $M_0 = M$ for $\phi = 1$, such that $\mathcal{A}(\hat{W}(t), M_0)$ is quadratically stable for $(\underline{\lambda}, \overline{\omega}) \in \kappa$ by Lemma 9.1 and

$$\mathcal{A}^T(\hat{W}(t), M_0)\mathcal{P} + \mathcal{P}\mathcal{A}(\hat{W}(t), M_0) \leq -\eta I_{n+m}, \tag{9A.9}$$

holds. It then follows that

$$\begin{aligned}\mathcal{A}^T(\hat{W}(t), M)\mathcal{P} + \mathcal{P}\mathcal{A}(\hat{W}(t), M) &= \begin{bmatrix} -R & Q_2(t) \\ Q_2^T(t) & -2\rho M \end{bmatrix} \\ &= \begin{bmatrix} -R & Q_2(t) \\ Q_2^T(t) & -2\rho M_0 \end{bmatrix} + \begin{bmatrix} 0_{n\times n} & 0_{n\times m} \\ 0_{m\times n} & -2\rho(M - M_0) \end{bmatrix} \\ &= \mathcal{A}^T(\hat{W}(t), M_0)\mathcal{P} + \mathcal{P}\mathcal{A}(\hat{W}(t), M_0) + \begin{bmatrix} 0_{n\times n} & 0_{n\times m} \\ 0_{m\times n} & -2\rho(M - M_0) \end{bmatrix} \\ &\leq -\eta I_{n+m} + \begin{bmatrix} 0_{n\times n} & 0_{n\times m} \\ 0_{m\times n} & -(\phi - 1)2\rho M_0 \end{bmatrix}.\end{aligned} \tag{9A.10}$$

Now, consider the Lyapunov function candidate given by

$$\mathcal{V}(\overline{e}) = \overline{e}^T \mathcal{P}\overline{e}. \tag{9A.11}$$

Differentiating Eq. (9A.11) and using Eq. (9A.3), and (9A.10) yields

$$
\begin{aligned}
\dot{\mathcal{V}}(\bar{e}(t)) &= 2\bar{e}^T(t)\mathcal{P}\dot{\bar{e}}(t) \\
&= \bar{e}^T(t)[\mathcal{A}^T(\hat{W}(t), M)\mathcal{P} + \mathcal{P}\mathcal{A}(\hat{W}(t), M)]\bar{e}(t) + 2\bar{e}^T(t)\mathcal{P}\mathcal{B}\omega_1(\cdot) \\
&\leq -\eta\|\bar{e}(t)\|_2^2 - (\phi - 1)2\rho x_c^T(t)M_0x_c(t) - 2\rho x_c^T(t)\omega_1(\cdot) \\
&\leq -\eta\|\bar{e}(t)\|_2^2 - (\phi - 1)2\rho\lambda_{\min}(M_0)\|x_c(t)\|_2^2 + 2\rho\|x_c(t)\|_2\|\omega_1(\cdot)\|_2.
\end{aligned} \tag{9A.12}
$$

Setting $\beta = 2\rho\lambda_{\min}(M_0)$ and noting that $\|\omega_1(\cdot)\|_2 \leq \omega^*$, Eq. (9A.12) can be written as

$$
\begin{aligned}
\dot{\mathcal{V}}(\bar{e}(t)) &\leq -\eta\|\bar{e}(t)\|_2^2 - (\phi - 1)\beta\|x_c(t)\|_2^2 + 2\rho\|x_c(t)\|_2\omega^* \\
&= -\eta\|e_r(t)\|_2^2 - \eta\|x_c(t)\|_2^2 - (\phi - 1)\beta\|x_c(t)\|_2^2 + 2\rho\|x_c(t)\|_2\omega^* \\
&= -\eta\|e_r(t)\|_2^2 - (\eta + (\phi - 1)\beta)\|x_c(t)\|_2\left(\|x_c(t)\|_2 - \frac{2\rho\omega^*}{\eta + (\phi - 1)\beta}\right).
\end{aligned} \tag{9A.13}
$$

It then follows that when $\|x_c(t)\|_2$ satisfies

$$
\|x_c(t)\|_2 > \frac{2\rho\omega^*}{\eta + (\phi - 1)\beta}, \tag{9A.14}
$$

then $\dot{\mathcal{V}}(\bar{e}(t)) < 0$. To analyze $\|e_r(t)\|_2$, first note that the right-hand side of Eq. (9A.13) is concave with a maximum at

$$
\|x_c(t)\|_2 = \frac{\rho\omega^*}{\eta + (\phi - 1)\beta}. \tag{9A.15}
$$

Using Eq. (9A.15) in Eq. (9A.13) gives the upper bound

$$
\begin{aligned}
\dot{\mathcal{V}}(\bar{e}(t)) &\leq -\eta\|e_r(t)\|_2^2 + (\eta + (\phi - 1)\beta)\frac{\rho\omega^*}{\eta + (\phi - 1)\beta}\left(\frac{\rho\omega^*}{\eta + (\phi - 1)\beta}\right), \\
&= -\eta\|e_r(t)\|_2^2 + \frac{(\rho\omega^*)^2}{\eta + (\phi - 1)\beta}.
\end{aligned} \tag{9A.16}
$$

So, when $\|e_r(t)\|_2$ satisfies

$$
\|e_r(t)\|_2 > \frac{\rho\omega^*}{\sqrt{\eta(\eta + (\phi - 1)\beta)}}, \tag{9A.17}
$$

$\dot{\mathcal{V}}(\bar{e}(t)) < 0$. Using Eq. (9A.14), and (9A.17), it follows that $\mathcal{V}(\bar{e})$ decreases outside the compact set

$$
\Omega = \left\{e_r \in \mathbb{R}^n, x_c \in \mathbb{R}^m : \|e_r(t)\|_2 \leq \frac{\rho\omega^*}{\sqrt{\eta(\eta + (\phi - 1)\beta)}} \text{ and } \|x_c(t)\|_2 \leq \frac{2\rho\omega^*}{\eta + (\phi - 1)\beta}\right\}. \tag{9A.18}
$$

Now, it follows that $\mathcal{V}(\bar{e})$ is upper- and lower-bounded as

$$\lambda_{\min}(\mathcal{P})\|\bar{e}(t)\|_2^2 \leq \mathcal{V}(\bar{e}) \leq \lambda_{\max}(\mathcal{P})\|\bar{e}(t)\|_2^2 = \lambda_{\max}(\mathcal{P})\left(\|e_r(t)\|_2^2 + \|x_c(t)\|_2^2\right), \tag{9A.19}$$

and noting $\|e_r(t)\|_2 \leq \|\bar{e}(t)\|_2$, it follows

$$\begin{aligned}
\lambda_{\min}(\mathcal{P})\|e_r(t)\|_2^2 &\leq \lambda_{\max}(\mathcal{P})\left(\|e_r(t)\|_2^2 + \|x_c(t)\|_2^2\right) \\
&\leq \lambda_{\max}(\mathcal{P})\left(\left(\frac{\rho\omega^*}{\sqrt{\eta(\eta+(\phi-1)\beta)}}\right)^2 + \left(\frac{2\rho\omega^*}{\eta+(\phi-1)\beta}\right)^2\right) \\
&= \lambda_{\max}(\mathcal{P})(\rho\omega^*)^2\left(\frac{\eta+(\phi-1)\beta}{\eta(\eta+(\phi-1)\beta)^2} + \frac{4\eta}{\eta(\eta+(\phi-1)\beta)^2}\right), \\
&= \lambda_{\max}(\mathcal{P})(\rho\omega^*)^2\left(\frac{5\eta+(\phi-1)\beta}{\eta(\eta+(\phi-1)\beta)^2}\right).
\end{aligned} \tag{9A.20}$$

From Please make Eq. (9A.20), we compute the bound for $\|e_r(t)\|_2$ as

$$\|e_r(t)\|_2 \leq \rho\omega^*\sqrt{\frac{\lambda_{\max}(\mathcal{P})}{\lambda_{\min}(\mathcal{P})}}\sqrt{\frac{5\eta+(\phi-1)\beta}{\eta(\eta+(\phi-1)\beta)^2}}. \tag{9A.21}$$

Since $\|\cdot\|_\infty \leq \|\cdot\|_2$, and this bound is uniform, then Eq. (9A.21) yields

$$\|e_r(t)\|_{\mathcal{L}_\infty} \leq \rho\omega^*\sqrt{\frac{\lambda_{\max}(\mathcal{P})}{\lambda_{\min}(\mathcal{P})}}\sqrt{\frac{5\eta+(\phi-1)\beta}{\eta(\eta+(\phi-1)\beta)^2}}, \tag{9A.22}$$

which is the bound given by Eq. (9.45).

Finally, note that $M - M_0 = \phi M_0 - M_0 = (\phi - 1)M_0$, which implies $\lambda_{\min}(M - M_0) = (\phi - 1)\lambda_{\min}(M_0)$. It follows that as M becomes very large, $\lambda_{\min}(M - M_0) \longrightarrow \infty$, which implies $\phi \longrightarrow \infty$, and hence from Eq. (9A.22) it follows that $\|e_r(t)\|_{\mathcal{L}_\infty} \longrightarrow 0$.

Proof of Theorem 9.3: Let $\omega_2(\cdot) \triangleq B\hat{W}^T(t)e(t) + BK_1e(t)$. Then, it follows from using Eqs. (9.11), (9.12), (9.17), and (9.26) in Eq. (9.29) that

$$\dot{x}_r(t) = (A + B\hat{W}^T(t))x_r(t) + BMx_c(t) + \omega_2(\cdot). \tag{9A.23}$$

In addition, let $\omega_3(\cdot) \triangleq -\hat{W}^T(t)e(t) - K_1e(t) + K_2c(t)$. Then, it follows from using Eqs. (9.11), (9.12), and (9.17) in Eq. (9.26) that

$$\dot{x}_c(t) = -Mx_c(t) - K_1x_r(t) - \hat{W}^T(t)x_r(t) + \omega_3(\cdot). \tag{9A.24}$$

Note that Eq. (9A.24) can be rewritten as

$$x_c(t) = M^{-1}[\omega_3(\cdot) - K_1x_r(t) - \hat{W}^T(t)x_r(t) - \dot{x}_c(t)], \tag{9A.25}$$

where it follows from Eq. (9A.23), and (9A.25) that

$$\dot{x}_r(t) = A_r x_r(t) + B_r(t)c(t) - B\dot{x}_c(t). \tag{9A.26}$$

Using Eq. (9.38), and (9A.26), one can write the reference model error dynamics as

$$\dot{e}_r(t) = A_r e_r(t) - B\dot{x}_c(t), \tag{9A.27}$$

which implies that if $\dot{x}_c(t) \to 0$ as $t \to \infty$, then $x_r(t)$ converges to $x_{r_i}(t)$.

Next, Eq. (9A.24) can be rewritten as

$$\dot{x}_c(t) = -Mx_c(t) + M\omega_4(\cdot), \tag{9A.28}$$

where $\omega_4(\cdot) \triangleq z(\cdot) + M^{-1}q(\cdot)$, $z(\cdot) \triangleq M^{-1}[-\hat{W}^T(t)x_{r_i}(t) - \hat{W}^T(t)e(t) - K_1 x_{r_i}(t) - K_1 e(t) + K_2 c(t)]$, and $q(\cdot) \triangleq -\hat{W}^T(t)e_r(t) - K_1 e_r(t)$. Note that Eq. (9A.28) can be equivalently represented as

$$\dot{x}_c(t) = z_1(t) - Mz_2(t) + q(\cdot), \tag{9A.29}$$

$$\dot{z}_1(t) = -Mz_1(t) + M\dot{z}(t), \tag{9A.30}$$

$$\dot{z}_2(t) = -Mz_2(t) + q(\cdot). \tag{9A.31}$$

Letting $x_0(t) = [e_r^T(t),\ z_2^T(t),\ z_1^T(t)]$, we have

$$\dot{x}_0(t) = \underbrace{\begin{bmatrix} A + B\hat{W}^T(t) & BM & -B \\ -K_1 - \hat{W}^T(t) & -M & 0 \\ 0 & 0 & -M \end{bmatrix}}_{\mathcal{A}_0(\cdot)} x_0(t) + \underbrace{\begin{bmatrix} 0 \\ 0 \\ M \end{bmatrix}}_{\mathcal{B}_0} \dot{z}(t). \tag{9A.32}$$

Note that since the upper left block of $\mathcal{A}_0(\cdot)$ is quadratically stable by Lemma 9.1, $-M$ is Hurwitz, and $\mathcal{A}_0(\cdot)$ is in an upper triangular form, then it follows that $x_0(t) \to 0$ as $t \to \infty$ if $\dot{z}(t) \to 0$ as $t \to \infty$.

Finally, $\dot{z}(t)$ can be written for the constant command case as

$$\dot{z}(t) = -M^{-1}[\dot{\hat{W}}^T(t)x_{r_i}(t) + \hat{W}^T(t)\dot{x}_{r_i}(t) + \dot{\hat{W}}^T(t)e(t) + \hat{W}^T(t)\dot{e}(t) + K_1\dot{x}_{r_i}(t) + K_1\dot{e}(t)]. \tag{9A.33}$$

From Theorem 9.1, it follows that $\dot{\hat{W}}(t) \to 0$ as $t \to \infty$. In addition, since $c(t)$ is constant, then it follows from Eq. (9.38) that $\dot{x}_{r_i}(t) \to 0$ as $t \to \infty$. Moreover, since $\ddot{e}(t)$ is bounded and $\dot{e}(t)$ is uniformly continuous as a direct consequence of Theorem 9.1, then $\dot{e}(t) \to 0$ as $t \to \infty$. This argument shows that $\dot{z}(t) \to 0$ as $t \to \infty$, and hence, $x_0(t) \to 0$ as $t \to \infty$, which shows that the error between the ideal reference model given by Eq. (9.38) and the modified reference model given by Eq. (9.29) vanishes as $t \to \infty$. Furthermore, from Theorem 9.1, we know $e(t) \to 0$ as $t \to \infty$, and hence, $x(t) - x_{r_i}(t) = e(t) + e_r(t) \to 0$ as $t \to \infty$.

Proof of Theorem 9.4: We first show the boundedness of the system error $e(t)$ and the weight error $\tilde{W}(t)$ for all $t \in \overline{\mathbb{R}}_+$. For this purpose, consider the Lyapunov function candidate given by

$$\mathcal{V}(e, \tilde{W}) = e^T P e + \gamma^{-1} \mathrm{tr}\, \tilde{W}^T \tilde{W}. \tag{9A.34}$$

Note that $\mathcal{V}(0, 0) = 0$ and $\mathcal{V}(e, \tilde{W}) > 0$ for all $(e, \tilde{W}) \neq (0, 0)$. Differentiating Eq. (9A.34) yields

$$\dot{\mathcal{V}}\left(e(t), \tilde{W}(t)\right) \leq -e^T(t)Re(t) \leq 0, \tag{9A.35}$$

which guarantees the Lyapunov stability, and hence the boundedness of the solution $(e(t), \tilde{W}(t))$ for all $t \in \overline{\mathbb{R}}_+$.

To show the boundedness of signals $x_r(t)$ and $x_c(t)$ (and hence $v(t)$) for all $t \in \overline{\mathbb{R}}_+$, consider the reference model given by Eq. (9.29) and the actuator dynamics given by Eq. (9.26) subject to Eq. (9.11), Eq. (9.12), and Eq. (9.17) as

$$\begin{aligned}
\dot{x}_r(t) &= A_r x_r(t) + B_r r(t) + B\left[Mx_c(t) + \hat{W}^T(t)\sigma(x(t)) + K_1 x(t) - K_2 c(t)\right] \\
&= A_r x_r(t) + BMx_c(t) + \hat{W}^T(t)\sigma(x(t)) + BK_1 x_r(t) + BK_1 e(t) \\
&= Ax_r(t) + BMx_c(t) + \hat{W}^T(t)\sigma(x(t)) + BK_1 e(t), \qquad (9A.36) \\
\dot{x}_c(t) &= -Mx_c(t) - \hat{W}^T(t)\sigma(x(t)) - K_1 x(t) + K_2 c(t) \\
&= -Mx_c(t) - K_1 x_r(t) - \hat{W}^T(t)\sigma(x(t)) - K_1 e(t) + K_2 c(t). \qquad (9A.37)
\end{aligned}$$

Then, Eq. (9A.36), and Eq. (9A.37) can be rewritten in a compact form as

$$\dot{\xi}(t) = \mathcal{A}(M)\xi(t) + \zeta(\cdot) + \omega(\cdot), \tag{9A.38}$$

with $\xi(t) \triangleq [x_r^T(t), x_c^T(t)]^T$,

$$\zeta(\cdot) = \begin{bmatrix} B\hat{W}^T(t)\sigma(x(t)) \\ -\hat{W}^T(t)\sigma(x(t)) \end{bmatrix}, \tag{9A.39}$$

and

$$\omega(\cdot) = \begin{bmatrix} BK_1 e(t) \\ -K_1 e(t) + K_2 c(t) \end{bmatrix}. \tag{9A.40}$$

Note that $\omega(\cdot)$ in Eq. (9A.38) is a bounded perturbation for all $t \in \overline{\mathbb{R}}_+$ as a result of the boundedness of the signals $e(t)$ and $\tilde{W}(t)$ for all $t \in \overline{\mathbb{R}}_+$. In the remainder of the proof, we consider two cases.

Case 1: We first consider the case when $\|x(t)\|_2 < \epsilon$. By Assumption 9.1, one can set $K(x(t)) = 0$, and hence, $\sigma(x(t)) = b(x(t))$, which is bounded for all $t \in \overline{\mathbb{R}}_+$. It then follows that Eq. (9A.38) can be rewritten as

$$\dot{\xi}(t) = \mathcal{A}(M)\xi(t) + \omega(\cdot), \tag{9A.41}$$

with

$$\omega(\cdot) = \begin{bmatrix} B\hat{W}^T(t)b(x(t)) + BK_1 e(t) \\ -\hat{W}^T(t)b(x(t)) - K_1 e(t) + K_2 r(t) \end{bmatrix}. \tag{9A.42}$$

Now, it follows that since $\omega(\cdot)$ is bounded for all $t \in \overline{\mathbb{R}}_+$ and $\mathcal{A}(M)$ is quadratically stable for $\underline{\lambda}_0 \in \kappa_0$ by Lemma 9.2, then $x_r(t)$ and $x_c(t)$ are also bounded for all $t \in \overline{\mathbb{R}}_+$ (see, for example, [12], additionally it follows from Eq. (9.26) that $v(t)$ is bounded for all $t \in \overline{\mathbb{R}}_+$.

Case 2: We next consider the case when $\|x(t)\|_2 \geq \epsilon$. By Assumption 9.1, one can set $b(x(t)) = 0$, and hence, $\sigma(x(t)) = K(x(t))x(t)$. We first show that $K(x(t))$ is bounded for all $t \in \overline{\mathbb{R}}_+$. Specifically, it follows from $\sigma(x(t)) = K(x(t))x(t)$ that

$$\begin{aligned}\frac{\|K(x(t))x(t)\|_2}{\|x(t)\|_2} &= \frac{\|\sigma(x(t))\|_2}{\|x(t)\|_2} \\ &= \frac{\|\sigma(x(t)) - \sigma_0 + \sigma_0\|_2}{\|x(t)\|_2} \\ &\leq \frac{\alpha\|x(t)\|_2}{\|x(t)\|_2} + \frac{\|\sigma_0\|_2}{\|x(t)\|_2},\end{aligned}$$

Taking the supremum of both sides yields

$$\sup_{\|x(t)\|_2 \neq 0} \frac{\|K(x(t))x(t)\|_2}{\|x(t)\|_2} \leq \sup_{\|x(t)\|_2 \neq 0} \left(\alpha + \frac{\|\sigma_0\|_2}{\|x(t)\|_2} \right),$$

which further implies

$$\|K(x(t))\|_2 \leq \alpha + \frac{\|\sigma_0\|_2}{\epsilon}, \tag{9A.43}$$

for $\|x(t)\|_2 \geq \epsilon$, and hence, $K(x(t))$ is bounded for all $t \in \overline{\mathbb{R}}_+$.

Next, the $\zeta(\cdot)$ matrix in Eq. (9A.38) can be rewritten as

$$\zeta(\cdot) = \begin{bmatrix} B\hat{W}^T(t)K(x(t))(x_r(t) + e(t)) \\ -\hat{W}^T(t)K(x(t))(x_r(t) + e(t)) \end{bmatrix}. \tag{9A.44}$$

Note that since $\hat{W}(t)$ is bounded for all $t \in \overline{\mathbb{R}}_+$, as a result of the projection-based weight update law, and $K(x(t))$ is bounded for all $t \in \overline{\mathbb{R}}_+$, $\hat{\theta}(t, x(t)) = \hat{W}^T(t)K(x(t))$ is bounded for all $t \in \overline{\mathbb{R}}_+$ and it follows from Eq. (9A.44)

$$\zeta(\cdot) = \begin{bmatrix} B\hat{\theta}(t, x(t))(x_r(t) + e(t)) \\ -\hat{\theta}(t, x(t))(x_r(t) + e(t)) \end{bmatrix}. \tag{9A.45}$$

Then, using Eqs. (9A.45), and (9A.38) can be rewritten as

$$\dot{\xi}(t) = \mathcal{A}(\hat{\theta}(t, x(t)), M)\xi(t) + \omega(\cdot), \tag{9A.46}$$

with

$$\omega(\cdot) = \begin{bmatrix} B\hat{\theta}(t, x(t))e(t) + BK_1e(t) \\ -\hat{\theta}(t, x(t))e(t) - K_1e(t) + K_2r(t) \end{bmatrix}. \tag{A.47}$$

Now, as in Case 1, it follows that since $\omega(\cdot)$ is bounded for all $t \in \overline{\mathbb{R}}_+$ and $\mathcal{A}(\hat{\theta}(t, x(t)), M)$ is quadratically stable for $(\underline{\lambda}, \overline{\omega}, \underline{\lambda}_1) \in \kappa_1$ by Lemma 9.2, then $x_r(t)$ and $x_c(t)$ are also bounded for all $t \in \overline{\mathbb{R}}_+$, and additionally it follows from Eq. (9.26) that $v(t)$ is bounded for all $t \in \overline{\mathbb{R}}_+$.

Finally, since $e(t)$ and $x_r(t)$ are bounded for all $t \in \overline{\mathbb{R}}_+$ for both cases, this implies $x(t)$ is bounded for all $t \in \overline{\mathbb{R}}_+$. In addition, it follows from Eq. (9.48) that $\dot{e}(t)$ is bounded for all $t \in \overline{\mathbb{R}}_+$, and hence $\ddot{\mathcal{V}}(e(t), \tilde{W}(t))$ is bounded for all $t \in \overline{\mathbb{R}}_+$. As a consequence of the boundedness of $\ddot{\mathcal{V}}(e(t), \tilde{W}(t))$ and Barbalat's lemma [12], $\lim_{t\to\infty} e(t) = 0$ is immediate.

Proof of Theorem 9.5: Consider the quadratic function given by

$$\mathcal{Q}(\overline{W}_{i_1,\ldots,i_l}, \Delta\lambda) = \mathcal{A}^T(\overline{W}_{i_1,\ldots,i_l}, \Delta\lambda)\mathcal{P}(\Delta\lambda) + \mathcal{P}(\Delta\lambda)\mathcal{A}(\overline{W}_{i_1,\ldots,i_l}, \Delta\lambda), \tag{A.48}$$

which is negative definite at the corners as a natural consequence of the LMI conditions given by Eqs. (9.64) through (9.66). Now, expanding $\mathcal{Q}(\overline{W}_{i_1,\ldots,i_l}, \Delta\lambda)$ yields

$$\begin{aligned}
\mathcal{Q}(\overline{W}_{i_1,\ldots,i_l}, \Delta\lambda) &= \left[\mathcal{A}_{0,i_1,\ldots,i_l} + \sum_{j=1}^{m}\Delta\lambda_j\mathcal{A}_j\right]^T\left[\mathcal{P}_0 + \sum_{j=1}^{m}\Delta\lambda_j\mathcal{P}_j\right] \\
&\quad + \left[\mathcal{P}_0 + \sum_{j=1}^{m}\Delta\lambda_j\mathcal{P}_j\right]\left[\mathcal{A}_{0,i_1,\ldots,i_l} + \sum_{j=1}^{m}\Delta\lambda_j\mathcal{A}_j\right] \\
&= \mathcal{A}_{0,i_1,\ldots,i_l}^T\mathcal{P}_0 + \mathcal{P}_0\mathcal{A}_{0,i_1,\ldots,i_l} \\
&\quad + \sum_{j=1}^{m}\Delta\lambda_j\left[\mathcal{A}_{0,i_1,\ldots,i_l}^T\mathcal{P}_j + \mathcal{P}_j\mathcal{A}_{0,i_1,\ldots,i_l} + \mathcal{A}_j^T\mathcal{P}_0 + \mathcal{P}_0\mathcal{A}_j\right] \\
&\quad + \sum_{j=1}^{m}\sum_{k=1}^{j-1}\Delta\lambda_j\Delta\lambda_k\left[\mathcal{A}_j^T\mathcal{P}_k + \mathcal{P}_k\mathcal{A}_j + \mathcal{A}_k^T\mathcal{P}_j + \mathcal{P}_j\mathcal{A}_k\right] \\
&\quad + \sum_{j=1}^{m}\Delta\lambda_j^2\left[\mathcal{A}_j^T\mathcal{P}_j + \mathcal{P}_j\mathcal{A}_j\right].
\end{aligned} \tag{9A.49}$$

Next, consider the quadratic function $x^T\mathcal{Q}(\overline{W}_{i_1,\ldots,i_l}, \Delta\lambda)x$ for any vector $x \neq 0$, which can be written as

$$x^T\mathcal{Q}(\overline{W}_{i_1,\ldots,i_l}, \Delta\lambda)x = \alpha_{0,1_{i_1,\ldots,i_l}} + \sum_{j=1}^{m}\Delta\lambda_j\alpha_{j,1_{i_1,\ldots,i_l}} + \sum_{j=1}^{m}\sum_{k=1}^{j-1}\Delta\lambda_k\Delta\lambda_j\beta_{k,j} + \sum_{j=1}^{m}\Delta\lambda_j^2\gamma_j, \tag{A.50}$$

where $\alpha_{0,1_{i_1,\ldots,i_l}} = x^T\left[\mathcal{A}_{0,i_1,\ldots,i_l}^T\mathcal{P}_0 + \mathcal{P}_0\mathcal{A}_{0,i_1,\ldots,i_l}\right]x$, $\alpha_{j,1_{i_1,\ldots,i_l}} = x^T\left[\mathcal{A}_{0,i_1,\ldots,i_l}^T\mathcal{P}_j + \mathcal{P}_j\mathcal{A}_{0,i_1,\ldots,i_l} + \mathcal{A}_j^T\mathcal{P}_0 + \mathcal{P}_0\mathcal{A}_j\right]x$, and $\beta_{k,j} = x^T\left[\mathcal{A}_j^T\mathcal{P}_k + \mathcal{P}_k\mathcal{A}_j + \mathcal{A}_k^T\mathcal{P}_j + \mathcal{P}_j\mathcal{A}_k\right]x$, and $\gamma_j = x^T[\mathcal{A}_j^T\mathcal{P}_j +$

$\mathcal{P}_j\mathcal{A}_j]x$ are fixed constants. It naturally follows that the corners of Eq. (9A.50) are negative whenever Eqs. (9.64) through (9.66) are satisfied. We now only need to guarantee that the maximums of Eq. (9A.50) occur at its corners. Therefore, a sufficient condition is the partial convexity constraint given by

$$\frac{\partial^2(x^T\mathcal{Q}(\overline{W}_{i_1,\ldots,i_l},\Delta\lambda)x)}{\partial\Delta\lambda_j^2} = x^T[\mathcal{A}_j^T\mathcal{P}_j + \mathcal{P}_j\mathcal{A}_j]x \geq 0, \tag{A.51}$$

for $j = 1, \ldots, m$. Since x is arbitrary, we obtain that

$$\mathcal{A}_j^T\mathcal{P}_j + \mathcal{P}_j\mathcal{A}_j \geq 0. \tag{A.52}$$

The results of Theorem 3.1 of [23] and Theorem 5.7 of [14] guarantee that the LMIs given by Eqs. (9.64) through (9.66) hold for $\Lambda = \text{conv}(\Lambda_0)$ since $x^T\mathcal{Q}(\overline{W}_{i_1,\ldots,i_l},\Delta\lambda)x$ always obtains its maximums at some corner of the parameter box Λ_0.

APPENDIX B: GHV STATE-SPACE MODEL

For the configuration with an altitude of 80,000 feet and a Mach number of 6, a linearized model under nominal conditions is obtained in the form of Eq. (9.70) with

$$A_p = \begin{bmatrix} -3.70\times10^{-3} & -7.17\times10^{-1} & 0 & -3.18\times10^{1} & -2.67\times10^{-4} \\ -5.35\times10^{-7} & -2.39\times10^{-1} & 1 & -2.95\times10^{-12} & 2.23\times10^{-7} \\ -2.79\times10^{-5} & 4.26 & -1.19\times10^{-1} & 0 & 3.94\times10^{-5} \\ -4.76\times10^{-8} & 1.31\times10^{-13} & 1 & -4.45\times10^{-14} & -1.33\times10^{-11} \\ -5.53\times10^{-10} & -5.87\times10^{3} & 0 & 5.87\times10^{3} & 0 \\ 5.99\times10^{-16} & -3.14\times10^{-11} & 0 & -3.04\times10^{-19} & -9.74\times10^{-16} \\ 1.47\times10^{-10} & -4.45\times10^{-6} & 0 & 0 & -1.00\times10^{-11} \\ -5.29\times10^{-12} & 3.98\times10^{-8} & 0 & 0 & 1.28\times10^{-12} \\ 8.08\times10^{-28} & 2.04\times10^{-22} & 1.01\times10^{-20} & 1.17\times10^{-16} & -1.73\times10^{-31} \end{bmatrix} \cdots$$

$$\cdots \begin{bmatrix} -8.81\times10^{-1} & 0 & 0 & -1.77\times10^{-15} \\ -1.06\times10^{-3} & 0 & 0 & -3.18\times10^{-21} \\ -1.47 & 0 & 0 & 0 \\ -1.08\times10^{-19} & 4.44\times10^{-16} & -9.58\times10^{-16} & -2.58\times10^{-18} \\ 0 & 0 & 0 & -3.26\times10^{-13} \\ -6.97\times10^{-2} & -1.04\times10^{-2} & -9.99\times10^{-1} & -5.35\times10^{-3} \\ -1.31\times10^{3} & -2.03 & -7.54\times10^{-3} & 0 \\ 2.07 & -1.55\times10^{-3} & -5.31\times10^{-2} & 0 \\ -2.38\times10^{-4} & 8.54\times10^{-1} & -8.84\times10^{-3} & -3.00\times10^{-6} \end{bmatrix} \tag{9B.1}$$

$$B_p = \begin{bmatrix} -6.53 \times 10^{-3} & -1.24 \times 10^{-13} & -2.98 \times 10^{-3} \\ -1.33 \times 10^{-4} & -2.44 \times 10^{-13} & 1.17 \times 10^{-7} \\ -1.84 \times 10^{-1} & -1.60 \times 10^{-13} & 2.48 \times 10^{-4} \\ 0 & 0 & 0 \\ 0 & 0 & 0 \\ -1.40 \times 10^{-16} & -2.47 \times 10^{-5} & -2.18 \times 10^{-4} \\ -5.90 \times 10^{-11} & -8.04 & 10.3 \\ 8.56 \times 10^{-14} & 3.17 \times 10^{-2} & 2.85 \times 10^{-1} \\ 0 & 0 & 0 \end{bmatrix} \tag{9B.2}$$

with the state vector being defined as $x_p(t) = [V(t), \alpha(t), q(t), \theta(t), h(t), \beta(t), p(t), r(t), \phi(t)]^T$, where $V(t)$ denotes the total velocity, $\alpha(t)$ denotes the angle-of-attack, $q(t)$ denotes the pitch rate, $\theta(t)$ denotes the pitch angle, $h(t)$ denotes the altitude, $\beta(t)$ denotes the sideslip angle, $p(t)$ denotes the roll rate, $r(t)$ denotes the yaw rate, and $\phi(t)$ denotes the roll angle. The control input vector is defined as $u(t) = [\delta_e(t), \delta_a(t), \delta_r(t)]^T$ where $\delta_e(t)$ denotes the elevator deflection, $\delta_a(t)$ denotes the aileron deflection, and $\delta_r(t)$ denotes the rudder deflection.

REFERENCES

[1] Narendra, K. S., and Annaswamy, A. M., *Stable Adaptive Systems*, Prentice Hall, Upper Saddle River, NJ, 1989.

[2] Ioannou, P. A., and Sun, J., *Robust Adaptive Control*, Prentice-Hall, Upper Saddle River, NJ, 1996.

[3] Lavretsky, E., and Wise, K. A., *Robust Adaptive Control*, Springer, New York, NY, 2013.

[4] Cao, C., and Hovakimyan, N., "$\mathcal{L}_1$ Adaptive Controller for Systems in the Presence of Unmodelled Actuator Dynamics," *IEEE Conference on Decision and Control*, New Orleans, LA, December 2007.

[5] Su, C.-Y., and Stepanenko, Y., "Backstepping-Based Hybrid Adaptive Control of Robot Manipulators Incorporating Actuator Dynamics," *International Journal of Adaptive Control and Signal Processing*, Vol. 11, No. 2, 1997, pp. 141–153.

[6] De La Torre, G., Yucelen, T., and Johnson, E., "Reference Control Architecture in the Presence of Measurement Noise and Actuator Dynamics," *American Control Conference*, Portland, OR, June 2014.

[7] Muse, J. A., "Frequency Limited Adaptive Control Using a Quadratic Stability Framework: Guaranteed Stability Limits," *AIAA Guidance, Navigation, and Control Conference*, National Harbor, MD, January 2014.

[8] Johnson, E. N., "Limited Authority Adaptive Flight Control," Ph.D. thesis, School of Aerospace Engineering, Georgia Institute of Technology, Atlanta, GA, 2000.

[9] Gruenwald, B. C., Wagner, D., Yucelen, T., and Muse, J. A., "Computing Actuator Bandwidth Limits for Model Reference Adaptive Control," *International Journal of Control*, Vol. 89, No. 12, 2016, pp. 2434–2452.

[10] Pomet, J. B., and Praly, L., "Adaptive Nonlinear Regulation: Estimation from the Lyapunov Equation," *IEEE Transactions on Automatic Control*, Vol. 37, No. 6, 1992, pp. 729–740.

[11] Haddad, W. M., and Chellaboina, V., *Nonlinear Dynamical Systems and Control, A Lyapunov-Based Approach*, Princeton University Press, Princeton, NJ, 2008.

[12] Khalil, H. K., *Nonlinear Systems*, Prentice Hall, Upper Saddle River, NJ, 1996.

[13] Boyd, S., El Ghaoui, L., Feron, E., and Balakrishnan, V., *Linear matrix inequalities in system and control theory*, SIAM Publishing House, Philadelphia, PA, 1994.

[14] Scherer, C., and Weiland, S., "Linear Matrix Inequalities in Control," *Lecture Notes, Dutch Institute for Systems and Control*, Vol. 3, 2000.

[15] Gahinet, P., Apkarian, P., and Chilali, M., "Affine Parameter-Dependent Lyapunov Functions and Real Parametric Uncertainty," Vol. 41, No. 3, March 1996, pp. 436–442.

[16] Lofberg, J., "Yalmip: A Toolbox for Modeling and Optimization in Matlab," *IEEE International Symposium on Computer Aided Control Systems Design*, New Orleans, LA, September 2004.

[17] Gahinet, P., Nemirovskii, A., Laub, A. J., and Chilali, M., "The LMI Control Toolbox," *IEEE Conference on Decision and Control*, Lake Buena Vista, FL, December 1994.

[18] Wagner, D., "A Linear Matrix Inequality-Based Approach for the Computation of Actuator Bandwidth Limits in Adaptive Control," Master's thesis, Department of Mechanical and Aerospace Engineering, Missouri University of Science and Technology, Rolla, MO, 2016.

[19] Lewis, F. L., and Syrmos, V. L., *Optimal control*, Wiley, Hoboken, NJ, 1995.

[20] Famularo, D. I., Valasek, J., Muse, J. A., and Bolender, M. A., "Observer-based Feedback Adaptive Control for Nonlinear Hypersonic Vehicles," *AIAA Guidance, Navigation, and Control Conference*, Grapevine, TX, January 2017.

[21] Wiese, D. P., Annaswamy, A. M., Muse, J. A., Bolender, M. A., and Lavretsky, E., "Adaptive Output Feedback Based on Closed-loop Reference Models for Hypersonic Vehicles," *Journal of Guidance, Control, and Dynamics*, Vol. 38, No. 12, 2015, pp. 2429–2440.

[22] Muse, J. A., "Frequency Limited Adaptive Control Using a Quadratic Stability Framework: Stability and Convergence," *AIAA Guidance, Navigation, and Control Conference*, National Harbor, MD, January 2014.

[23] Gahinet, P., Apkarian, P., and Chilali, M., "Affine Parameter-dependent Lyapunov Functions and Real Parametric Uncertainty," *IEEE Transactions on Automatic Control*, Vol. 41, No. 3, 1996, pp. 436–442.

INDEX

Note: Page numbers followed by *f* or **t** (indicating figures and tables).

SUPPORTING MATERIALS

A complete listing of titles in the Progress in Astronautics and Aeronautics series is available from AIAA's electronic library, Aerospace Research Central (ARC) at arc.aiaa.org. Visit ARC frequently to stay abreast of product changes, corrections, special offers, and new publications.

AIAA is committed to devoting resources to the education of both practicing and future aerospace professionals. In 1996, the AIAA Foundation was founded. Its programs enhance scientific literacy and advance the arts and sciences of aerospace. For more information, please visit www.aiaafoundation.org.